Elevated Carbon Dioxide

Elevated Carbon Dioxide

Impacts on Soil and Plant Water Relations

M. B. Kirkham

CRC Press
Taylor & Francis Group
Boca Raton London New York

CRC Press is an imprint of the
Taylor & Francis Group, an **informa** business

CRC Press
Taylor & Francis Group
6000 Broken Sound Parkway NW, Suite 300
Boca Raton, FL 33487-2742

First issued in paperback 2019

ISBN-13: 978-1-4398-5504-1 (hbk)
ISBN-13: 978-0-367-38299-5 (pbk)

Library of Congress Cataloging-in-Publication Data

Kirkham, M. B.
    Elevated carbon dioxide : impacts on soil and plant water relations / author: Mary Beth Kirkham.
        p. cm.
    Includes bibliographical references and index.
    ISBN 978-1-4398-5504-1 (alk. paper)
    1. Plants--Effect of atmospheric carbon dioxide on. 2. Plant-water relationships. 3. Plant-soil relationships. 4. Growth (Plants) I. Title.

QK753.C3K57 2011
632'.19--dc22
                                             2010039437

Visit the Taylor & Francis Web site at
http://www.taylorandfrancis.com

and the CRC Press Web site at
http://www.crcpress.com

To the memory of my mother who suggested the book

and

To the rest of my family for support

Contents

Preface

Water and carbon dioxide are the two most important compounds affecting plant growth. In introductory botany textbooks, we have seen the familiar equation for photosynthesis, which shows carbon dioxide (CO_2) joining with water (H_2O), in the presence of light and chlorophyll, to form sugar ($C_6H_{12}O_6$) and oxygen (O_2), as follows:

$$6CO_2 + 6H_2O \xrightarrow{\text{light+chlorophyll}} C_6H_{12}O_6 + 6O_2$$

Life on earth would not be possible without photosynthesis. We survive because of the oxygen and food (sugars) produced by photosynthesis. Therefore, it is of critical importance to look at the plant water relations under elevated CO_2, because the CO_2 concentration in the atmosphere is increasing.

The CO_2 concentration in the atmosphere was first recorded by Charles D. Keeling (1928–2005) of the Scripps Institution of Oceanography at the University of California at San Diego. He monitored it beginning in 1957 at Mauna Loa in Hawaii and in Antarctica at the South Pole. In the 50 year period between 1958 and 2008, the CO_2 concentration in the atmosphere increased from 316 to 385 ppm. This book aims to put the information in one source as no books document plant water relations under elevated CO_2.

This book was developed from research conducted in the Evapotranspiration Laboratory at Kansas State University between 1984 and 1991 with field-grown sorghum, winter wheat, and rangeland plants under elevated CO_2. Such experiments had not been done before in the semiarid Great Plains of the United States. The rising levels of CO_2 in the atmosphere were of interest to the U.S. Department of Energy, which funded our work. It has been 27 years since we started our first experiments. We can thus make some predictions, based on our early results, about how plants are responding to elevated CO_2, which was 330 ppm in 1984 when we started our studies. I present some of these predictions in this book.

This book is not a literature review. It describes experiments that appear in peer-reviewed journal articles. Thousands of papers have been written on the effects of elevated levels of atmospheric CO_2 on plants. It is impossible to review the entire literature. I have thus picked selected articles as examples and then discuss each one, often providing an illustration from the paper. For each paper that I illustrate, I mention the full name (common and scientific) of the plant under investigation. When this is not mentioned in the original article, I have referred to Fernald (1950) and Bailey (1974). For each experiment, I also provide the type of soil used (if it is given in the original article) and the general conditions of the experiment [greenhouse, growth chamber, open-top or enclosed chambers, or FACE (free-air carbon dioxide enrichment) facility]. All information in the book has been taken from hard copy sources (books and journal articles). I have carefully documented the source of the information. When it comes from a lengthy article or a book, I have provided the exact page on which I found the information. In this way, the interested reader can easily find the source.

This book has much instructive material, which I use to teach my graduate-level class. When a new scientific concept is raised, I provide a detailed explanation.

The book deals only with water and elevated CO_2. It does not deal with temperature, nutrients, or other factors, such as the greenhouse effect (warming of the atmosphere by trace gases), that affect plant growth under elevated CO_2, although in the last chapter I mention temperature briefly.

The book is organized as follows. I start with an introductory chapter (Chapter 1) dealing with drought, because it is predicted that the central Great Plains, where Kansas is located, will become drier as the CO_2 concentration in the atmosphere increases. In this chapter, I provide a preliminary overview of the three types of photosynthesis: C_3, C_4, and Crassulacean acid metabolism. The book then describes water as it moves from the soil through the plant and out into the atmosphere. This is the way that water moves through the soil–plant–atmosphere continuum. Chapters 2 through 5 deal with soil. Chapter 2 discusses the composition of the soil atmosphere. Chapter 3 deals with the interaction of elevated CO_2 in the soil with the physical factors in the soil, such as compaction, that affect root growth. Because oxygen is a key factor for root growth, Chapter 4 deals with variable oxygen concentration of the soil along with elevated CO_2 in the soil. Carbon dioxide can be elevated both in the soil and in the atmosphere. Therefore, I had to distinguish the effects of elevated CO_2 in the soil from elevated CO_2 in the atmosphere. While Chapters 2 through 4 focus on elevated CO_2 in the soil, Chapter 5 deals with soil when the atmosphere above it is elevated with CO_2.

After the discussion of soil and elevated CO_2, I then consider the root. Chapter 6 deals with elevated CO_2 and root growth. In this chapter, studies with roots have been done with carbon isotopes, and I discuss carbon isotope ratios and how they are used in plant science in detail. Chapter 7 deals with the effects of elevated CO_2 on plant water, osmotic, and turgor potentials. Chapters 8 and 9 deal with stomata under elevated CO_2. Chapter 8 discusses stomatal conductance. In this chapter, I present material about the resistances in leaves, the units used to measure stomatal conductance, and the physiological factors affecting stomatal opening and closing. Chapter 9 deals with stomatal density. In this chapter, the geological timescale is presented as we learn from the geological record that stomatal density is affected by CO_2. Next, I take the water out of the plant into the atmosphere and discuss the effects of elevated CO_2 on transpiration and evapotranspiration in Chapter 10. In this chapter, I discuss ethylene, because it is a gas similar to CO_2, and I distinguish the two gases and their effects on plants. I also cover the general principles of evapotranspiration and how it is determined. Next, I discuss water use efficiency in Chapter 11. In this chapter, I present material about the historical aspects of water use efficiency. Chapter 12 compares C_3 and C_4 plants under elevated CO_2 and provides a detailed account of C_4 photosynthesis and its advantages, and how it has evolved. Chapter 13 deals with plant anatomy and focuses on the xylem (including wood)—the tissue that carries water in plants. In this chapter, I discuss the three variations of C_4 photosynthesis, because it is necessary to know this material to understand one of the topics described in this chapter. Chapter 14 deals with phenology and how this is affected by elevated CO_2. In this chapter, I also cover the Q_{10} value (a value related to the rate of reactions), the degree-day concept, and heat units as these are necessary to know to understand how a plant progresses through its different phenological stages. I also provide sample problems to help students calculate the Q_{10}. Finally, Chapter 15 deals with the growth and yield of many different kinds of plants under elevated CO_2 and well-watered conditions. In any work dealing with plant water relations, some measure of plant growth (e.g., height, biomass, or leaf area) should be provided, because growth integrates all factors affecting a plant. Therefore, I have devoted an entire chapter to this.

Each chapter ends with a summary. Because the humanistic side of science is usually overlooked in scientific books, I have appended to several chapters the biographies of people who have developed the concepts discussed in those chapters.

The units for CO_2 in the different chapters are expressed as ppm, μmol/mol, μL/L (or μl/l—liter can either be capitalized or not capitalized when it is abbreviated), cm^3/m^3, or in Pascal (Pa). When I worked with the Tri-Societies (American Society of Agronomy, Crop Science Society of America, and Soil Science Society of America) on the book that we (Allen et al. 1997) edited, it suggested the common unit of μmol/mol. Therefore, in this book I have converted the units as much as possible to μmol/mol. I have left the unit of Pascal (pressure unit) unchanged, as one needs to know the temperature and elevation of the place where the research was done to convert it to a concentration unit. As I did not have this information, I retained the unit as Pascal.

A key part of the book are the figures. They have been redrawn from the originals by Eldon J. Hardy, who was my draftsman at the College of Engineering when I was in the Department of Agronomy at Oklahoma State University. He is now retired. This book would not have been possible without his help. His drawings are unparalleled for uniformity, clearness, and precision. I extend my profound thanks to him for working closely with me during the last three years to prepare these drawings.

I am grateful to the publisher, CRC Press, Taylor & Francis, for publishing this book. I would especially like to thank Randy Brehm, editor, Chemical and Life Sciences Group, Taylor & Francis, for her prompt and peerless assistance during all stages of publication of this book. I would like to thank the four reviewers that she contacted and who supported the publication of this book. I would also like to thank John Edwards, project coordinator, Editorial Project Development, Taylor & Francis, for his help during the production of this book.

REFERENCES

Allen, L.H., M.B. Kirkham, D.M. Olszyk, and C.E. Whitman (Eds.). 1997. *Advances in Carbon Dioxide Effects Research*. ASA Spec. Pub. No. 61. Madison, WI: American Society of Agronomy, Crop Science Society of America, Soil Science Society of America. 228pp.

Bailey, L.H. 1974. *Manual of Cultivated Plants*. Revised edn. Fourteenth printing. New York: Macmillan Publishing Company. 1116pp.

Fernald, M.L. 1950. *Gray's Manual of Botany*. New York: American Book Company. 1632pp.

1 Elevated Atmospheric Carbon Dioxide: Drought

INTRODUCTION

In this introductory chapter, we consider the effect of elevated levels of CO_2 on the growth of plants under drought. Drought is our first topic, because it causes most (40.8%) of the crop losses in the United States (Boyer 1982), and it is predicted that as CO_2 concentration increases in the atmosphere, large areas, especially the crop-producing latitudes of the central United States, will become drier. We know as a fact that the CO_2 concentration in the atmosphere is increasing, as documented by the careful measurements of Keeling (1970). (See Appendix for a biography of Charles David Keeling.) The increase is thought to be due to the increased burning of fossil fuels, which has occurred since the middle of the last century (Figure 1.1). The CO_2 concentration in the atmosphere has risen from 316 ppm in 1958 to 385 ppm in 2008 (Dlugokencky 2009). Figure 1.2 shows the CO_2 concentration between 1959 and 1979 (Idso 1982), and Figure 1.3 shows the concentration between 1980 and 2005 (Schnell 2005).

We first look at the predictions. We then consider plants with the C_3- and C_4-type of photosynthesis and how elevated CO_2 will affect their photosynthesis, and consequently growth, differently under drought. (In Chapter 12, we shall investigate C_3 and C_4 plants under elevated CO_2 in more detail.) We then look at experimental data, starting with the studies we (in the Evapotranspiration Laboratory at Kansas State University) did in the 1980s in the field in the semiarid central Great Plains of the United States We follow with studies done in controlled environments using a variety of plants, including crops, trees (both angiosperms and gymnosperms), and a cactus-family plant. Finally, we mention the effects of elevated CO_2 on plants grown under saline conditions, because drought and salinity often occur together.

Many studies have been done concerning elevated CO_2 and drought. Here, we look at only a few examples and focus on growth. Growth is critical to monitor in any experiment, because it integrates all stresses such as drought. Of the four soil physical factors that affect plant growth (water, temperature, aeration, and mechanical impedance), water is the most important (Kirkham 2005, pp. 1, 6). In the final chapter (Chapter 15), we return to growth of plants under elevated CO_2, but under well-watered conditions.

PREDICTIONS

Manabe and Wetherald (1980) predicted the effects on precipitation and evaporation, and the difference between the two, if the CO_2 concentration is doubled or if it is quadrupled (Figure 1.4). This figure demonstrates that the effects of elevated levels of CO_2 will depend on latitude. It is projected that between 37° and 47°N latitude, there will be a decrease in precipitation, an increase in evaporation, and a decrease in the excess of precipitation over evaporation (Manabe and Wetherald 1980; Rosenberg 1982).

Even a small difference between precipitation and evaporation can make a big difference in irrigation needs for agriculture. Although only about one-seventh of the cropland in the United States is irrigated, it produces a disproportionately large share of the market value of the crops (Waggoner 1984). Irrigation accounts for about half of all water consumption in the United States.

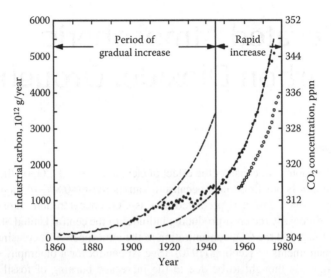

FIGURE 1.1 The global rate of production of industrial CO_2 carbon and the mean global atmospheric CO_2 concentration as a function of time. The curves through the industrial CO_2 carbon data are exponential fits to the intervals 1860–1914 and 1945–1978. Adapted from Keeling (1982, Fig. 6). (Reprinted with permission from Idso, S.B., *Carbon Dioxide: Friend or Foe?* IBR Press, A division of the Institute for Biospheric Research, Inc., Tempe, AZ, pp. 92, Fig. V-1; Keeling, C.D., The oceans and terrestrial biosphere as future sinks for fossil fuel CO_2, in Reck, R.A. and Hummel, J.R., eds., *Interpretation of Climate and Photochemical Models, Ozone and Temperature Measurements*, AIP Conference Proceedings No. 82, American Institute of Physics, New York, 1982, pp. 47–82, Fig. 6. Copyright 1982, American Institute of Physics.)

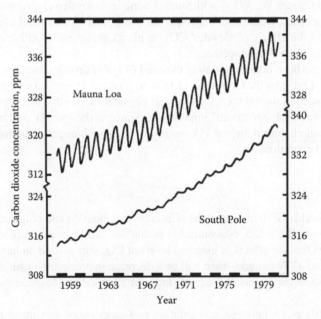

FIGURE 1.2 Continuous trends of atmospheric CO_2 concentration at Mauna Loa, Hawaii and at the South Pole. (Adapted from Idso, S.B., *Carbon Dioxide: Friend or Foe?* IBR Press, A division of the Institute for Biospheric Research, Inc., Tempe, AZ, pp. 92, Fig. I-1; Keeling, C.D., The oceans and terrestrial biosphere as future sinks for fossil fuel CO_2, in Reck, R.A. and Hummel, J.R., eds., *Interpretation of Climate and Photochemical Models, Ozone and Temperature Measurements*, AIP Conference Proceedings No. 82, American Institute of Physics, New York, 1982, pp. 47–82, Figs. 1 and 2. Copyright 1982, American Institute of Physics.)

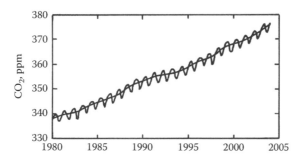

FIGURE 1.3   Globally averaged trace gas (CO_2) mole fraction determined from samples collected as part of the NOAA/CMDL Global Cooperative Air Sampling Network (data courtesy of T.J. Conway, NOAA/CMDL). (NOAA stands for National Oceanic and Atmospheric Administration, and CMDL stands for Climate Modeling and Diagnostics Laboratory.) (Reprinted from Schnell, R.C., *Bull. Am. Meteorol. Soc.*, 86(6), S20, 2005, Fig. 3.1, top. With permission from the American Meteorological Society.)

Because irrigation uses runoff and runoff is often only the small difference between precipitation and evaporation, changes in precipitation can cause relatively great changes in the supply of water for irrigation. Thus, irrigation and the valuable crops that it produces seem particularly susceptible to a decrease in precipitation or an increase in evaporation that may occur as a result of increased levels of CO_2 (Waggoner 1984).

The area between 37° and 47°N latitude includes the major grain-growing areas of the Midwest of the United States that are both in the humid [e.g., Iowa and Illinois, where corn (*Zea mays* L.) dominates] and semiarid parts of the country [e.g., Kansas and Nebraska, where wheat (*Triticum aestivum* L.) dominates]. Soybean [*Glycine max* (L.) Merr.], a legume, is also a major crop in Iowa and Illinois.

PHOTOSYNTHESIS OF C_3 AND C_4 PLANTS

Corn has the C_4-type of photosynthesis, and wheat and soybean have the C_3 type. So we must consider the effect of drought on crops with C_4 and C_3 photosynthesis. As we know from introductory botany, the concentration of CO_2 in the air directly affects plants, because photosynthesis depends on CO_2, which is taken up by plants, and, in the presence of light, water, and chlorophyll, it is converted to sugar, and oxygen is given off. In all plants, photosynthesis involves the C_3 process that converts CO_2 into molecules with three carbons (phosphoglyceric acid, a three-carbon compound). In some species, called C_4 plants, CO_2 is first converted into molecules with four carbons (oxaloacetic acid, a four-carbon molecule), which are then transported within the leaf to sites (bundle-sheath cells) (Kirkham 2005, pp. 357–362) where CO_2 is released, providing a high concentration of CO_2 for the C_3 process. We shall return to the difference in C_3 and C_4 photosynthesis in more detail in Chapters 6, 12, and 13.

The same enzyme that catalyzes the first step in the C_3 process (ribulose-1,5-bisphosphate carboxylase/oxygenase or Rubisco) (Parry et al. 1993) can also catalyze an oxidation that leads to photorespiration, which is respiration that occurs in the light. Photorespiration consumes as much as a third of the CO_2 that the plant has absorbed in the light (Waggoner 1984). Photorespiration is slowed by high concentrations of CO_2 and, hence, occurs more rapidly in C_3 plants, such as wheat, than in C_4 plants, such as corn (also called maize; both wheat and corn are in the grass family or Poaceae). When atmospheric CO_2 is relatively low, net photosynthesis is faster in C_4 than in C_3 plants (Figure 1.5). But at higher levels of CO_2, the changes in photorespiration per change in CO_2 lead to greater increases in net photosynthesis for C_3 plants than for C_4 plants. Therefore, increasing levels of CO_2 will affect C_3 and C_4 plants differently. C_3 plants should benefit more from an increase in CO_2 concentration than C_4 plants (Figure 1.5).

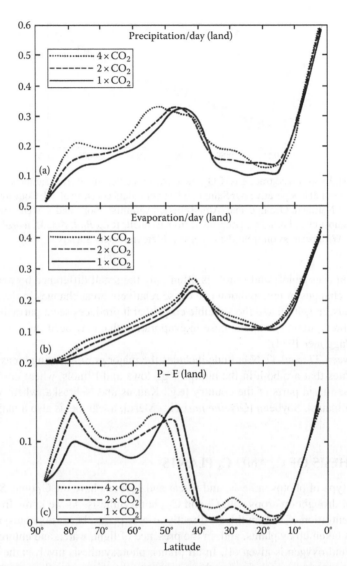

FIGURE 1.4 Zonal-mean values of (a) precipitation rate, P; (b) evaporation rate, E; and (c) precipitation rate minus evaporation rate, (P – E), over the continent. Units are cm/day. (Reprinted from Manabe, S. and Wetherald, R.T., *J. Atmos. Sci.*, 37, 99, 1980, Fig. 14. With permission from the American Meteorological Society, Copyright 1980.)

Now, let us look briefly at the effect of elevated levels of CO_2 on transpiration, because this is going to affect how well a plant can survive under drought. The water in plants can be directly affected by CO_2, because CO_2 closes stomata. (In Chapters 8 and 10, we return to study in detail the effect of elevated CO_2 on stomata and transpiration, respectively.) It is essential that the stomata let CO_2 into the leaf for photosynthesis, but at the same time, open stomata also allow water to escape. It is well known that CO_2 is an antitranspirant. That is, as the CO_2 level is increased, stomata close and transpiration is decreased. The closing of the stomatal pores conserves water. Again, C_3 and C_4 plants are affected differently (Figure 1.6). In light, the stomata of maize, a C_4 plant, have been shown to narrow when the CO_2 concentration is increased from 300 to 600 μmol/mol (0.03%– 0.06%), and transpiration is decreased by about 20%. Transpiration from wheat, a C_3 plant, however,

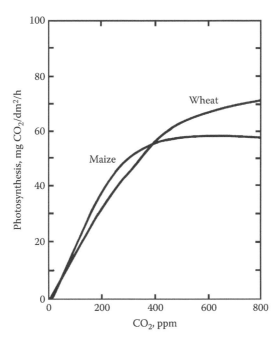

FIGURE 1.5 Although the predicted increase in the amount of CO_2 in the atmosphere will lead to higher temperatures and less rainfall, its effects on agriculture may not be altogether negative. Photosynthesis, which reduces CO_2 to carbohydrates (calculated here per square decimeter of leaf surface exposed to bright light per hour), increases with the amount of CO_2 present. Maize (corn) shows the rapid photosynthesis of C_4 plants (those that first convert CO_2 to a four-carbon compound) at 300 ppm CO_2 (300 μmol/mol CO_2), and wheat shows the greater increase in photosynthesis at high concentrations of CO_2 of C_3 plants (which convert CO_2 to a compound with three-carbon atoms). (Reprinted from Waggoner, P.E., *Am. Sci.* 72, 179, 1984, Fig. 1. With permission from the Crop Science Society of America.)

is decreased by only about 5%. The contrast is assumed to be typical of the difference between C_3 and C_4 plants (Rosenberg 1981).

PHOTOSYNTHESIS OF CAM PLANTS

In addition to plants that have the C_3 and C_4 photosynthetic systems, there is a third group of plants that keep their stomata open at night and fix CO_2 into organic acids, especially malic acid (Salisbury and Ross 1978, p. 145). Their metabolism of CO_2 is unusual, and because it was first investigated in members of the Crassulaceae (or orpine family), it is commonly called Crassulacean acid metabolism, often abbreviated CAM. CAM has been found in 18 families, including the Cactaceae (cactus family), Orchidaceae (orchid family), Bromeliaceae (pineapple family), Liliacea (lily family), and Euphorbiacea (spurge family). Many CAM plants are succulents, but not all. Some succulents, like the halophytes (plants that can grow in saline soils), do not possess CAM. Species with CAM usually lack a well-developed palisade layer of cells, and most of the leaf cells are spongy mesophyll. Bundle-sheath cells are present, but in contrast to those of C_4 plants, they are similar to the mesophyll cells (Salisbury and Ross 1978, p. 145). The only commercially important plant with CAM metabolism is pineapple (*Ananas comosus* Merr.; Bromeliaceae). Little research has been done with CAM plants and elevated CO_2, but we shall discuss one paper at the end of this chapter.

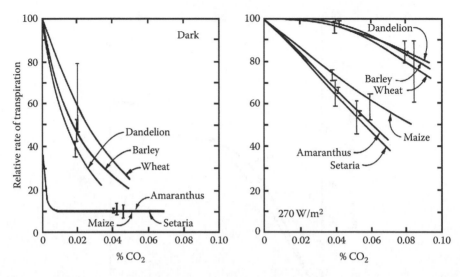

FIGURE 1.6 Effect of CO_2 concentration on transpiration of C_3 and C_4 species in light (right) and darkness (left). Wheat (*Triticum* sp.; Poaceae or grass family), barley (*Hordeum vulgare* L.; Poaceae), and dandelion (*Taraxacum officinale* Weber; Compositae or composite family) are C_3 plants, and maize (*Zea mays* L.; Poaceae), amaranthus (*Amaranthus* sp.; Amaranthaceae or amaranth family), and setaria (*Setaria* sp.; *Setaria italica* Beauv. is foxtail millet; Poaceae) are C_4 plants. (With kind permission from Springer Science+Business Media: *Climatic Change*, The increasing CO_2 concentration in the atmosphere and its implication on agricultural productivity. I. Effects on photosynthesis, transpiration and water use efficiency, 3, 1981, 265–279, Rosenberg, N.J., Fig. 3.)

FIELD STUDIES WITH CROPS

Now, let us turn to the experiments we did under field conditions with elevated CO_2 in Kansas in the 1980s. Chaudhuri et al. (1990) found that elevated CO_2 compensated for reductions in yield of winter wheat (*T. aestivum* L. cv. Newton), a C_3 crop, due to low water supply. We grew the wheat under ambient (340 μmol/mol) and elevated levels (485, 660, and 825 μmol/mol) of CO_2 during three growing seasons in 16 underground boxes (77 cm long, 37 cm wide, and 180 cm deep) containing a silt loam soil. Water in half of the boxes was maintained at 0.38 m³/m³ (high water level; this was field capacity) and in the other half between 0.14 and 0.25 m³/m³ (low water level or about half field capacity). Boxes were weighed to determine the amount of water used by transpiration. Plastic chambers (121 × 92 × 168 cm) covered the boxes to maintain different CO_2 levels. Grain yield of the high-water-level wheat grown under ambient CO_2 was about the same as the grain yield of low-water-level wheat grown at the highest level of CO_2 (825 μmol/mol) (3 year means: 725 and 707 g/m² for the 340 and 825 μmol/mol levels, respectively) (Figure 1.7). Similar results were obtained for yield components (spike number, spike weight, kernels/spike, and kernel weight).

Even though leaf area was not measured by Chaudhuri et al. (1990), it probably was increased under elevated levels of CO_2, because transpiration increased as the level of CO_2 increased. Consequently, Chaudhuri et al. (1990) found no reduction in total water used by wheat grown under elevated CO_2 and drought. However, water use efficiency (WUE) was increased. WUE is the amount of dry matter (grain yield or vegetative yield) divided by the amount of water needed to produce that dry matter. The amount of water lost can be based on evapotranspiration or transpiration (Kirkham 2005, pp. 469–484). The inverse of WUE is called the water requirement. In semiarid regions like Kansas, it is important to increase the WUE (or decrease the water requirement). This means that more grain can be obtained for the same amount of water. We shall return to the effects of elevated levels of CO_2 on WUE and water requirement in Chapter 11. Chaudhuri et al. (1990) found that the

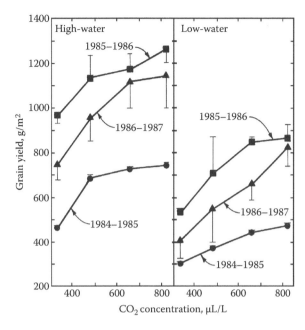

FIGURE 1.7 Grain yield of well-watered (high water level) and drought-stressed (low water level) winter wheat as affected by CO_2 concentration during a 3 year study. Vertical bars = ±standard deviation. Only half of the bars is drawn for clarity. (Reprinted from Chaudhuri, U.N. et al., *Agron. J.*, 82, 637, 1990. With permission from the American Society of Agronomy.)

water requirement for grain production decreased as the CO_2 concentration increased. The results are given in Chapter 11 and showed that the water requirement of wheat was reduced by about 30% when the CO_2 concentration was increased 2.4 times (from 340 to 825 μmol/mol) CO_2. This decrease in water requirement would be beneficial in semiarid regions.

In another early field experiment, Nie et al. (1992a) compared the photosynthetic rate of a C_3 grass (Kentucky bluegrass; *Poa pratensis* L.; Poaceae) and a C_4 grass (big bluestem; *Andropogon gerardii* Vitman; Poaceae) growing in a silty clay loam soil in the spring in a tall-grass prairie under two levels of CO_2 (ambient and twice ambient or 356 and 715 μmol/mol CO_2, respectively) in closed top chambers and watered with two different levels of water (field capacity or 0.38 m^3/m^3 and no water added). The prairie is a rangeland and these two grasses are dominant on it. Kentucky bluegrass grows in the cooler months of the spring and fall, and big bluestem grows in the hot summer months. The rangelands occur in the extensive mid-continental area of North America, where it is predicted that rainfall will be reduced as the CO_2 in the atmosphere increases. The area forms a band of dry land 790–1580 km wide that stretches from Mexico to Canada (Kirkham et al. 1991).

Under well-watered conditions, elevated CO_2 increased the photosynthetic rate of Kentucky bluegrass by 48% (from 9.2 to 17.8 μmol/m^2/s) but did not affect the photosynthetic rate of big bluestem, the C_4 grass (average photosynthetic rate of about 22 μmol/m^2/s) (Figure 1.8, top) (Nie et al. 1992a). When no water was added, photosynthetic rates of both grasses were decreased, but, again, elevated CO_2 had little effect on the photosynthetic rate of big bluestem (Figure 1.8, bottom). However, the average photosynthetic rate of Kentucky bluegrass was increased from 8.9 to 16.3 μmol/m^2/s, so it had a photosynthetic rate similar or slightly less than that of big bluestem under drought (average photosynthetic rate of big bluestem under the dry conditions and high CO_2 was 20.8 μmol/m^2/s). With elevated CO_2 and high water, Kentucky bluegrass had a photosynthetic rate almost the same to that of big bluestem under both high moisture (Figure 1.8, upper right) and low moisture (Figure 1.8, lower right). The results showed that the photosynthetic rate of the C_3 grass under elevated CO_2 could be increased to a rate about as high as that of the C_4 grass under ambient CO_2. However,

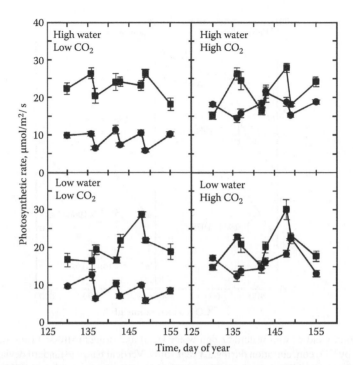

FIGURE 1.8   Net photosynthetic rate of Kentucky bluegrass (circles) and big bluestem (squares) grown under a high (715 μmol/mol) and a low (354 μmol/mol) atmospheric CO_2 concentration and a high (field capacity) and a low (no water added) soil moisture. Means of 8–16 measurements. Vertical bars = ±standard error. (With kind permission from Springer Science+Business Media: *Photosynthetica*, Photosynthesis of a C_3 grass and a C_4 grass under elevated CO_2, 26, 1992a, 189–198, Nie, D., He, H., Kirkham, M.B., and Kanemasu, E.T., Fig. 1.)

even though the photosynthetic rate of Kentucky bluegrass was increased under the elevated CO_2, its average photosynthetic rate was still less than that of big bluestem. The average photosynthetic rates during the season for big bluestem and Kentucky bluegrass grown under the elevated CO_2 and high water level were 21.8 and 17.8 μmol/m²/s, respectively. These values suggest that if Kentucky bluegrass is to have a photosynthetic rate similar to that of big bluestem, the ambient level of CO_2 must be more than doubled.

In another field study, double-the-ambient level of CO_2 (658 vs. 337 μmol/mol) did not affect the rate of photosynthesis of big bluestem under either a high water level (field capacity) (Figure 1.9, top) or a low water level (half field capacity (Figure 1.9, bottom) (Kirkham et al. 1991). Other studies show that the photosynthetic rate of C_4 plants does not increase when the CO_2 concentration is augmented (Patterson and Flint 1990; Ziska et al. 1990). Reviews (Morison 1993; Idso and Idso 1994) indicate that the increase in growth caused by elevated CO_2 is usually greater in drought-stressed plants than in well-watered ones. However, it appears that it will be the C_3 plants that will benefit most from augmented CO_2 under dry conditions.

A 2 year (1993–1994) free-air CO_2 enrichment (FACE) experiment was done in the desert region of the southern United States (Maricopa, Arizona) on a clay loam soil to see the effects of elevated CO_2 and drought on spring wheat (*T. aestivum* L. cv. Yecora Rojo) (Wall et al. 2006). (We describe how FACE experiments are set up in the first section of Chapter 6.) Plants were exposed to ambient levels of CO_2 (370 μmol/mol or ambient + 180 μmol/mol; i.e., 550 μmol/mol) (FACE plants) under ample and reduced water supplies (100% and 50% replacement of evapotranspiration, respectively). Photosynthetic rate varied according to the stage of development (Figure 1.10). When inflorescences were emerging (75 days after emergence; Figure 1.10a), the FACE plants and the watered control plants had similar photosynthetic rates. Elevated CO_2 compensated for reductions in photosynthetic

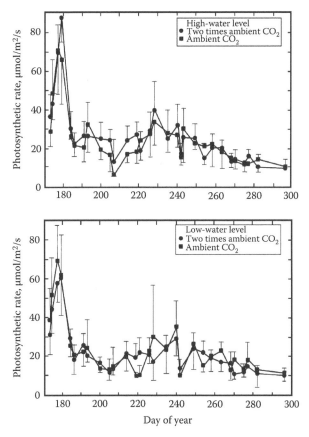

FIGURE 1.9 Photosynthetic rate of big bluestem leaves with two atmospheric CO_2 levels and two soil–water levels. (Reprinted from Kirkham, M.B. et al., *Crop Sci.*, 31, 1589, 1991, Fig. 4. With permission from the American Society of Agronomy.)

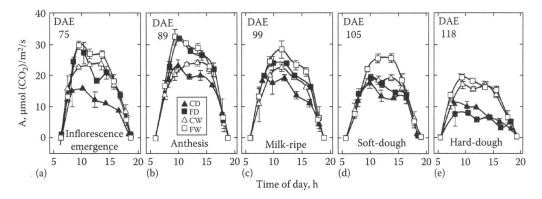

FIGURE 1.10 Dawn to dusk trends in mean leaf net assimilation rate (A) of fully expanded sunlit wheat leaves for days after 50% emergence (DAE) and five developmental stages during a 1993 experiment in Maricopa, Arizona. Plants were exposed to ambient (370 μmol/mol) or free-air CO_2 enrichment (FACE: 550 μmol/mol CO_2) under ample (wet) and reduced (dry) water supplies (100% and 50% replacement of evapotranspiration, respectively). Symbols: Open squares: CO_2-enriched plants, wet; closed squares: CO_2-enriched plants, dry; open triangles: ambient level CO_2, wet; closed triangles: ambient level CO_2, dry. (Reprinted from Wall, G.W. et al., *Agron. J.*, 98, 354, 2006, Fig. 6. With permission from the American Society of Agronomy.)

rate due to drought. The FACE wet and dry plants had the same photosynthetic rate. At the milk-ripe stage (99 days after emergence; Figure 1.10c) and soft-dough stage (105 days after emergence; Figure 1.10d), the photosynthetic rate of the FACE wheat was higher than that of plants under the other treatments. But by the hard-dough stage (118 days after emergence; Figure 1.10e), plants under the well-watered conditions had a higher photosynthetic rate than plants under dry conditions, and CO_2 enrichment had no effect on increasing the photosynthetic rate under drought.

In a later 2 year (1998–1999) study at the same FACE facility in Arizona with elevated CO_2 (561 μmol/mol; ambient CO_2 concentration was 368 μmol/mol), sorghum plants (*Sorghum bicolor* (L.) Moench 'Dekalb DK54') were grown under well-watered and dry conditions (Ottman et al. 2001). During each season, the well-watered plots received an average of 1132 mm of water, and the dry plots received 383 mm. At final harvest, the elevated CO_2 increased yield from 999 to 1151 g/m^2 in the dry plots, but it had no effect in the wet plots. Ottman et al. (2001) conclude that as CO_2 concentration in the atmosphere increases, sorghum yield is likely to be higher in the future in areas where water is limited.

CONTROLLED ENVIRONMENT STUDIES WITH CROPS

Studies done under controlled environment conditions have confirmed the field studies done by Chaudhuri et al. (1990), Kirkham et al. (1992), Nie et al. (1992a), and Wall et al. (2006). Kaddour and Fuller (2004) grew three Syrian cultivars of durum wheat (*T. durum* Desf.) under controlled environmental conditions with either 400 μmol/mol CO_2 or 1000 μmol/mol CO_2 at two levels of water availability. Plants grew in a loam-based compost. The fully irrigated pots were given water to reestablish 90% of available water content (about pot capacity) once the soil water availability had reached 60% available water capacity. Water in the pots in the drought-stressed treatment was restored to 70% available water capacity when the soil water availability had reached 45% available water content. All cultivars responded in a similar manner to both the CO_2 and water stress. Raising the CO_2 level enabled plants to compensate for the low water availability, so that plants at high CO_2 with drought stress had the same ear number as those under ambient CO_2 and irrigated conditions (Figure 1.11). Leaf area index at ear emergence was depressed by low water availability but was unaffected by CO_2 level (Figure 1.12). Biomass production followed the same pattern as for ear number, with CO_2 improving dry weight under both water-stressed and irrigated treatments (Figure 1.13).

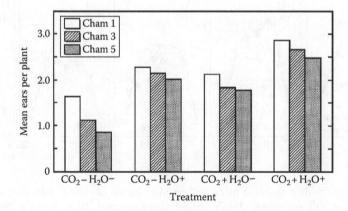

FIGURE 1.11   Effect of elevated CO_2 (1000 μmol/mol) and water stress on ear production of three durum wheat cultivars registered in Syria (Cham 1, Cham 3, and Cham 5). Control plants grew at an atmospheric CO_2 concentration of 400 μmol/mol CO_2. CO_2-=control CO_2; CO_2+=elevated CO_2; H_2O-=water restricted; H_2O+=full irrigation. (Reprinted from Kaddour, A.A. and Fuller, M.P., *Cereal Res. Commun.*, 32, 225, 2004, Fig. 2. With permission from Akadémiai Kiadó.)

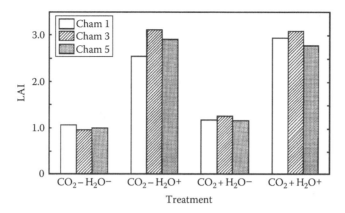

FIGURE 1.12 Effect of elevated CO_2 (1000 μmol/mol) and water stress on leaf area index (LAI) at ear emergence of durum wheat cultivars registered in Syria (Cham 1, Cham 3, and Cham 5). Symbols are the same as in Figure 1.11. (Reprinted from Kaddour, A.A. and Fuller, M.P., *Cereal Res. Commun.*, 32, 225, 2004, Fig. 3. With permission from Akadémiai Kiadó.)

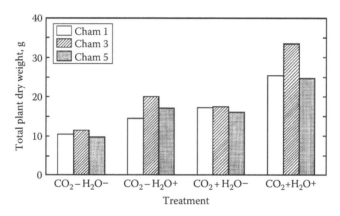

FIGURE 1.13 Effect of elevated CO_2 (1000 μmol/mol) and water stress on vegetative biomass production at ear emergence of three durum wheat cultivars registered in Syria (Cham 1, Cham 3, and Cham 5). Symbols are the same as in Figure 1.11. (Reprinted from Kaddour, A.A. and Fuller, M.P., *Cereal Res. Commun.*, 32, 225, 2004, Fig. 4. With permission from Akadémiai Kiadó.)

Müller (1993) studied the influence of two CO_2 concentrations (ambient or 340 μmol/mol and elevated or 800 μmol/mol) and of drought stress on dry matter production of two winter wheat cultivars (Alcedo and Norman) grown in pots with a mixture of sand and a sandy loam soil, which were placed in two climate chambers. Soil with the well-watered and drought-stressed plants was maintained between 60%–80% and 40%–50% of the maximum water holding capacity of the soil, respectively. As in other studies (Chaudhuri et al. 1990; Kaddour and Fuller 2004; Wall et al. 2006), he saw that elevated CO_2 compensated for reductions in growth due to drought stress. Similarly, André and Du Cloux (1993) also found that doubling of CO_2 (from 330 to 660 μmol/mol) compensated for water-stress-induced inhibition of photosynthesis of wheat (*T. aestivum* L. cv. Capitole) grown in closed chambers.

Another C_3 grass, rice, also has increased photosynthesis under drought when the CO_2 level is raised. Widodo et al. (2003) measured the midday photosynthetic rate of rice (*Oryza sativa* L. cv. IR-72) grown in sunlit, controlled-environment chambers at two CO_2 concentrations (ambient or 350 μmol/mol and elevated or 700 μmol/mol). Four water-management regimes were imposed: continuously flooded; drought stress imposed during panicle initiation; drought stress imposed during anthesis; and drought stress imposed at both stages, that is, panicle initiation and anthesis. When rice plants reached the developmental stage required for initiating drought treatments, the paddy water was drained from the bottom of the soil bins of the chambers. Drought was terminated on the day when rice plants desiccated to the point of near-zero photosynthetic rate in high light at midday, and water was restored in the afternoons. At elevated CO_2 concentration, photosynthetic rates were high at most sampling dates. Under flooded conditions, average photosynthetic rates of plants grown under 700 and 350 μmol/mol CO_2 averaged about 35 and 25 μmol/m²/s. Near the end of drought periods, water deficit caused decreases in the photosynthetic rate. These drought-induced effects were more severe for plants grown at the ambient than at the elevated concentration of CO_2. Plants grown under elevated CO_2 concentration were able to maintain midday leaf photosynthesis longer into the drought period than plants grown at ambient CO_2 concentration. By 76 days after planting, leaves of the plants grown with 700 μmol/mol CO_2 with drought imposed at panicle initiation and at both stages (panicle initiation and anthesis) maintained moderate photosynthetic rates (15 and 11 μmol CO_2/m²/s, respectively), while leaves of the plants grown with 350 μmol/mol CO_2 with drought imposed at panicle initiation and at both stages had low midday photosynthetic rates (1 and 4 μmol CO_2/m²/s, respectively). In addition, midday leaf photosynthetic rates recovered from water deficit more rapidly in the elevated CO_2 treatment. While doubled CO_2 did not compensate fully for reductions in photosynthesis due to drought, the elevated CO_2 did increase the photosynthetic rate of water-stressed rice. Thus, in the absence of other potential climate stresses, rice grown under future increases in atmospheric CO_2 concentration may be better able to tolerate drought (Widodo et al. 2003).

As noted, soybean, a legume, is a major crop grown in the humid Midwest of the United States. Allen et al. (1994) determined photosynthetic rate and transpiration rate of soybean [*Glycine max* (L.) Merr. cv. Bragg; Fabacea (formerly called Leguminosae) or the pulse family], grown under two moisture regimes in controlled-environment, outdoor chambers at 330 and 660 μmol/mol CO_2. Well-watered plants were kept at field capacity, and drought-stressed plants received no water during a 13 day period. Leaflets at high CO_2, either water stressed or well watered, had higher photosynthetic rates (Figure 1.14a) and lower transpiration rates (Figure 1.14b), and, therefore, higher water use efficiencies (calculated by dividing the photosynthetic rate by the transpiration rate; Figure 1.14c) than those at control CO_2 levels. Photosynthetic rates of soybean grown under drought stress and elevated CO_2 were higher than those of soybean grown under well-watered, ambient CO_2 during the first three days of the water-stress treatment (Figure 1.14a). At the end of the experiment, soybean grown with the elevated CO_2 under drought had a photosynthetic rate similar to that of well-watered soybean under ambient levels of CO_2. So, in this experiment with soybean (Allen et al. 1994), elevated CO_2 only compensated for reductions in photosynthesis due to water stress during early stages of drought stress.

TREES

We shall first consider experiments in which gymnosperm trees and angiosperm trees have been studied separately. In gymnosperms, the ovules are borne in an exposed position on the sporophyll (a modified leaf that bears sporangia; a sporangium is a spore case or a single cell giving rise to spores) (Friend and Guralnik 1959). In contrast, the angiosperms or flowering plants have their ovules and seeds within a closed ovary. In angiosperms, the sporophyll refers to the stamens and carpels (Esau 1977, p. 527). In Greek, the word gymnosperm means naked seed and the word angiosperm means a seed in a case. We remember that gymnosperms are the more primitive group, and the flowering plants are highly evolved. Gymnosperms include ancient lines of seed-bearing plants. Extinct lines, recognized as gymnosperms, were conspicuous in the Paleozoic era (520–185 million years ago; see Figure 9.10 for the geological time chart) and are regarded as the ancestral stock from which the modern conifers may have originated (Foster and Gifford 1959, p. 320). Angiosperms constitute the dominant and most ubiquitous vascular plants of modern floras on the earth. The fragmentary

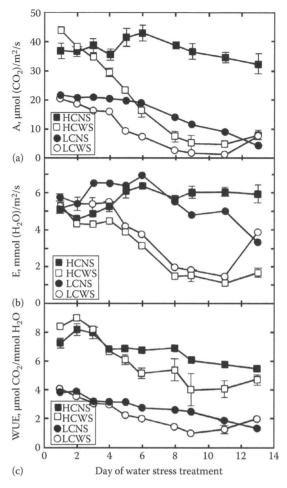

FIGURE 1.14 Midday averages of (a) assimilation rate (A), (b) transpiration rate (E), and (c) water use efficiency (WUE) of soybean leaflets grown at CO_2 concentrations of 660 µmol/mol (high) and 330 µmol/mol (low) under non-stressed and water-stressed conditions. The legends HCNS and HCWS label the high CO_2, non-stressed and water-stressed treatments, respectively, and LCNS and LCWS label the low CO_2, non-stressed and water-stressed treatments, respectively. (Reprinted from Allen, L.H. et al., *Agron. J.* 86, 625, 1994, Fig. 3. With permission from the American Society of Agronomy.)

evidence that we have concerning the evolution of angiosperms is provided by fossilized wood and especially by impressions of leaf form and venation. It shows that by the Middle Cretaceous, angiosperms had reached a high stage of morphological specialization. The angiosperms apparently suddenly appeared in the Cretaceous period (150–60 million years ago; Figure 9.10) and evolved at a much faster rate than that of the gymnosperms (Foster and Gifford 1959, p. 444).

Townend (1993) grew 2-year-old Sitka spruce [*Picea sitchensis* (Bongard) Carrière] plants, a gymnosperm in the Pinaceae or pine family, from four clones in boxes filled with peat under two moisture regimes in naturally lit growth chambers for 6 months at either ambient (350 µmol/mol) or augmented (600 µmol/mol) CO_2 concentrations to elucidate the plant's response to drought under elevated CO_2. Plants grew in boxes that were kept well watered or allowed to dry out slowly over the summer. Severe drought was not obtained, because the lowest soil matric potential (measured with tensiometers) under dry conditions was −0.05 MPa. The matric potential under the well-watered conditions was maintained at −0.005 MPa. Plants growing in elevated CO_2 showed a 6.9% increase in mean relative growth rate compared to controls in the drought treatment and a 9.8% increase

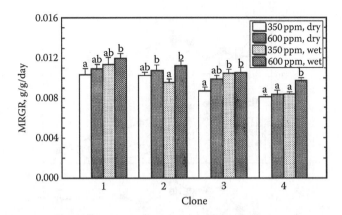

FIGURE 1.15 Mean relative growth rates (MRGR) over a 6 month growing period for four clones of Sitka spruce plants. Bars without a letter in common signify a significant difference ($P < 0.05$). Comparisons apply within clones only. Error bars indicate ± one standard deviation. (Reprinted from Townend, J., *Tree Physiol.*, Fig. 6, 1993 by permission of Oxford University Press.)

compared to controls in the well-watered treatment (Figure 1.15). However, only one clone (Clone 2; Figure 1.15) had increased growth under elevated CO_2 and drought compared to plants grown with ambient CO_2. Clone 2 plants that grew with 600 µmol/mol CO_2 and drought had a higher growth rate than Clone 2 plants that grew with 350 µmol/mol CO_2 and well-watered conditions. This result indicated that elevated CO_2 could compensate for reductions in growth of spruce due to a mild drought, at least in one clone.

Liang and Maruyama (1995) assessed the effects of atmospheric CO_2 enrichment and drought stress on photosynthetic rate, leaf conductance, transpiration, and WUE in 2-year-old alder (*Alnus firma* Sieb. et Zucc.), an angiosperm in the Corylaceae or hazelnut family. It is a common pioneer tree species in the cool-temperate mountain zone in Japan. Measurements were conducted in a controlled environment laboratory at three CO_2 concentrations: 350 (ambient), 600, and 900 µmol/mol. Plants were grown in a mixture of loam, gravel, and vermiculite. Five watering regimes were imposed. Plants were watered at the following leaf-water potentials (measured with a pressure chamber): water potential of higher than −0.3 MPa (well watered); −0.5; −0.8, −1.0 MPa; and lower than −1.2 MPa (serious drought stress). Water was added daily to the well-watered control to maintain the soil at pot capacity. Under well-watered conditions, rates of photosynthesis increased with increasing CO_2 concentrations and leaf conductance decreased. With drought stress, photosynthesis, leaf conductance, and transpiration decreased. Nevertheless, under drought, photosynthesis was stimulated by elevated CO_2, while CO_2 had no effect on transpiration. Transpiration of CO_2-enriched plants was similar to that of plants grown at the ambient level of CO_2. The overall result was an increase in WUE with drought. They concluded that if a doubling of atmospheric CO_2 concentration occurs by the mid-twenty-first century, then the photosynthetic rate of *Alnus firma* in drought-affected regions may be expected to increase. But a reduction in water use is not indicated (Liang and Maruyama 1995). Kerstiens et al. (1995) pointed out that the total water consumption of a tree, which is dependent on leaf area, is a critical factor, in determining a plant's response to elevated CO_2. However, in semiarid regions, WUE is equally as important as total water used, as noted for wheat in Kansas (Chaudhuri et al. 1990).

Tschaplinski et al. (1995) found that elevated CO_2 compensated for drought-induced reductions in growth of three tree species, all angiosperms. They grew 1-year-old seedlings of American sycamore (*Platanus occidentalis* L.; Platanaceae or plane-tree family), sweet gum (*Liquidambar styraciflua* L.; Hamamelidaceae or witch-hazel family), and sugar maple (*Acer saccharum* Marsh.; Aceraceae or maple family) in pots placed in four open-top chambers containing either ambient air

(343 μmol/mol CO_2) or ambient air enriched with 300 μmol/mol CO_2 (652 μmol/mol average during the study). Two soil moisture regimes were included within each chamber: a well-watered treatment with plants watered daily and a drought treatment in which plants were subjected to a series of drought cycles. Duration and depth of the drought cycles were determined by soil matric potential. Mean soil water potentials at re-watering for the water-stressed seedlings under ambient CO_2 for sugar maple, sweet gum, and sycamore were −0.5, −0.7, and −1.8 MPa, respectively, compared with greater than −0.1 MPa for well-watered plants.

Under well-watered conditions, elevated CO_2 increased the growth rate of sugar maple by 181% resulting in a 4.3-fold increase in total plant dry weight after 81 days (Figure 1.16a), compared with 1.6- and 1.4-fold increases for sycamore (Figure 1.16b) and sweet gum (Figure 1.16c), respectively, after 69 days (Tschaplinski et al. 1995). Root/shoot ratios were not affected by elevated CO_2 in sycamore (Figure 1.16e) or sweet gum (Figure 1.16f). But the CO_2-induced increase in dry weight of sugar maple was preferentially allocated to shoots rather than roots under well-watered conditions

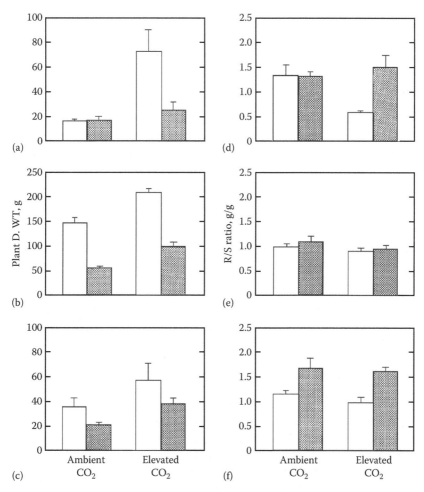

FIGURE 1.16 Effects of drought and atmospheric CO_2 concentration on total plant dry weight (a–c) and root/shoot (R/S) ratios (d–f) of sugar maple (top), American sycamore (middle), and sweet gum seedlings (bottom). Seedlings were grown in pots under either ambient (343 μmol/mol CO_2) or elevated (652 μmol/ mol CO_2), and under periodic drought cycles (filled bars) or were kept moist (open bars). Data are means and one standard error of the means of five or six plants per treatment. (Reprinted from Tschaplinski, T.J. et al., *New Phytol.* 129, 63, 1995, Fig. 1. With permission from New Phytologist, Lancaster University.)

(Figure 1.16d). This result contrasts with what is typically observed. Root growth usually increases under high CO_2. We shall study roots under elevated CO_2 in detail in Chapter 6.

Sugar maple under the drought-stressed, elevated CO_2 conditions grew as well as the sugar maple under the ambient CO_2 (either drought stressed or well watered) (Figure 1.16a). However, elevated CO_2 compensated, partially or fully, for drought-caused reductions in growth in sycamore and sweet gum. The compensation was full for sweet gum (Figure 1.16c). For sycamore, the compensation was only partial (Figure 1.16b). Tolley and Strain (1984b, 1985) reported that elevated CO_2 compensated fully for drought-induced reductions in growth of sweet gum. Drought-stressed sweet gum grown in 1000 µmol/mol CO_2 exhibited a photosynthetic rate similar to well-watered plants grown at ambient CO_2 (Tolley and Strain 1985). Sionit et al. (1985) found only a partial compensation for drought in sweet gum seedlings grown at 675 µmol/mol CO_2. Double the ambient level of CO_2 did not affect the photosynthetic rate of sweet cherry (*Prunus avium* L.; Rosaceae or rose family) under drought (Centritto 2005). These results suggest that the degree of compensation of photosynthetic rate to drought depends on the degree of drought experienced and the concentration of CO_2 (Tschaplinski et al. 1995).

Extrapolation of short-term responses of tree seedlings grown in pots to long-term responses of mature trees growing in a forest is risky (Tschaplinski et al. 1995). But results of such short-term studies may suggest how competitive relationships among trees under elevated CO_2 will change. Competition may be different on mesic versus xeric sites. The short-term growth stimulation of sugar maple by elevated CO_2 was greater than that of sycamore and sweet gum when plants were well watered, but this advantage was eliminated by drought. The three species studied by Tschaplinski et al. (1995) would rarely, if ever, be found together in the same forest. Longer-term studies with mature trees of species that typically coexist are essential to determine the impact of drought on forests, as the CO_2 concentration in the atmosphere increases. Sugar maple may benefit the most of the three species studied in a CO_2-enriched atmosphere, but productivity will be limited if frequent droughts occur.

To determine if high CO_2 concentrations may ameliorate the negative effects of drought, Miao et al. (1992) grew two angiosperms, gray birch (*Betula populifolia* Marsh.; Corylaceae) and red maple (*A. rubrum* L.; Aceraceae), exposed to four watering regimes with ambient (350 µmol/mol) and elevated (700 µmol/mol) atmosphere CO_2 levels. Plants grew under greenhouse conditions in pots filled with a mixture of soil, sand, and Turface. Turface is an artificial medium, somewhat like "kitty litter," used in greenhouse production (Jaeger and Hellmers 1981). Plants received one of four moisture treatments designated "very dry," "dry," "moist," and "flooded." Plants in very dry pots were watered only when they wilted. Dry pots were watered when the upper 2 cm of soil dried. Moist pots received water when the top soil appeared dry. Flooded pots were kept in standing water maintained to a height 2 cm below the soil surface.

The magnitude of growth enhancement due to CO_2 was dependent on soil moisture and species. Red maple showed the greatest response to elevated CO_2 under moist soil conditions (Figure 1.17b), while gray birch showed the greatest response under dry soil conditions (Figure 1.17a). Under "very dry" and "dry" conditions, doubled CO_2 increased the growth of gray birch more than that of red maple. However, the elevated CO_2 concentration increased growth of red maple, too, under dry conditions. The results showed that the magnitude of the response of gray birch and red maple seedlings to elevated CO_2 will be contingent on the availability of soil moisture. We shall return to soil moisture under elevated atmospheric CO_2 in Chapter 5.

Xiao et al. (2005) studied a leguminous tree (*Caragana intermedia* Kuanget H.C. Fu; Fabaceae) exposed to drought and elevated CO_2. The plant is a deciduous shrub that occurs in Inner Mongolia, China. It is typically about 0.7–3.0 m tall and has plumose, compound leaves. It reduces erosion and acts as a windbreak. Seedlings were grown in sand with a slow release fertilizer and placed in two greenhouses exposed to two CO_2 concentrations (350 or 700 µmol/mol) under three watering regimes: well watered (watered to 60%–70% of pot capacity); moderately watered (45%–55% of pot capacity); and drought stressed (30%–40% of pot capacity). Height, basal diameter, and

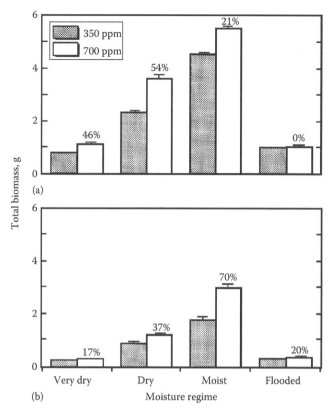

FIGURE 1.17 Total biomass of (a) gray birch and (b) red maple grown under two levels of CO_2 (350 and 700 μmol/mol) and four levels of moisture. Numbers on the top of lighter columns represent percent increase in total biomass of plants grown in 700 μmol/mol CO_2 relative to plants grown in 350 μmol/mol CO_2. Error bars refer to one standard error. (Reprinted with kind permission from Springer Science+Business Media: *Oecologia*, Elevated CO_2 differentially alters the responses of cooccurring birch and maple seedlings to a moisture gradient, 90, 1992, 300–304, Miao, S.L., Wayne, P.M., and Bazzaz, F.A., Fig. 1.)

number of leaves of plants grown under the well-watered conditions and with the 350 μmol/mol CO_2 concentration were similar to those of plants grown under the two water-stressed treatments and 700 μmol/mol CO_2. So elevated CO_2 compensated for reductions in growth due to drought. The authors concluded that elevated CO_2 may weaken the effect of drought stress on growth of *C. intermedia* seedlings.

GYMNOSPERM VERSUS ANGIOSPERM TREES

While a tremendous amount of work has been done studying and modeling trees under elevated CO_2 (Thornley and Cannell 1996; Medlyn et al. 1999; Valentini et al. 2000), few studies apparently have compared specifically gymnosperms and angiosperms under elevated CO_2 and drought, even though models have considered them separately (Whitehead et al. 2001, 2004). Gymnosperms dominate the cold regions, because of their vascular system. The two conducting cells in the xylem tissue are tracheids and vessel members. Angiosperms, the most highly evolved plants, have both tracheids and vessel members, while gymnosperms have only tracheids (Kirkham 2005, p. 318). Water in tracheids is less likely to cavitate because they are narrower than the vessel elements, which can be wide and long at maturity. Imagine a northern tree thawing in the spring. As the ice melts, the tracheids become filled with liquid containing the many bubbles of air that had been forced out by freezing. As melting continues and transpiration begins, tension begins to develop in the xylem tissue. Because of the small dimensions of the tracheids involved, the pressure difference across the curved air–water interface bounding the bubbles would be considerable, resulting in much higher pressure in an air bubble than would exist in the water. Any bubbles that form should dissolve fairly readily, restoring the integrity of the water column (Kirkham 2005, p. 327). This process could not occur in the wide and long vessel members.

In addition to the difference in vascular cells between gymnosperms and angiosperms, the distribution and positioning of stomata differs between the two classes of plants. The epidermis of the conifer leaf of pine (*Pinus*), for example, is heavily cuticularized and has such thick walls that the spaces between the cells are almost obliterated (Esau 1965, p. 446). The fiber-like cells of the hypodermis (a layer or layers of cells beneath the epidermis and distinct from the underlying ground tissue cells; Esau 1977, p. 512) also have thick walls and form a compact layer interrupted only beneath the stomata. The epidermis bears numerous stomata on one or all sides in different conifers. The front cavity of the stoma is typically filled with a whitish or dark, granular or alveolate (honeycombed) occluding material (Esau 1965, p. 446). Since this material is porous and the pores are filled with air, the stomata appear white superficially. The guard cells are sunken and are overtopped by the subsidiary cells. The shape of stomata in gymnosperms differs from that in C_4 plants like monocotyledonous grasses, which are dumbbell-shaped, and C_3 dicotyledonous plants, which are elliptical (Kirkham 2005, p. 382). In the angiosperms, the stomata are distributed all over the leaves, even though leaves of tree species often have stomata only on the lower (abaxial) surface (Kirkham 2005, p. 380; see also Table 9.1). This arrangement of stomata in angiosperms contrasts with that in gymnosperms. In many genera, including *Pinus*, the stomata occur in longitudinal rows parallel with the vascular bundles (Esau 1965, p. 446).

Because of the difference in vascular and stomatal anatomy between gymnosperm and angiosperms, one would expect that the two classes of plants might respond differently to elevated CO_2 under drought. Tolley and Strain (1984b, 1985) grew an angiosperm (*L. styraciflua* L., sweet gum) and a gymnosperm (*Pinus taeda* L., loblolly pine) under elevated CO_2 and drought, even though they made no comparison specifically between the two classes of plants. The two species co-occur in forests in the Piedmont of the United States (Sionit et al. 1985). The Piedmont is a plateau between the Atlantic coast and the Appalachian Mountains, covering parts of Alabama, Georgia, South Carolina, North Carolina, and Virginia (Friend and Guralnik 1959). The two plants are early successional tree species, which commonly invade fields of the North Carolina Piedmont (Tolley and Strain 1985). However, the drier microsites of abandoned fields in the North Carolina Piedmont may frequently be dominated by loblolly pine as the first arboreal species (Tolley and Strain 1984b).

In a study without water stress (Tolley and Strain 1984a), the two species were exposed to three levels of CO_2 (350 or ambient, 675, and 1000 µmol/mol). Elevated CO_2 did not increase growth of loblolly pine, while it increased height, leaf area, basal stem diameter, and total dry weight of sweet gum. In a subsequent study, Tolley and Strain (1984b) subjected the two tree species to water stress using the same three CO_2 levels (350, 675, and 1000 µmol/mol). Plants grew in pots in walk-in growth chambers in the Duke University Phytotron in Durham, North Carolina. The substrate was mixture of gravel, vermiculite, and Turface. Half of the seedlings in each CO_2 treatment had water withheld until plant water potentials reached about −2.5 MPa, while the remaining plants were well watered. At the end of the drying cycle, stressed plants were returned to well-watered conditions for a 14 day recovery period. After the 2 week recovery period, stressed loblolly pine and sweet gum seedlings grown at 1000 µmol/mol CO_2 had total dry weights similar to corresponding control values. For sweet gum seedlings in particular, the reduction of early seedling growth following exposure to a period of drought under normal atmospheric CO_2 concentration was ameliorated by growing plants under elevated CO_2.

Tolley and Strain (1985) also measured gas exchange rates of the two species. Carbon dioxide enrichment changed little the photosynthetic rates of either water-stressed or well-watered loblolly pine seedlings (Figure 1.18, right). In contrast, the effect of water stress was delayed for

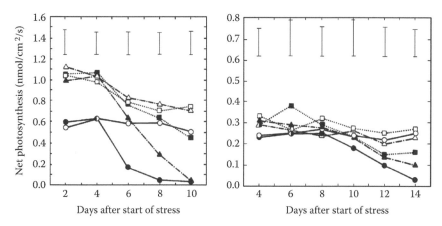

FIGURE 1.18 Net photosynthesis over a drying cycle for well-watered (open symbols) and water-stressed (closed symbols) seedlings grown at 350 (open and closed circles), 675 (open and closed triangles), and 1000 (open and closed squares) μmol/mol CO_2. Left: sweet gum; right: loblolly pine. Each value is the average of three measurements. Error bars are Tukey's HSD (honestly significant difference test) values at $P \leq 0.05$. Note the difference in magnitude of the photosynthetic scale for sweet gum and pine. (Reprinted with kind permission from Springer Science+Business Media: *Oecologia*, Effects of CO_2 enrichment and water stress on gas exchange of *Liquidambar styraciflua* and *Pinus taeda* seedlings grown under different irradiance levels, 65, 1985, 166–172, Tolley, L.C. and Strain, B.R., Fig. 2.)

sweet gum seedlings grown at elevated CO_2, particularly at 1000 μmol/mol CO_2 (Figure 1.18, left). Development of internal plant water deficits was slower for stressed sweet gum seedling grown at elevated CO_2. As a result, these seedlings maintained higher photosynthetic rates over the drying cycle than stressed sweet gum seedlings grown at 350 μmol/mol CO_2 and stressed loblolly pine seedlings grown at ambient and enriched CO_2 levels. Tolley and Strain (1984b, 1985) concluded that with the future increases in atmospheric CO_2 concentrations, sweet gum seedlings should be able to tolerate longer exposure to low soil moisture, resulting in a greater survival of seedlings on dry sites of abandoned fields in the North Carolina Piedmont, now dominated by loblolly pine.

Growth and transpiration efficiency (biomass accumulation/water consumption) of two temperate-forest tree species exposed to ambient and elevated CO_2 concentrations (350 and 700 μmol/mol, respectively) were investigated under well-watered and drought-stressed conditions by Guehl et al. (1994). The trees were *Quercus petraea* Lieblein (Durmast oak; Fagaceae or beech family) and *P. pinaster* Aiton (cluster pine; Pinaceae). The plants grew in a mixture of sphagnum peat and sand. Half of the plants were well watered by restoring soil water content to pot capacity twice a week, and half of the plants were subjected to a soil drought by restoring soil water content to 40% of pot capacity twice a week. Under well-watered conditions, doubling the CO_2 concentration for one growing season increased biomass by 138% in *Q. petraea* and 68% in *P. pinaster* (Figure 1.19, top). In contrast, under drought, elevated CO_2 increased biomass by only 47% in *Q. petraea* and had no significant effect on biomass growth in *P. pinaster*. Transpiration efficiency was higher in *Q. petraea* than in *P. pinaster* in all treatments (Figure 1.19, bottom). The CO_2-promoted increase in transpiration efficiency was higher in *Q. petraea* (about 80%) than in *P. pinaster* (about 50%). The results showed that the elevated CO_2 fully compensated for reductions in growth due to drought of *Q. petraea*, an angiosperm, but had no effect in alleviating growth reduction in *P. pinaster*, a gymnosperm.

Taken together, the results indicate that several angiosperm trees [e.g., sycamore (Tschaplinski et al. 1995); sweet gum (Tolley and Strain 1984a,b; Tschaplinski et al. 1995); gray birch and red maple (Miao et al. 1992); *C. intermedia*, a leguminous tree (Xiao et al. 2005); Durmast oak (Guehl et al. 1994)] will grow better than gymnosperm trees [e.g., Sitka spruce (Townend 1993); loblolly pine (Tolley and Strain 1984b, 1985); cluster pine (Guehl et al. 1994)] under drought as the CO_2

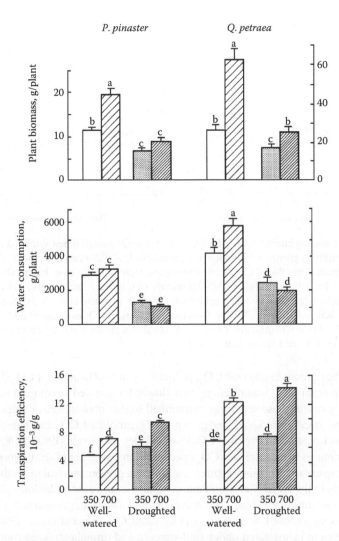

FIGURE 1.19 Growing season integrated whole-plant biomass accumulation, water consumption, and transpiration efficiency for *P. pinaster* (cluster pine) and *Q. petraea* (Durmast oak) grown under two CO_2 concentrations (350 and 700 µmol/mol) and two water treatments (well-watered or pot capacity and drought-stressed or watered when soil water content reached 40% of pot capacity). Mean values (± standard errors) not sharing common letters are significantly different ($P < 0.05$). (Reprinted from Guehl, J.M. et al., *Tree Physiol.*, 14, Fig. 1, 1994 by permission of Oxford University Press.)

concentration in the atmosphere increases. Townend (1993) did observe an increase in growth of one clone of Sitka spruce, out of the four he studied, due to elevated CO_2 when the plants were exposed to a mild drought. When the two classes of plants (angiosperms and gymnosperms) have been compared in the same experiment (Tolley and Strain 1984b, 1985; Guehl et al. 1994), elevated CO_2 has not affected the photosynthetic rate or growth of gymnosperms under drought, but angiosperms have responded. I do not know the reason for the nonresponsiveness of gymnosperms to elevated CO_2 under drought. Gymnosperms often have evergreen leaves that are needle shaped. In contrast, angiosperm trees usually produce new, broad leaves in the spring and must grow fast to produce seeds before winter and dormancy. The flat leaves with stomata spread out over the surfaces apparently are well adapted to take up CO_2 for fast growth. The gymnosperms are not under such constraints and can grow slowly, yet survive.

The lack of response of gymnosperms, C_3 plants, to elevated CO_2 under drought is like that of C_4 plants. However, C_4 plants have a physiological reason to be nonresponsive to elevated CO_2. They recycle CO_2 within their leaves, unlike C_3 plants, so C_4 plants have enough CO_2 for maximum photosynthesis and growth even without elevated levels of CO_2 in the atmosphere. Concentrations of CO_2 above about 300 μmol/mol often appear not to affect the photosynthetic rate of C_4 plants (Figure 1.5) (Waggoner 1984). However, this cannot be considered as a general rule, because C_4 plants do increase their rate of photosynthesis at concentrations above 300 μmol/ mol (see Chapter 12).

CAM PLANT

Luo and Nobel (1993) studied the response of a CAM plant (*Opuntia ficus-indica* Mill.; Indian fig; Cactaceae) to elevated CO_2. They did not expose the plants to drought under elevated CO_2. But it is important to see how plants in the cactus family might respond to elevated CO_2, because they grow in arid regions. As noted, CAM plants keep their stomata open at night, unlike C_3 and C_4 plants that close their stomata at night. Cladodes (35 cm long, 20 cm wide, 2 cm thick) were planted with one-third of their surface area below the soil surface in pots containing a loamy sand. A cladode is a flattened leaf-like stem (Allen 1990). The pots were placed in two growth chambers with CO_2 concentrations of 360 or 720 μmol/mol. Plants were well watered, and the soil water potential in the root zone was above −0.3 MPa. Relative growth rates of the plants were comparable to values for productive C_3 and C_4 plants. Under the elevated CO_2 concentration, biomass of newly initiated cladodes was 7% higher (Figure 1.20), and area was 8% less (Figure 1.21), compared to plants grown with 360 μmol/mol CO_2. The combination of more biomass and less area caused the specific cladode mass to be 16% higher for the doubled CO_2 concentration. (Specific mass is the biomass divided by the area.) Luo and Nobel (1993) noted that the basal cladodes of *O. ficus-indica* had a large capacity for storage of water, which could buffer environmental stresses such as drought and thereby reduce their effects on growth of daughter cladodes. The results indicate that cacti, which

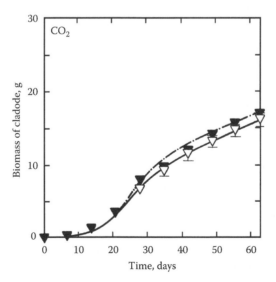

FIGURE 1.20 Biomass of cladodes of *O. ficus-indica* grown under two CO_2 levels: 360 μmol/mol (open triangles) or 720 μmol/mol (closed triangles). (Reprinted from Luo, Y. and Nobel, P.S.: Growth characteristics of newly initiated cladodes of *Opuntia ficus-indica* as affected by shading, drought and elevated CO_2, *Physiol. Plant.*, 1993, 87, 467–474, Fig. 1. Copyright Wiley-VCH Verlag GmbH & Co. KGaA. Reproduced with permission.)

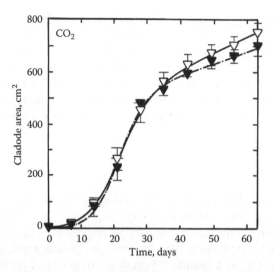

FIGURE 1.21 Cladode area of *O. ficus-indica* (Indian fig) grown under two CO_2 levels: 360 μmol/mol (open triangles) or 720 μmol/mol (closed triangles). (Reprinted from Luo, Y. and Nobel, P.S.: Growth characteristics of newly initiated cladodes of *Opuntia ficus-indica* as affected by shading, drought and elevated CO_2, *Physiol. Plant.*, 1993, 87, 467–474, Fig. 2. Copyright Wiley-VCH Verlag GmbH & Co. KGaA. Reproduced with permission.)

naturally live under aridity in deserts, may grow better as the CO_2 concentration in the atmosphere increases. They also suggest that the time of day that a plant takes up its CO_2 (night for the CAM plants) does not affect its responsiveness to elevated CO_2.

SALINITY

Because drought and salinity often occur together, it is important to consider plants exposed to salinity and elevated CO_2. Zeroni and Gale (1989) in the Negev, a southern region in Israel of partially reclaimed desert (Friend and Guralnik 1959), studied rose (*Rosa hybrida* cv. Sonia; Rosacea), a salt-sensitive plant, grown under salt stress in growth chambers simulating conditions of winter in greenhouses in the desert in Israel. The roses were placed in chambers with one of three concentrations of CO_2 (320, 600, and 1200 μmol/mol). Plants grew in pots containing volcanic cinder and were fertilized with a nutrient solution. The plants were irrigated with water containing one of four levels of salt. Local, brackish well water was used that contained a total soluble salt level of 3436 g/m³. This is equivalent to an electrical conductivity of about 5 millimhos/cm or 5 deciSiemens/m (dS/m) (Richards 1954, p. 11). The concentrations used in the experiment were 0%, 25%, 50%, 75%, and 100% of the well water concentrations, or 0, 859, 1718, 2577, and 3436 g/m³. Plants died when treated with 3436 g/m³ salt. Subsequently, the highest concentration used was 2577 g/m³ total soluble salt. Salinity tolerance increased at the two high concentrations of CO_2 (Figure 1.22).

In addition, raising the salt concentration in the presence of either 600 or 1200 μmol/mol CO_2 caused a decrease in blindness compared with either the controls without salt or the plants grown at 320 μmol/mol CO_2. Blindness is a problem in rose production (Laurie et al. 1969, pp. 327–351). The normal flowering shoot on a greenhouse rose possesses fully expanded sepals, petals, and reproductive parts (pistils and stamens). The failure to develop a flower on the apical end of the stem is a common occurrence; such shoots are termed "blind." The sepals and petals are present, but the reproductive parts are absent or aborted. The cause of blind shoots is not known. Severe defoliation will often temporarily increase blindness (Laurie et al. 1969, p. 340). Zeroni and Gale

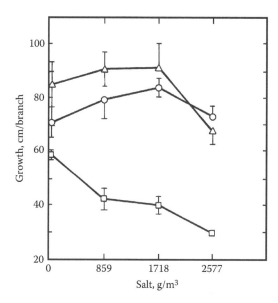

FIGURE 1.22 Branch length of "Sonia" roses grown at four levels of salinity and three levels of CO_2: 320 μmol/mol (squares); 600 μmol/mol (circles); and 1200 μmol/mol (triangles). (Reprinted from Zeroni, M. and Gale, J., *J. Hort. Sci.*, 64, 503, 1989, Fig. 5. With permission from the Headley Brothers, Ltd., The Invicta Press.)

(1989) observed that the decrease in blindness was accompanied by an increase in cane length and, in some cases, increases in flower diameter and fresh weight. The results demonstrated that growers of roses in desert regions could irrigate the plants with local water that may be brackish if the CO_2 is supplemented, as can be done in climate-controlled greenhouses. In addition, the results showed that like water stress, salt stress can be ameliorated by elevated levels of CO_2 (Zeroni and Gale 1989).

Nicolas et al. (1993) also observed that salinity tolerance of two types of wheat (*T. aestivum* L. cv. Matong and *T. durum* Desf. cv. Modoc) was improved under an elevated level of CO_2. Modoc is more salt-sensitive than Matong. The wheat plants were grown for 6 weeks in two greenhouses in Canberra, Australia, one with ambient CO_2 (350 μmol/mol) and the other with double the ambient level (700 μmol/mol CO_2). The plants grew in polyvinyl tubes filled with a potting mix consisting of river loam, peat moss, and river sand with fertilizer added. A solution of 150 mol/m³ NaCl + 10 mol/m³ $CaCl_2$ was applied to the salt-treated plants, while the nonsaline controls were irrigated with water. Using the conversion chart of Richards (1954, p. 11), 150 mol/m³ NaCl has an electrical conductivity of 18 millimhos/cm, or 18 dS/m and 10 mol/m³ $CaCl_2$ has an electrical conductivity of 2.3 millimhos/cm or 2.3 dS/m. Doubling the CO_2 concentration increased shoot and root growth of each cultivar under both saline and nonsaline conditions (Figure 1.23). Under salinity, high CO_2 increased dry matter accumulation of the salt-sensitive Modoc to a greater extent (+104%) than that of the more salt-tolerant Matong (+73%). In the salt treatment, shoot growth was increased to a greater extent than root growth (Figure 1.23, right). Leaf area showed similar trends to shoot dry weight (Figure 1.24). Salinity reduced tillering, but high CO_2 partly reversed the effects of salinity (Figure 1.25). Their results showed that high CO_2 increased growth by stimulating the development of tiller buds that would otherwise have been inhibited. Growth may also be increased under salinity stress in an atmosphere with augmented CO_2, because oxidative damage is mitigated by elevated CO_2 (Pérez-López et al. 2009). For a discussion of oxidative stress, see Bray et al. (2000, pp. 1189–1197).

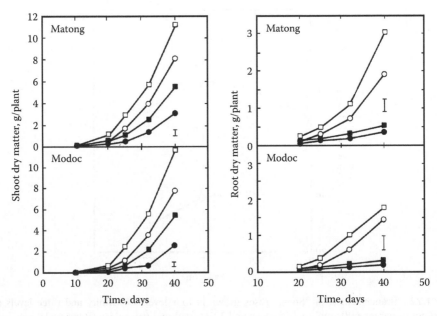

FIGURE 1.23 Effects of salinity and CO_2 level on shoot and root dry matter for two wheat cultivars "Matong" and "Modoc." Closed circles: salt, ambient CO_2 (350 μmol/mol); closed squares: salt, high CO_2 (700 μmol/mol); open circles: no salt, ambient CO_2; open squares: no salt, high CO_2. Bars represent the least significant difference ($P = 0.05$) in all figures. (Reprinted from Nicolas, M.E. et al., *Austral. J. Plant Physiol.*, 20, 349, 1993, Fig. 1. With permission from CSIRO Publishing.)

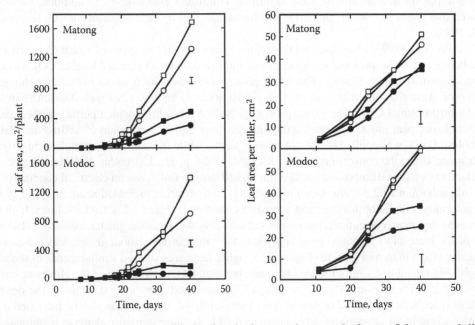

FIGURE 1.24 Effects of salinity and CO_2 level on leaf area and average leaf area of the emerged tillers for two wheat cultivars "Matong" and "Modoc." The symbols are the same as those defined in the legend of Figure 1.23. (Reprinted from Nicolas, M.E. et al., *Austral. J. Plant Physiol.*, 20, 349, 1993, Fig. 3. With permission from CSIRO Publishing.)

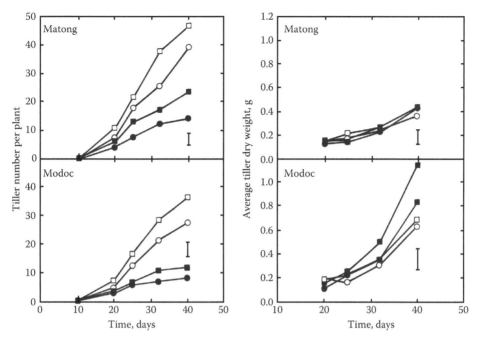

FIGURE 1.25 Effects of salinity and CO_2 level on the total number of tillers per plant and the average dry weight of the emerged tillers for two wheat cultivars "Matong" and "Modoc." The total tiller number includes all tiller buds visible to the naked eye. The symbols are the same as those defined in the legend of Figure 1.23. (Reprinted from Nicolas, M.E. et al., *Austral. J. Plant Physiol.*, 20, 349, 1993, Fig. 2. With permission from CSIRO Publishing.)

SUMMARY

Elevated CO_2 has either completely or partially compensated for reductions in growth and photosynthetic rate of the C_3 plants, wheat, soybean, and rice due to drought. The degree of compensation of drought-induced reduction in growth depends upon the level of CO_2, the species, the stage of plant growth, and the severity of the moisture stress. The data indicate that in a CO_2-increased world, C_3 plants will survive better under drought compared to a low-CO_2 world. Results of research cited in this chapter indicated that elevated CO_2 does not increase the photosynthetic rate of C_4 plants under drought. Thus, in the area that is predicted to be most affected by elevated CO_2 (latitudes between 37° and 47° N), C_3 plants, such as wheat, which is widely grown in the region, should benefit more from elevated CO_2 than C_4 plants, if the climate becomes drier. However, at ambient CO_2 levels, C_4 plants have a higher photosynthetic rate than C_3 plants, both under well-watered and drought-stressed conditions. In general, the overall effect of elevated CO_2 is to make C_3 plants appear to respond the same as C_4 plants, as far as their photosynthetic rate is concerned. Data indicate that as CO_2 concentration in the atmosphere increases, yield of C_4 plants will be maintained under drought, and yield of C_3 plants will increase such that it will become more similar to that of C_4 plants under drought.

Elevated CO_2 has compensated partially or fully for reductions in growth and photosynthetic rate of several angiosperm trees due to drought. However, experiments show that some gymnosperm trees do not respond to elevated levels of CO_2 under drought. If gymnosperms and angiosperms grow in the same forest, angiosperms may be better able to compete as the CO_2 level in the atmosphere increases, if conditions become drier. The biomass of cladodes of a CAM plant in the cactus family (Indian fig) increased with elevated CO_2. Therefore, the fact that CAM plants keep their stomata open at night, in contrast to C_3 and C_4 plants, does not seem to affect their responsiveness to elevated CO_2. Elevated CO_2 has compensated for reductions in growth of roses and wheat due to salinity. In addition, elevated CO_2 reduced blindness (failure to form flowers) in roses. Thus, the effects of salt stress can be ameliorated by elevated levels of atmospheric CO_2.

APPENDIX

BIOGRAPHY OF CHARLES DAVID KEELING

Charles David Keeling, the first person to measure accurately the rising levels of CO_2 in the atmosphere, was born on April 20, 1928, in Scranton, Pennsylvania, United States (Brown 2005). He was the son of Ralph Franklin and Grace Noerr (Sherburne) Keeling. He married Louise Barthold and they had five children: Andrew, Ralph, Emily, Eric, and Paul (Marquis Who's Who 2004). He graduated from the University if Illinois in 1949 (Marquis Who's Who 2004), majoring in chemistry (Brown 2005). He got a PhD from Northwestern University in 1954 (Marquis Who's Who 2004), where he studied chemistry and isotope geochemistry (Heimann 2005). He was a research fellow in geochemistry at the California Institute of Technology in Pasadena between 1953 and 1956 (Marquis Who's Who 2004). Then he joined Scripps Institution of Oceanography of the University of California at San Diego in La Jolla, California, where he was an assistant research chemist (1956–1960), associate research chemist (1960–1964), and associate professor of oceanography (1964–1968). He became professor of oceanography in 1968 (Marquis Who's Who 2004).

When he moved to the California Institute of Technology in 1953, he became interested in the problem of measuring levels of CO_2 in the atmosphere (Heimann 2005). Before that time, the prospect that levels of atmospheric CO_2 could be altered on a global scale because of the burning of coal, gas, and oil was largely theoretical. At the end of the nineteenth century, Svante Arrhenius (see Chapter 14 for a biography of Arrhenius) and Arvid Högbom had suggested that temperature might be affected by greenhouse gases. But the possibility of increasing atmospheric CO_2 was of no concern because the ocean was believed to absorb most of any increase (Heimann 2005). In the 1950s, however, Roger Revelle and Hans Suess at the Scripps Institution of Oceanography in California realized that the sink capacity of the ocean is limited, because of the slow mixing with deeper waters. Measurement of atmospheric CO_2 thus, became a compelling topic of scientific interest. A job offer from Revelle brought Keeling to Scripps, where he remained his entire career (Heimann 2005), except for sabbatical leaves. During a sabbatical stay in Stockholm with Bert Bolin in the early 1960s, they demonstrated that fossil-fuel emissions contributed to the global distribution of atmospheric CO_2 (Heimann 2005). [Bert Bolin died on December 30, 2007, at the age of 82. For his obituary, see Watson (2008).]

Keeling looked for alternatives to the existing, questionable methods of determining variations in CO_2 and eventually resorted to using an infrared analyzer (Heimann 2005). Because this is a relative measurement, he had to devise a system of calibration to make the measurements accurate and comparable between different sites and over long time spans. In conjunction with the "International Geophysical Year 1957" (this was a series of global geophysical activities spanning 1957–1958), systems were set up on Mauna Loa in Hawaii and in Antarctica, supplemented by laboratory measurements on air flasks regularly sampled at the South Pole. The choice of such remote places was to ensure that they were as far away from local sources of CO_2 as possible, thus allowing reliable detection of changes in the background atmosphere (Heimann 2005).

Keeling produced a classic paper showing that the atmospheric CO_2 concentration was rising at a rate consistent with estimates of fossil-fuel emissions (Keeling 1970). In this paper, he shows a figure that has become known as the "Keeling Curve" (Koepp 2005). The figure has "Concentration Atmospheric CO_2 (ppm)" on the y-axis and "Year" on the x-axis, beginning with 1958. He also showed that the concentration at Mauna Loa exhibits a characteristic seasonal variation. Using concurrent measurements of the $^{13}C/^{12}C$ isotope ratio in CO_2, which differs according to the exchange processes with the land and ocean, Keeling was able to demonstrate that this seasonal variation is probably driven by the annual cycle of vegetation in the Northern Hemisphere (Heimann 2005). The records also show interannual variations, which can be related to climate fluctuations such as those of the El Niño-Southern Oscillation. And using concurrent $^{13}C/^{12}C$ isotope measurements, Keeling's group could estimate the relative contributions of land and ocean to these phenomena (Heimann 2005).

After making these measurements of CO_2, many would have moved on to other scientific investigations (Heimann 2005). However, Keeling anticipated the power of long-term observations, and, despite many funding crises, relentlessly pursued the measurements. Keeling was not always an easy colleague to work with (Heimann 2005). He was meticulous and demanded the same standards from his coworkers. But it is thanks to this attention to detail that we now have two highly accurate records of atmospheric CO_2 (Mauna Loa and South Pole) spanning half a century (Heimann 2005). Between the late 1950s and today, levels of CO_2 have risen from about 315 ppm (Heimann 2005) to 385 ppm (Dlugokencky 2009).

Charles Keeling's son, Ralph Keeling, works not only in the same field as his father but also at the same institution, the Scripps Institution of Oceanography (Kerr 2003). Ralph Keeling makes precise measurements—precise to six significant figures—of the oxygen in the atmosphere. That allows him to track the carbon passing through atmosphere, oceans, and plants, a prime concern of greenhouse-warming researchers. For Ralph Keeling, early signs of scientific interest came at age 2, when he was just learning to talk (Kerr 2003). On a summer-time trip to the Cascade Mountains [a mountain range in western Oregon and Washington and British Columbia (Friend and Guralnik 1959)], the family was admiring Nisqually Glacier when Ralph piped up, "How do you make glaciers?"

In college, Ralph Keeling had been considering theoretical physics, not his father's work with analytical instruments (Kerr 2003). But after one semester of graduate school at Harvard, he joined his parents in Switzerland, where his father was on sabbatical leave. With his father's encouragement, he became a member of a project of the late Hans Oeschger, who was developing a technique for extracting gases trapped in glacial ice over the millennia. Ralph said, "That galvanized my interest in being an instrumentalist" (Kerr 2003). He went on to develop a novel interferometric technique for measuring atmospheric oxygen so precisely that its combination with carbon could be followed (Kerr 2003). His work, for example, has analyzed ocean and land-biota sinks for CO_2 (Keeling et al. 1996) and has allowed precise measurements of anthropogenic CO_2 in the ocean (Keeling 2005).

Charles Keeling won many awards. In 2001, the President of the United States, George W. Bush, selected him to receive the U.S. Medal of Science, the highest award in the United States, for lifetime achievement in scientific research (Brown 2005; Marquis Who's Who 2004). He got a Special Achievement Award from the Vice President of the United States in 1997. He was recipient of the Half Century Award of the American Meteorological Society in 1981 and the Blue Planet Prize of the Asahi Glass Foundation in 1993 (Marquis Who's Who 2004), which is given by the Science Council of Japan (Heimann 2005). He also received the Tyler Prize for Environmental Achievement (Heimann 2005). He was a fellow of the John Simon Guggenheim Memorial Foundation in 1962. He was named fellow of the American Association for the Advancement of Science and the American Geophysical Union, from which he also received the Ewing Medal in 1991. He was a fellow of the American Academy of Arts and Sciences and a member of the National Academy of Science, as well as a member of Sigma Xi (Marquis Who's Who 2004). He was a member of the commission on global pollution of the International Association of Meteorology and scientific director of the Central CO_2 Calibration Laboratory of the World Meteorological Organization (Brown 2005).

Charles Keeling was involved in the political life of his hometown. He also conducted a choir, and, after a hard day at work, he enjoyed playing the grand piano, which he did with proficiency (Heimann 2005). In his biography in Marquis Who's Who (2004), Keeling lists his avocations as musical performance and off-road walking. He had great interest in foreign cultures. During a sabbatical stay in Switzerland, he took on the challenge of taking lessons in the Bernese

Swiss–German language with success. But his major love was nature, and especially the mountains (Heimann 2005).

He died on June 22, 2005 (Brown 2005), of a heart attack in Hamilton, Montana (Koepp 2005). He was survived by his widow, Louise, and his five children (Brown 2005). His fame was such that his death was recognized in popular publications. According to his obituary in *Time* magazine, Keeling's "precise measurements of carbon dioxide in the atmosphere over five decades became the undisputed basis for global-warming concerns" (Koepp 2005). His final resting place is near his summer house in the wilderness of Montana (Heimann 2005).

2 Elevated Carbon Dioxide in the Soil: Composition of the Soil Atmosphere

INTRODUCTION

As we noted in Chapter 1, the four soil physical factors that affect root growth, and consequently plant growth, are water, aeration, temperature, and mechanical impedance (Kirkham 2005, p. 1). In Chapter 1, we looked at the effect of lack of water on plants under elevated atmospheric CO_2. In this chapter, we turn to soil aeration and focus on elevated CO_2 in the soil atmosphere. Most biological reactions that occur in soil consume oxygen (O_2) and produce CO_2. The effects of soil aeration on growth of plants can be divided into two categories (Taylor and Ashcroft 1972, p. 352). First, reduction of soil constituents occurs, which in turn influences growth. In the absence of adequate O_2 in the soil, sulfide, ferrous, and manganous compounds form; organic matter is reduced to alcohols, aldehydes, or methane, and hydrogen gas may appear; and essential nitrates are reduced to toxic nitrites or to free nitrogen gas (Taylor and Ashcroft 1972, p. 352). We shall not be concerned about these soil processes in this chapter. Second, aeration directly affects the physiological condition of plants. They require adequate soil aeration for the proper development of their root systems. Because the greatest influence of aeration on root growth is attributable to the concentrations of O_2 and CO_2 in the soil air (Taylor and Ashcroft 1972, p. 353), we shall consider both gases in this chapter.

COMPOSITION OF THE SOIL ATMOSPHERE

Boussingault and Lévy (1853) were early researchers who studied the composition of the atmosphere of the soil. (See Appendix for a biography of Boussingault.) Their research, along with that of others after them, established five general characteristics of the soil atmosphere (Taylor and Ashcroft 1972, p. 365):

1. The sum of the CO_2 and O_2 in the soil air is nearly the same as the sum of these two components in the air above the soil.
2. The CO_2 content usually increases with depth.
3. The CO_2 content of the soil shows a marked seasonal trend.
4. The proportion of CO_2 in the soil air depends upon the amount of organic matter in the soil.
5. Rainfall has a strong influence upon the CO_2 content of soil air.

Let us consider each one in more detail.

CO_2 AND O_2 IN SOIL AND AIR

As we shall discuss further in this chapter, as well as in Chapter 4, it is generally felt that diffusion is the main way that gases move in the soil. Thus, the amount of CO_2 in the air above the soil should be about the same as the CO_2 in the soil, unless there are pockets of high respiration, such as those that occur when manure is put on the soil or when a pathogen is localized in a particular part of the soil (Don Kirkham, Iowa State University, personal communication, c. 1985). However, even in healthy soil without manure, respiration of roots will cause somewhat elevated levels of CO_2 around roots. Because of the difficulty of measuring concentration of CO_2 and O_2 right at the root surface, exact concentrations are hard to determine. But some general conclusions can be made about the concentration of CO_2 and O_2 in the soil air. Even though it is felt that the sum of CO_2 and O_2 is about the same above the soil and in the soil (Taylor and Ashcroft 1972, p. 365), the concentrations of CO_2 and O_2 in the above-ground air differ from those in the below-ground air. There tends to be an inverse relation between the concentrations of these two components in the soil (Taylor and Ashcroft 1972, p. 353). The amount of CO_2 is usually greater in the soil, and the amount of O_2 less, than in the atmosphere above. Acceptable average figures for soil air might be 0.25% CO_2 and 20.73% O_2 compared to 0.03% CO_2 (but rising) and 20.95% O_2 in the atmosphere (Taylor and Ashcroft 1972, p. 365). [The concentration of CO_2 in the atmosphere in 2008 was 0.0385% (Dlugokencky 2009).] Cropped soils generally have more CO_2 and less O_2 than bare or fallow soils, and grass-covered soils are generally higher in CO_2 than cultivated ones (Taylor and Ashcroft 1972, p. 365).

When aeration is restricted, the soil processes use up O_2 and produce CO_2 faster than they can be exchanged with the atmosphere. The net result is a reduction of O_2 and a buildup of CO_2 in the soil. There is often a question as to which factor causes reduced root growth. Is it a result of O_2 deficiency or CO_2 toxicity (Taylor and Ashcroft 1972, p. 353)? As we shall see Chapters 3 and 4, apparently the CO_2 concentration has to become high before it restricts root growth. So it appears that O_2 concentration in the soil is the critical factor affecting root growth. Diffusion of gases in soil practically stops when the fraction of air-filled pores is less than 10%. Therefore, roots need at least 10% by volume air space in the soil to survive (Kirkham 1994). Taylor and Ashcroft (1972, p. 354) state that a deficiency of O_2 may be said, in general, to exist when O_2 comprises 15% or less of the soil air if the remainder is nitrogen. A maximum concentration of CO_2 that roots can tolerate is not established.

The concentration of CO_2 in the soil has to be extremely high to kill plants, as demonstrated by CO_2 from volcanoes. Areas of dead and dying trees have long been observed near a dormant volcano at Mammoth Mountain in eastern, central California, United States, which last erupted about 500 years ago (Farrar et al. 1995; Williams 1995). The damage was initially considered to be an effect of 4 years of drought (Farrar et al. 1995). But it was noted that all trees, regardless of age or species, were equally affected within dead or dying areas, which gradually increased in number and size. The soils in the region have normal temperatures and show no signs of acid alteration or other lethal agents such as sulfur gas, radon, or mercury. Also, biological pests were not the cause of the tree kill.

The death of the trees was a mystery until Farrar et al. (1995) did their study. In July 1994, they did a reconnaissance soil-gas survey in the tree-kill areas and found that samples from the 15 cm depth contained >20% CO_2. Then, in September and October of 1994, they did a thorough study and analyzed CO_2 in the soil from samples taken along two transects near Reds Lake (RL) and Horseshoe Lake (HSL), which covered both the tree-kill areas and control sites in a healthy forest. The types of trees in the forests were not named by them. The CO_2 concentrations ranged from less than 1% in the healthy forest to more than 90% at several locations within tree-kill areas. Where CO_2 concentrations exceeded 30%, most trees were dead (Figure 2.1). [The cover of the August 24, 1995, issue of *Nature* with the articles by Williams (1995) and Farrar et al. (1995) also shows the dead trees.] Farrar et al. (1995) conclude that the trees were being killed by large volumes of CO_2 emitted through the soil from magma deep within the mountain. The forest die-off was the most conspicuous surface manifestation of the magmatic processes at Mammoth Mountain. While volcanologists (Farrar et al. 1995) are interested in the massive vegetation kills as obvious indicators of possible magmatic unrest at depth, we as plant and soil scientists are interested in the data, which show that extremely high concentrations of CO_2 in the soil are needed to kill tree roots (greater than 30%).

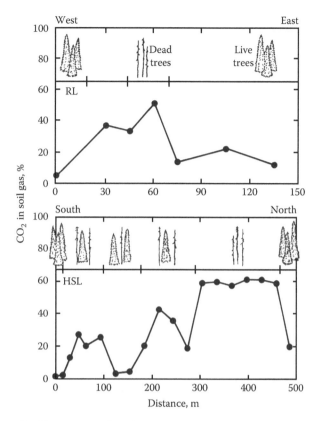

FIGURE 2.1 Carbon dioxide concentrations in September 1994 at 0.6 m depth along two transects at Mammoth Mountain in eastern California, United States. Top panel: Transect RL near Reds Lake; bottom panel: Transect HSL near Horseshoe Lake. Symbols at the top of each panel denote relative degree of tree mortality along the transects. (Reprinted by permission from Macmillan Publishers Ltd., *Nature*, Farrar, C.D., Sorey, M.L., Evans, W.C., Howle, J.F., Kerr, B.D., Kennedy, B.M., King, C.Y., and Southon, J.R., Forest-killing diffuse CO_2 emission at Mammoth Mountain as a sign of magmatic unrest, 376, 675–678, Fig. 2, Copyright 1995.)

High concentrations of CO_2 also are needed to be lethal to humans. Farrar et al. (1995) did a survey of confined spaces in areas of known high CO_2 discharge near HSL and found CO_2 concentrations greater than 1% in campsite lavatories and small tents, 25% in a small cabin, and 89% in a utility vault of 0.6 m diameter and 1 m depth. These values exceed U.S. health standards (Joseph 1985, pp. 155–156), described in the next paragraph, and the last two would be quickly lethal (Farrar et al. 1995). The CO_2 leakage may have contributed to the death of a cross country skier in 1998 (Altevogt and Celia 2004).

The U.S. Occupational Safety and Health Administration (OSHA) provides information to employers and employees about the safety of different chemicals (Joseph 1985, p. xv). Carbon dioxide is not regulated as a hazardous substance or hazardous waste. Overexposure by inhalation may lead to rapid breathing, rapid heart rate, headache, sweating, shortness of breath, dizziness, mental depression, visual disturbances, shaking, unconsciousness, and death (Joseph 1985, p. 155). Skin or eye contact with solid CO_2 may cause frostbite. The time-weighted average (TWA) for the permissible exposure limit set by OSHA is 5000 ppm (0.5%). The short-term exposure limit (STEL) set by the American Conference of Governmental Industrial Hygienists (ACGIH) is 15,000 ppm (1.5%). The STELs are limits established by the ACGIH that represent the maximum concentrations workers can be exposed to for 15 min periods without suffering adverse effects with certain excursion limits (Joseph 1985, p. xxiv).

Mineral or geothermal springs also produce large amounts of CO_2. They are being used as experimental areas to study the effects of elevated CO_2 on plants. Two mineral springs in New Zealand, near Kamo and Kaeo in the North Island, release large volumes of CO_2 resulting in concentrations near the vents of $5000\,\mu mol/mol$ (0.5%) and $2000\,\mu mol/mol$ (0.2%), respectively (Newton et al. 1996a). In central-western Italy, more than 100 geothermal CO_2 springs exist, 8 of which were studied by Miglietta et al. (1993a). Carbon dioxide concentrations at these springs were as high as $10,000\,\mu mol/mol$ (1%). The springs offer the chance to study plants that have been exposed to elevated CO_2 for lengthy periods.

CO_2 WITH DEPTH IN THE SOIL

It has been known for a long time that the CO_2 concentration increases with depth below the soil surface. Lundegårdh (1927, pp. 423–424) reported that in the top 15 cm depth, the CO_2 concentration in the soil air averages about 0.3% for lighter soils but may rise to 1% or more, especially in wet soils rich in humus. Even in the same field, the variation from plot to plot may be prominent, particularly in soils that are not well cultivated and contain lumps of decomposing organic matter. Lundegårdh (1927, p. 424) suggested that by determining the CO_2 concentration at the 15 cm depth, one can detect poor aeration of the soil that can be removed by cultivation or drainage.

The values for CO_2 concentration in the soil air are important to estimate growth conditions for roots. Lundegårdh (1927, pp. 419–420) said that a CO_2 concentration higher than 1% in the soil atmosphere may become dangerous to crop growth. But the resistance of different plants to elevated levels of CO_2 in soil varies greatly. It is well known that wet-land plants, like rice (*Oryza sativa* L.), appear to tolerate low O_2 in the root zone better than mesophytes. [A mesophyte is a plant adapted to grow under medium conditions of moisture (Friend and Guralnik 1959).] However, an internal supply of O_2 via the aerial organs appears to supply O_2 to the roots of bog plants (Leyton and Rousseau 1958, p. 467), and the distance O_2 may need to move to roots deep in soil may affect their survival. (In Chapter 4, we talk of movement of gases up and down a plant.) Differences in tolerance have also been reported among tree species. Willow (*Salix* sp.), alder (*Alnus* sp.), mangrove (*Rhizophora* sp. or *Avicennia* sp.), and black spruce (*Picea mariana* Mill.) are associated with waterlogged soils, and Norway spruce (*P. abies* Karst), Sitka spruce (*P. sitchensis* Carr.), Scots pine (*Pinus sylvestris* L.), and jack pine (*P. banksiana* Lamb) are species associated with well-aerated soils (Leyton and Rousseau 1958, p. 467). Cannon and Free (1925, p. 162) report the O_2 content at a depth of 38 cm (15 in.) in a soil in India (type not given) immediately about the roots of three types of plants: *Crotolaria* (common name: rattle-box, which is a large genus especially prevalent in Asia with annual and perennial erect herbs and shrubs), 2.23%–9%; *Indigofera* (indigo), 3.33%–5.24%; and *Zea* (the genus of corn, *Zea mays* L.), 6.25%–13.82%.

The CO_2 concentration increases with depth, while soil and root respiration decreases with depth. Lundegårdh (1927, p. 424) measured CO_2 production in the profile of a sandy soil (uncropped) and found the following values for soil respiration (in $g/m^2/h$): 0–10 cm depth, 0.271; 10–20 cm depth, 0.084; 20–30 cm depth, 0.016; and 30–40 cm depth, 0.009. The data showed that the soil microbiological activity was concentrated in the surface layer and decreased rapidly with depth.

Bhupinderpal-Singh et al. (2005) labeled pasture plants, growing in columns with a silt loam soil, with $^{14}CO_2$ to see the distribution of label with depth. The plants were ryegrass (*Lolium perenne* L.) and white clover (*Trifolium repens* L.). They analyzed the carbon in the root exudates at the soil–root interface, which would be an indication of respiration by the roots. They also analyzed the carbon in the roots. Seven days after pulse labeling, most of the total below-ground ^{14}C at the soil–root interface was recovered from the 0–20 mm depth (55%), followed by 20–40 mm (22%), 40–80 mm (15%), 80–120 mm (4%), and 120–200 mm (4%). Similarly, of the total below-ground ^{14}C in roots, 68%, 17%, 9%, 3%, and 3% were recovered from the 0–20, 20–40, 40–80, 80–120, and 120–200 mm soil depths, respectively. Decreased ^{14}C activity in roots with increasing soil depth followed the pattern of decreasing root mass with depth, as follows: 13.2 g in the 0–20 mm depth;

4.5 g at 20–40 mm; 1.9 g at 40–80 mm; 0.8 g at 80–120 mm; 0.8 g at 120–160 mm; and 0.4 g in the 160–200 mm depth. The data indicated that CO_2 production by these pasture roots decreased rapidly with depth, because their mass decreased rapidly with depth. Their data supported an earlier observation that surface CO_2 fluxes are proportional to root activity (Ben-Asher et al. 1994).

Even though CO_2 evolution from roots decreases with depth, CO_2 generation in the deep subsurface of the soil (beyond the root zone) has long been recognized (Keller 1991). Keller (1991) studied an 18-m-thick surface deposit of clayey calcareous till in the Western Glaciated Plains of Canada. He found large increases in the CO_2 concentration with depth. The subsoil CO_2 production appeared to be due to respiration by aerobic microbial communities that utilize the soil- and sediment-derived organic carbon as substrates.

However, most CO_2 from respiration comes from the soil surface (Hunt et al. 2004), because that is the location of roots and microorganisms that break down roots and residue. Hunt et al. (2004) studied two sandy loam soils where grassland plants grew on the South Island of New Zealand. The dominant plants were native fescue [*Festuca novae-zelandiae* (Hack.) Cockayne] and the introduced flat-weed (*Hieracium pilosella* L.). The soils were well drained and formed two distinct phases, determined by past fluvial processes. The "stony soils" had a high stone content (58% by volume), and the "shallow soils" had few stones (6%). Carbon-dioxide production in the stony soils at the 0.0–0.1 and 0.1–0.2 m depths was 49 and 18 mg/kg, respectively. For the shallow soils, these values were 75 and 25 mg/kg, respectively, for the two depths. Under well-watered conditions, 26% of the soil surface CO_2 efflux came from soil microbial activity. Soil respiration declined by a factor of four as the soil water content declined from field capacity (0.21 m³/m³ for the stony soils; 0.27 m³/m³ for the shallow soils) to the driest value obtained (0.04 m³/m³). Hunt et al. (2004) estimated that 85% of the ecosystem respiration emanated from the soil surface.

Hendry et al. (1999) studied CO_2 production at a field site 10 km south of Saskatoon, Canada (52° 09′ N; 106° 39′ W), to calculate the rates of in situ CO_2 production with depth through an unsaturated zone over 550 days. Maximum CO_2 gas concentrations occurred during the summer (0.85%–1.22%), and minimum concentrations occurred during the winter (0.04%–0.24%). They found that CO_2 gas concentrations decreased with increasing depth during the summer and increased with depth during the winter. Respiration rates ranged from 5 µg carbon/g/day in the soil (top) horizon in the summer to zero in the winter. Below 3 m in unsaturated sections in the C horizon, the rate remained relatively constant throughout the year at about 10^{-4}–10^{-5} µg carbon/g/day. [C horizons or layers are little affected by pedogenic processes. The material of C horizons may be either like or unlike that from which the set of horizons formed. In general, C horizons are above bedrock but below organic or mineral layers (Soil Science Society of America 1997).] Roots contributed about 75% of the CO_2 in the summer months. They concluded that although microorganisms are present in subsurface environments, their in situ activity in the unsaturated zone of northern soils may be low.

Dueñas et al. (1995) measured emissions of CO_2 from soils in the semi-urban area surrounding Málaga, Spain. The soil at one site, labeled M1, had 32% clay, 13% fine sand, 55% coarse sand, and 2.85% organic matter. The soil at another site, labeled M2, had 88% clay, 4% fine sand, 8% coarse sand, and 3.92% organic matter. (The three soil textural classes, clay, sand, and silt, always add to 100% when the soil texture is determined. Organic matter is excluded in this 100%.) The soil at site M1 was bare, and the soil at site M2 was in a forest with *Ficus* (fig) and *Eucalyptus* (gum tree) trees. Measurements of CO_2 were made by inserting tubes into the soil and taking samples of soil air. The samples were injected into a gas chromatograph to measure CO_2. The concentration of CO_2 decreased with depth (Figure 2.2). Even though one soil was bare and the other was vegetated, the change in CO_2 concentration with depth was similar. At the 0.6 m depth (maximum depth measured for the vegetated soil), the CO_2 concentration in the soil was about 0.008 ppm on a volumetric basis, and at the 0.5 m depth in the bare soil, the CO_2 concentration was about 0.010 ppm on a volumetric basis. Their data confirm the general observation that CO_2 concentration in the soil increases with depth.

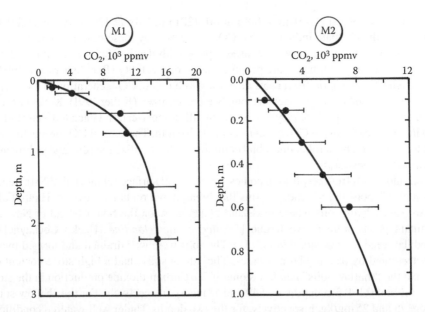

FIGURE 2.2 Concentration of CO_2 with depth in two soils in the semi-urban area near Malaga, Spain. Site M1 was a bare soil and site M2 was a forested soil. Horizontal lines show the standard deviations. Note the different depth scales on the ordinate. (Reprinted from *Chemosphere*, 30, Dueñas, C., Fernañdez, M.C., Carretero, J., Liger, E., and Pérex, M., Emissions of CO_2 from some soils, 1875–1889, Copyright 1995, Fig. 2a, with permission from Elsevier.)

The rate of emission with depth depends upon the depth of the water table. Moore and Dalva (1993) in Montreal, Canada, used laboratory columns (80 cm long; 10 cm diameter), which they filled with three different types of peat soils, to determine the effect of water table depth on CO_2 flux from the surface of the soils. The samples of peat were collected from three sites: an open bog in southern Ontario, Canada; a poor fen in subarctic Quebec; and a swamp with trees in Quebec. The base of each column had a water-tight cap and the top was sealed with a removable cap that contained a septum, through which headspace air was sampled with a syringe. The concentration of CO_2 in the headspace air was determined by gas chromatography. The water table in columns was lowered to a depth of about 50 cm beneath the peat surface in each column, at a rate of 2 cm/day. After a period of 15 days to allow the columns to equilibrate with the lowered water table, the water table was raised at a rate of 2 cm/day until saturation was reached, and sampling was continued for 10 days.

Lowering of the water table created gas-filled pore space, which averaged 26%, 36%, and 25% for the bog, fen, and swamp columns, respectively. All three peat types showed a strong increase in CO_2 emission rates as the water table was lowered from the surface to a depth of 50 cm (Figure 2.3a through c for bog, fen, and swamp, respectively). A notable pattern of the falling limb (Figure 2.3, solid circles) was large CO_2 emissions compared to the rising limb (Figure 2.3, open circles). Emission rates at the 50 cm depth for the falling water table were about 10 times those when the water table was at the peat surface. Note that some values in Figure 2.3 are off scale and have been written as numbers on the right-hand side of Figure 2.3a through c. The results of Moore and Dalva (1993) show that more CO_2 was emitted from a peat soil with a falling water table than a peat soil with a rising water table.

Dinsmore et al. (2009) found that depth of water table affected CO_2 emissions from a peatland soil near Edinburgh, Scotland. Northern peatlands are considered to contain approximately a third of the estimated total global soil carbon pool (Dinsmore et al. 2009). Vegetation on the soil was a patchy mix of coarse grasses and soft rush covering a *Sphagnum* base layer. They subjected soil

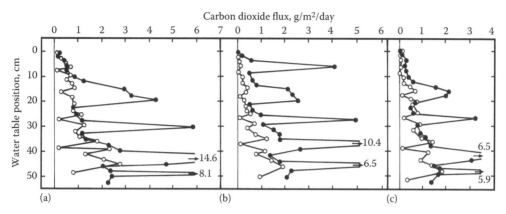

FIGURE 2.3 The relation between CO_2 emission rates and water table position on the falling (closed circles) and rising (open circles) limbs of an experiment to examine the effect of a dynamic water table on gas emissions from three types of peat-land soils: (a) bog, (b) fen, and (c) swamp. Data show the mean values ($n = 5$). (Moore, T.R. and Dalva, M.: The influence of temperature and water table position on carbon dioxide and methane emissions from laboratory columns of peatland soils. *J. Soil Sci.* 1993, 44, 651–664, Fig. 3. Copyright Wiley-VCH Verlag GmbH & Co. KGaA. Reprinted with permission.)

cores with vegetation to two water table treatments: 0–5 and 30–35 cm. Emissions of CO_2 were highest in the cores subjected to the low water table treatment. That is, when the water table was put at a depth 30 cm lower in the soil, the flux of CO_2 out of the soil into the atmosphere was greatly increased. Lowering of the water table in organic soils by drainage often increases the release of soil carbon as CO_2 (Smith et al. 2003). Even though CO_2 emissions from peatlands, which are often near marshes, can be reduced by putting the water table lower in the soil, emissions of CO_2 from them are high. In a survey of CO_2 emissions from different soil types in China, highest CO_2 emission rates were from marshland sites and lowest CO_2 emission rates were from sand dunes (Yao et al. 2010). If a soil is too dry (as in a sand dune), plants and microbes grow with difficulty due to lack of water, and there is little CO_2 production from roots or microbial activity. Reduced soil moisture lowers CO_2 efflux (Garten et al. 2009). However, variability of water content along a 100 m slope of an evergreen broadleaved forest in southern China did not result in any difference in CO_2 emission from the soil, as determined by Fang et al. (2009). The soil was a lateritic red earth formed from sandstone. Soil CO_2 release was independent of slope position in their study.

SEASONAL AMOUNTS OF CO_2 AND O_2

There is a marked seasonal fluctuation in O_2 content of the soil atmosphere, as shown in Figure 2.4 (Boynton and Reuther 1938). This figure shows three soils in an apple (*Malus* sp.) orchard in New York: sandy loam (high O_2 throughout the year); silty clay loam (low O_2 in spring); silty clay (low O_2 in winter and spring). The sandy loam soil is well aerated and produces many apples; the silty clay is poorly aerated and trees are short lived with low productivity. During the period from late December to April in New York, the soil is frequently frozen on the surface. This produces a sheet of ice that prohibits the movement of gases, so that O_2 is gradually used and cannot be replaced by diffusion (Taylor and Ashcroft 1972, pp. 365–366).

Crusting over of the soil will have the same effect as an ice sheet, and the crust prevents diffusion of O_2 to roots. Tillage corrects this problem (Kirkham 1994). As noted earlier, diffusion appears to be the main method of exchange of gases between the soil and the air. For soil aeration, it is generally assumed that convection is negligible, and that the major mechanism of gas transport is diffusion within the soil space (Kirkham 1994), even though the supposition has been challenged, as reviewed by Tanner and Simonson (1993). The assumption of gas transport by diffusion is based

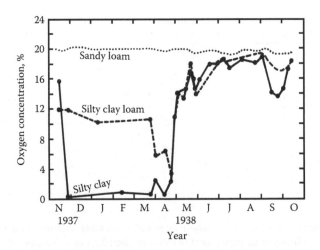

FIGURE 2.4 Seasonal trend of oxygen concentration at 2 ft (61 cm) depth in three orchard soils of western New York. The sandy loam soil is well aerated and produces many apples; the silty clay is poorly aerated and trees are short-lived with low productivity. (Reprinted from Boynton, D. and Reuther, W., *Proc. Am. Soc. Hort. Sci.* 36, 1, 1938, Fig. 1, bottom left. With permission. Copyright 1938, American Society for Horticultural Science.)

in large part on data obtained by Buckingham (1904) and Romell (1922). The velocity of diffusion is large (Buckingham 1904, p. 44).

The seasonal trend is also illustrated in Figure 2.5 (Furr and Aldrich 1943) for an irrigated date (*Phoenix* sp.) garden on a calcareous very fine sandy loam soil. Respiration of roots and soil organisms reduced the O_2 concentration and increased the CO_2 concentration. Because respiration increases with temperature, these changes were more marked in the summer than in the fall and winter. An increase in soil moisture tension (decrease in soil water content) (Figure 2.5, bottom) was accompanied by an increase in O_2 content (Figure 2.5, top). [Soil moisture tension is the old term for matric potential (Kirkham 2005, p. 56).] The changes in gas concentrations, as shown in Figure 2.5, would be more noticeable in soil high in organic matter, because of the greater microbial activity, than in soil containing little organic matter (Kramer 1969, p. 138). We shall talk about organic matter later in this chapter.

We have been discussing seasonal changes in CO_2 concentration in the soil. We digress here briefly to see how CO_2 concentration in the air above the soil changes with season. Rosenberg (1981) presented information about the annual cycle of CO_2 concentration just above the soil surface for typical agricultural land in the eastern Great Plains of North America (Mead, Nebraska; 41° 09′ N; 96° 30′ W). The mean daily CO_2 concentration at 16 m above ground level varied from about 340 ppm in winter to a minimum of about 328 ppm in early August of 1972. The annual range for daytime concentrations was greater. It varied from about 339 ppm in winter to about 305 ppm in late July and early August when photosynthetic activity in this region is greatest. Figure 2.6 shows the change in CO_2 concentration at two levels above the soil (16 and 1.0 m) during 24 h on July 17, 1973. During the day at 16 m, the midday concentrations were about 300 ppm. At 1.0 m above the soil, the midday concentrations were even lower, because the growing plants extracted CO_2 from the air. These data are typical of many days in July and August. They demonstrate that agricultural lands provide a strong sink for CO_2, and this sink strength is easily detected 16 m above the ground. Maximal CO_2 fluxes above the soil surface at Mead, Nebraska, measured between 5.6 and 16 m above ground, occurred in mid to late July (Figure 2.7), when photosynthetic activity was at its maximum (Rosenberg 1981).

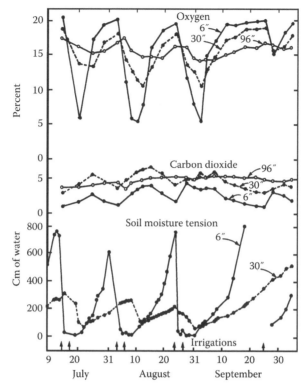

FIGURE 2.5 Variations in oxygen, carbon dioxide, and soil moisture tension of soil at various depths (6, 30, and 96 in. depths or 15, 76, or 244 cm depths). An increase in soil moisture tension indicated a decrease in water content and was accompanied by an increase in oxygen content. The data come from an irrigated date garden on calcareous very fine sandy loam soil. (Reprinted from Furr, J.R. and Aldrich, W.W., *Proc. Am. Soc. Hort. Sci.*, 42, 46, 1943, Fig. 1. With permission from American Society for Horticultural Science.)

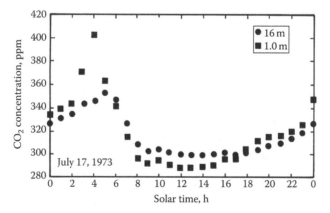

FIGURE 2.6 Typical diurnal pattern of CO_2 concentration at 1 and 16 m above ground during the growing season at Mead, Nebraska, United States. (Reprinted with kind permission from Springer Science + Business Media: *Climat. Change*, The increasing CO_2 concentration in the atmosphere and its implications on agricultural productivity. I. Effects on photosynthesis, transpiration and water use efficiency, 3, 1981, 265–279, Rosenberg, N.J., Fig. 6.)

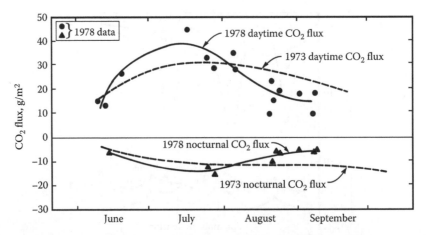

FIGURE 2.7   Growing season CO_2 flux in 1973 and 1978 at Mead, Nebraska, United States. (Reprinted with kind permission from Springer Science+Business Media: *Climat. Change*, The increasing CO_2 concentration in the atmosphere and its implications on agricultural productivity. I. Effects on photosynthesis, transpiration and water use efficiency, 3, 1981, 265–279, Rosenberg, N.J., Fig. 7.)

CO_2 Emissions as Affected by Tillage

Conservation agriculture in its version of permanent raised bed planting with crop residue retention increases yields and improves soil characteristics such as aggregate distribution and organic matter content (Patiño-Zúñiga et al. 2009). The objective of the study of Patiño-Zúñiga et al. (2009) was to investigate how conservation agriculture with permanently raised beds and residue affected CO_2 emissions. The experiment was carried out at El Batán in Central Mexico, which has an annual rainfall of 600 mm per year. The soil was a Cumulic Haplustoll, which is used for farming in the region, and it has good physical and chemical properties. [Cumulic Haplustolls are moderately extensive in the Great Plains of the United States, and nearly all are cultivated (Soil Survey Staff 1975, p. 305).] The experiment had tilled and non-tilled beds, and each type of bed either had residue (wheat or maize straw) or no residue (straw was removed). With residue retention, the CO_2 emission was 1.2 times larger in soil under tilled beds compared to non-tilled beds. They found that permanently raised bed planting with crop residue retention and not tilled decreased emissions of CO_2 compared to soil under conventionally tilled raised beds.

Fortin et al. (1996) showed that CO_2 fluxes from the soil surface vary not only with season, but also with the type of cultivation. During a 2 year period (1992–1993), they cultivated a silt loam soil in Ottawa, Ontario, Canada in two different ways: conventional tillage and no-tillage. Conventional plots were moldboard plowed in the fall to a depth of 180 mm and disked once in the spring of each year. Barley (*Hordeum vulgare* L.) was planted on May 8, 1992 (day of year 129), and oat (*Avena sativa* L.) was planted on May 5, 1993 (day of year 125), with a no-till drill. Following planting, rings (soil collars) for CO_2-emission measurements and time domain reflectometry probes, used to get soil water content, were installed for the season. Carbon dioxide fluxes were measured using a closed-chamber system. Soil temperatures were measured with thermocouples. In general, soil CO_2 emissions were lower in the no-till plots than in the conventionally tilled plots until about day 168 (Figure 2.8), a period during which soil temperature in no-till plots was lower (the average of the plots was 12.7°C) than in the conventionally tilled plots (average: 15.9°C). The lower soil temperatures were due to the 100% residue cover. During this period, soil water in the no-till plots was higher than in the conventionally tilled plots (Figure 2.9). From mid-June (June 15 = day of year 166) until plowing in August (day of year 239 = August 27), soil temperature differences between treatments were less than 1°C. During this time, soil CO_2 emissions were higher in the no-till plots than in the conventionally tilled plots (Figure 2.8). Yet the soil water content was still higher in the

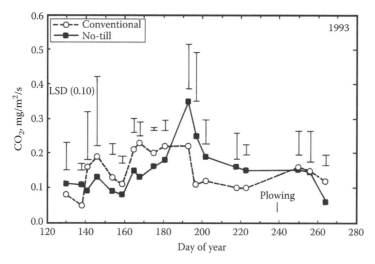

FIGURE 2.8   Soil CO_2 emissions from a conventional-tillage and a no-tillage Typic Endoaquoll planted to oat in Ottawa, Ontario, in 1993. Harvest and plowing were done within the same week. Least significant difference (LSD) values are shown. (Reprinted from Fortin, M.-C. et al., *Soil Sci. Soc. Am. J.* 60, 1541, 1996, Fig. 4. With permission from the Soil Science Society of America.)

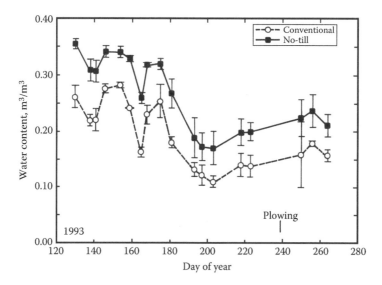

FIGURE 2.9   Volumetric soil water content in conventional-tillage and a no-tillage Typic Endoaquoll planted to oat in 1993. Values (with standard deviations) are averaged for the 150 mm depth. (Reprinted from Fortin, M.-C. et al., *Soil Sci. Soc. Am. J.*, 60, 1541, 1996, Fig. 5B. With permission from the Soil Science Society of America.)

no-till plots (Figure 2.9). This suggested that soil respiration, as reflected by the CO_2 emissions, is inhibited at low soil water contents. Emissions of CO_2 from the two treatments were similar after plowing on day 239. The results of Fortin et al. (1996) showed that no-tilled soil does not necessarily have a higher or lower CO_2 emission than conventionally tilled soil. The amount of CO_2 released will depend upon soil moisture and temperature, and these are variable during a season.

Reicosky and Lindstrom (1993) also showed that the amount of CO_2 emitted from the soil surface depends upon tillage. They determined the effect of no-tillage and four fall tillage methods on

CO_2 flux from a clay loam soil in Minnesota, a state in the northern Corn Belt of the United States. The four tillage treatments were as follows: moldboard plow only, moldboard plow plus disk harrow twice, disk harrow once, and chisel plow once. The depths of tillage for these four methods were 0.250, 0.250, 0.075, and 0.150 m. Wheat (*Triticum aestivum* L.) was in the plots before tillage. The wheat was harvested on August 22, 1991 and tilled into the soil on September 4, 1991 (day of year 247). No-tillage had 0% residue incorporation. The CO_2 flux was measured with a portable system with an infrared gas analyzer. Measurements of CO_2 flux started within 5 min after the completion of tillage and continued for 19 days. Moldboard plow had the roughest soil surface and the highest initial CO_2 flux (29 g/m²/h) and maintained the highest flux throughout the study. Moldboard plow plus disking twice and chisel plow had similar initial rates (7 and 6 g/m²/h, respectively), which were greater than disk harrow and no-tillage (flux ranged from 0.2 to 0.7 g/m²/h). The high values probably reflected a flush of microbial CO_2 and CO_2 released from large voids generated by the tillage event. The high initial CO_2 fluxes were more related to depth of soil disturbance, which resulted in a rougher surface and larger voids, than to residue incorporation. Measurement of the CO_2 flux from the tilled surfaces was continued on day of year 253 after a major rainfall event (49 mm received in 3 days—day of year 250, 251, and 252). Results were consistent with earlier findings (Figure 2.10). Moldboard plow showed the highest CO_2 flux and no-tillage showed the lowest. The results of Reicosky and Lindstrom (1993) contrasted with those of Fortin et al. (1996), who showed that no-till plots emitted more CO_2 during summer months than conventionally tilled plots (Figure 2.8). This is because the no-till plots had more water in the soil than the conventionally tilled plots (Figure 2.9). Lack of water was probably limiting microbial activity and root respiration in their conventionally tilled plots. Reicosky and Lindstrom (1993) made their measurements after a heavy rainfall, and water was not limiting microbial activity in any of their treatments. They noted that even though it was not part of their study, the variation in soil temperature was probably the major factor affecting CO_2 emissions under their different tillage treatments.

In passing, we note that the methods to measure CO_2 flux from the soil are becoming easier to make. Until recently, the CO_2 flux from the soil has often been measured using a closed-chamber system and an infrared gas analyzer (Reicosky and Lindstrom 1993; Fortin et al. 1996). Such methods can cause air disturbances, which may alter CO_2 concentration in the soil, and they also are laborious (Young et al. 2009). However, recent advances in sensor technology provide robust capability for continuous measurement of soil gases. Young et al. (2009) evaluated the performance of solid-state CO_2 sensors (Model GMM220 series, Vaisala, Inc., Helsinki, Finland) in silt loam soil

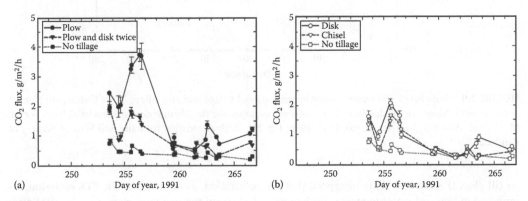

FIGURE 2.10 Intermediate-term effect of fall tillage method on CO_2 flux versus time following a major rainfall event: (a) moldboard plow, moldboard plow plus disk harrow, and no till; (b) disk harrow, chisel plow, and no till. Tillage was done on September 4, 1991 (day of year 247). Each data point is the mean of three replications. Error bars represent ±1 standard deviation. The sign convention is positive, indicating CO_2 flux from the soil surface to the atmosphere. (Reprinted from Reicosky, D.C. and Lindstrom, M.J., *Agron. J.*, 85, 1237, 1993, Fig. 4. With permission from the American Society of Agronomy.)

with winter wheat (*T. aestivum* L.). Measurements were compared with those taken with an infrared gas analyzer. In a field of irrigated winter wheat, soil CO_2 concentration exceeded the 10,000 µmol/mol limit of the GMM222 sensor. Alternatively, the GMM221 sensor, designed to measure between 0 and 20,000 µmol/mol, showed that soil CO_2 concentrations were between 14,000 and 16,000 µmol/mol. They concluded that the GMM222 accurately measured soil CO_2 concentrations under field conditions that were within the sensor detection limit. However, periods of high biological soil activity required the GMM221 sensor with a higher detection limit.

Advancements are also being made in the measurement of the carbon isotope ratio ($\delta^{13}C$). (In Chapter 6, we state how $\delta^{13}C$ is determined using values of ^{13}C and ^{12}C and its importance in root studies.) Marron et al. (2009) tested a tunable diode laser spectrophotometer (TDLS) to determine the seasonal and daily variability in $\delta^{13}C$ of soil CO_2 efflux. This method, based on the measurement of the absorption of an infrared laser emission at specific wave lengths of $^{13}CO_2$ and $^{12}CO_2$, allows the continuous monitoring of the two isotopes. The TDLS was installed in the Hesse forest in the northeastern part of France early during the 2007 growing season to determine the seasonal and daily variability in $\delta^{13}C$. Measurements were compared to those taken with an isotope ratio mass spectrometer. The two methods gave similar values of $\delta^{13}C$, showing the reliability of the TDLS system. The system was able to record pronounced daily as well as seasonal variations in $\delta^{13}C$ (up to 1.5‰). The $\delta^{13}C$ observed in the forest fluctuated between values of litter and of root respiration recorded during incubation, suggesting that temporal variations were associated with changes in the relative contribution of the two compartments during the day and during the season.

Diurnal CO_2 Emissions

Not only do CO_2 emissions from the soil differ with seasons, they also differ diurnally. In a study to measure root respiration, Nobel and Palta (1989) determined soil O_2 and CO_2 concentrations for two species of shallow-rooted cacti grown in the field: barrel cactus, *Ferocactus acanthodes* (Lemaire) Britton, and the widely cultivated platyopuntia, *Opuntia ficus-indica* (L.) Miller (also called Indian-fig). [Opuntia is the old Latin name used by Pliny and probably is derived from Opus, a town in Greece (Bailey, 1974, p. 703).] The *F. acanthodes* was obtained from Agave Hill (850 m elevation) about 10 km south of Palm Desert, California, where the soil was a loamy sand. The *O. ficus-indica* was collected at a cactus farm (395 m elevation) about 15 km west of San Bernardino, California, where the soil was a sandy clay loam. To measure the concentrations of O_2 and CO_2 in the center of the root zone of *F. acanthodes* at Agave Hill, a cylindrical airstone (commercially available for the aeration of aquaria) was placed in the soil on April 3, 1989. The depth was 10 cm, which is the center of the rooting zone of *F. acanthodes*. The soil was extremely dry with a water potential of less than −6 MPa. Another airstone was placed at a similar location for another *F. acanthodes*, and the water equivalent to 50 mm of rainfall was applied from 12:00 to 16:00 h on April 3, 1989, raising the soil water potential in the root zone to 0.00 MPa by 08:00 h on April 4, 1989, the time when soil gas phase measurements were commenced. A syringe was used to draw gases from a tygon tube that led from the airstone to the soil surface. Concentrations of CO_2 gas in the tube were determined with a portable photosynthesis system (Li-Cor 6200, Li-Cor, Inc., Lincoln, Nebraska). Similar techniques were used to determine the gas phase composition at a soil depth of 10 cm in the rooting zone of *O. ficus-indica* at the cactus farm on April 11, 1989, when the soil water potential was −2 MPa under natural conditions and 0.00 MPa after watering.

The CO_2 concentration in the center of the root zone of *F. acanthodes* tended to increase during the daytime and to decrease at night (Nobel and Palta 1989) (Figure 2.11). Maximum values were 3860 µL/L at 16:00 h under wet conditions and 635 µL/L at 18:00 h under dry conditions, when the soil at a 10 cm depth achieved maximum temperatures of 28°C and 32°C, respectively. Minimum soil CO_2 concentrations of 1230 µL/L under wet conditions and 450 µL/L under dry conditions occurred in the early morning (Figure 2.11), when soil temperatures at 10 cm were about 19°C and 21°C, respectively. The CO_2 concentration 10 cm below the soil surface averaged 540 µmol/mol for *F. acanthodes* under dry conditions and 2400 µL/L under wet conditions. In contrast to the more

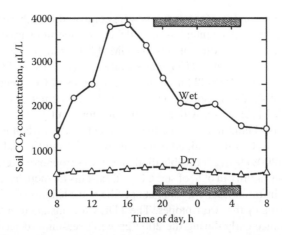

FIGURE 2.11 Daily patterns of CO_2 concentration in the root zone of *F. acanthodes* at Agave Hill (in California, United States) for a soil depth of 10cm under dry conditions (dry) and the day after applying a depth of 50mm of water in a circular pattern 50cm in radius centered on the plant (wet). Stippled areas indicate night. (Reprinted with kind permission from Springer Science+Business Media: *Plant Soil*, Soil O_2 and CO_2 effects on root respiration of cacti, 120, 1989, 263–271, Nobel, P.S. and Palta, J.A., Fig. 1.)

than twofold daily increase in soil CO_2 concentration under wet conditions for *F. acanthodes*, the soil O_2 concentrations in the rooting zone in the field were indistinguishable from that in the ambient air (21% by volume). The soil O_2 concentration at 10cm was essentially constant, the value average 99.5%±2.2% (mean and standard deviation for 12 measurements) of that of the ambient air. For the dry soil, the soil O_2 concentration averaged 99.0%±2.1% (n =12) of the ambient air value.

For *O. ficus-indica* growing in the commercial plantation, the soil was shaded and hence did not experience large daily oscillations in temperature at 10cm below the surface (Nobel and Palta 1989). The soil at the center of the root zone had a CO_2 concentration of 1120±60µL/L (standard deviation for $n=4$) in the late afternoon under natural (dry) conditions and 4240±260µL/L ($n=4$) when wet. The CO_2 concentrations in the early morning were 1040±50µL/L ($n=4$) under natural conditions and 4090±410µL/L ($n=3$) when wet. The CO_2 concentration at 10cm averaged 1080µL/L under natural conditions and 4170µL/L under wet conditions. As for the soil with *F. acanthodes*, no daily change was discernible for the soil O_2 concentrations at a depth of 10cm, which averaged 100.0%±0.1% (standard deviation for $n=10$) of the ambient air value under natural conditions and 99.8%±0.1% ($n=8$) under wet conditions.

Nobel and Palta (1989) measured the respiration rates of roots by growing the plants in pots filled with soil from the Agave Hill site (loamy sand). For both species, the respiration rate in the laboratory was zero at 0% O_2 and increased to its maximum value at 5% O_2 for rain roots (roots induced by watering) and 16% for established roots. Established roots of *O. ficus-indica* were slightly more tolerant of elevated CO_2 than were those of *F. acanthodes*; 5000µL/L inhibited respiration by 35% and 46%, respectively, for the two species. For both species, root respiration was reduced to zero at 20,000µL/L (2%) CO_2. Inhibition by 2% CO_2 was irreversible and led to the death of cortical cells in established roots in 6h. The results showed that the root respiration of these two cacti was inhibited by elevated soil CO_2 concentrations and that 2% in the soil was lethal.

ORGANIC MATTER

Lundegårdh (1927) found that large amounts of organic matter had to be added to soil to have an effect on soil respiration. When he added small to medium quantities of farmyard manure to sands or loams not too poor in humus, the effect upon soil respiration was small. The manured plots at first evolved less CO_2 than the control plots, and then the respiration increased again.

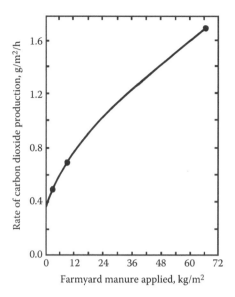

FIGURE 2.12   The influence of large amounts of farmyard manure on the production of carbon dioxide in the soil. (Reprinted from Taylor, S.A. and Ashcroft, G.L., *Physical Edaphology. The Physics of Irrigated and Nonirrigated Soils*, W.H. Freeman and Company, San Francisco, CA, 533 pp., Fig. 12.11. With permission; Data from Lundegårdh, H., *Soil Sci.*, 23, 417, 1927, Fig. 9.)

However, large quantities of manure always increased soil respiration (Figure 2.12). [Lundegårdh's (1927) data have been redrawn by Taylor and Ashcroft (1972, p. 366).] Although not stated by Lundegårdh, we can estimate how much manure he added (see *X*-axis, Figure 2.12). Let us assume he had a silt loam soil, which to the depth of the plow layer (17 cm) weighs 2,300,000 kg/ha (Higgs et al. 1981, p. 69). We remember that 1 ha = 10,000 m^2. To the depth of the plow layer, then we have 23,000 kg/100 m^2 soil. Dividing the largest amount of manure he added (Figure 2.12), which was 6,500 kg/100 m^2, by 23,000 kg/100 m^2, we see he added about 28% manure to the plow layer— indeed this is a large amount. Lundegårdh (1927, p. 430) concluded that the native microorganisms of the farmyard manure persist only when large quantities are given, so that the manure is composed of coherent lumps. If the manure is uniformly spread out among the soil particles, the original environment of the manure organisms is destroyed, and it will take some time for them to invade the new substrate.

Cannon and Free (1925, p. 162) cited research on manured and unmanured soils in India. The O_2 content varied greatly, so much so that it appeared to reach the critical concentration for root growth. [Cannon and Free (1925, pp. 161–162) did not state what this critical concentration of O_2 was but noted that the critical concentration varies among species. The value appears to be between 2% and 15% O_2. See the comments above, as well as Kirkham (1994) and Chapter 4.] In unmanured lands in India in the spring, there was from 13.5% to 18.6% O_2 at depths down to 1.5 m (5 ft). In October, and in manured land, there was from 11.4% to 17.8% O_2, but in swamp land, there was only 0.54% O_2.

Not only does the amount of organic matter in the soil affect the rate of CO_2 evolution, but also the atmospheric CO_2 concentration under which the organic matter was produced affects CO_2 evolution. Gorissen et al. (1995) grew rye grass plants (*L. perenne* L. cv. Bartlett) under two atmospheric CO_2 concentrations of 350 and 700 μmol/mol in growth chambers. After 4 weeks, the plants were harvested, divided into shoots and roots, and dried; the roots were then ground. Pots were filled with 30 g of a loamy sand soil, and the soils were amended with 1 mg root material per g dry soil. Decomposition of roots cultivated at 700 μmol/mol showed an increase in decomposability during the first 2 days compared to roots cultivated at 350 μmol/mol CO_2. Increased amounts of easily decomposable C compounds in the roots, for example, sucrose, under elevated CO_2 might have been responsible for this observation. After 8 days, decomposition of the roots grown at $700\,\mu mol/mol$ CO_2 was decreased. At the end of the incubation (64 days), the decrease in decomposition increased to 24% compared to the roots cultivated at $350\,\mu mol/mol$ CO_2. The decrease was possibly due to a change in root quality, such as a limited nitrogen supply by the root material caused by an increased carbon-to-nitrogen ratio, which was 18 and 32 in the roots grown at 350 and $700\,\mu mol/mol$, respectively. Their results indicated that decreased decomposability of roots grown under elevated CO_2 concentrations can potentially annul the additional CO_2 release from soil organic matter.

Ross et al. (1996) also found that the release of CO_2 from decomposition of harvested herbage depended upon the CO_2 concentration under which the plants grew. They took samples of plants and the associated soil (a Pukepuke sand) from a 30-year-old pasture near Bulls, New Zealand. The principal plants in the samples were *L. perenne* L. (ryegrass, a C_3 plant) and *Paspalum dilatatum* Poir. (paspalum or dallis-grass, a C_4 plant). They placed the turves in controlled environmental rooms that had atmospheric CO_2 concentrations of 300 or $700\,\mu mol/mol$. Seasons were artificially imposed (spring, summer, and autumn). The photoperiods and temperatures in the rooms were appropriate for each season. Water was withheld completely from half of the turves in each CO_2 treatment during the final 2 months of the "summer" period; the other half of the turves in each treatment were kept continuously moist. Herbage for yield was harvested every 3 weeks (except during "winter" when samples were taken every 6 weeks). Samples for decomposition were dried and ground. Decomposition of herbage, harvested at the end of each "season," was determined by incubating plant material (1.0% on a dry weight basis) with 10.0 g of soil (at 60% of water-holding capacity) in flasks at 25°C for 21 days and measuring CO_2–C production at intervals. Under both wet and dry conditions, the release of CO_2 from the decomposition of "summer"-harvested herbage was greater in the $700\,\mu mol/mol$ treatment than from the $350\,\mu mol/mol$ treatment (Figure 2.13, top). Treatment differences were less marked in the "autumn" samples (Figure 2.13, bottom). They concluded that elevated CO_2 may not necessarily lead to greater carbon sequestration, but it will likely result in greater carbon turnover. Their data also confirmed that the atmospheric CO_2 concentration under which organic matter is produced affects CO_2 evolution.

The amount of CO_2 emitted from manure will depend upon its water content. Cabrera et al. (1994) monitored CO_2 emissions from poultry litter, which is a mixture of excreta, bedding material, feathers, and waste feed. They studied both a fine fraction, which can be recycled as fertilizer, and unfractionated litter (whole litter), which can be recycled as bedding material. The litter came from three poultry houses. Houses 1 and 3 had dirt floors and House 2 had a cement floor. The fine fraction was obtained after sieving through a 0.83 mm sieve. After fractionation, the fractions and the whole litters were stored in wooden bins for 0, 2, 4, 8, or 16 weeks of storage, during which time CO_2 emission from samples was monitored. Samples were incubated at their natural water content (0.17 kg H_2O/kg dry litter) and at a water content of 0.50 kg H_2O/kg dry litter. The CO_2 emission was measured, because it indicates the development of anaerobic microsites adequate for denitrification. Denitrification is the biological reduction of nitrate or nitrite to gaseous nitrogen—either molecular nitrogen or the oxides of nitrogen. The process results in the escape of nitrogen as a gas into the air and hence is undesirable in agriculture. Denitrification results in the reduction of fertilizer value of the litter. Increasing the water content to 0.5 kg H_2O/kg increased the CO_2 emission in both the whole litter and the fine fraction (Figure 2.14; open and closed triangles show the wet treatment). The increase in water content increased respiration and production of CO_2, which in turn led to anaerobic conditions appropriate for denitrification. The results of Cabrera et al. (1994) suggest that more CO_2 will be evolved from manure at higher water contents than at lower water contents.

RAINFALL, IRRIGATION, AND FLOODING

Rains saturate the soil and limit aeration. Irrigations have the same effect. Early investigators of the soil air (e.g., Boussingault and Léwy 1853) did not study irrigation, but if they had considered

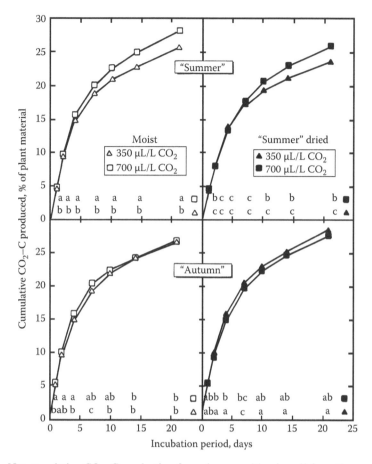

FIGURE 2.13 Net cumulative CO_2–C production from decomposition in soil from the same treatment, of herbage harvested from moist or dried soil at the end of "summer" and from moist soil at the end of "autumn." At each incubation period and "season," values not marked with the same letter differ significantly ($p < 0.05$). (Reprinted with kind permission from Springer Science+Business Media: *Plant Soil*, Elevated CO_2 effects on carbon and nitrogen cycling in grass/clover turves of a Psammaquent soil, 182, 1996, 185–198, Ross, D.J., Saggar, S., Tate, K.R., Feltham, C.W., and Newton, P.C.D., Fig. 2.)

it, they would have concluded that irrigation as well as rainfall has a marked influence upon the CO_2 content of the soil air. Drainage and any other factor that influence the water relations of soil also influence the aeration status (Taylor and Ashcroft 1972, p. 365). As we saw from the data of Moore and Dalva (1993) (Figure 2.3), the depth of the water table affects the rate of CO_2 emission. They used peat soils and saw that lowering of the water table increased the CO_2 flux. Opening of the soil pores to air allowed CO_2 to escape. It is expected that a high water table retards gaseous diffusion and interchange of soil gases with atmospheric gases (Taylor and Ashcroft 1972, p. 366). In such situations, CO_2 might build up rapidly but cannot be readily exchanged. As a consequence, aeration becomes inadequate, with undesirable consequences (Taylor and Ashcroft 1972, p. 366), such as limitation of root growth. Let us now look at changes in CO_2 in the soil with irrigation.

Argo et al. (1996) measured CO_2 and O_2 partial pressures at three locations (3.8, 7.5, and 10.3 cm from the top) in 15-cm-tall pots containing flowering chrysanthemums [*Dendranthema × grandiflorum* (Ramat.) Kitamura] grown in one of three root media. They were 100% Canadian sphagnum peat, 50% sphagnum peat and 50% vermiculite, and 100% vermiculite. Before differential irrigation treatments began, average ambient CO_2 and O_2 partial pressures in the media were 63 Pa and 21 kPa,

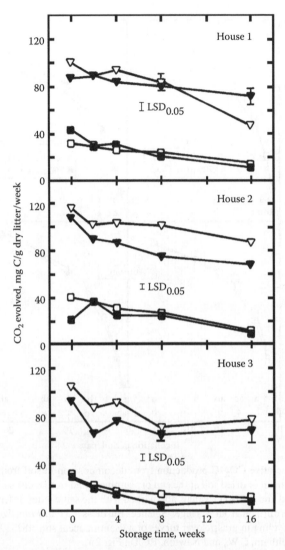

FIGURE 2.14 Emission of CO_2 from a fine fraction of poultry litter (open triangles and open squares) and whole poultry litter (closed triangles and closed squares) from three broiler houses during 16 weeks of storage. The litter was incubated at an unamended water content of 0.17 kg H_2O/kg (open and closed squares) and at an amended water content of 0.50 kg H_2O/kg (open and closed triangles). (Reprinted from Cabrera, M.L. et al., *Commun. Soil Sci. Plant Anal.*, 25, 2341, 1994, Fig. 3. With permission.)

respectively, and were similar in the three sampled layers in root media with an average moisture content of 50%–60% of container capacity. The water used for both irrigations was an alkaline well water with 320 mg $CaCO_3$/L. Within 190 min after a surface (drip)-irrigation application, the partial pressure CO_2 in the media increased to as much as 1600 Pa and O_2 in the media decreased to 20.5 kPa. With subirrigation, CO_2 partial pressures in the media increased to a maximum value of only 170 Pa and O_2 in the media remained at 21 kPa. The high CO_2 partial pressures in the media measured after surface irrigation were not persistent. Within 180 min, CO_2 partial pressures returned to levels averaging 45% higher (100 Pa) than that measured before the irrigation. Partial pressures of O_2 in the media also returned to ambient levels 180 min after the surface irrigation. The results showed that subirrigation kept CO_2 concentrations in the root media low. Applications of subirrigation water will not result in the same amount of moisture in soil as will applications of

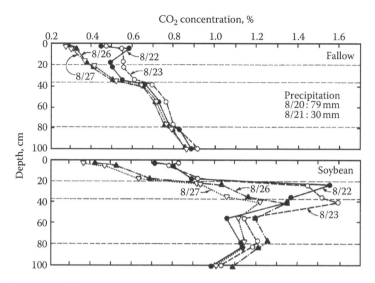

FIGURE 2.15 Changes in CO_2 concentration in soil after heavy rainfall. (Reprinted from Osozawa, S. and Hasegawa, S., *Soil Sci.*, 160, 117–124, 1995, Fig. 6. With permission.)

surface water. With subirrigation, super-capillary-sized pores are not filled with water while they are in surface irrigation. Subirrigation provides more soil aeration than surface addition of water (Kirkham 2005, p. 79).

Osozawa and Hasegawa (1995) measured changes in the concentrations of CO_2 to a depth of 100 cm in an Andisol under both fallow and when soybean [*Glycine max* (L.) Merr.] grew for 1 year (June 24, 1991 to June 24, 1992). [Andisols are mineral soils that are dominated by soil properties related to volcanic origin of materials (Soil Science Society of America 1997).] Gas sampling tubes were installed in the soil at depths of 2.5, 5, 17.5, 22.5, 34.5, 39.5, 55, 76.5, 81.5, and 100 cm below the soil surface. The concentration of CO_2 in the soil depended on rainfall and season. Heavy rainfall, which closed the air pathways to the atmosphere, resulted in a higher CO_2 concentration in the shallow soil layers. Figure 2.15 shows the changes in the CO_2 concentration profile in soil under fallow and under soybean after heavy rainfalls of 79 mm on August 20 and 30 mm on August 21. The CO_2 peak under fallow appeared at the 5 cm depth on August 22 but disappeared by August 23. However, the CO_2 concentrations in the deeper layers on August 23 were higher than on the previous day. Thereafter, the CO_2 concentration decreased greatly at shallow depths. Under soybean, the CO_2 concentration at depths greater than 55 cm was the highest on August 26, 5 days after the rainfall, although the CO_2 concentrations in the shallower layers had started to decrease. The changes in CO_2 concentration with time and depth may be explained by the following processes. First, the CO_2 concentration increases in the shallow layers because of the lack of a continuous air-void pathway to the atmosphere. Second, with the progression of the drying process (drainage and evapotranspiration), CO_2 in the shallow layers begins to diffuse both upwards and downwards, resulting in a decrease of the CO_2 concentration in the shallow layers and an increase in the deeper layers. Thereafter, the CO_2 concentration throughout the soil profile decreases to the initial condition.

The seasonal changes in the CO_2 concentration are shown in Figure 2.16 (Osozawa and Hasegawa 1995). At the beginning of the experiment on June 24, 1991, the CO_2 concentration profiles under fallow and soybean were the same. The CO_2 concentrations under fallow increased with depth throughout the year and showed maximum values toward the end of July and in August. Thereafter, the CO_2 concentration began to decrease. In contrast, the CO_2 concentrations under soybean showed maximum values at depths that shifted downward gradually with soybean growth from 20 to 40 cm

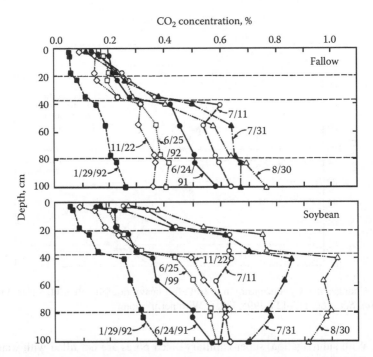

FIGURE 2.16 Seasonal changes in CO_2 concentration. (Reprinted from Osozawa, S. and Hasegawa, S., *Soil Sci.*, 160, 117–124, 1995, Fig. 7. With permission.)

on July 11 to 40 to 80 cm on August 30, suggesting the effect of root elongation. Although the root length density decreased with depth, some roots reached the 80 cm depth on August 30, 1991. The CO_2 concentrations under soybean were higher than those under fallow, except on June 24, 1991. The data of Osozawa and Hasegawa (1995) showed that changes in CO_2 in soil are greatly influenced by soil moisture. Heavy rainfall resulted in a high CO_2 concentration in shallow layers at first, then a decrease in the concentration at shallow depths accompanied by an increase in the concentration in deeper layers with the progression of drying.

CO_2-Amended Irrigation Water

In contrast to poor aeration due to water-saturated soil from rains and irrigation, CO_2 has actually been added to irrigation water to try and increase growth. Mauney and Hendrix (1988) irrigated cotton (*Gossypium hirsutum* L.), one of the most responsive crops to CO_2 enrichment (Kimball 1983), with carbonated water. ["Carbonate" means to charge with carbon dioxide (Friend and Guralnik 1959).] They grew the plants in buckets containing a potting mix that was allowed to drain. They placed the buckets in rows in a greenhouse, so the CO_2 escaping into the atmosphere would enhance the growth of both control and treated rows. The carbonated water was produced by commercially available carbonators and was saturated with CO_2 at the time of irrigation and carried 1.5–1.8 g/L. This concentration of CO_2 was about 3×10^3 times greater than the control water. The half-life of this concentration was about 24 h. The pH of the carbonated irrigation water was 3.6 and the pH of the control water was 4.3. They repeated the experiment two times. In the first experiment, growth (i.e., boll number, leaf area, and dry weight per plant) was not increased due to the carbonated water. In the second experiment, boll number per plant was increased from 11.3 (control) to 17.3 (plants with carbonated water); leaf area increased from 0.52 m² per plant (control) to 0.86 m² per plant; and total dry weight increased from 125.5 to 282.9 g per plant. In the second experiment, they measured the carbon-dioxide exchange rate (CER) (photosynthetic rate) of the plants. At 50 days after planting, there was no increase in CER due to the carbonated water. But at 110 days, the CER increased from 18 (control) to 25 μmol/m^2/s. By analyzing carbon isotope ratios in the lint of the cotton plants, they determined that none of the carbon fixed by photosynthesis came from the irrigation water. They attributed the increased photosynthesis and growth with the carbonated water to uptake of Zn and Mn, which were deficient in the control leaves. The decrease in pH due to the irrigations with carbonated water apparently made these elements more available than they were with the control water.

Nakayama and Bucks (1980) also found positive effects of carbonated water on plant growth. They applied carbonated water to soils via a subsurface drip irrigation system, and it produced a 20% increase in wheat (*T. aestivum* L.) yield and a decrease in pH of 1.5 units in the calcareous soil that they used in their experiments. Hartz and Holt (1991) conducted a study to determine the effects of CO_2-amended irrigation water on production of tomato (*Lycopersicon esculentum* Mill. cv. Bingo) and cucumber (*Cucumis sativus* L. cv. Dasher II) grown under field conditions in a fine sandy loam at Riverside, California, or a sandy loam at Irvine, California. Beginning 3 weeks after planting, bottled CO_2 was injected continuously into the irrigation stream at 0, 0.5, or 1.0 g/L. Carbon dioxide amendment acidified the irrigation water, causing a drop in pH of about 2.0 units (from about 7.3 to 5.3 at 1.0 g/L CO_2). At Riverside, CO_2 amendment of the irrigation water had no effect on marketable yield of tomato or cucumber. At Irvine, yields of tomato and cucumber decreased linearly with increasing amounts of CO_2 injection. Hartz and Holt (1991) suggested that the detrimental effect of CO_2 treatment at Irvine may have been related to the use of plastic mulch. The mulch increased CO_2 concentration in the soil atmosphere. They hypothesized that further addition of CO_2 through water injection may have resulted in supra-optimal concentrations of CO_2 that were detrimental to yield. Why this hypothesis did not cause a decrease in yield at Riverside is not stated. The results of the field studies at Riverside and Irvine were in contrast to earlier reports on other crops, which showed beneficial results from CO_2 injection into irrigation water. Hartz and Holt (1991) concluded that a better understanding of the mechanisms of action of CO_2-amended irrigation water on plant growth is required before root-zone enrichment can be advocated as a commercial production practice.

During a 2 year (1988–1989) field experiment in Colorado, Novero et al. (1991) applied carbonated water to tomato plants. The carbonated water was produced using a diffuser-type commercial carbonator. The pH of this water was 4.5 in both years, whereas, the pH of the control water was 6.5 in 1988 and 7.3 in 1989. Irrigation water applied to the plots was delivered through flexible polyethylene drip tubing placed on top of the plant beds. In 1988, half the plots were mulched with a thin layer of black polyethylene, and in 1989, all the plots were mulched. They sampled the CO_2 concentration in the air above the plots and also took soil-air samples at the 10 cm depth. Fruit yield was increased by 7.8% by application of the carbonated water in the mulched study in 1988, but the carbonated water had no effect in the unmulched study. In 1989, marketable yields were increased by 16.4% in the carbonated water experiment. In both years, concentration of CO_2 in the soil air under the mulched plots was increased due to the irrigation with carbonated water. For example, in 1989, 1 day after an irrigation, the concentration of CO_2 at the 10 cm depth in the control plot was 3,200 μmol/mol and in the plot with the carbonated water, it was 7,100 μmol/mol. The CO_2 concentrations in the air above the soil for control plots at three different times, that is, during, 1 h after, and 1 day after irrigation at the 1 cm height were 363, 377, and 346 μmol/mol, respectively. However, for the plots that were irrigated with carbonated water, the CO_2 concentrations at these three times and height (1 cm) were 1268, 493, and 370 μmol/mol. At the 15 cm height, control plots at these three times had 335, 346, and 351 μmol/mol. Treated plots at the 15 cm height had 420, 372, and 353 μmol/mol CO_2 for these three times, respectively. The carbonated water increased the atmospheric CO_2 concentration to at least the 15 cm height. But the effect was short-lived, and, by 1 day after the irrigation, the CO_2 concentration in the atmosphere had decreased to control levels. However, carbonated water application decreased the soil pH for periods of up to 5 days after an irrigation and increased uptake of phosphorus, potassium, calcium, magnesium, zinc, iron, manganese, copper, and boron. They concluded that because of the limited duration of enrichment relative to the entire growing season for the carbonated water treatments, the yield increase could not be due only to atmospheric enrichment of CO_2. The yield increase resulted from the combined effects of a small amount of atmospheric CO_2 enrichment and increased nutrient uptake due to the changed soil pH.

Basile et al. (1993) saw that carbonated water containing 2600 µmol/mol CO_2 decreased soil pH about 1.5 units, and the temporary reduction in soil pH induced nutrient mobility. The soil was a clay loam with sand and municipal sludge added. They had no plants in their study, so the reduced pH on plant uptake was not reported by them. During a 2 year field experiment in New Jersey (1991 and 1992), Storlie and Heckman (1996) found that carbonated irrigation water had no effect on yield and plant nutrient status of bell pepper (*Capsicum annuum* L. cv. Gator Belle in 1991 and cv. Bell Captain in 1992) grown in a sandy loam soil. Carbon dioxide was injected into the irrigation water at 0, 0.33, 0.67, or 1.0 the base rate of 0.0273 mol/L (1.2 g/L). In 1991, CO_2 injection acidified the irrigation water and caused a reduction in pH from 4.9 to 4.4, 4.2, and 4.1 for the three CO_2 treatments, respectively. Canopy CO_2 concentrations were measured at four heights (0, 15, 30, and 45 cm). Air CO_2 concentration in the lowest portion of the canopy increased by 115 µmol/mol during irrigation, but returned to the ambient CO_2 concentration within 30 min after an irrigation ceased. At higher points in the canopy (15, 30, and 45 cm above the plant base), no enrichment was detected due to the carbonated water irrigation. Carbon dioxide can be directly absorbed through the plant roots and used in plant metabolism (e.g., respiration; Stolwijk and Thimann 1957). But direct absorption is probably not a major contributing source to increased productivity (Stolwijk and Thimann 1957; Storlie and Heckman 1996).

CARBONATION AND HENRY'S LAW

A factor that affects the efficacy of carbonated irrigation water is Henry's law. Carbonated water is under pressure. Carbonated beverages have a long history and were developed from European attempts to imitate the popular and naturally effervescent waters in famous springs. Jan Baptista van Helmont (1577–1644) first used the term "gas" in his reference to the CO_2 content. Numerous reports of experiments with carbonated water were included in the *Transactions* of the Royal Society of London in the late 1700s, including the studies of Joseph Priestley (Riley and Korab 1971). (See Appendix for a biography of Joseph Priestley.) Duplication of the waters of Bad Pyrmont, a spa in Germany near the city of Hamelin (spelled Hameln in German; of "Pied Piper" fame), was one of the major objectives, probably because of the popularity of the spa and also because of the relatively simple nature of the mineral content of the waters there. Priestley, who obtained "fixed air" (carbon dioxide) for his experiments from a brewery near his house, published *Directions for Impregnating Water with Fixed Air* in 1772. Priestley was awarded the Copley medal by the Royal Society of London for his reports concerning "fixed air" and the mixture of water with it. He felt that medicinal value existed in artificially carbonated water. While it has not been proved that carbonation offers a direct medical benefit, carbonated beverages are used to alleviate postoperative nausea when no other food can be tolerated, as well as to ensure adequate liquid intake. Antoine Lavoisier in France identified the "fixed air" of Priestley as a combination of carbon and oxygen, which he called *gas acide carbonique* (acidified carbonated gas). (See Appendix for a biography of Antoine Lavoisier.) In 1781–1783, the English chemist Thomas Henry described apparatus for the production of carbonated waters on a commercial scale (Riley and Korab 1971). (I believe that Thomas Henry may be the father of William Henry, whose work we describe in this chapter, even though *Encyclopedia Britannica*, my source of information, does not state this.)

Today, CO_2 is supplied to the soft drink manufacturer in either liquid or solid ("dry ice") form, maintained under high pressure in heavy steel containers. As the pressure is released, the material changes into the gaseous form, in which it can be dissolved in water in quantities that vary with temperature and pressure (Riley and Korab 1971). The amount of gas that the water will absorb increases as the pressure is increased and as the temperature is decreased. Carbonation of the water is effected by chilling the liquid and cascading it in thin layers over a series of plates in an enclosure containing CO_2 gas under pressure. Soft drink producers use bottles and metal cans to keep the beverages under pressure. Still (non-effervescent) beverages, such as fruit juices, are not under pressure and may be packaged not only in bottles and cans but also in cardboard cartons (Riley and Korab 1971). We experience CO_2 on our tongues in carbonated beverages through physical and chemical sensation (Frommer 2010). The feel of fizz relies in part on the detection of CO_2 bubbles that stimulate somatosensory receptors on the tongue. The response to CO_2 depends on the sour sensor cells. A gene encoding a carbonic anhydrase is specifically expressed in these cells. In experiments with mice, this enzyme was essential for detection of CO_2 (Frommer 2010).

William Henry (1775–1836) described experiments on the quantity of gases absorbed by water under different pressures. (See Appendix for a biography of William Henry.) The conclusion he reached was that "water takes up of gas condensed by one, two or more additional atmospheres, a quantity which, ordinarily compressed, would be equal to twice, thrice, etc., the volume absorbed under the common pressure of the atmosphere" (Oesper 1971). Stated in other words, Henry's law is "The solubility (in terms of mass dissolved) of a gas in a liquid, at constant temperature, is directly proportional to the pressure of the gas" (Gregg 1963, p. 375). If there is a mixture of gases, the solubility of each gas is directly proportional to its partial pressure. Henry's law applies to dilute solutions (Newton 1971). In a very dilute solution, a solute molecule will (with rare exceptions) have only solvent molecules as near neighbors, and the probability of escape of a particular solute molecule into the gas phase is expected to be independent of the total concentration of solute molecules. In this case, the rate of escape of solute molecules will be proportional to their concentration in the solution, and solute will accumulate in the gas until the return rate is equal to the rate of escape. With a very dilute gas, this return rate will be proportional to the partial pressure of the solute. Thus, we expect that for a solution very dilute in solute, in equilibrium with a gas at low pressure (Daniels and Alberty 1966, p. 140; Newton 1971), the following equation holds:

$$p_2 = X_2 K_2 \qquad\qquad (2.1)$$

where
  p is the partial pressure
  X is the mole fraction
  K is the Henry's-law constant

The subscript 2 indicates that the solute (in this case, simply the component at lower concentration) is being considered. The value of Henry's-law constant, K_2, is obtained by plotting the ratio p_2/X_2 versus X_2 and extrapolating to $X_2 = 0$. Such a plot is shown in Figure 2.17 for acetone in ether–acetone solution at 30°C.

Thus, the quantity of gas absorbed by a liquid is dependent upon the pressure. It seems that for irrigation water to hold CO_2, it needs to be maintained under pressure. Because the irrigation water is not under pressure, the extra CO_2 added by artificial injection of CO_2 into irrigation water probably will not last long enough to have an effect on plant growth. Waters high naturally in CO_2 from springs could provide CO_2-enriched irrigation water. As noted earlier in this chapter, CO_2 emissions from mineral springs are being used to study the effects of elevated levels of CO_2 on plant growth in New Zealand (Newton et al. 1996a; Ross et al. 2002) and Italy (Miglietta et al. 1993a).

Henry's law says nothing about time—that is, how long it takes CO_2 to escape once the pressure is removed. As noted, Mauney and Hendrix (1988) said that the half-life of the carbonated water that they used to irrigate cotton was about 24 h. After the pressure is released, the CO_2 in the carbonated

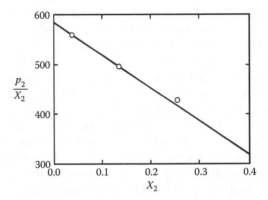

FIGURE 2.17   Evaluation of the Henry's-law constant, K_2, for acetone in ether–acetone solutions at 30°C. (Daniels, F. and Alberty, R.A., *Physical Chemistry*, 3rd edn., 767pp, 1966, Fig. 5.10. Copyright Wiley-VCH Verlag GmbH & Co. KGaA. Reprinted with permission.)

water added to the soil will equilibrate with the water and air in the soil and return to the ambient levels normally found in the soil.

SALINITY

Salinity is a big problem on irrigated soils. Waters act as sources of salts when used for irrigation (Richards 1954, p. 3). Unless they are leached through the soil, they build up in the soil and limit plant growth. In the first chapter, we discussed the effects of elevated levels of atmospheric CO_2 on plants grown under saline conditions. Little information exists for plants supplied with CO_2 in a saline root zone. Cramer and Lips (1995) grew tomato seedlings (*L. esculentum* (L.) Mill. cv. F144) in hydroponic culture with 0 or 100 mM NaCl and aerated the root solution with either ambient (360 μmol/mol CO_2) or CO_2-enriched air (5000 μmol/mol). Under the ambient CO_2 conditions, the dry weights of plants grown with 0 or 100 mM NaCl were 1.30 and 0.75 g per plant, respectively. When the root solution was enriched with 5000 μmol/mol CO_2, the dry weights of plants grown with 0 or 100 mM NaCl were 1.16 and 0.91 g per plant, respectively, a significant difference. The elevated CO_2 counteracted some of the growth reductions due to salinity. A fan was used to remove any CO_2 that escaped from the root zone into the air. So the increase in growth under salinity was not due to elevated levels of CO_2 in the atmosphere. Cramer and Lips (1995) felt that the enriched CO_2 allowed plants to increase their supply of carbon for amino acid synthesis in roots, thus allowing them to grow better under salinity.

FLOODING

Hook et al. (1971) studied the effect of flooding on swamp tupelo [*Nyssa sylvatica* var. *biflorac* (Walt.) Sarg.], a flood-tolerant tree. It grows both on sites that are inundated during most of the growing season and in drained, moist soils. Its growth is reduced by about 50% by flooding in stagnant as compared with moving water. Sustained rates of anaerobic respiration occur in *Nyssa* under low oxygen concentrations or in the absence of oxygen. It is assumed that this anaerobic respiration permits the roots to grow in an environment devoid of oxygen. During anaerobic carbohydrate metabolism, either lactic acid is formed (often called *glycolysis*) or ethanol and CO_2 are produced (termed *alcoholic fermentation*) (Conn and Stumpf 1963, pp. 163–175). Glycolysis and alcoholic fermentation involve the same reactions, except for the initial and final stages. In glycolysis, the initial reactant is a polysaccharide (Conn and Stumpf 1963, p. 164). The plant starches and animal glycogens are examples of polysaccharides (Conn and Stumpf 1963, pp. 45–46). The fundamental

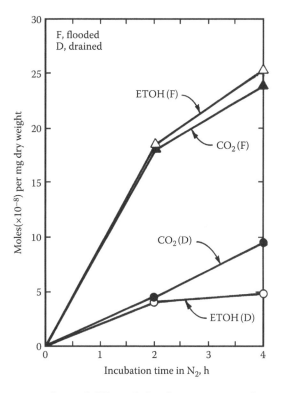

FIGURE 2.18 Ethanol concentration and CO_2 evolution from swamp tupelo root tips from drained and flooded pots after incubation in nitrogen. Each point is the average of six seedlings. (Reprinted from Hook, D.D., Brown, C.L., and Kormanik, P.P., *J. Exp. Bot.* 1971, Fig. 1 by permission of Oxford University Press.)

unit of starch and glycogen is glucose (Conn and Stumpft 1963, p. 46). In alcoholic fermentation, the initial reactant is glucose. In glycolysis, the final product is lactic acid; and in alcoholic fermentation, the final products are CO_2 and ethanol.

Hook et al. (1971) transplanted the swamp tupelo seedlings into drained or flooded pots in the greenhouse. The soil was a mixture of sand and peat moss. After 31 days in flooded pots, the seedlings were removed from the pots, and tests were made on terminal 1 cm segments of lateral roots (root tips) after 2 and 4 h incubation in nitrogen gas (N_2) (no oxygen) to monitor CO_2 evolution. Roots were macerated in an ice bath, and the root extracts were centrifuged and analyzed for ethanol. Under flooding, the roots of swamp tupelo produced ethanol (Figure 2.18). The ethanol pathway (alcoholic fermentation) appeared to be the primary source of CO_2 in swamp tupelo roots under anaerobic conditions.

Hook et al. (1971) also investigated the tolerance of roots of swamp tupelo and sweet gum (*Liquidambar styraciflua* L.) to CO_2 in the root medium. Two-month-old seedlings were grown with their roots in three different levels of CO_2 (2%, 10%, and 31%) with 1% O_2 in a closed liquid-culture system. Stems of the seedlings were sealed airtight with rubber stoppers and silicon rubber. The tops of the seedlings were exposed to atmospheric conditions and the roots to selected gas mixtures. Swamp tupelo roots tolerated much higher concentrations of CO_2 than sweet gum roots without adverse effects. Sweet gum seedlings survived 15 days in 2% CO_2 with only slight chlorosis. However, sweet gum seedlings died within 10 days in 31% CO_2 and within 15 days in 10% CO_2. In contrast, growth of swamp tupelo was not hindered by either 2% or 10% CO_2, but 31% CO_2 retarded root development, growth in height, and rate of O_2 uptake by roots. Rapid growth by swamp tupelo and slow growth by sweet gum with only 1% oxygen in the root environment suggested that differences in CO_2 tolerance of these two species were closely tied to the O_2 requirement of their roots. As we shall see in Chapters 3 and 4, the ability of plants to survive appears to be more dependent on the O_2 concentration in the root zone rather than the CO_2 concentration.

SUMMARY

The following conclusions can be made about CO_2 in the soil atmosphere. The amount of CO_2 is usually greater in the soil, and O_2 less, than in the atmosphere above. Average figures for soil air might be 0.25% CO_2 and 20.73% O_2 compared to 0.0385% CO_2 and 20.95% O_2 in the atmosphere. The CO_2 concentration in the soil usually increases with depth. A seasonal fluctuation in the CO_2 and O_2 content of the soil atmosphere occurs. Respiration of roots and soil organisms reduces the O_2 concentration and increases the CO_2 concentration. Because respiration increases with temperature, these changes are more marked in the summer than in the fall and winter. No-tilled soil does not have a higher or lower CO_2 emission than conventionally tilled soil. The amount of CO_2 released under different tillage systems depends upon soil moisture and temperature, and these are variable during a season. Organic matter added to the soil, like manure, increases the CO_2 evolved from the soil. The decomposability of organic matter grown under elevated levels of CO_2 varies from that grown under ambient levels of CO_2. When organic matter from vegetation grown under elevated CO_2 decomposes, it can either produce more, less, or the same amount of CO_2 compared to vegetation grown under ambient CO_2. Rainfall and irrigation increase the CO_2 concentration in soil, because water closes the air pathways to the atmosphere. Carbonated irrigation water generally has no effect on growth of plants. When growth is increased with carbonated water, it is usually due to a lowering of pH of the soil that makes essential nutrients more available. The CO_2 in the carbonated water added to soil dissipates quickly.

APPENDIX

BIOGRAPHY OF JEAN BAPTISTE BOUSSINGAULT

Jean Baptiste Boussingault (1802–1887), French chemist and pioneer in agricultural research, was born in Paris and spent his early manhood in South America. On his return to France, he held the chair of chemistry at Lyons and then the chair of agricultural and analytical chemistry in Paris. His most important work, which was translated into several European languages, was *Agronomie, chimie agricole, et physiologie* (*Agronomy, agricultural chemistry, and physiology*) (1860–1874; second edition, 1884). His writings included papers on the quantity of nitrogen in different foods, the amount of gluten in different wheats, investigations on the question of whether plants can assimilate free nitrogen from the atmosphere (which he answered in the negative), the respiration of plants, the function of their leaves, and the value of manures (*Encyclopaedia Britannica* 1971).

BIOGRAPHY OF JOSEPH PRIESTLEY

Joseph Priestley (1733–1804), discoverer of oxygen (Friend and Guralnik 1959), is widely known for his contributions to the chemistry of gases. A dissenting English minister, he published extensively not only his experimental investigations and the history of science, but also on philosophy, religion, education, and political theory. He is equally well known in his native land and in the United States, where he spent the last 10 years of his life. Born at Birstall, near Leeds in England, on March 13, 1733, the son of strict Calvinists, he soon rejected the authority of the established churches and became a dissenting minister at 22, and 5 years later, a teacher of classics and literature at a dissenting academy in Warrington (near Manchester). While there, Priestley wrote essays on education and English grammar. He earned a doctor of laws degree from the University of Edinburgh (Van Klooster and Passmore 1971).

The early experiments he did dealt with electricity. His work at Warrington left him free to visit London for a month each year, where he met Benjamin Franklin (1706–1790), who encouraged him to complete his first major scientific work, *The History and Present State of Electricity* (1767). Meanwhile, he had been admitted in 1766 as a fellow of the Royal Society (Van Klooster and Passmore 1971).

In 1767, he became pastor of Mill Hill chapel, near Leeds. Living next to a brewery, he became interested in the "fixed air" (carbon dioxide) that lay over the liquids in the fermentation vats. On moving to another location, he continued his studies with this gas, which he made by pouring acid on chalk. He dissolved the gas in water and obtained "an exceedingly pleasant sparkling water, resembling Seltzer water." For this application, which was his first contribution to the chemistry of gases, the Royal Society awarded him the Copley Medal in 1773. About this time, Priestley was fortunate in meeting a fellow sympathizer with the cause of the American colonies, Lord Shelburne. He became librarian to this wealthy nobleman, which gave him leisure time to indulge in scientific pursuits (Van Klooster and Passmore 1971).

With the aid of a newly acquired burning glass, he obtained on August 1, 1774, a new gas from mercuric oxide that was five or six times as pure as ordinary air. He even foresaw its future uses: "it may be peculiarly salutary for the lungs in certain cases," which greatly increased his fame and led to a meeting in Paris with Antoine Lavoisier and other noted scientists to whom he related his findings (Van Klooster and Passmore 1971). Lavoisier at once repeated Priestley's experiments with the new gas, which he named oxygen, from the Greek for "acid-maker," because of its acid-forming properties. It gave Lavoisier the clue to demolish the phlogiston theory of which, nevertheless, Priestley remained a lifelong adherent. [Phlogiston is an imaginary element formerly believed to cause combustion and to be given off by anything burning (Friend and Guralnik 1959).] Meanwhile, the theological heresies of Priestley so strained his relations with Shelburne that he resigned in 1783 to become minister to a dissenting congregation in Birmingham (Van Klooster and Passmore 1971).

In Birmingham, where Priestley spent the happiest decade of his life, he wrote the final two of his six volumes *On Different Kinds of Air*. An abridged edition in three volumes appeared in 1790. In addition to being a scientist, philosopher, political theorist, and dissenting minister, Priestley was also an educator. He was probably the first ever to teach experimental science to school children (Van Klooster and Passmore 1971).

Priestley's nontechnical writing, particularly his *History of the Corruptions of Christianity* (1782), and his belief in the cause of the American colonists and later of the French Revolution made him unpopular with the public. The resentment against Priestley and his friends came to a head on the second anniversary (July 14, 1791) of the fall of the Bastille. Mob rule existed in Birmingham for 3 days. Priestley's church, his house, and his laboratory were burned, and he himself escaped in disguise to Worcester. Later, he settled in London, where he spent 3 unhappy years. On April 7, 1794, he left for America to join his three sons, who had already found new homes there. Arriving in New York City in June 1794, he was offered a professorship and a ministry, both of which he declined. He built himself a house and laboratory in the town of Northumberland, Pennsylvania, and there he died on February 6, 1804 (Van Klooster and Passmore 1971).

He was timid about his scientific abilities, saying he was "not apt to be very confident" about conclusions from his observations (Van Klooster and Passmore 1971). Even so, he had the attributes of a born investigator: an intense curiosity, a passion for experimentation, and a native honesty and frankness. He was ingenious at developing experimental methods. Being the first to collect gases over mercury, he discovered, in addition to oxygen, several other gases, including nitrogen, ammonia, nitrogen dioxide, nitrous oxide, hydrogen chloride, and sulfur dioxide. Some of these are highly soluble in water and thus had escaped the attention of contemporary chemists. Another significant contribution to chemistry was Priestley's work on "the purification of air by plants and the influence of light on that process," which provided the stimulus for subsequent studies by Jan Ingenhousz (Dutch–English physician and scientific investigator who discovered the principles underlying photosynthesis, 1730–1799), Jean Senebier (Swiss pastor known for his contributions to the understanding of photosynthesis, 1742–1809), and others on respiration and photosynthesis (Van Klooster and Passmore 1971).

BIOGRAPHY OF ANTOINE LAURENT LAVOISIER

Antoine Laurent Lavoisier (1743–1794), French chemist and father of modern chemistry, was a brilliant experimenter and many-sided genius. He was born in Paris on August 26, 1743. His father, an *avocat au parlement* (literal translation: a lawyer in a legislative assembly), gave him an excellent education at the Collège Mazarin, and he studied mathematics and astronomy with Nicolas de Lacaille, chemistry with the elder Rouelle, and botany with Bernard de Jussieu (Duveen 1971). In 1766, he received a gold medal from the Academy of Sciences for an essay on the best means of lighting a large town. Among his early works were papers on the analysis of gypsum, on thunder, on the aurora, and a refutation of the prevalent belief that water, by repeated distillation, is converted into earth. He also accompanied J.E. Guettard (1715–1786) on a long geological trip and assisted him in preparing his mineralogical atlas of France. In 1768, he was nominated *adjoint chimiste* (associate chemist) to the academy. He passed through all the grades in the academy, was director in 1785, and was made treasurer in 1791. His father bought him a title of nobility in 1772 (Duveen 1971).

He became a member of the *ferme générale* (general farming). Appointed *régisseur des poudres* (manager of gunpowder) in 1775, he abolished the annoying search for saltpeter (potassium nitrate or sodium nitrate) in the cellars of private houses. He also increased the production of salt and improved the manufacture of gunpowder (Duveen 1971). In 1778, he started a model farm at Fréchines, where he demonstrated the advantages of scientific agriculture. In 1785, he was nominated to the committee on agriculture and, as its secretary, drew up reports and instructions on the cultivation of crops and promulgated various agricultural schemes. Chosen a member of the provincial assembly of Orléans in 1787, he planned the improvement of the social and economic conditions of the community by means of savings banks, insurance societies, and canals. He advanced money without interest to the towns of Blois and Romorantin for the purchase of barley during the famine of 1788. He was associated with committees on hygiene, coinage, the casting of cannon, and public education. He was secretary and treasurer of the commission appointed in 1790 to secure uniformity of weights and measures throughout France. This led to the establishment of the metric system (Duveen 1971).

Lavoisier's name will always be associated with the overthrow of the phlogistic doctrine that had dominated the development of chemistry for more than a century, and with the establishment of the foundations upon which modern science rests. On November 1, 1772, Lavoisier deposited with the Academy of Sciences a note that stated that sulfur and phosphorus when burned increased in weight because they absorbed "air." However, the metallic lead formed from litharge (a yellowish-red oxide of lead, PbO, made by heating lead in a current of air) by reduction with charcoal weighed less than the original litharge because it had lost "air." He did not explain the exact nature of the airs concerned in the processes until after the preparation of "dephlogisticated air" (oxygen) by Joseph Priestley in 1774. He concluded that Priestley's new "air" was what was absorbed by burning sulfur and phosphorus, "nonvital air" or nitrogen remaining behind (Duveen 1971).

In a memoir presented to the academy in 1777, read in 1779 but not published until 1781, he assigned to dephlogisticated air the name oxygen, or "acid producer." Combustion, as explained by Lavoisier, was not the result of the liberation of hypothetical "phlogiston," but the result of the combination of the burning substance with oxygen. On June 25, 1783, he announced to the academy that water was the product formed by the combination of hydrogen and oxygen. He burned alcohol and other combustible organic compounds in oxygen and, from the weight of water and carbon dioxide produced, calculated their composition (Duveen 1971).

In addition to his purely chemical work, Lavoisier, mostly in conjunction with Pierre Laplace (French astronomer and mathematician, 1749–1827), devoted attention to physical problems, especially those connected with heat. The two carried out some of the earliest thermochemical investigations and did a series of experiments to determine specific heats. Lavoisier also studied fermentation, respiration, and animal heat, and considered the processes as essentially chemical in nature (Duveen 1971).

Lavoisier's membership of the *ferme générale* (general farming) made him an object of suspicion to the authorities during the French Revolution. Marat accused him of putting Paris in a prison and of stopping the circulation of air in the city by the wall erected at his suggestion in 1787. [Jean Paul Marat (1743–1793) was one of the leaders of the French Revolution, and he was assassinated by Charlotte Corday (Friend and Guralnik 1959).] In August 1792, Lavoisier had to leave his house and laboratory. In November 1792, the people belonging to the farmers-general were ordered arrested, and in May 1794, they were sent to be tried by the revolutionary tribunal. The trial lasted only part of a day, and Lavoisier and 27 others were condemned to death. That same afternoon, May 8, 1794, he and his companions, including his father-in-law, were guillotined at the Place de la Révolution (now Concorde). His body was thrown into a common grave (Duveen 1971). The following day, Joseph Louis Lagrange (1736–1813), a French–Italian mathematician and astronomer who served with Lavoisier on the commission to reform the weights and measures, said, "It required only a moment to sever that head, and perhaps a century will not be sufficient to produce another like it" (Duveen 1971). Lavoisier's apparatus is preserved in the Musée du Conservatoire National des Arts et Métiers (Museum of the National Academy of Arts and Trades) in Paris. The famous painting of Lavoisier and his wife by Jacques Louis David (French artist, 1748–1825), done in 1788, is now at Rockefeller University in New York City (Duveen 1971).

BIOGRAPHY OF WILLIAM HENRY

William Henry (1775–1836), English physician and chemist, is the author of "Henry's law," which states that the amount of gas absorbed by a liquid is proportional to the pressure of the gas upon the liquid. He was the son of Thomas Henry (1734–1816), an apothecary and writer on chemistry, who translated Lavoisier's *Opuscules*. William Henry was born in Manchester, England, on December 12, 1775, and was educated at Edinburgh, Scotland (MD, 1807). He devoted himself to industrial and scientific studies. One of his best-known papers (*Philosophical Transactions of the Royal Society*, 1803) describes experiments on the quantity of gases absorbed by water under different pressures. His *Elements of Experimental Chemistry* went through 11 editions. In a fit of melancholia, he committed suicide at Pendlebury (in the Greater Manchester, England, area) on September 2, 1836 (Oesper 1971).

3 Elevated Carbon Dioxide in the Soil: Interaction with the Soil Physical Factors That Affect Root Growth

INTRODUCTION

In Chapter 2, we studied the composition of the soil atmosphere. We noted that, in addition to aeration, the three soil physical factors that affect root growth are water, mechanical impedance (compaction), and temperature, but did not consider each factor individually. In this chapter, we discuss the interaction of each factor with oxygen or CO_2 in the soil and how this interaction affects root growth.

SOIL WATER

Roots need at least 10% by volume air space in the soil to survive (Kirkham 1994; Kirkham 2005, p. 85). If the soil pores are filled up with water and have no air, then roots lack oxygen. Air contains 20.946% oxygen (Weast 1964, p. F-88) or rounding, 21%. Therefore, the minimum amount of oxygen necessary in the soil for roots is about 2.1% ($0.21 \times 0.10 = 0.021$). This value agrees with that published by Gliński and Stępniewski (1985, p. 83) for the critical concentration of oxygen at the root surface necessary for survival (2%). Because of the vital importance of oxygen, most studies have been done concerning the effect of soil oxygen on root growth rather than CO_2. Unlike oxygen, the limiting (maximum) concentration of CO_2 in the soil for root growth has not been established.

Gingrich and Russell (1956) studied the influence of oxygen on the growth of corn (*Zea mays* L.) in a silt loam soil from Illinois, United States, held at different soil matric potentials. Soil matric potentials were measured using a pressure membrane apparatus. Different oxygen concentrations (0.26%, 2.10%, 5.25%, 10.5%, and 21.0% by volume) were obtained by mixing air, which had 20.9% oxygen, with nitrogen gas. Root growth was greater at high soil matric potentials (low tensions) than at lower ones (Figure 3.1). Root growth increased when oxygen concentration increased, except at the two lowest soil matric potentials (−900 and −1200 J/kg or −9 or −12 bar; remember 1 bar = 100 J/kg). When the soil matric potential was not limiting growth, the optimum oxygen concentration in the aerating gas was above 10.5% (Gingrich and Russell 1956; Taylor and Ashcroft 1972, p. 315).

To obtain a general relation between aeration status and soil matric potential, Wiegand and Lemon (1958) introduced the concept of a *critical value*. This is the critical concentration of dissolved oxygen at the root surface that determines whether or not there is enough oxygen in the soil for normal root functioning. They reasoned that if the oxygen in the soil solution at the surface of the root were below this value, growth was limited by aeration. If it were higher than this value, there was no reduction in the growth. They could determine the critical value by measuring an aeration parameter with a platinum electrode. [This aeration parameter is now called the oxygen diffusion rate. See Kirkham (2005, pp. 129–141) for an explanation of it and how it is measured.]

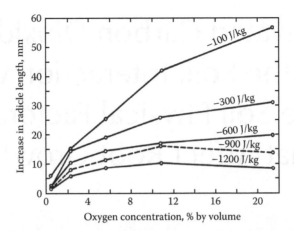

FIGURE 3.1 Influence of oxygen concentration on the growth of the root (radicle) of corn seedlings at several values of soil matric potential. The one line is dashed only to distinguish it from the neighboring lines. (Reprinted from Gingrich, J.R. and Russell, M.B., *Agron. J.*, 48, 517, 1956, Fig. 2A. With permission from the American Society of Agronomy.)

They calculated the critical value to be $4.45\,\mu g/cm^3$ oxygen, which was equivalent to 12% oxygen. They studied two soils: a Miller clay at College Station, Texas, and an Amarillo fine sandy loam at Big Spring, Texas. The relation between the concentration of dissolved oxygen at the root surface and the matric potential in the moist Miller clay at matric potentials greater than −40 J/kg (greater than −0.40 bar) is shown in Figure 3.2. The critical value is shown by the arrow in this figure. The measured values (dots on the solid line) all fell below the critical value, showing that the clay soil

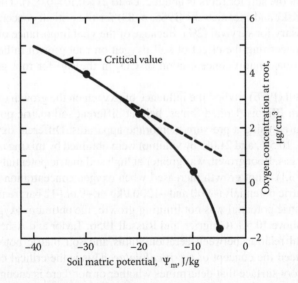

FIGURE 3.2 Calculated concentration of oxygen at the root surface as a function of matric potential (replotted from Wiegand and Lemon 1958). The negative values of oxygen concentration at the root surface result from the assumption that all roots consume oxygen at the same rate regardless of aeration status, whereas an assumption that the consumption rate would decrease below the critical value results in a shape like the dotted line (added by Taylor and Ashcroft). (Reprinted from Taylor, S.A. and Ashcroft, G.L., *Physical Edaphology. The Physics of Irrigated and Nonirrigated Soils*, W.H. Freeman and Company, San Francisco, CA, 1972, p. 358, Fig. 12.6; Redrawn from Wiegand, C.L. and Lemon, E.R., *Soil Sci. Soc. Am. Proc.*, 22, 216, 1958, Fig. 2. With permission from the Soil Science Society of America.)

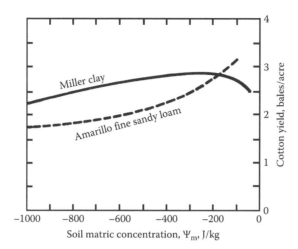

FIGURE 3.3 The influence of matric potential upon the yield of cotton for two soils. Aeration was limiting in the Miller clay at high matric potentials (between 0 and about –200 J/kg), and water was limiting at lower potentials (drier soil). Aeration was adequate in the Amarillo fine sandy loam at all matric potentials, and hence, yield was limited only by soil water. (Reprinted from Taylor, S.A. and Ashcroft, G.L., *Physical Edaphology. The Physics of Irrigated and Nonirrigated Soils*, W.H. Freeman and Company, San Francisco, CA, 1972, p. 359, Fig. 12.7; Redrawn from Wiegand, C.L. and Lemon, E.R., *Soil Sci. Soc. Am. Proc.*, 22, 216, 1958, Figs. 4 and 5. With permission from the Soil Science Society of America.)

did not have enough oxygen for normal root functioning at matric potentials between about –5 and –30 J/kg (–0.05 to –0.30 bar). Due to their technique, they found a negative value of oxygen concentration at the root surface, which is meaningless (Taylor and Ashcroft 1972, p. 359). The curve should be like the dotted line, as supplied by Taylor and Ashcroft (1972, p. 358).

Because a relation exists between soil matric potential and aeration, it is logical that the reduction in cotton (*Gossypium hirsutum* L.) yield found by Wiegand and Lemon (1958) at high matric potentials in the clay (Figure 3.3) resulted from restricted aeration (Taylor and Ashcroft 1972, p. 359). The reduction shown at progressively lower soil matric potentials resulted from water deficiency. Restricted aeration was not evident in the Amarillo fine sandy loam, and the entire reduction in growth was attributed to decreasing soil matric potential.

In the above paragraphs, we have been citing the book by Taylor and Ashcroft (1972). Their chapter on soil aeration gives important information about the influence of oxygen on plant growth. It was written by an expert on the topic, Sterling A. Taylor. See Appendix for a biography of him. In a key paper (Taylor 1949), he described a method to measure partial pressure of oxygen in the soil, which is difficult to do. Because of this difficulty, measurement of the oxygen diffusion rate (e.g., the work of Wiegand and Lemon 1958), rather than measurement of the actual oxygen concentration in the soil, is the standard method to evaluate the aeration status of the soil (Don Kirkham, Iowa State University, personal communication, c. 1985).

SOIL COMPACTION

The amount of aeration required for root growth depends upon the amount of work that a root must do in expanding against the forces in the soil (Kirkham 2005, p. 23). Compaction greatly reduces aeration and results in low oxygen concentration and high CO_2. High mechanical resistance in a soil is due to many factors, ranging from natural causes, such as the presence of endemic claypans (Grecu et al. 1988) to man-made ones, such as tillage with heavy farm equipment (van der Ploeg et al. 2006), trampling, or tamping soil to fill in holes.

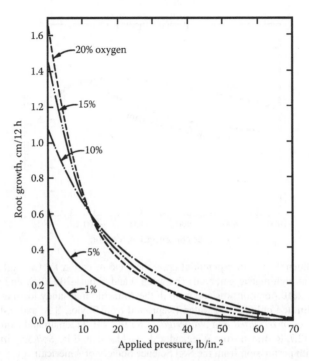

FIGURE 3.4 Rate of root growth as a function of applied pressure at various oxygen percentages (1 lb/in.^{2} = 6896 N/m^2). (Reprinted from Gill, W.R. and Miller, R.D., *Soil Sci. Soc. Am. Proc.*, 20, 154, 1956, Fig. 3. With permission from the Soil Science Society of America.)

Gill and Miller (1956) simultaneously controlled the mechanical resistance and oxygen environment of the roots of corn (*Zea mays* L.) seedlings. The length of the corn roots was measured at various times while the seedlings were under different mechanical pressures and oxygen concentrations. The seedlings grew in "soil," which consisted of 50 µ glass beads. They controlled mechanical impedance by subjecting the soil surface (where the roots grew) to a uniform load by means of a thin rubber diaphragm (membrane) forced against the soil surface by gas under controlled pressure. Mixtures of nitrogen and oxygen gases were used to obtain five levels of oxygen: 20%, 15%, 10%, 5%, and 1%. The gases diffused through the rubber membrane to the roots. There was an interaction between the influence of mechanical pressure and the aeration status on the growth of the corn roots (Figure 3.4). Under full oxygen availability as provided by 20% oxygen, the growth rate was reduced by increasing mechanical pressure so that the roots subjected to a pressure of 10 lb/in.^{2} (68,963 N/m^2; 1 lb/in.^{2} = 6,896.3 N/m^2) grew only about one-half as fast as those subjected to no pressure (Taylor and Ashcroft 1972, p. 323). When little energy was expended against mechanical forces, low concentrations of oxygen (1% and 5%) gave some growth. As the applied pressure increased, the amount of oxygen necessary for a given growth rate also increased. Thus, a growth rate of 0.2 cm in 12 h was achieved with only 1% oxygen when the mechanical work required was only that necessary to overcome about 5 lb/in.^{2} of pressure. But when the work required was increased to that necessary to overcome about 20 lb/in.^{2}, the oxygen concentration had to be raised to about 5% to maintain this growth rate. At 35 lb/in.^{2}, oxygen concentrations in excess of 10% were required (Taylor and Ashcroft 1972, pp. 355–356). Oxygen concentrations above 10% appeared to give about the same growth rate of corn seedling roots no matter what the pressure was.

The results demonstrated that roots can exert less mechanical pressure, and hence, do less mechanical work, in a medium that is poorly aerated than they can in a medium that is well supplied with oxygen (Taylor and Ashcroft 1972, p. 323). When soil is compacted, there is a concomitant increase in mechanical impedance and a decrease in porosity, the latter resulting in reduced aeration. The influence of mechanical impedance and reduced aeration are difficult to separate in field experiments, but the non-limiting water range allows us to see the effect of each (Letey 1985). Mechanical resistance can be measured with a penetrometer (Kirkham 2005, pp. 117–128), and aeration can be measured by the oxygen diffusion rate method (Kirkham 2005, pp. 129–144). Measurement of the oxygen status in the environment has become so important that a new branch of science has been established called "oxygenology," which is defined as the "scientific discipline related to the presence and role of oxygen in nature on Earth" (Stępniewski and Stępniewska 1998). The European Union is actively engaged in studies of oxygenology (Stępniewski et al. 2005).

Tackett and Pearson (1964a) looked at the effect of different bulk densities and oxygen concentrations in the soil on the root growth of cotton (*G. hirsutum* L.). They put soil taken from the B1 horizon (18–25 cm) of a Norfolk sandy loam in aluminum cylinders, 7.6 cm long, and compacted it to five bulk densities: 1.3, 1.5, 1.6, 1.7, and 1.9 g/cm^3. Over this compacted soil, they placed another 7.6 cm of soil that was a control soil. They did not give the bulk density of the control soil, but it probably was 1.3 g/cm^3. They called the soil compacted to different densities the "subsoil." Between the control soil and the test soil (subsoil), they placed a paraffin membrane through which the roots could grow. They had an aerating chamber at the bottom of each cylinder to which they applied four levels of oxygen gas in nitrogen gas. The oxygen levels were 1.2%, 5.0%, 10.0%, and 21.0% oxygen with the difference due to nitrogen gas. They planted four cotton seeds in the surface and then took measurements for 92.5 h after planting. The moisture content was constant and remained at −1/3 bar. Gas samples were taken by inserting a hypodermic needle through a port in the center of the subsoil core and determining the content on a gas chromatograph.

The minimum oxygen level that could be approached before reducing root length depended upon the density. At a subsoil density of 1.3 g/cm^3, it was necessary to reduce the oxygen level to approximately 5% before a decrease in root length occurred. At a density of 1.5 g/cm^3, reduction in root length occurred when the oxygen level was decreased below 10%. As soil density increased, however, mechanical impedance became the dominant factor limiting root growth at all levels of oxygen. This was especially apparent at a density of 1.9 g/cm^3, where roots were able to penetrate only a few millimeters of the soil. These roots were short, thick, and distorted.

The amounts of CO_2 produced by unit weight of respiring roots for the duration of the experiment at the different oxygen and density levels are shown in Figure 3.5. Respiration of roots grown in soil of low density was not affected appreciably by changes in oxygen concentration. However, at the higher bulk densities, the roots produced more CO_2 and the amount of CO_2 produced increased with increases in oxygen concentration. The high amount of CO_2 produced at the highest bulk density reflects the extra energy that these roots probably were exerting to push through the soil.

In a second experiment, Tackett and Pearson (1964b) varied the level of CO_2 and of subsoil density in a growth chamber to study the interaction of these two factors on root growth of cotton seedlings at constant oxygen concentration. They used the same Norfolk sandy loam that they used previously (Tackett and Pearson 1964a) and compacted the soil to bulk densities of 1.3, 1.5, 1.6, and 1.7 g/cm^3. They supplied CO_2 at four levels (0.03%, 6.1%, 11.9%, and 24.0% by volume). Oxygen concentration in the gases was maintained constant at 21% and nitrogen gas made up the difference. The time of the experiment was 92.5 h.

When results of all densities were averaged, CO_2 did not affect the rate of root elongation until the concentration of CO_2 reached a level of about 12%. Beyond this concentration, there was a decline with increasing CO_2 concentration. A difference in response occurred at different densities (Figure 3.6). Rate of root elongation decreased as CO_2 was increased beyond 12% at all bulk densities except for the highest tested. At this density of 1.7 g/cm^3, mechanical impedance dominated root growth and there was no effect of CO_2. When the bulk density was reduced to 1.6 g/cm^3, the impedance ceiling on root penetration was raised somewhat (Tackett and Pearson 1964b). As bulk density was decreased further, mechanical impedance became still less important and any increase in CO_2 concentration resulted in reduction in root growth. The absence of extreme toxicity at high

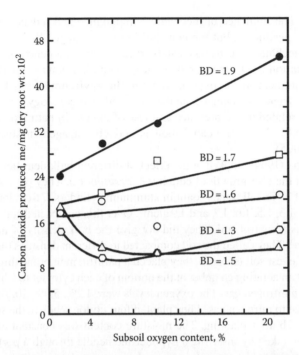

FIGURE 3.5 The relationship between subsoil O_2 content and bulk density and the amount of CO_2 produced by cotton roots. (Reprinted from Tackett, J.L. and Pearson, R.W., *Soil Sci. Soc. Am. Proc.*, 28, 600, 1964a, Fig. 8. With permission from the Soil Science Society of America.)

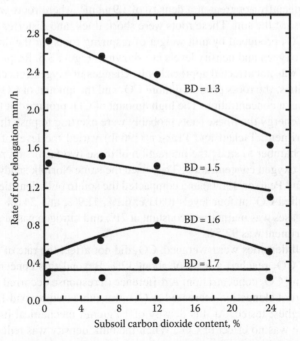

FIGURE 3.6 Effect of CO_2 level on the rate of cotton seedling root penetration of subsoils compacted to different bulk densities. Oxygen was maintained at 21%. (Reprinted from Tackett, J.L. and Pearson, R.W., *Soil Sci. Soc. Am. Proc.*, 28, 741, 1964b, Fig. 3. With permission from the Soil Science Society of America.)

levels of CO_2 (up to 24% in the soil) indicates that CO_2 would rarely, if ever, limit root growth in a Norfolk sandy loam. Mechanical impedance to root penetration would overshadow aeration effects.

SOIL TEMPERATURE

Temperature has a direct influence on the growth rate of plant roots (Cannon and Free 1925, p. 2). It also directly affects the rate of soil respiration, that is, the amount of CO_2 that is evolved from the soil as a result of respiration by microbes, soil fauna, and roots. This rate of respiration is dependent on moisture. But it is also dependent on temperature through the Q_{10} value, which reflects the rate of a reaction or reactions in a living organism. We discuss the Q_{10} value in detail in the chapter on phenology (Chapter 14). Because in this book, we are focusing on the effects of elevated CO_2 on water relations rather than the effects of elevated CO_2 on temperature, we shall not further discuss direct effects of temperature on root growth. We do note that underground heating has been studied for decades to see its effects on crop growth (Willis et al. 1957). Now underground heating, such as that at Harvard Forest in Petersham, Massachusetts (Rustad 2001), as well as aboveground heating with infrared radiators (Luo et al. 2001; Kimball et al. 2008), is being done with the specific purpose to see how possible global warming may affect biological processes.

Temperature also has an indirect effect on root growth. A decrease in temperature may cause a reduction in the minimum concentration of oxygen necessary for full growth rate. Because respiration of roots and soil organisms is influenced by temperature, the effect of temperature is more pronounced in summer. With lower temperatures, plants can grow at lower oxygen concentrations because there is less metabolic activity. Cannon and Free (1925) grew several different plants under varying temperature and oxygen concentrations. In all experiments, the soil used was a sand from a wash on the grounds of the Desert Laboratory, Tucson, Arizona (Cannon and Free 1925, p. 16). The sand was put in glass culture tubes where the aeration status of the roots could be varied (Cannon and Free 1925, p. 14). Results for cotton (*G. barbadense* L.) are shown in Figure 3.7, where we see that if the temperature of the soil was about 18°C, oxygen concentrations as low as 2.2% produced maximum growth rate. With 21% oxygen, the maximum growth rate occurred at about 30°C (Taylor and Ashcroft 1972, p. 354).

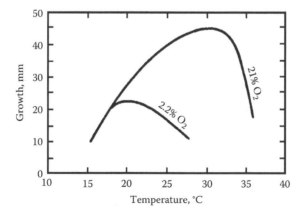

FIGURE 3.7 Sketch of the growth of roots as a function of temperature at two different oxygen percentages. The sketches are based upon measurements of *G. barbadense* L. Maximum growth rate at a given oxygen percentage is highly dependent upon temperature. (From Free, E.E., Differences between nitrogen and helium as inert gases in anaerobic experiments on plants, in Cannon, W.A. (ed.), *Physiological Features of Roots, with Especial Reference to the Relation of Roots to Aeration of the Soil*, Carnegie Institution of Washington Publication No. 368, Intelligencer Printing Company, Lancaster, PA, 1925, 168pp, Fig. 20. With permission.)

SUMMARY

Roots need at least 10% by volume air space in the soil to survive, which means that the minimum oxygen concentration in the soil needs to be about 2% because air contains 21% oxygen. No such limiting (maximum) concentration has been established for CO_2. Because oxygen is more important for root growth than CO_2, most studies concerning the effects of soil water content and soil compaction on root growth have focused on oxygen in the soil rather than CO_2. The absence of extreme toxicity at high levels of CO_2 (24%) in one study done with a sandy loam soil indicated that CO_2 would rarely, if ever, limit root growth in this soil. Excessive moisture and mechanical impedance apparently are more important in decreasing root growth than elevated levels of CO_2 in the soil.

APPENDIX

BIOGRAPHY OF STERLING ANGUS TAYLOR

Sterling A. Taylor, soil physicist and pioneer in the study of oxygen movement in the soil, was born on February 16, 1918, in Salem, Utah. He got the bachelor of science degree at Utah State College in 1941. As he finished the BS degree, the United States was getting involved in World War II, and he was among the first of the graduates to be called into the service (Thorne 1972). He married his college sweetheart, Frances Glassett, and went immediately into the Army, where he served between 1942 and 1946 (Cattell 1955). He attained the rank of major (Thorne 1972).

He sought and obtained a graduate assistantship in 1946 at Cornell University in Ithaca, New York. He chose the field of soil physics and studied with Morell Belote (M.B.) Russell (1914–1992) on problems of water relationships of soils. When he finished his doctorate degree in 1949, Utah State University was looking for a person to replace Willard Gardner (1883–1964) (Gardner 1977), considered to be the father of soil physics in the United States. Taylor was offered the job and accepted. He became the leader in a major effort at Utah State to evaluate the combinations of factors that would produce the highest yields of the most important crops. His efforts, along with those of his colleagues, resulted in a series of papers showing how irrigation, fertilizers, and plants could be coordinated to produce maximum crop yields. The principles resulting from these studies served as an important base in the modern agricultural revolution that has doubled and trebled crop yields. His findings were widely applied in solving the problem of food supply for the hungry world (Thorne 1972).

He also became deeply involved in the arid regions of the world. He and his students set out to find how plants remove water from soils and how this process is related to climate. The success of such studies required accurate methods for measuring water stress in plants, water movement in soils and plants, evaporation, transpiration, and aspects of meteorology. Out of these studies came the first steps toward the development of thermocouple psychrometers (e.g., Campbell et al. 1966; see also Taylor and Ashcroft 1972, pp. 265–267). He also clarified the terminology used in soil–plant–water relations (Taylor 1968).

An unusual feature of Taylor's laboratory was the close fellowship and teamwork that he achieved with his students. His laboratory became a featured place for visitors interested in seeing how graduate studies and research programs can complement one another (Thorne 1972). One of his most outstanding qualities was his dedication. He had definite goals and clearly defined patterns of work. He did not let unimportant extraneous details interfere with his major objectives. Sometimes his immediate supervisors were frustrated by his refusal to give time to social affairs and to time-consuming committee activities that abound in academia. But, as Thorne (1972) pointed out, the running of errands and similar activities are sometimes a substitute for the more difficult course of study, thinking, teaching, experimenting, and writing. Taylor did not take coffee breaks. This was not because he was antisocial, but rather because he found his professional activities more interesting. He saw in nature relationships few others could see. He knew the deep personal satisfaction that comes from generating ideas and discovering new truths (Thorne 1972).

The professional associates who knew Taylor best included soil physicists and meteorologists. They were located in practically every state of the union and in foreign countries, especially England, Belgium, the former U.S.S.R., Italy, Australia, Egypt, Spain, and Venezuela. He was in great demand for advice on problems associated with the development of irrigated agriculture. He counseled on the water plan for California, and on similar plans for Egypt and Venezuela, and on the organization of research in Puerto Rico (Thorne 1972).

Don Kirkham (1908–1998), in his oral history of his years as an assistant professor of mathematics and physics at Utah State University (then Utah State College) (1937–1940) (Kirkham 1995–1996), says the following about Sterling A. Taylor, who was a student in both classes Don Kirkham taught at Utah State: physics (which included meteorology) and mathematics. "Sterling Taylor went into soil physics and became an outstanding soil physicist. Sterling Taylor got all A's and became valedictorian. Later on, I helped Sterling Taylor get a Guggenheim." A Guggenheim is a prestigious fellowship supported by the John Simon Guggenheim Memorial Foundation, New York, New York.

Sterling Taylor died of cancer on June 8, 1967. He was near completion of his book just before he died (Thorne 1972). The book, with coauthor Gaylen L. Ashcroft (Taylor and Ashcroft 1972), is still widely used by teachers and has some of the best information available for students of soil physics (Loyd R. Stone, Kansas State University, personal communication, c. 2005). Taylor's son, Sterling Elwynn Taylor, professor of agricultural meteorology in the Department of Agronomy at Iowa State University, was hired by the university in 1979 as the first extension climatologist in the nation. He is well known across Iowa by farmers who depend on his radio programs for weather information (Don Kirkham, personal communication, c. 1997).

4 Elevated Carbon Dioxide in the Soil: Variable Oxygen Concentration and Root Growth

INTRODUCTION

In Chapter 3, we focused on three of the four soil physical factors that control plant growth (water, compaction, and temperature) and how they interact with the fourth factor (aeration) under elevated CO_2 and reduced oxygen. As noted in that chapter, most studies consider the lack of oxygen rather than elevated CO_2 in the soil, because roots need at least 10% by volume air space in the soil to survive (or about 2% oxygen as a minimum). This is because diffusion of gases in the soil practically stops when the fraction of air-filled pore spaces is less than 10% (Kirkham 2005, p. 129). When roots lack oxygen, they cannot respire and CO_2 collects in the soil. In this chapter, we expand our discussion beyond the four soil physical factors that affect plant growth to consider soil aeration in more detail. First, we focus on how variable concentrations of oxygen in the soil affect root growth. We then consider responses of different species to oxygen in the soil, followed by a comparison of soil and root production of CO_2. Next, we take a look again (as we did briefly in Chapter 3) at limiting concentrations of oxygen in the soil and maximal concentrations of CO_2, which permit root growth. Finally, we consider movement of oxygen down a plant, which is one way a plant can survive low levels of oxygen in the soil.

VARIABLE OXYGEN CONCENTRATION IN THE SOIL

We remember that air contains 20.9% oxygen, 78.1% nitrogen, and 0.934% argon (to three significant figures) (Kirkham 2005, p. 132). These percentages add up to 99.934%. The amount of CO_2 in the air is a trace amount, and that is why it is called a "trace gas" when its concentration in the atmosphere is reported each year in the State of the Climate report published by the American Meteorological Society in the *Bulletin of the American Meteorological Society* [e.g., see Figure 1.3 from Schnell (2005)]. Even though the concentration of CO_2 in the atmosphere has been increasing, the *Handbook of Chemistry and Physics* does not note this increase. In my library, I have copies of 40th edition (Hodgman 1959), the 45th edition (Weast 1964), and the 75th edition (Lide 1994). These editions list the CO_2 concentration in the atmosphere as follows: 0.03% (Hodgman 1959, p. 3371); 0.033% ± 0.001% (Weast 1964, p. F-88); and 0.000314 (fractional volume, dimensionless) (Lide 1994, pp. 14–14). The concentration given by Lide (1994) (0.0314% or 314 ppm) is even less than the concentration given by Weast (1964) (0.033% or 330 ppm). The concentration of CO_2 in the atmosphere in 2008 was 384.9 ppm (Dlugokencky 2009). So the *Handbook of Chemistry and Physics* is not keeping up with the increasing CO_2 concentration in the atmosphere. These concentrations of CO_2 in the atmosphere, as listed in the *Handbook of Chemistry and Physics* and in the State of the Climate in the *Bulletin of the American Meteorological Society*, are small compared to the concentrations of the three main gases in the atmosphere (oxygen, nitrogen, and argon). Because it is generally felt that diffusion is the main way that gases move in the soil [see Kirkham (1994) for a review of the literature], the amount of CO_2 in the air above the soil should be about the same as the CO_2 in the soil, unless there are pockets of high respiration, such as those that occur when manure is put on the soil. When it is stated that "the sum of CO_2 and O_2 in the soil air is 21% by volume" (Acock and Allen 1985, p. 70), this means that most of the 21% is oxygen (20.9%) and only a tiny amount (0.0385%) is CO_2. Because of the importance of oxygen in the soil air, we now turn to studies in which its concentration has been varied.

Huck (1970) investigated effects of short-term fluctuations in soil oxygen status on radicle elongation rates of cotton (*Gossypium hirsutum* L.) and soybean [*Glycine max* (L.) Merr.] (Taylor et al. 1972). He measured elongation rates while oxygen content of a gas stream that passed through the soil surrounding the roots was varied. Elongation ceased completely within 2 or 3 min after all oxgyen was purged from the system with 100% nitrogen gas, and it returned to normal shortly after 21% oxygen was reapplied to the system, if the period of complete anaerobiosis did not exceed 30 min (Figure 4.1 for cotton). Periods of anaerobiosis longer than 30 min caused increasing proportions of taproot death until all were killed at 3 h for cotton and 5 h for soybeans. This tissue death occurred within the region of elongation and did not extend to older tissue. Setting the oxygen level at 3% for 5 h resulted in an initial reduction in the rate of taproot extension (Figure 4.2 for cotton), but elongation rates gradually returned to near those at 21% oxygen. In his experiments, Huck (1970) found that lateral root initiation and development after a period of anaerobiosis occurred immediately above the killed portion. He concluded that distribution of roots through a volume of soil could be influenced by anaerobiosis of even a few hours. This short-term condition might happen in flooded soils or in soils with perched water tables caused by the presence of tillage pans (Taylor et al. 1972).

In the above paragraph, we have cited the work of Taylor et al. (1972), which includes the authorship of Elizabeth ("Betty") L. Klepper. She along with Morris Huck, another coauthor on the paper, was the first to film the growth of cotton roots under the variable oxygen concentrations, as described above in words. The film, now transferred to electronic versions, is available from the Tri-Societies [American Society of Agronomy (ASA), Crop Science Society of America

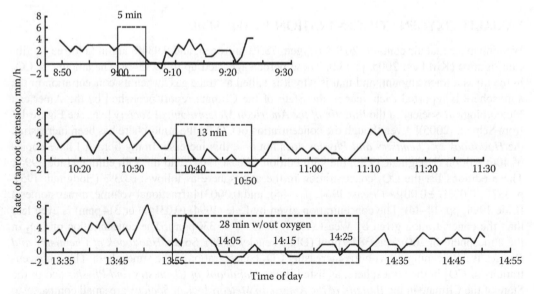

FIGURE 4.1 Elongation rate of a single cotton taproot approximately 2 days after initiation of germination. Times shown are actual clock time on day of treatment, so the tracing represents successive periods of anaerobic treatment of the same root on the same day. (Reprinted from Huck, M.G., *Agron. J.*, 62, 815, 1970, Fig. 2. With permission from the American Society of Agronomy.)

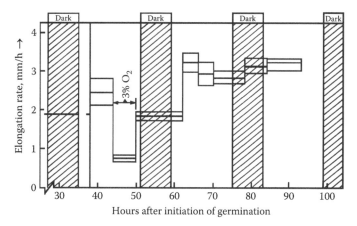

FIGURE 4.2   Elongation rate of a single cotton taproot as affected by passage through the growth medium of a gaseous mixture containing 3% O_2 and 97% nitrogen gas. (Reprinted from Huck, M.G., *Agron. J.*, 62, 815, 1970, Fig. 3. With permission from the American Society of Agronomy.)

(CSSA), and Soil Science Society of America (SSSA) 1999]. See Appendix for a biography of Betty Klepper.

The importance of aeration is shown in Figure 4.3, in which the elongation of the roots of *Impatiens balsamina* L. (garden balsam) is plotted as a function of time (Hunter and Rich 1925). The growth rate in the beginning was rapid, because there was adequate oxygen and a low level of CO_2 (Taylor and Ashcroft 1972, pp. 353–354). When the oxygen was used up and the CO_2 level increased in the vicinity of the roots, growth stopped for a time, probably while the processes of gaseous exchange could effect replacement of the CO_2 near the root with oxygen. When the aeration was adequate, growth started again and the entire cycle was repeated. Artificial aeration of the soil with 500 cm³ of air extended the length of the active growth period fourfold. Prior to aeration,

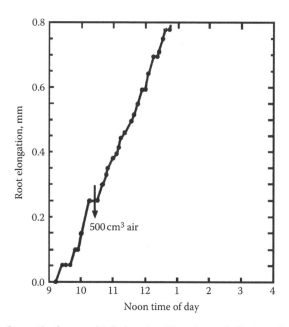

FIGURE 4.3   The rate of growth of roots of *I. balsamina*. Note the periodic growth; the length of the period is greatly increased by aeration. (From Hunter, C. and Rich, E.M., *New Phytol.*, 24, 257, 1925, Fig. 1. With permission.)

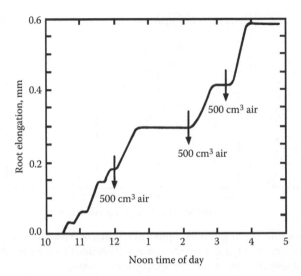

FIGURE 4.4 Roots of *I. balsamina* that had ceased to elongate resume active elongation after aeration. (From Hunter, C. and Rich, E.M., *New Phytol.*, 24, 257, 1925, Fig. 2. With permission.)

the period of rapid growth was about 15–20 min followed by 5–15 min of inactivity (Taylor and Ashcroft 1972, p. 354). Following each addition of 500 cm³ of air to the soil, there was a period of rapid growth that extended to about 80 min, after which the growth cycle was the same as before aeration (Figure 4.4). The increases in root elongation following the artificial aerations (Figures 4.3 and 4.4) may have been caused by the removal of CO_2, which may have become high enough locally to have a narcotic effect on the roots (Taylor and Ashcroft 1972, p. 354). Without artificial aeration, the activity of the root cells may have been temporarily suspended until the CO_2 accumulation was reduced by diffusion. Alternatively, a deficiency of oxygen developed, thus inhibiting respiration until the concentration of oxygen near the roots could be increased by exchange or artificial aeration (Taylor and Ashcroft 1972, p. 354).

An abrupt reduction in soil aeration will cause a growing plant to wilt, due to a reduction in the rate of water absorption. If the soil air is suddenly replaced by CO_2, transpiration is reduced. Because the same effect occurs if a de-topped plant is attached to a vacuum, it is concluded that at least part of the effect of CO_2 is to reduce the permeability of the root tissues to water (Taylor and Ashcroft 1972, p. 360).

VARIATION IN SPECIES

Growth of roots is affected by the aeration status of the soil. Cannon and Free (1925, p. 150) showed the influence of varying concentrations of oxygen (0.6%–10%) in the soil on root-growth rate of 11 species of plants: *Allium cepa* L. (onion); *Citrus medica* var. *limon* L. (lemon); *C. sinensis* Osbeck (sweet orange); *G. barbadense* L. (sea-island cotton); *G. hirsutum* L. (upland cotton); *Opuntia versicolor* (the genus *Opuntia* is prickly pear or Indian fig); *Pisum sativum* L. (garden pea); *Potentilla anserina* L. (silverweed); *Prosopis velutina* (common name unknown); *Salix lasiolepis* (the genus *Salix* is willow); and *Zea mays* L. (corn or maize). There was variation in their response to soil air containing different amounts of oxygen. However, root growth of most species ceased at 1% oxygen in the soil air (Taylor and Ashcroft 1972, p. 356). Corn roots were the only ones grown at 10% oxygen in the soil. The other species grew with less oxygen. When grown with 3.6% oxygen, growth of corn roots was about half of what it was at 10% oxygen. Figure 4.5 is a cross section of a corn root showing the characteristic air spaces that develop in the cortex with inadequate aeration (McPherson 1939). When oxygen is removed entirely from the soil air, no root growth of any

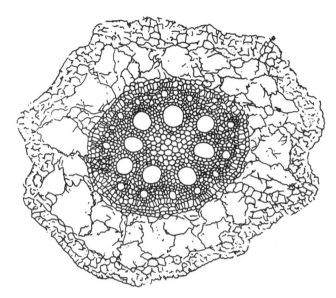

FIGURE 4.5   Cross section of a corn root showing characteristic air spaces developed with inadequate aeration. (From McPherson, D.C., *New Phytol.*, 38, 190, 1939, Fig. 1. With permission.)

crop species can take place, except that of hydrophytes such as rice (*Oryza sativa* L.) (Taylor and Ashcroft 1972, p. 356).

In general, the same conditions that influence root growth also affect top growth in a medium in which aeration is restricted (Taylor and Ashcroft 1972, p. 357). When aerated with CO_2, all upland plants soon wilt and further growth is prevented. Aeration with nitrogen gas greatly reduces top growth. For tomato (*Lycopersicon esculentum* Mill.), the reduction is as much as 90%; for barley (*Hordeum vulgare* L.), it is less; and rice is unaffected (Taylor and Ashcroft 1972, p. 357). At a given temperature, the rate of increase of growth as related to oxygen content in the aeration stream follows a curve similar to the one diagrammed in Figure 4.6 (Gingrich and Russell 1956).

SOIL VERSUS ROOT EVOLUTION OF CO_2

Soil respiration is the total of all soil processes in which CO_2 is produced. Pure chemical oxidation will probably always occur, but, as a rule, only to a small extent. The microflora (bacteria and fungi) may produce the chief part of the CO_2, but plant roots also give off considerable amounts of CO_2 (Lundegårdh 1927, p. 421). In Chapter 2 on the composition of the soil atmosphere, we looked at experiments that separated CO_2 evolved from the microorganisms in the soil versus CO_2 from roots, as the CO_2 concentration changed with depth. We here also look at some of experiments that have compared the two sources of CO_2 (microorganisms and roots). The separation of CO_2 from roots and that from soil organisms has always been difficult.

Lundegårdh (1927, p. 421) measured the respiration from a bare soil and that from the soil covered with oats (*Avena sativa* L.). The bare soil showed a respiration of $0.268 \, g/m^2/h$ and the latter a respiration of $0.399 \, g/m^2/h$, a difference of $0.131 \, g/m^2/h$. The root respiration, thus, formed about 30% of the total soil respiration. Further experiments indicated, however, that a great deal of what he called root respiration was due to bacteria that inhabit the root surface and probably are fed with the organic substances that are delivered from the roots (Lundegårdh 1927, p. 421). He then cultivated oats and wheat (*Triticum* sp.) in sterilized soil and in non-sterilized soil. The roots from the non-sterilized soil evolved about 45% more CO_2 than the roots from the sterilized soil.

Lundegårdh (1927, pp. 439–440) saw that the variations in the CO_2 concentration among the leaves of sugar beet (*Beta vulgaris* L.) in the free atmosphere above the plants were related to

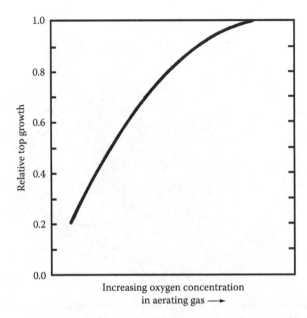

FIGURE 4.6 Diagram of the relative top growth of plants (top growth with given oxygen content)/(top growth with 21% oxygen) as influenced by oxygen content of the aerating gas. (Reprinted from Taylor, S.A. and Ashcroft, G.L., *Physical Edaphology. The Physics of Irrigated and Nonirrigated Soils*, W.H. Freeman and Company, San Francisco, CA, 1972, p. 358, Fig. 12.5; Drawn from Gingrich, J.R. and Russell, M.B., *Agron. J.* 48, 517, 1956, Fig. 2C. With permission from the American Society of Agronomy.)

corresponding variations in soil respiration. The correlation was especially strong, if plots had received fertilizer. When the soil respiration was stimulated by applications of mineral fertilizers, the CO_2 concentration among the leaves was raised, too. If the sugar beet leaves were growing rapidly, however, leaves were absorbing more CO_2 than the soil could produce. As long as the CO_2 in the soil was lower than the CO_2 concentration in the free air (which he called the "standard value" and measured 5 m above the ground), there was a diffusion of CO_2 downward, as illustrated by the arrows at the top of the left-hand side of Figure 4.7. All CO_2 that came from the soil in this case was absorbed by the plants. When the CO_2 evolved from the soil exceeded the standard value, there was a diffusion stream from the level of the leaves up to the free atmosphere (see the arrows at the top of the right-hand side of Figure 4.7). Part of the CO_2 evolved from the soil in this case escaped between the leaves and disappeared into the free atmosphere.

Perhaps the most challenging phase of the terrestrial carbon cycle occurs below ground (Chapin and Ruess 2001). Although the above-ground plant production through photosynthesis is relatively well documented from field measurements and globally distributed satellite observations, the quantity of carbon that plants transfer below ground is not well known. Micro-video cameras provide direct observations of the growth, longevity, and decomposition of fine roots. However, other large components of below-ground plant production, such as exudation of organic compounds by roots and transfer of carbohydrates to associated mycorrhizal fungi, are difficult to study nondestructively and to estimate. As noted above, the separation of CO_2 from roots and that from soil organisms has long been difficult. Estimates of the contribution of root respiration to total CO_2 efflux from soil have ranged from 10% to 90%, with methodological uncertainties accounting for most of this variation (Chapin and Ruess 2001).

Experiments carried out by Högberg et al. (2001) in northern Sweden illustrated the large influence of roots and associated fungi on the carbon cycle. They developed a new approach to determine root respiration, which provided an integrated estimate of the proportion of soil respiration derived from roots and their symbiotic mycorrhizae. Their experiments involved girdling all trees

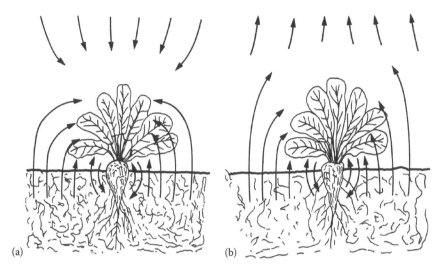

FIGURE 4.7 (a) Diagram of the diffusion currents when the carbon dioxide factor in the soil is below the standard value. (From Lundegårdh, H., *Soil Sci.*, 23, 417, 1927, Fig. 15. With permission.) (b) Diagram of the diffusion currents when the CO_2 factor in the soil exceeds the standard value. (From Lundegårdh, H., *Soil Sci.*, 23, 417, 1927, Fig. 16. With permission.) [Lundegårdh (1927) defines the standard value as the CO_2 concentration in the free air measured 5 m above the ground.]

in experimental plots of Scots pines (*Pinus sylvestris* L.) in a boreal forest. In girdling, bark 1.5 m above the ground was stripped to a depth that allowed upward water flux but prevented downward transport of carbohydrates to roots and mycorrhizae. Soil respiration was measured as the rate of CO_2 accumulation in closed chambers inserted into the soil surface. By comparing soil respiration in girdled plots, in which root and mycorrhizal respirations had been eliminated, with control plots where they had not, Högberg et al. (2001) concluded that over half of the soil respiration in the forest derived from the current year's photosynthesis. However, their work did not determine the ratio of plant-associated respiration derived from roots to that from their symbiotic mycorrhizae, and this ratio remained unknown.

LIMITING CONCENTRATION OF OXYGEN

Many contradictory reports exist concerning levels of oxygen and CO_2 that limit root growth (Kramer 1969, p. 140). This is not surprising because experiments have been conducted on different species of plants in various stages of development, growing in a variety of media from fine-textured soils to water cultures. The methods used to aerate root systems vary widely and produce different results. The actual oxygen concentrations at the root surfaces are different in water culture, in soil aerated by forced circulation of a gas mixture, and in soil aerated by diffusion. Furthermore, measurements of samples of gas removed from the soil do not necessarily indicate the concentrations available at the root surfaces. These are usually covered with films of water through which oxygen diffuses much more slowly than through air. The diffusion coefficient of oxygen in water is approximately 2.6×10^{-5} cm^{-2} s^{-1} compared with 1.9×10^{-1} cm^{-2} s^{-1} in air. Therefore, diffusion through water films covering roots is likely to be the limiting factor in oxygen supply (Kramer 1969, p. 140).

Attempts have been made to study the oxygen supply to roots by measuring the oxygen diffusion rate to a platinum electrode inserted in the soil to simulate a root (Kramer 1969, p. 140; Kirkham 2005, pp. 129–144). (We mentioned the oxygen diffusion rate in Chapter 3.) In general, it appears that oxygen diffusion rates of less than 0.20 μg/cm²/min are inadequate, and values greater than 0.40 μg/cm²/min are high enough for optimum growth. However, good yields of several crops at oxygen diffusion rates of 0.15 μg/cm²/min have been reported (Kramer 1969, p. 140). These plants were grown in much larger containers than those used by most workers. This suggests that crowded root systems require a higher rate of oxygen diffusion than those in large volumes of soil. (In Chapter 6, we consider the effects of root restriction on plants grown under elevated CO_2.) Kramer (1969, p. 141) states that oxygen concentrations above 10% are adequate for growth. He even suggested that high concentrations maintained by energetic aeration with air containing 20% oxygen can be inhibitory (Kramer 1969, p. 141). As noted in Chapter 3, the minimum (critical) concentration at a root surface is often considered to be about 2% (Gliński and Stępniewski 1985, p. 83; Kirkham 1994).

MAXIMUM CO_2 IN SOIL THAT ALLOWS CROP GROWTH

An interaction occurs between CO_2 and oxygen in the soil air. Taylor and Ashcroft (1972, p. 354) state that a deficiency of oxygen may be said, in general, to exist when oxygen comprises 15% or less of the soil air if the remainder is nitrogen. CO_2 might be tolerated to an extent of about 50% of the soil air if there is a normal oxygen supply, but probably much less than 50% if the amount of oxygen is low. A number of experiments indicate that in the range of concentrations found in natural field conditions, oxygen deficiency is more serious in reducing absorption than is CO_2 accumulation (Taylor and Ashcroft 1972, p. 360). If the oxygen content is high, there is little decrease in water uptake as a result of high CO_2 content. When there is an excess of CO_2 in the absence of oxygen, transpiration is reduced little more than by oxygen deficiency alone.

Considerable uncertainty exists concerning the effects of high CO_2 on root growth (Kramer 1969, p. 141). Work in New York orchards indicates that the CO_2 concentration rarely rises above 12% and it is usually much lower (Kramer 1969, p. 141). Stolwijk and Thimann (1957) found that a concentration of 0.5% CO_2 stimulated root growth of pea (*P. sativum* L.) and a concentration of 1.5% CO_2 inhibited its growth. But other workers have reported no injury at concentrations of 20% for cotton, 20% or more for corn and soybean, and at least 6.8% for tomato (Kramer 1969, p. 141). Castonguay et al. (2009) assessed plant survival of annual bluegrass (*Poa annua* L.) and creeping bentgrass [*Agrostis stolonifera* L. var. *palustris* (Huds.)] exposed to combinations of O_2 and CO_2 concentrations at low temperatures (1°C or −2°C). Plants were exposed to four different atmospheric compositions: low O_2 (0%) and low CO_2 (0%); low O_2 (0%) and high CO_2 (15%); high O_2 (20%) and high CO_2 (15%); and a control treatment at 21% O_2 and ~0.04% CO_2. Their results indicated that lack of O_2 (anoxia) rather than high CO_2 was the source of plant damage. It appears that the CO_2 concentration usually found in the soil is not high enough to cause injury, but the concentration of oxygen is often low enough to be inhibitory (Kramer 1969, p. 141).

Huang et al. (1997) reviewed the effect of high CO_2 concentration in the root zone on the growth of different crops. They used the pressure unit Pascal (Pa) to express the amount of CO_2 and oxygen. They said that in well-aerated soils, soil CO_2 levels are often near 1 kPa and oxygen levels are about 20 kPa. They reported that root elongation of soybean [*Glycine max* (L.) Merr.] and maize (*Zea mays* L.) is inhibited at CO_2 levels above 20 kPa in the soil atmosphere. Root growth of sunflower (*Helianthus annuus* L.), bean (*Phaseolus vulgaris* L.), pea (*P. sativum* L.), and alfalfa (*Medicago sativa* L.) is completely inhibited by a CO_2 concentration of 6.5 kPa in the atmosphere of the rooting medium, while root growth of oat (*A. sativa* L.) and barley (*H. vulgare* L.) is unaffected. [For the data for sunflower, bean, pea, oat, and barley, Huang et al. (1997) used data from Stolwijk and Thimann (1957) as the source for their information. Stolwijk and Thimann (1957) express CO_2 in concentration units. Stolwijk and Thimann (1957) said that 6.5% CO_2 was the root medium concentration that completely inhibited growth of sunflower, bean, and pea. So Huang et al. (1997) are equating 6.5 kPa CO_2 with 6.5% CO_2. A direct conversion cannot be made, because one needs to know temperature and elevation to convert concentration units to pressure units.] Huang et al. (1997) grew in nutrient culture two genotypes of wheat (*Triticum aestivum* L.), which differed in tolerance to hypoxia, under high CO_2. When plants grew with 10 kPa CO_2 (c. 10%) in the root zone and ambient oxygen, root growth was reduced in both genotypes.

MOVEMENT OF GASES UP AND DOWN A PLANT

One way a plant can tolerate lack of oxygen in the soil (aside from the species being tolerant) is the movement of oxygen down from the shoot to the root zone. Oxygen produced in shoots during photosynthesis moves down to roots. This movement has been studied in *Nuphar* sp. (yellow pond lily). A gradient of oxygen concentration from leaves to rhizomes of *Nuphar* occurs on sunny days, and at the same time, there is a gradient of CO_2 concentration from rhizomes to leaves (Kramer 1969, p. 139). When leaves are darkened, oxygen concentration in distant parts of rhizomes falls to less than 1%. Experiments have shown that a considerable amount of oxygen moves from shoots to roots in rice (Kramer 1969, p. 139). Such movement has been demonstrated in corn and barley, although to a lesser extent than in rice. In rice, sufficient oxygen to meet needs can be supplied to roots only to a depth of 10 cm by diffusing through the internal air spaces of the roots (Kramer 1969, p. 140). Downward movement of oxygen in rice appears to be restricted after the stems elongate, when oxygen is supplied to the deeper roots from a mat of fine roots formed at the surface of the water. One of the most remarkable examples of movement of oxygen from shoots to roots is in pineapple (*Ananas comosus* Merr.), where gas containing up to 80% oxygen has been observed to escape from roots in a culture solution (Kramer 1969, p. 140). The oxygen apparently comes from photosynthesis and is observed only when leaves are illuminated. It probably occurs because the stomata of pineapple are tightly closed most of the day and oxygen formed in photosynthesis cannot escape by the usual pathway. (Pineapple has Crassulacean acid metabolism.) In addition to downward movement of oxygen in plants, upward movement of CO_2 from roots to shoots may be a significant factor in photosynthesis of certain plants (Kramer 1969, p. 140).

Taylor et al. (1972) reported experiments that determined the effect of longitudinal internal diffusion of oxygen on pea (*P. sativum* L.) radicle elongation. When air with 21% oxygen was maintained around the shoot, sufficient oxygen diffused through the plant pathways for radicles in humidified nitrogen gas (no oxygen) to elongate at a rate 20% of the control. When the radicles were at 21% oxygen, but the shoots were in nitrogen, the radicle elongation rate was 50% of the control value. These data indicate that oxygen moves both up and down root tissue. Similar results have been obtained for CO_2 and ethylene gas (Taylor et al. 1972).

Grosse et al. (1992) used tracer gas studies (ethane) and oxygen uptake measurements to study pressurized gas transport. They explained pressurized gas transport as follows: Oxygen supply to roots by diffusion is only effective over short distances. Over long gas transport pathways, as, for example, in trees that provide oxygen to their root systems from above-ground tissues, diffusion alone may be insufficient. However, tree species growing in wet or frequently flooded sites have adaptations that improve the oxygen supply to their roots. The well-known "knees" of bald cypress (*Taxodium distichum* L. Rich.) and the prop roots of mangrove (*Rhizophora* sp. or *Avicennia* sp.) have long been recognized as such adaptations. In the case of mangroves, air roots project above mud and are provided with minute openings called lenticels into which air diffuses. The air then passes through the soft spongy tissue to roots beneath the mud (Encyclopaedia Britannica 1971b). Another method of gas transport, called pressurized gas transport, has been recognized as an additional mechanism by which an increased flow of air can be supplied to submerged plant organs. The thermo-osmotically active partition required for pressurized gas transport has been localized within the phellogen layer of the lenticellular meristem of black alder [*Alnus glutinosa* (L.) Gaertn.], which has intercellular pores with a nearly optimal average diameter of 14 nm. When the stem is warmed relative to the surrounding air through the absorption of radiant energy by the darkly pigmented bark, a flow of gas from the atmosphere into the stem occurs in accordance with a physical effect that has been described by many different terms, but which is defined now as thermo-osmosis of gases. This gas flow generates a positive pressure in the stem's intercellular system causing a flow of air to the roots. This flow increases not only the oxygen partial pressure in the roots but also the rate of oxygen loss to the rhizosphere (Grosse et al. 1992).

Grosse et al. (1992) found that pressurized gas transport improved oxygen supply to roots in the wetland tree species bald cypress (Taxodiaceae family), *Betula pubescens* J.F. Ehrh. (birch; Betulaceae family), and *Populus tremula* L.(European aspen; Salicaceae family), but not in *Acer pseudoplatanus* L. (sycamore maple; Aceraceae family) and *Ilex aquifolium* L. (English holly; Aquifoliaceae family), which are found in drier habitats. In the deciduous tree species *B. pubescens* and *P. tremula*, pressurized gas transport was most evident during the resting period, which is characterized by soil anoxia following waterlogging of the natural habitat. Because pressurized gas transport is found in species of distantly related families, Grosse et al. (1992) hypothesized that it helps wetland species survive the initial period of soil flooding before acclimation to waterlogging occurs.

A method used to measure oxygen movement through plants is carried out in introductory plant physiology laboratories. The experiment is done to monitor the rate of photosynthesis, but it also shows how fast oxygen is moving through the plant. The setup is shown in Figure 4.8. An elodea plant (*Elodea canadensis* Michx., an aquatic plant) is cut and placed upside down in a cylinder of water. The cut end is attached to a water-filled, graduated pipette, which is clamped off at the other end. As the oxygen moves through the plant, it moves into the pipette and displaces the water, and the rate of oxygen production can be measured by noting how fast the water falls in the pipette (Newcomb et al. 1964, pp. 42–46). This water plant (elodea) is favorable for such measurements, because oxygen produced in the leaves and stem collects in the large intercellular spaces in the stem,

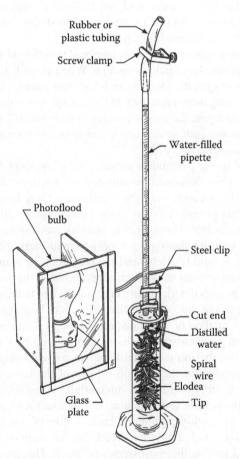

FIGURE 4.8 Arrangement of apparatus for determining the rate of oxygen release from photosynthesizing *Elodea*. (Reprinted from Newcomb, E.H. et al., *Plants in Perspective. A Laboratory Manual of Modern Biology*, W.H. Freeman and Company, San Francisco, CA, 1964, p. 43, Fig. 9.1. With permission from Eldon H. Newcomb, Department of Botany, University of Wisconsin, Madison, WI.)

passes through them, and then bubbles up through the water at the cut end of the plant (Newcomb et al. 1964, p. 43).

SUMMARY

Short periods when oxygen is lacking in the soil (e.g., 3 min or less for cotton) can immediately reduce root growth. But growth returns to normal if oxygen is reapplied. Periods of anaerobiosis of longer than 30 min for cotton caused root death. The concentration of CO_2 in the soil that inhibits root growth varies. Some studies indicate that if the oxygen content is adequate, CO_2 concentrations of 20%–50% have a negligible effect on growth. Other studies show that concentrations of 6.8%–10% are injurious. A concentration as low as 1.5% CO_2 reduced the growth of pea in one study. The influence of poor soil aeration on plants seems to be primarily from oxygen deficiency rather than from CO_2 toxicity. Oxygen can be transported from the shoots to the roots to help improve oxygen supply to roots under anaerobic conditions.

APPENDIX

BIOGRAPHY OF ELIZABETH ("BETTY") L. KLEPPER

Betty Klepper, the first woman to be named Fellow of the SSSA (in 1985) and who elucidated root growth under field conditions, was born on March 8, 1936, in Memphis, Tennessee, where she received her elementary and secondary education (Mortvedt 1986). Her father was a lawyer and her mother was a school teacher and housewife (McIntosh and Simmons 2008). She holds a BA degree in chemistry from Vanderbilt University, Nashville, Tennessee, where she was elected to Phi Beta Kappa and graduated *summa cum laude* in 1958. Under a Marshall Scholarship from the British government, she studied chemistry and botany at Exeter University in England in 1958 and 1959. After 1 year as a high school science teacher, she returned to graduate school in botany and received the MA in 1963 and the PhD in 1966 from Duke University. Her graduate research on plant–water relations was done under the supervision of Paul J. Kramer, sometimes called the father of plant–water relations in the United States. She spent 18 months with the Commonwealth Scientific and Industrial Research Organization at Griffith, New South Wales, Australia, working on irrigation research problems. From 1968 to 1972, she served as assistant professor in the Department of Botany and Microbiology at Auburn University (Alabama), where she taught the undergraduate plant physiology course and a graduate-level course in plant–water relations and mineral nutrition. At Battelle Northwest Laboratories, Richland, Washington, Dr. Klepper studied the radioecology of long-lived actinides and fission products from 1972 to 1976. From 1976 until her retirement, she was a plant physiologist at the United States Department of Agriculture, Agricultural Research Service, Columbia Plateau Conservation Research Center at Pendleton, Oregon, where she was appointed research leader in 1985 (Mortvedt 1986). Her scientific and writing skills were sought after by journals, and she served on the editorial boards of *Crop Science*, *Plant Physiology*, and *Field Crops Research*. In 1989, she spent a month in the Ivory Coast as a consultant for the International Atomic Energy Commission (personal communication, May 16, 1989).

Dr. Klepper's research has concerned plant and soil-water relations, especially the function of roots. She was one of the first to use a pressure chamber to document the diurnal cycle of plant water potential (Klepper 2000). She along with Dr. Howard M. Taylor helped to develop the rhizotron, an underground box with soil that has a glass side through which root growth can be observed. These rhizotrons can be seen in a classic film that she and her colleague, Morris G. Huck, made for the XI International Botanical Congress held in Seattle, Washington, between August 24 and September 2, 1969, "Time-Lapse Photography of Root Growth." It was originally produced on 16 mm film by the Auburn University Television Department (Auburn, Alabama), but it is now available in DVD (digital versatile disc or digital video disc) and VCR (videocassette recorder) formats (ASA, CSSA, and SSSA 1999).

Betty Klepper gives freely of her time to help others, especially women scientists. She encouraged me in my career. She visited the Evapotranspiration Laboratory of Kansas State University on October 1–2, 1982, to help advise one of my students, Dr. Susan J. Grecu, who was studying root growth into claypan soils using minirhizotrons (Grecu et al. 1988a,b). She explained to us the proper equation to use to determine root length density. At that time, I remember the good advice she gave me: "If you don't believe in your own research, no one will." We have to be our own best cheerleaders.

In 1997, when she was retired, she told me that she spends at least 3 h/day working on her investments (personal communication, October 28, 1997). She donates money to a foundation at a small college in Missouri to fund a scholarship, in the name of her father, for a law student. She said she was proud that she was able to make this scholarship available. Through wise financial planning, she has been able to endow a lectureship for the CSSA, the Betty Klepper Endowed Lectureship. Each year, an outstanding crop scientist is brought to the annual meeting of the society to give a lecture. The first lecture was in 2001.

She has won many awards, and was the first woman to win many of them. In 1985, when she was named the first woman to become Fellow of the SSSA, she also became Fellow of the ASA, the second woman to attain such an honor. Rosalind Morris, a wheat geneticist at the University of Nebraska, Lincoln, was the first in 1979. Dr. Klepper became Fellow of the CSSA in 1989, the first woman to attain Fellowship of all three of the Tri-Societies (ASA, CSSA, and SSSA). She was the first woman to be elected President of any of the Tri-Societies when she became President of CSSA in 1997. In 2004, she received the Monsanto Crop Science Distinguished Career Award, the first woman to receive it. (This award was discontinued in 2009.) In 2006, she received the CSSA Presidential Award (McIntosh and Simmons 2008).

Since retirement, Dr. Klepper has been involved with environmental education. She coordinates the Stewards of the Umatilla River Environment (S.U.R.E.), a citizens' group working to improve the environment along this river in Pendleton, Oregon. The group plants native riparian trees for wildlife habitat and flood prevention. They provide nesting boxes for wood ducks and song birds, which are helpful for her hobby of bird watching. She also has led workshops for local farmers and developed a "River School" for fifth graders (McIntosh and Simmons 2008). Pendleton, Oregon, has honored her for her environmental activities.

5 Elevated Carbon Dioxide in the Atmosphere: Interaction with the Soil Physical Factors That Affect Root Growth

INTRODUCTION

In Chapter 3, we considered the effects of elevated CO_2 in the soil on three of the four soil physical factors that affect root growth: soil water, soil compaction, and soil temperature. The fourth factor is soil aeration, which includes the effects of both oxygen and CO_2, and we saw that lack of oxygen in the soil limits root growth more than elevated levels of CO_2. We now turn to the same three soil physical factors as they are affected by elevated CO_2 in the air above the soil.

SOIL WATER

We first consider soil water. Much research has been done to consider the possibility of using soil to sequester carbon, as the concentration of CO_2 in the air increases (Lal et al. 1999). Less known is the effect of elevated atmospheric CO_2 on soil water content. But we need to know this to manage soil water as the CO_2 concentration in the atmosphere increases. Allen, L.H., Jr. (1990) said that the amount of CO_2 in the atmosphere might double within 100 years. As noted in Chapter 1, models predict that, in response to a doubling of the atmospheric CO_2 concentration, soil moisture will be reduced throughout an extensive mid-continental area of North America (Manabe and Wetherald 1987; Mitchell and Warrilow 1987; Kellogg and Zhao 1988), where temperate grasslands are important agriculturally because they provide natural pastures for grazing animals.

The models are based on predicted increased temperatures and reduced rainfall. As an example of such a model, we look at the predictions of Valdés et al. (1994), who used predictions of temperature and precipitation changes from general-circulation-model simulations (Kellogg and Zhao 1988) to determine soil moisture at Amarillo, Texas, and San Antonio, Texas. They stated that the most likely scenario [a *scenario* is a postulated sequence of future events (Allen, R.E. 1990)] is the one representing an increase of 2°C in the mean annual temperature. Precipitation scenarios, they said, are more variable, but a decrease in the mean annual precipitation of about 10% is the most used one. To represent the climate scenarios, both with and without global warming, synthetic traces of hourly values of precipitation and evapotranspiration potential were generated. When these two values (+2°C increase in temperature; −10% decrease in precipitation) were applied to the synthetically manufactured weather data, they got the results shown in Figure 5.1 for Amarillo, Texas, and in Figure 5.2 for San Antonio, Texas. The modeled data showed that the monthly means for soil moisture at both locations will decrease as the CO_2 concentration in the atmosphere doubles.

Soil moisture is also predicted to decrease in agricultural regions of California. Gleick (1987) modeled climatic impacts on a major watershed in northern California using climate-change scenarios from general circulation models. He concluded that changes in temperature and precipitation,

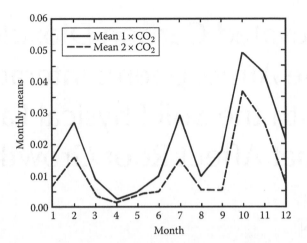

FIGURE 5.1 Soil moisture concentration monthly means for Amarillo, Texas, for ambient ($1 \times CO_2$) and doubled ($2 \times CO_2$) atmospheric CO_2 concentrations. The doubled CO_2 values are based on the scenario of a 10% decrease in precipitation and a 2°C rise in temperature. (Reprinted from *J. Hydrol.*, 161, Valdés, J.B., Seoane, R.S., and North, G.R., A methodology for the evaluation of global warming impact on soil moisture and runoff, 389–413, Copyright 1994, Fig. 9, with permission from Elsevier.)

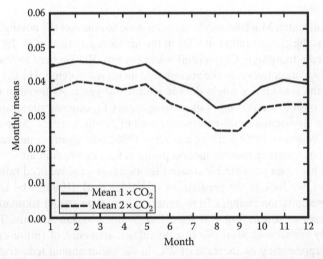

FIGURE 5.2 Soil moisture concentration monthly means for San Antonio, Texas, for ambient ($1 \times CO_2$) and doubled ($2 \times CO_2$) atmospheric CO_2 concentrations. The doubled CO_2 values are based on the scenario of a 10% decrease in precipitation and a 2°C rise in temperature. (Reprinted from *J. Hydrol.*, 161, Valdés, J.B., Seoane, R.S., and North, G.R., A methodology for the evaluation of global warming impact on soil moisture and runoff, 389–413, Copyright 1994, Fig. 10, with permission from Elsevier.)

caused by increases in atmospheric trace-gas concentrations, could have major impacts on soil moisture in important agricultural areas. Of particular importance were predicted patterns of summer soil-moisture drying that were consistent across the entire range of tested scenarios. The decreases in summer soil moisture ranged from 8% to 44% (Figure 5.3). The data in Figure 5.3 are average summer soil-moisture values in the agricultural portion of the Sacramento Basin for 10 different scenarios: precipitation changes of 0% and plus and minus 10% or 20% and average monthly temperature changes of an increase of 2°C and 4°C.

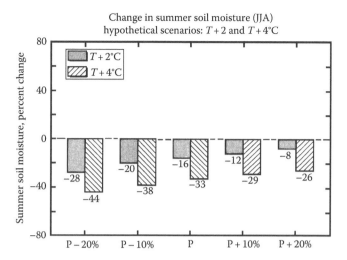

FIGURE 5.3 Percent change in average summer (June, July, and August) soil moisture between the base case and the hypothetical scenarios with an increase in average monthly temperature of 2°C and 4°C ($T+2$°C and $T+4$°C) and precipitation (P) changes of zero and plus and minus 10% and 20%. Note that all these scenarios show decreases in summer soil moisture. (Reprinted with kind permission from Springer Science+Business Media: *Climatic Change*, Regional hydrologic consequences of increases in atmospheric CO_2 and other trace gases, 10, 1987, 137–161, Gleick, P.H., Fig. 9.)

Because apparently no measurements of the soil water content under augmented atmospheric CO_2 had been made, we measured over a 2 year period (1989–1990) the soil water content in a tall grass prairie in Kansas exposed to ambient (337 μmol/mol) and twice ambient (658 μmol/mol) levels of CO_2 (Kirkham et al. 1993). The dominant species on the grassland is big bluestem (*Andropogon gerardii* Vitman), a warm-season grass. About 70% of the botanical composition of the grassland is big bluestem (Kirkham et al. 1991). The plants grew in closed-top chambers (1.5 m in diameter and 1.8 m tall). Half of the plots (chambers) were kept at field capacity (high water level), and half of the plots were kept at half field capacity (low water level). We did not control the temperature. However, daytime air temperatures at 100 cm above ground in the chambers averaged 2.7°C warmer than outside during the two growing seasons (Nie et al. 1992b). The soil at the site was a Tully silty clay loam (Order: Mollisol; fine, mixed, mesic Pachic Arguistoll) with a field capacity of 0.38 m³/m³. Soil water content was measured weekly during the two growing seasons with a neutron probe between the 100 and 2000 mm depth. Measurements were taken in 100 mm increments to a depth of 600 and then in 200 mm depth increments to a depth of 2000 mm. Soil water contents in the 0–50 and 50–100 mm depths were determined gravimetrically. They were put on a volumetric basis by multiplying by the bulk density (1.03 and 1.12 Mg/m³ at the 0–50 and 50–100 mm depths, respectively.) Total water in the soil profile from the 0–2000 mm depth then was calculated. Soil water content at the midpoint of each depth was plotted in figures showing the water content on different dates during each season.

The water contents in May, June, July, August, and September in each season (1989 and 1990) are shown in Figures 5.4 through 5.8, respectively. At the beginning of the 1989 and 1990 seasons (May) and under the high-water treatment, the soil water content was greater in the top 0.075 m with the elevated level of CO_2 than with the ambient level (Figure 5.4, left-hand side). For the low-water treatment, the standard-error bars overlapped between the two CO_2 treatments. Nevertheless, more water was in the soil at each depth with the elevated level of CO_2 than with the ambient level (Figure 5.4, right-hand side). Under the low-water treatment, in the total 2000 mm profile, soil exposed to elevated CO_2 had 26 mm (624–598 mm) and 17 mm (736–719 mm) more water in the profile on the May dates in 1989 and 1990, respectively, than soil exposed to ambient CO_2. Under the high

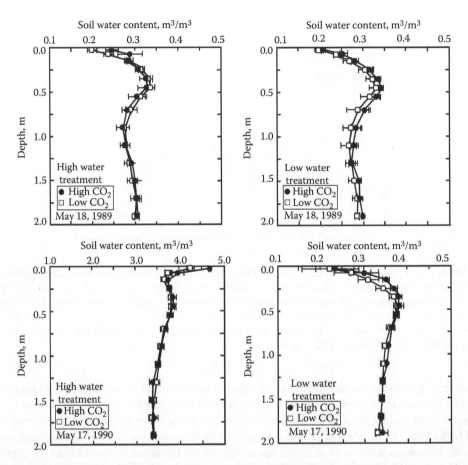

FIGURE 5.4 Water content on May 18, 1989 (top) and May 17, 1990 (bottom) at different depths in a soil of a grassland with a high (twice ambient) and a low (ambient) atmospheric CO_2 concentration and a high (field capacity) and a low (half field capacity) water level. Horizontal bars=±standard error; n=4. (Reprinted from Kirkham, M.B. et al., Responses of plants to elevated levels of carbon dioxide, in *Proceedings of the Symposium on Plant Growth and Environment*, October 16, 1993, Suwon, Korea, Korean Agricultural Chemical Society, Suwon, Korea, pp. 130–161, Fig. 1. With permission from the Korean Agricultural Chemical Society (now called the Korean Society for Applied Biological Chemistry, Seoul National University, Seoul, Republic of Korea).)

water level, these differences were 11 mm (628–617 mm) and 4 mm (758–754 mm) in 1989 and 1990, respectively.

One month later on June 14, 1989 and June 14, 1990 (Figure 5.5), soil under the elevated level of CO_2 again had more water than soil under the ambient level. Differences were especially evident with the low-water treatment, where, in 1989, differences were significant between the 0.7 and 1.3 m depths, and, in 1990, between the 0.025 and 0.35 m depths. As in May, under the low water level, the differences between the high- and low-CO_2 treatments were greater (689 – 649 mm=40 mm and 677 – 656 mm=21 mm for 1989 and 1990, respectively) than those between the high- and low-CO_2 treatments under the high water level (725 – 702 mm=23 mm and 730 – 725 mm=5 mm for 1989 and 1990, respectively). In July 1989 under the high water level, the plots with elevated CO_2 had more water only at the deepest depth (1.9 m depth; Figure 5.6, top left). In July 1990 under the high water level, there was more water in the plots with elevated CO_2 than with ambient CO_2 (715 – 698 mm=17 mm) (Figure 5.6, bottom left). In July 1989 and July 1990 under the low water level, there was more water in the plots with elevated CO_2 than in the plots with ambient CO_2 (662 – 625 mm=37 mm and 636 – 605 mm=31 mm for 1989 and 1990, respectively) (Figure 5.6, right-hand

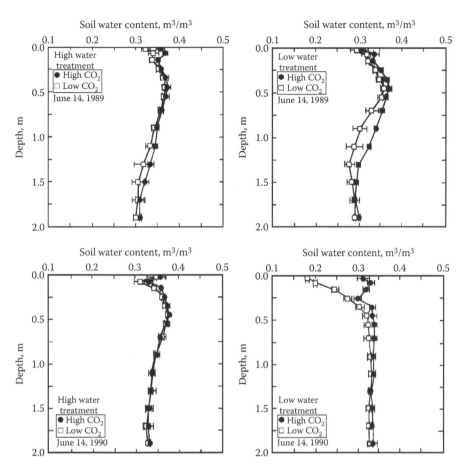

FIGURE 5.5 Water content on June 14, 1989 (top) and June 14, 1990 (bottom) at different depths in a soil of a grassland with two atmospheric CO_2 concentrations and two soil water levels. For details, see legend of Fig. 5.4. (Reprinted from Kirkham, M.B. et al., Responses of plants to elevated levels of carbon dioxide, in *Proceedings of the Symposium on Plant Growth and Environment*, October 16, 1993, Suwon, Korea, Korean Agricultural Chemical Society, Suwon, Korea, pp. 130–161, Fig. 2. With permission from the Korean Agricultural Chemical Society (now called the Korean Society for Applied Biological Chemistry, Seoul National University, Seoul, Republic of Korea).)

side). The August soil-moisture measurements each year also showed more water in the soil under the elevated CO_2 atmosphere than in the soil under the ambient level of CO_2 for both water treatments (Figure 5.7). For the high water level, the total amount of soil water in the profile in August 1989 was 21 mm more (724 – 703 mm) under elevated CO_2, and in August 1990 it was 27 mm more (802 – 775 mm) under elevated CO_2. For the low water level, the total amount of soil water in the profile in August 1989 was 46 mm more (622 – 576 mm) under elevated CO_2, and in August 1990 it was 4 mm more (756 – 752 mm) under elevated CO_2. Under the high water level in September 1989 and September 1990, there was more water in the plots with elevated CO_2 only at the surface of the soil (Figure 5.8, left). Under the low water level in September 1989 and September 1990, the total amount of soil water in the profile was higher under the elevated CO_2 treatment than under the ambient CO_2 treatment (698 mm – 680 mm = 18 mm in 1989; 653 – 621 mm = 32 mm in 1990).

The total profile soil water contents, determined weekly for the 1989 season, were published by Kirkham et al. (1991) and are shown in Figure 5.9 for the high-water treatment and Figure 5.10 for the low-water treatment. For the elevated and ambient concentrations of CO_2, the average amounts of water in the soil during the 1989 season were 722 and 705 mm, respectively, for the high water

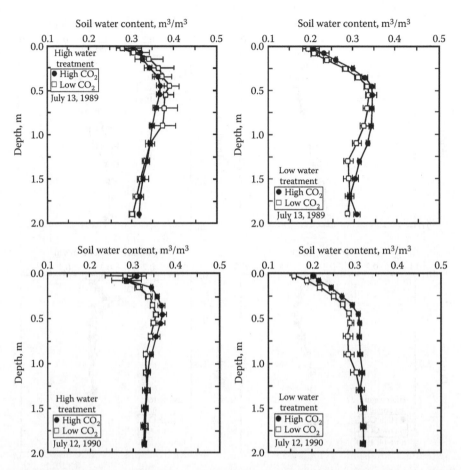

FIGURE 5.6 Water content on July 13, 1989 (top) and July 12, 1990 (bottom) at different depths in a soil of a grassland with two atmospheric CO_2 concentrations and two soil water levels. For details, see legend of Figure 5.4. (Reprinted from Kirkham, M.B. et al., Responses of plants to elevated levels of carbon dioxide, in *Proceedings of the Symposium on Plant Growth and Environment*, October 16, 1993, Suwon, Korea, Korean Agricultural Chemical Society, Suwon, Korea, pp. 130–161, Fig. 3. With permission from the Korean Agricultural Chemical Society (now called the Korean Society for Applied Biological Chemistry, Seoul National University, Seoul, Republic of Korea).)

level. For the low water level, these values were 669 and 638 mm, respectively. Therefore, with twice-ambient CO_2, the soil contained an average of 17 and 31 mm more water for the field-capacity and half-field-capacity water regimes, respectively.

The higher soil water content in the plots with elevated CO_2 was probably the result of a lower transpiration rate (Kirkham et al. 1993). Under the high- and low-water treatments, doubled CO_2 reduced transpiration of leaves by 25% and 35%, respectively, in 1989, and by 24% and 20%, respectively, in 1990 (Figure 5.11). Transpiration was reduced with elevated CO_2 because stomatal resistance was increased (Figure 5.12). In 1989 and 1990, elevated CO_2 increased stomatal resistance by averages of 1.6 and 1.0 s/cm, respectively, for both water treatments. It has been known for a long time that CO_2 is an antitranspirant because it closes stomata (van Bavel 1974). Because elevated CO_2 closed stomata, canopy temperatures increased under high CO_2, as transpirational cooling was reduced (Figure 5.13). For the low- and high-water treatments, elevated CO_2 increased average, seasonal canopy temperature by 1.7°C and 0.8°C, respectively, in 1989, and by 1.4°C and 1.5°C, respectively, in 1990. These values are in fairly good agreement with data of Idso et al. (1987), who found that a doubling of atmospheric CO_2 (from 330 to 660 μmol/mol) resulted in a midday temperature increase

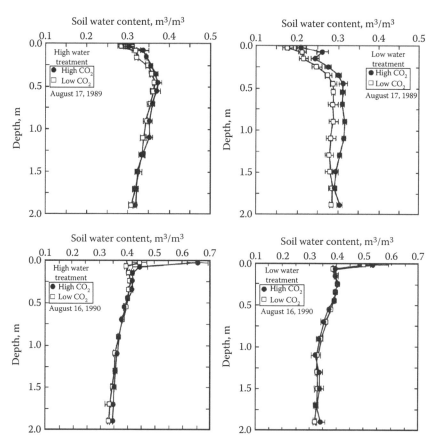

FIGURE 5.7 Water content on August 17, 1989 (top) and August 16, 1990 (bottom) at different depths in a soil of a grassland with two atmospheric CO_2 concentrations and two soil water levels. For details, see legend of Figure 5.4. (Reprinted from Kirkham, M.B. et al., Responses of plants to elevated levels of carbon dioxide, in *Proceedings of the Symposium on Plant Growth and Environment*, October 16, 1993, Suwon, Korea, Korean Agricultural Chemical Society, Suwon, Korea, pp. 130–161, 1993, Fig. 4. With permission from the Korean Agricultural Chemical Society (now called the Korean Society for Applied Biological Chemistry, Seoul National University, Seoul, Republic of Korea).)

of 1.1°C for well-watered cotton (*Gossypium hirsutum* L.) in Phoenix, Arizona. We shall discuss stomatal resistance and transpiration under elevated CO_2 in detail in Chapters 8 and 10, respectively.

In 1989 and 1990, for both water levels, photosynthetic rate of big bluestem did not differ with CO_2 level, except under the low-water treatment during the mid-season of 1990 (June 11 to July 16 or day of year 162–197, Figure 5.14, lower right). During this period, the photosynthetic rate of high-water big bluestem with ambient CO_2 (open squares, Figure 5.14, upper right) was about the same as that of low-water big bluestem with doubled CO_2 (closed circles, Figure 5.14, lower right). The augmentation by CO_2 of the photosynthetic rate of big bluestem under dry conditions was not evident in 1989, perhaps because the surface of the soil did not become as dry in 1989 during this period as it did in 1990. During day of year 162–197 in 1990 in the low-water treatment, the surface (0–100 mm) soil water content in the ambient plots was less than $0.20 \, m^3/m^3$ (Figures 5.5 and 5.6, lower right), but in the plots with elevated CO_2, it was more than $0.20 \, m^3/m^3$ (between 0.21 and $0.34 \, m^3/m^3$, Figures 5.5 and 5.6, lower right). The higher surface soil water content in the plots with elevated CO_2 perhaps permitted the higher photosynthetic rate of big bluestem exposed to doubled CO_2. In 1989, the soil water content never fell below $0.20 \, m^3/m^3$, except for the top layer of the low-water, low-CO_2 plots on August 17, 1989 (Figure 5.7, upper right). The results suggested that

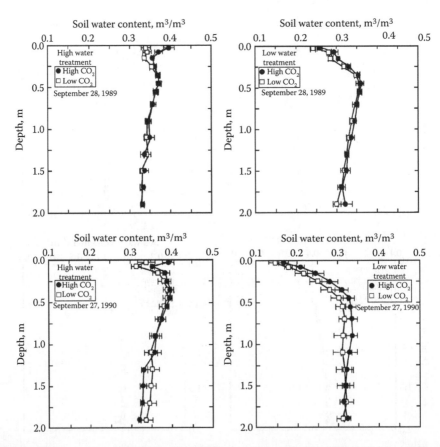

FIGURE 5.8   Water content on September 28, 1989 (top) and September 27, 1990 (bottom) at different depths in a soil of a grassland with two atmospheric CO_2 concentrations and two soil water levels. For details, see legend of Figure 5.4. (Reprinted from Kirkham, M.B. et al., Responses of plants to elevated levels of carbon dioxide, in *Proceedings of the Symposium on Plant Growth and Environment*, October 16, 1993, Suwon, Korea, Korean Agricultural Chemical Society, Suwon, Korea, pp. 130–161, 1993, Fig. 5. With permission from the Korean Agricultural Chemical Society (now called the Korean Society for Applied Biological Chemistry, Seoul National University, Seoul, Republic of Korea).)

under dry conditions (i.e., soil water content in the silty clay loam of less than $0.20\,m^3/m^3$), elevated CO_2 may compensate for reductions in the photosynthetic rate of the C_4 plant, big bluestem, due to drought. In Chapter 1, we discussed other studies that showed that elevated CO_2 can compensate for reductions in growth due to drought.

Data obtained after the 1989–1990 experiment (Kirkham et al. 1993) have confirmed that soil water content in the mid-continental United States is greater under elevated CO_2 than under ambient CO_2 (Knapp et al. 1993; LeCain et al. 2003; Morgan et al. 2004; Nelson et al. 2004). Knapp et al. (1993) did their research in the field in 1991 and 1992 on the same tall-grass prairie in Kansas that Kirkham et al. (1993) studied in 1989 and 1990, but in a different location. Knapp et al. (1993) do not report the soil type, but Morgan et al. (2004) in their summary of different experiments, including the one by Knapp et al. (1993), give it as in the Fine Tully series soils, the same type of soil that Kirkham et al. (1991, 1993) had in their experiments (Tully silty clay loam). LeCain et al. (2003) and Nelson et al. (2004) worked in a semiarid short-grass steppe in Colorado. LeCain et al. (2003) reported that the soil moisture results under elevated atmospheric CO_2 for the Colorado studies. In their study, open-top chambers were used to investigate responses of *Pascopyrum smithii* (Rydb.) A. Love (western wheatgrass) (C_3) and *Bouteloua gracilis* (H.B.K.)

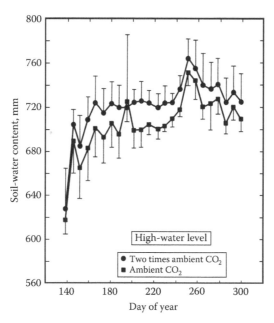

FIGURE 5.9 Soil-water content from the 0–2000 mm depth in a tallgrass prairie with a high (twice ambient) and a low (ambient) atmospheric CO_2 concentration and a high (field capacity) water level. Vertical bars = ± standard deviation. Only half of each bar is drawn for clarity. (Reprinted from Kirkham, M.B. et al., *Crop Sci.*, 31, 1589, 1991, Fig. 1, Copyright 1991. With permission from the Crop Science Society of America.)

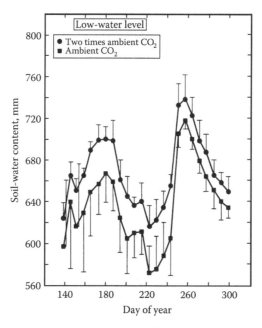

FIGURE 5.10 Soil-water content from the 0–2000 mm depth in a tallgrass prairie with a high (twice ambient) and a low (ambient) atmospheric CO_2 concentration and a low (half field capacity) water level. Vertical bars = ± standard deviation. Only half of each bar is drawn for clarity. (Reprinted from Kirkham, M.B. et al., *Crop Sci.*, 31, 1589, 1991, Fig. 1, Copyright 1991. With permission from the Crop Science Society of America.)

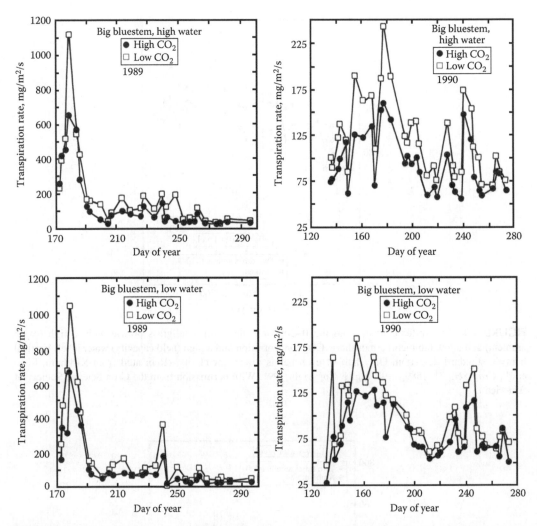

FIGURE 5.11 Transpiration rate of big bluestem in 1989 (left) and 1990 (right) on a grassland with two atmospheric CO_2 concentrations and two soil water levels. For details, see legend of Figure 5.4. (Reprinted from Kirkham, M.B. et al., Responses of plants to elevated levels of carbon dioxide, in *Proceedings of the Symposium on Plant Growth and Environment*, October 16, 1993, Suwon, Korea, Korean Agricultural Chemical Society, Suwon, Korea, pp. 130–161, 1993, Fig. 6. With permission from the Korean Agricultural Chemical Society (now called the Korean Society for Applied Biological Chemistry, Seoul National University, Seoul, Republic of Korea).)

Lag. (blue grama) (C_4) grown at ambient CO_2 (360 µmol/mol) and elevated CO_2 (720 µmol/mol) in the steppe during 4 years. The experiment was established in the spring of 1997 on a 6 ha field of native grassland. The soil was a Remmit fine sandy loam (field capacity not given). Soil water content was measured weekly using a neutron probe. Total growing season precipitation was 480, 302, 523, and 247 mm in 1997, 1998, 1999, and 2000, respectively (long term average = 280 mm). On many dates, soil water content was higher in the chambers with elevated CO_2 compared to those under the ambient CO_2 (Figure 5.15). When averaged over the 4 years of the study, soil water content in the upper meter was 129.4 mm in elevated CO_2 chambers compared with 113.6 mm in ambient CO_2 chambers, an increase of 14%. Ambient CO_2 chambers often had a lower soil water content than the un-chambered plots owing to higher air temperatures in the chambers. Because of reduced stomatal conductance and transpiration, soil water content was increased. The authors

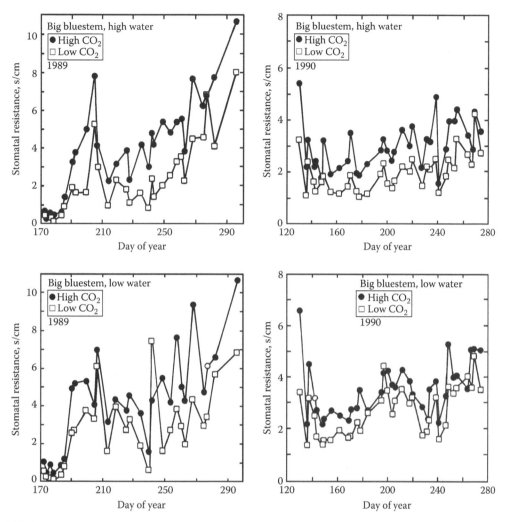

FIGURE 5.12 Stomatal resistance of big bluestem in 1989 (left) and 1990 (right) on a grassland with two atmospheric CO_2 concentrations and two soil water levels. For details, see legend of Figure 5.4. (Reprinted from Kirkham, M.B. et al., Responses of plants to elevated levels of carbon dioxide, in *Proceedings of the Symposium on Plant Growth and Environment*, October 16, 1993, Suwon, Korea, Korean Agricultural Chemical Society, Suwon, Korea, pp. 130–161, 1993, Fig. 7. With permission from the Korean Agricultural Chemical Society (now called the Korean Society for Applied Biological Chemistry, Seoul National University, Seoul, Republic of Korea).)

concluded that, in this semiarid ecosystem, responses of plants will be driven primarily by the effect of elevated CO_2 on soil water relations.

In contrast to the results obtained on grasslands in Kansas and Colorado in the United States, Nowak et al. (2004a) found that soil water content was not increased under elevated CO_2 in a desert ecosystem. They measured soil water content from September 1997 to September 2002 at the Nevada (United States) Desert Free Air CO_2 Enrichment (FACE) Facility in the Mojave Desert. The objective was to determine if elevated atmospheric CO_2 (550 µmol/mol CO_2) conserved soil water for a desert scrub community. Ambient CO_2 was 370 µmol/mol. The soil type and its field capacity were not stated. They measured soil water content in the top 0–0.2 or 0–0.5 m (depending upon the length of probe that they had) using time-domain reflectometry (see Kirkham 2005, pp. 187–205, for an explanation of how time-domain reflectometry can be used to obtain volumetric

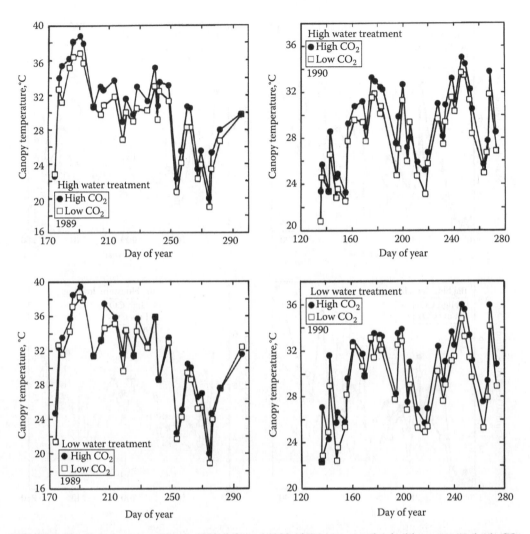

FIGURE 5.13 Canopy temperature in 1989 (left) and 1990 (right) on a grassland with two atmospheric CO_2 concentrations and two soil water levels. For details, see legend of Figure 5.4. (Reprinted from Kirkham, M.B. et al., Responses of plants to elevated levels of carbon dioxide, in *Proceedings of the Symposium on Plant Growth and Environment*, October 16, 1993, Suwon, Korea, Korean Agricultural Chemical Society, Suwon, Korea, pp. 130–161 1993, Fig. 8. With permission from the Korean Agricultural Chemical Society (now called the Korean Society for Applied Biological Chemistry, Seoul National University, Seoul, Republic of Korea).)

soil water content). The plant in the plots where they installed the probes was *Larrea tridentata* (DC.) Cov., an evergreen shrub. They measured soil water content to the 1.85 m depth using a neutron probe. Over all sampling dates, soil water in the top 0.2 m of soil was not different between the elevated and ambient CO_2 treatments. Similarly, soil water in the top 0.5 m of soil was not different between the two treatments. Finally, as with the shallower layers, soil water in the top 1.85 m of soil was not different between treatments. They gave no plant-physiological data. It could be that under the dry conditions of the desert, the stomata of the desert plants usually remain closed, so that large amounts of water are not transpired. In dry periods, desert plants may not be transpiring at all, and the water lost from the soil might be similar to barren land. Studies are needed in which soil water content is measured on bare soil under ambient and elevated atmospheric CO_2 concentrations.

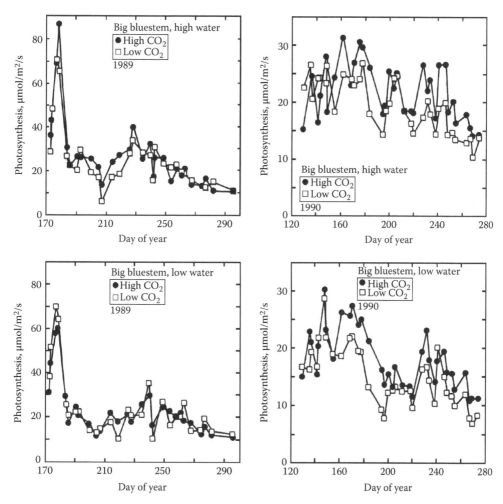

FIGURE 5.14 Photosynthesis of big bluestem in 1989 (left) and 1990 (right) on a grassland with two atmospheric CO_2 concentrations and two soil water levels. For details, see legend of Figure 5.4. (Reprinted from Kirkham, M.B. et al., Responses of plants to elevated levels of carbon dioxide, in *Proceedings of the Symposium on Plant Growth and Environment*, October 16, 1993, Suwon, Korea, Korean Agricultural Chemical Society, Suwon, Korea, pp. 130–161, 1993, Fig. 9. With permission from the Korean Agricultural Chemical Society (now called the Korean Society for Applied Biological Chemistry, Seoul National University, Seoul, Republic of Korea).)

In another FACE study done under irrigation in an arid region (Arizona), Kimball and Bernacchi (2006) found that the soil dried more slowly with elevated atmospheric CO_2 due to a reduction in evapotranspiration. (See Chapter 10 for a discussion of the effects of elevated CO_2 on evapotranspiration.) Therefore, the water content of the soil in FACE plots was often higher than that in plots under ambient atmospheric CO_2 concentration.

While the data in the mid-continental United States are consistent in showing that elevated CO_2 increases soil moisture content (Kirkham et al. 1993; Knapp et al. 1993; LeCain et al. 2003), Newton et al. (1996b) in New Zealand found no effect of elevated CO_2 on soil moisture content in temperate pastures. They took samples of turf with soil attached from a pasture and placed them in steel bins in two growth rooms, one at ambient CO_2 concentration (350 μmol/mol CO_2) and one at elevated CO_2 concentration (700 μmol/mol CO_2). Each sample had a dimension of 1×0.5×0.25 m, and the soil was a Pukepuke sand. Soil moisture content was measured using time-domain reflectometry

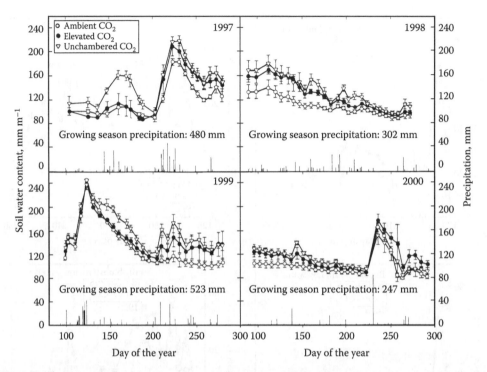

FIGURE 5.15 Soil moisture (to 1 m) and growing season precipitation for 4 years in open-top chambers and ambient (360 µmol/mol) and elevated (720 µmol/mol) CO_2 and in plots without chambers on the Colorado short-grass steppe. Data are means of three replications ± standard error. (Reprinted from LeCain, D.R., Morgan, J.A., Mosier, A.R., and Nelson, J.A., *Ann. Bot.*, 92, 41, Fig. 1, 2003 by permission of Oxford University Press.)

with probes 150 mm long. Plants grew under well-watered or drought-stressed conditions. Drought decreased soil water content from 0.60 to 0.10 g/g. But under both wet and dry conditions, the soil water content was the same under the two CO_2 treatments.

The difference in results between the New Zealand study and the studies done in the United States could be due to several factors. First, the soil types differed. Perhaps differences due to CO_2 treatment are difficult to observe in a fast-draining sandy soil, such as the one that Newton et al. (1996b) used. [Newton et al. (1996b) do not state if the soil water was allowed to drain out of the steel bins.] Second, the difference could also be due to confinement of the roots. In the studies in Kansas and Colorado, roots grew in the field and their depth of penetration was not limited. In the experiment of Newton et al. (1996b), the steel bins were 1 m in depth, and if the roots had been grown in the field, they might have penetrated to a deeper depth. Third, Newton et al. (1996b) did not present soil water content at different depths, and this can vary under elevated and ambient CO_2, as shown by Kirkham et al. (1993). For example, on June 14, 1989 (Figure 5.5, top right), Kirkham et al. (1993) observed differences in soil moisture at the 0.70–1.30 m depth, but not between the 0 and 0.70 m depth and the 1.30–1.90 m depth. Fourth, it could be that Newton et al. (1996b) did not take deep enough measurements to see differences due to CO_2 treatment. Even though their steel bins were 1 m in depth, they measured soil water content only to the 0.150 m depth. The moisture content between the 0.150 and 1 m depth might have been changing due to CO_2 treatment. Unlike neutron probes, which can take measurements to the 3 m depth, time-domain reflectometry measurements are limited to 1 m (Kirkham 2005, pp. 197–198). In Kansas, we need to measure deeper than 1 m to observe depth of depletion of water by plant roots. Depth of depletion of water by sunflower (*Helianthus annuus* L.) roots can be as deep as 2.7 m (Rachidi et al. 1993a).

Similar to the results of Newton et al. (1996b), Kammann et al. (2005) found no differences in soil moisture due to CO_2 treatment. They carried out their studies during a 5 year period (1998–2002) at the FACE site in Giessen, Germany, managed as a meadow. Giessen is in the central, western part of Germany, north of the city of Frankfurt. They enriched the plots with CO_2 year-round during daylight hours to +20% above ambient conditions. They did not give the ambient concentrations for the different years, but they stated in the introduction to their paper that the current concentration of CO_2 in the air was $375 \mu mol/mol$. Like Newton et al. (1996b), they measured soil moisture using time-domain reflectometry to the 0.150 m deep. The plots were not irrigated, and the annual mean precipitation at the site was 645 mm. The annual precipitation in 1998, 1999, 2000, 2001, and 2002 was 704.9, 565.1, 596.5, 619.5, and 750.3 mm, respectively. Despite the variable amount of rain in the soil due to rain (they called 1998 and 2002 wet years), they found no difference in soil moisture at the 0–150 mm depth in plots due to CO_2 during the 5-five study. However, they did see that when the top soil dried out (in July and August periods), the soil in the plots with elevated atmospheric CO_2 became wetter compared to the soil with the ambient CO_2 concentration. Kammann et al. (2005) stated that the "Giessen grassland does have a better ground water access than other grasslands under CO_2 enrichment," such as the Colorado short-grass steppe (Morgan et al. 2004) and the California desert scrub community (Nowak et al. 2004a). The average ground water table depth in the summer at the Giessen site is 84 cm and in the winter it is 53 cm. Morgan et al. (2004) do not give depth of ground water at their site. On the native tall-grass prairie near Manhattan, Kansas [called the Konza Prairie (Reichman 1987)], where Knapp et al. (1993) did their study, the depth to the water table is 100–150 ft (30.5–45.7 m) (Loyd R. Stone, Kansas State University, personal communication, September 22, 2008). Although Kammann et al. (2005) do not state this, water could have been moving up from the ground water to replace the water used by the plants in the ambient plots. Morgan et al. (2004), who summarize studies done under elevated CO_2, including the ones in Colorado, cite modeling data. These data predict that more water may percolate to deep depths when CO_2 in the atmosphere is increased. However, this deep percolation is predicted to occur only in wet years. We note that in wet years, stomata will be open under ambient CO_2 but more closed under elevated CO_2, which means water not used by the plants under elevated CO_2 can move to depth. In dry years, or dry parts of the season, the water can move up from the ground water in places like Giessen and replenish the water used by the plants grown under ambient CO_2. Because water is moving up and down in the soil profile differently in plots under elevated atmospheric CO_2 compared to ambient CO_2, measurements of soil water need to be taken during the season at different depths in the soil, such as Kirkham et al. (1993) did (Figures 5.4 through 5.8) to document these differences.

The studies by Nowak et al. (2004a), Newton et al. (1996b), and Kammann et al. (2005) showed no differences in soil moisture in plots under elevated atmospheric CO_2 and ambient CO_2. It is interesting to note that they reported no instances of soil moisture being less under elevated atmospheric CO_2.

SOIL COMPACTION

Little information exists concerning the effect of elevated atmospheric CO_2 on soil physical properties other than soil water. However, Prior et al. (2004) did measure how elevated atmospheric CO_2 altered soil bulk density, a measure of compaction. They collected soil samples from the surface layer (0–6 cm) after 5 years of growth of a legume [soybean; *Glycine max* (L.) Merr.] or non-legume [grain sorghum; *Sorghum bicolor* (L.) Moench] under two CO_2 levels (360 and 720 $\mu mol/mol$) using open-top field chambers in Auburn, Alabama, United States. The soil was a Blanton loamy sand that had been fallow for more than 25 years before the study began. The plants grew under no-tillage conditions. There were three replications. Six chambers had soybean (three at ambient CO_2; three at elevated CO_2) and six chambers had sorghum (three at ambient CO_2; three at elevated CO_2). In addition to bulk density, Prior et al. (2004) also measured soil carbon content, saturated hydraulic conductivity, and water stable aggregates. Soil carbon content was increased by 16% and 29% under CO_2-enriched conditions in the sorghum and soybean cropping systems, respectively. With soybeans, soil bulk density decreased (Figure 5.16) whereas saturated hydraulic conductivity (Figure 5.17) and aggregate stability (Figure 5.18) increased as a result of elevated CO_2. Concentration of CO_2 had little

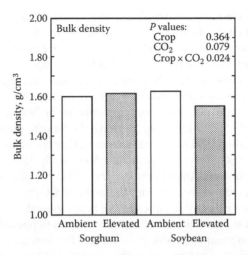

FIGURE 5.16 Effect of atmospheric CO_2 concentration (ambient or elevated) on soil bulk density in sorghum and soybean cropping systems. Probability values for main effect and interaction terms are shown. Bars represent means ($n=3$). (Reprinted from Prior, S.A., Runion, G.B., Torbert, H.A., and Rogers, H.H., Elevated atmospheric CO_2 in agroecosystems: Soil physical properties, *Soil Sci.*, 169, 434–439, 2004, Fig. 1. With permission.)

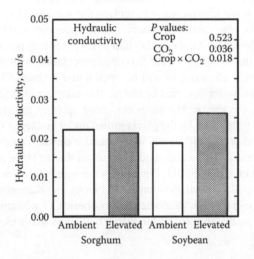

FIGURE 5.17 Effect of atmospheric CO_2 concentration (ambient or elevated) on saturated soil hydraulic conductivity in sorghum and soybean cropping systems. Probability values for main effect and interaction terms are shown. Bars represent means ($n=3$). (Reprinted from Prior, S.A., Runion, G.B., Torbert, H.A., and Rogers, H.H., Elevated atmospheric CO_2 in agroecosystems: Soil physical properties, *Soil Sci.*, 169, 434–439, 2004, Fig. 2. With permission.)

effect on soil properties with sorghum. Prior et al. (2004) concluded that soil physical properties may be altered in a future CO_2-enriched environment. But CO_2-induced benefits may favor legume-based systems such as soybean over grasses like sorghum.

SOIL TEMPERATURE

Apparently, little or no information exists concerning the effect of elevated atmospheric CO_2 on soil temperature. The thermal properties of soil will be changed as water is differentially lost due to different transpiration rates under ambient and elevated CO_2 concentrations. Temperature probes, such as the dual-probe heat-pulse probes (Song et al. 1998, 1999), need to be placed in plots under

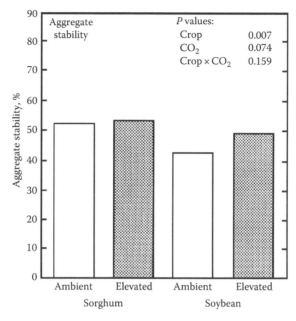

FIGURE 5.18 Effect of atmospheric CO_2 concentration (ambient or elevated) on soil aggregate stability in sorghum and soybean cropping systems. Probability values for main effect and interaction terms are shown. Bars represent means ($n = 3$). (Reprinted from Prior, S.A., Runion, G.B., Torbert, H.A., and Rogers, H.H., Elevated atmospheric CO_2 in agroecosystems: Soil physical properties, *Soil Sci.*, 169, 434–439, 2004, Fig. 3. With permission.)

ambient and elevated CO_2 concentrations to see how soil temperature changes with CO_2 concentration. These probes can measure both volumetric soil water content and soil temperature.

SUMMARY

Our results showed that, during two growing seasons, soil in a grassland of mid-continental North America under doubled CO_2 was wetter than soil under ambient CO_2, and the difference in moisture was greater under dry conditions than under wet conditions. Less water was lost from the soil under elevated CO_2 than from the soil under ambient CO_2, because elevated CO_2 closed stomata and reduced the transpiration rate of the plants. Other studies done in grasslands in the mid-continental United States, as well as under irrigated conditions in an arid part of the United States (Arizona), have confirmed these results. However, field studies done in a desert region of the western United States and in a meadow in Germany have shown no differences in soil moisture under elevated and ambient concentrations of atmospheric CO_2. Soil moisture may not differ under ambient and elevated CO_2 in deserts, perhaps because of limited transpiration by desert plants under dry conditions and small water loss from the soil. In Germany, differences in soil water were probably not observed because water could move up into the root zone from the ground water table. These results suggest that deep-draining soils are needed to document differences in soil moisture under elevated and ambient CO_2. The papers reported in this chapter show no instances of soil moisture being less under elevated atmospheric CO_2 compared to ambient CO_2. Soil moisture is either increased or the same under elevated CO_2 compared to ambient CO_2. Studies are needed to compare the soil-water content of bare soil under ambient and elevated atmospheric CO_2 concentrations. Climate modelers, who predict that soil will become drier during the summer in the grasslands of the mid-continental area of North America in response to a doubling of atmospheric CO_2, must consider the effect of CO_2 on transpiration in their models. More data are needed to determine how elevated atmospheric CO_2 affects soil physical properties other than soil water (i.e., soil compaction and soil temperature).

6 Elevated Atmospheric Carbon Dioxide: Root Growth

INTRODUCTION

In Chapters 3 and 4, we looked at elevated CO_2 in the soil and its effects on roots. We now turn to the effects of elevated CO_2 in the atmosphere on roots. The major portal for entry of water into a plant is the root, and the major exit is the stoma. Consequently, we take a detailed look at roots in this chapter, and in Chapters 8 and 9 we focus on stomata.

FIELD STUDIES WITH SORGHUM AND WHEAT

We first consider the studies that we in the Evapotranspiration Laboratory did in the 1980s in Kansas, located at the semiarid central Great Plains of the United States. During the summer of 1984, Chaudhuri et al. (1986a) grew sorghum [*Sorghum bicolor* (L.) Moench; hybrid Asgro Correl] in 16 underground boxes containing a silt loam soil. Over the boxes were placed closed-top chambers where the CO_2 concentration was either ambient (330 μmol/mol CO_2) or elevated (485, 660, and 795 μmol/mol CO_2). We described the experimental setup also in Chapter 1. The boxes, which were 77 cm × 37 cm at the top and 180 cm deep, were kept well watered, and no moisture stress was imposed on the sorghum. The boxes (called rhizotrons) could be pulled out of the ground for observation of roots, and on one side of the underground box was glass on which a grid was etched. The number of intersections that the roots made with the grid was used to determine root length using the method of Newman (1965). Chaudhuri et al. (1986a) put the length of roots, determined by the Newman method, on a per area basis to get units of cm/cm^2.

We note in passing that even though the Newman (1965) method is a relatively simple one to determine the total length of roots in a sample, the method developed by Marsh (1971) is even simpler. A glass grid is etched with squares 1.27 cm (one-half inch) on a side to use the Marsh (1971) method. One counts the intersections that the roots make with the grid. Based on the mathematics developed by Newman (1965), the counts give estimations of root length directly in centimeters, when the grid is one-half inch on a side. The method also can be used in the laboratory. A sample of roots, after extraction, is teased out in a Petri plate with water. The grid is placed under the Petri plate, and the intersections that the roots make with the grid are counted. The number of intersections equals the root length in centimeters. The roots in the Petri plate are dried, and then their dry weight is determined. One then knows the relation between root weight and root length.

Sorghum grown at elevated concentrations of CO_2 yielded more roots than plants grown with 330 μmol/mol. Figure 6.1 shows the root length on four different days after emergence (23, 36, 51, and 72 days) at four depths. Although the values presented underestimated the total root length because minute root hairs were not visible through the thick glass (8 mm), they provided a good comparison of the root length with increased CO_2 levels. The measurements (Figure 6.1) indicated that enriched CO_2 treatment increased the root length compared to the ambient treatment on all days of measurement. At harvest (97 days after planting), root weight of sorghum plants grown with 795 μmol/mol of CO_2 was greater down to the 120 cm depth compared to that of roots subjected to lower CO_2 concentrations (Figure 6.2). The total dry weights at harvest of 10 sorghum plants (the number in each rhizotron box) grown at 330, 485, 660, and 795 μmol/mol CO_2 were 42, 45, 56, and 68 g, respectively (Burnett et al. 1985, p. 31).

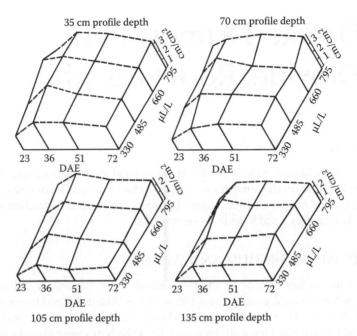

FIGURE 6.1 Sorghum root length (cm) per unit area (cm^2) as affected by increased CO_2 at different soil depths and at different growth stages. (Reprinted from *Agric. Forest Meteorol.*, 37, U.N. Chaudhuri, Burnett, R.B., Kirkham, M.B., and Kanemasu, E.T., Effect of carbon dioxide on sorghum yield, root growth, and water use, 109–122, Copyright 1986a, Fig. 7, with permission from Elsevier.)

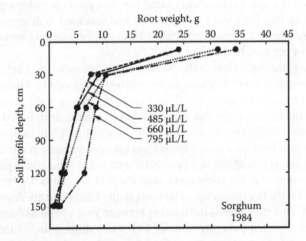

FIGURE 6.2 Sorghum root weight (g) as affected by increased CO_2 at different soil profile depths at maturity. (Reprinted from *Agric. Forest Meteorol.*, 37, U.N. Chaudhuri, Burnett, R.B., Kirkham, M.B., and Kanemasu, E.T., Effect of carbon dioxide on sorghum yield, root growth, and water use, 109–122, Copyright 1986a, Fig. 8, with permission from Elsevier.)

There was an increase in root length (Figure 6.1) and root weight (Figure 6.2) between the ambient level of CO_2 in 1984 (330 μmol/mol) and the first increment (485 μmol/mol) (a difference in CO_2 concentration of 155 μmol/mol). In 2008, the CO_2 concentration in the atmosphere was 385 μmol/mol (Dlugokencky 2009). The CO_2 concentration increased 55 μmol/mol (330–385 μmol/mol) in these 24 years (1984–2008), a rate of 2.3 μmol/mol/year. The 385 μmol/mol CO_2 in the atmosphere in 2008 is 100 μmol/mol CO_2 less than the first increment that Chaudhuri et al.

(1986a) used in 1984. So we already should be seeing an increase in sorghum root growth in the central Great Plains as a result of increased atmospheric CO_2 concentration. We can calculate this increase in dry weight. With an increase of 155 μmol/mol CO_2, root weight increased 3 g per 10 sorghum plants (42–45 g). We can put these values on an area basis. The area of each rhizotron box was 2849 cm² or 0.28 m². The 42 and 45 g weights per box for the roots at 330 and 485 μmol/mol CO_2 treatments, respectively, are equivalent to 150 and 161 g/m², an increase of 11 g/m² when the CO_2 concentration was increased 155 μmol/mol. Therefore, with an increase of 55 μmol/mol CO_2 in the atmosphere (increase between 1984 and 2008), root weight of sorghum has increased 4 g/m².

Elevated CO_2 also increased the shoot weight of sorghum (Chaudhuri et al. 1986a), and, as a result, root/shoot (R/S) ratios were similar under all four CO_2 levels and averaged 0.09.

Winter wheat (*Triticum aestivum* L. cv. Newton) was grown in the same underground boxes (where the sorghum had grown) for 3 years (1984–1985; 1985–1986; and 1986–1987) (Chaudhuri et al. 1990). The wheat was exposed to four CO_2 levels: 340 (ambient), 485, 660, and 825 μmol/mol. [The ambient CO_2 concentration at the beginning of the wheat experiment in 1984 was similar to that of the CO_2 concentration for sorghum, that is, 330 μmol/mol (Chaudhuri et al. 1986b). But as the experiment progressed, it increased. So Chaudhuri et al. (1990) plotted the 1984–1985 data with an average atmospheric CO_2 concentration for the ambient treatment of 340 μmol/mol.] The wheat plants grew under two moisture regimes. Half of the 16 boxes were maintained at field capacity (0.38 m³/m³). The plants grown in this treatment were called the well-watered or not-stressed plants. And half of the boxes were maintained between 0.14 and 0.25 m³/m³ (the drought-stressed plants). Roots of the plants grown under elevated levels of CO_2 and well-watered conditions penetrated to the maximum depth of observation (176 cm) before roots of plants grown under ambient levels of CO_2 (Figure 6.3; data for fall of 1986, which are representative of all 3 years). The distribution of roots with depth obtained at harvest in 1987 under well-watered and drought-stressed conditions is presented in Figure 6.4. The data again are representative of those obtained during the

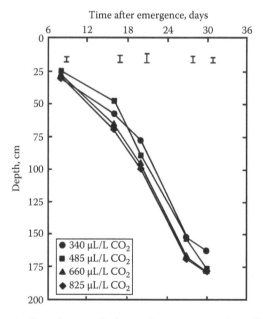

FIGURE 6.3 Depth of penetration of roots of winter wheat, grown under well-watered conditions in the fall of 1986, as affected by four CO_2 concentrations in the atmosphere. Bars indicate the maximum standard deviation that occurred among the four treatments on each measurement day. (Reprinted from Chaudhuri, U.N. et al., *Crop Sci.*, 30, 853, 1990, Fig. 1. With permission from the Crop Science Society of America.)

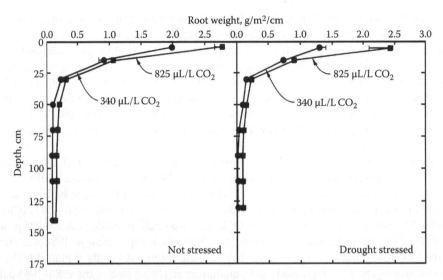

FIGURE 6.4 Root dry weight at different depths upon harvest in 1986 for well-watered or drought-stressed winter wheat grown under ambient or elevated CO_2. Horizontal bars = standard deviation. Only half of each bar is drawn, for clarity. Most standard-deviation bars fell within the symbol and are not shown. (Reprinted from Chaudhuri, U.N. et al., *Crop Sci.*, 30, 853, 1990, Fig. 2. With permission from the Crop Science Society of America.)

3 year study. Data for only two concentrations of CO_2 (ambient and the maximum concentration of 825 μmol/mol) are shown to improve the readability of the figure. Data for the 485 and 660 μmol/mol CO_2 concentrations fell between the two curves for each half of the figure. Under the well-watered (Figure 6.4, left) and drought-stressed (Figure 6.4, right) conditions, weight of roots of plants grown under 825 μmol/mol CO_2 was greater at all depths, compared with roots of plants grown under ambient CO_2. The difference in root growth under elevated and ambient levels of CO_2 was most pronounced at the top level (0–10 cm). Root weight remained relatively constant below about 50 cm for all treatments. This is in contrast to sorghum, where the highest concentration of CO_2 (795 μmol/mol) increased root growth down to the 120 cm depth (Figure 6.2). Root weight of well-watered wheat plants grown with ambient CO_2 generally was similar to that of drought-stressed plants grown with 825 μmol/mol CO_2. These results suggested that in the 0 to 10 cm zone, elevated levels of atmospheric CO_2 can compensate for restrictions in wheat root growth by drought (Chaudhuri et al. 1990).

Data for wheat root weight at harvest for all 3 years of the study are presented in Figure 6.5. Root dry weight increased as the CO_2 concentration increased. Dry weights of well-watered roots grown at ambient CO_2 (3 year mean: 118 g/m²) were similar to those of drought-stressed roots grown at the highest level of CO_2 (3 year mean: 123 g/m²). These results again suggest that high CO_2 (825 μmol/mol) can compensate for drought-stressed conditions. As for sorghum, the R/S ratio for wheat generally was similar among the different CO_2 treatments. But the ratio was higher under drought (averaging 0.13) than under the well-watered conditions (averaging 0.11), which indicated that shoot growth was reduced more by drought than root growth (Chaudhuri et al. 1990).

As for sorghum, we can calculate how much wheat root weight has increased in the past 24 years due to an increase in CO_2 concentration of the atmosphere of 55 μmol/mol (between 1984 and 2008). For well-watered conditions, the average root weight at harvest under 340 μmol/mol CO_2 (control) was 118 g/m² (as noted earlier), and the average root weight under the 485 μmol/mol treatment was 139 g/m². An increment of 155 μmol/mol CO_2 (remembering that CO_2 in the atmosphere was 330 μmol/mol when we started our wheat experiments) resulted in an increase

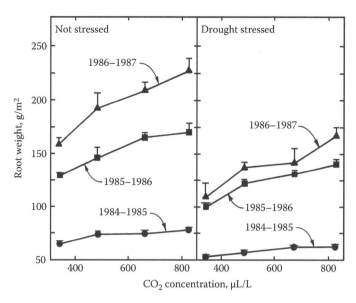

FIGURE 6.5 Root weight at harvest of well-watered or drought-stressed winter wheat as affected by CO_2 concentration during a 3 year study. Vertical bars = standard deviation. Only half of each bar is drawn, for clarity. (Reprinted from Chaudhuri, U.N. et al., *Crop Sci.*, 30, 853, 1990, Fig. 3. With permission from the Crop Science Society of America.)

of 21 g/m². Thus, an increase of 55 µmol/mol CO_2 in the atmosphere between 1984 and 2008 has resulted in an increase in wheat root weight of 7 g/m².

A free-air CO_2 enrichment (FACE) experiment done in Maricopa, Arizona, United States, with spring wheat (*T. aestivum* L. cv. Yecora Rojo) by Wechsung et al. (1995) confirmed the findings of Chaudhuri et al. (1990) for winter wheat. Before we consider the results of Wechsung et al. (1995), let us briefly describe FACE experiments. In these experiments, plants are exposed to elevated levels of CO_2 under natural conditions. No chambers are used. They were developed in the late 1980s (Kimball et al. 1999). Kimball (2006) reviewed research done with cotton (*Gossypium hirstum* L.), wheat, and sorghum at the FACE facility in Maricopa, Arizona. Lewin et al. (1994) described the FACE system that was used by Wechsung et al. (1995). A ring of spigots that released CO_2 into the air surrounded the plot where CO_2 was elevated. The diameter of the ring (diameter of the plot) at the Maricopa, Arizona, site was 22 m (Hendry and Kimball 1994). Such systems require large amounts of CO_2, because it is released freely into the air. A picture of a FACE experiment, carried out in Champaign, Illinois, United States, is shown on the cover of the November 2006, issue of *Plant, Cell and Environment* (Bernacchi et al. 2006). Nowak et al. (2004b) summarize results from 16 FACE sites carried out in the United States, Europe, Australia, and New Zealand. In Chapter 15, we compare growth of plants grown under elevated CO_2 in FACE experiments with those grown in chambers.

We now return to the experiment of Wechsung et al. (1995). The spring wheat grew on a clay loam soil under irrigated conditions (no water stress) with 550 µmol/mol CO_2 (the FACE plants) or with 370 µmol/mol CO_2 (the control plants). Root dry weight was higher in the FACE plots compared to the control plots for all growth stages (i.e., 34% greater at the three-leaf stage; 21% greater at tillering; 23% greater at stem elongation; 28% greater at dough development; and 19% greater at harvest). Growth rate of roots exposed to the elevated CO_2 was higher compared to that of the control roots. Also, roots grown with the elevated CO_2 explored a greater proportion of the soil profile earlier in the season than did the control roots. However, plants grown in the FACE plots did not have a deeper root system than control plants.

CONTROLLED-ENVIRONMENT STUDIES WITH WHEAT

Fitter et al. (1996) and van Vuuren et al. (1997) studied the effect of double the ambient level of CO_2 (700 µmol/mol) on spring wheat (*T. aestivum* L. cv. Tonic) grown in containers 120 cm deep and placed in plant growth chambers. They found that root biomass increased with CO_2 concentration. Elevated CO_2 changed the vertical distribution of the roots, with a greater stimulation of root growth in the top layers of the soil. In another experiment with spring wheat (cultivar Florence Aurore) grown under laboratory conditions for 28 days, CO_2 enrichment (700 µmol/mol) increased root growth by 23% compared to root growth under ambient CO_2 (350 µmol/mol CO_2) (Billes et al. 1993).

Derner et al. (2004) measured intergenerational above- and below-ground responses to elevated CO_2 (700 µmol/mol) of two genotypes of semi-dwarf, hard red, spring wheat (*T. aestivum* L.) grown under greenhouse conditions in a sandy loam soil. Control plants grew at 360 µmol/mol CO_2. The two genotypes were a rust resistant wheat (AZ-MSFRS-82RR) and a hybrid (DK49, Douglas King). These plants were progeny of seeds produced from a previous generation of plants grown at elevated CO_2 (700 µmol/mol). Because neither genotype in the first generation exhibited increased shoot growth with CO_2 enrichment (root growth was not measured in the first generation), the objective of Derner et al. (2004) was to assess if exposure to elevated CO_2 in subsequent generations resulted in increased above- and below-ground plant growth. They found that root and shoot growth of the two genotypes increased under elevated CO_2 to a similar extent in both the second and third generations. In the second generation, CO_2 enrichment increased root weight at final harvest by 88% (from 0.441 g/pot for the control treatment to 0.829 g/pot for the plants grown with 700 µmol/mol CO_2), and root length increased from 2088 cm/pot for the control treatment to 3005 cm/pot for the elevated CO_2 treatment. In the third generation, elevated CO_2 increased root weight at harvest by 28% (from 0.120 g/pot for the controls to 0.154 g/pot for the elevated CO_2 treatment), and root length increased from 1228 cm/pot for the control plants to 1509 cm/pot for the plants grown under the elevated CO_2. Derner et al. (2004) suggested that the CO_2 enrichment affected seed composition (a higher carbon:nitrogen ratio), so that seeds produced under high CO_2 subsequently produced seedlings with greater growth. These plants continued to respond to elevated CO_2 throughout the growth period, so, by final harvest, they had the increased growth observed during both generations.

SOYBEANS, TEPARY BEAN, BUSH BEAN, BARLEY, AND COTTON

Del Castillo et al. (1989) studied roots of soybean [*Glycine max* (L.) Merr. cv. Forrest] grown in outdoor, sunlit plant-growth chambers at CO_2 concentrations of 330, 450, 600, and 800 µmol/mol. The soil bins were filled with a sand and vermiculite mixture. They measured roots down to the 80 cm depth. They found that root weight was 26%–31% higher in the CO_2-enriched chambers compared with the 330 µmol/mol treatment, and that cumulative root length was approximately proportional to the CO_2 concentration. Rogers et al. (1992) grew soybean plants (cultivar Lee) in fertilized perlite under controlled environmental conditions at 350 or 700 µmol/mol CO_2. They measured roots down to the 50 cm depth. They found an increase in root dry weight and root length for high CO_2-grown plants (see Figure 6.6a, for root length). However, unlike length, numbers of roots at specific depths did not increase due to CO_2 enrichment (Figure 6.6b). The results of Rogers et al. (1992) suggested that elevated CO_2 does not affect meristematic activity of the roots (i.e., new roots are not produced), but it does increase root length of existing roots.

Salsman et al. (1999) grew in nutrient solution two species of bean: *Phaseolus acutifolius* Gray (common name: tepary bean) and *P. vulgaris* L. (common names: kidney bean, string bean, or bush bean) placed under three atmospheric CO_2 concentrations of 385 µmol/mol (ambient), 550 µmol/mol, or 700 µmol/mol in a greenhouse. Elevated CO_2 increased root and shoot growth of *P. actuifolius*, but not *P. vulgaris*. Root mass was increased by nearly 60% in *P. acutifolius*. They noted that same genus had responded differently to CO_2 in other studies (Bazzaz 1990), and the reasons for the

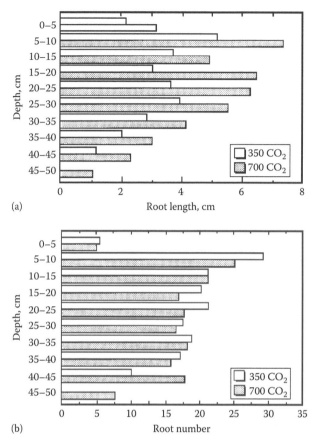

FIGURE 6.6 (a) Length of first-order lateral roots with depth ($n = 12$). (b) Number of roots for the same 12 soybean plants with depth. (Rogers, H.H., Binghan, G.E., Cure, J.D., Smith, J.M., and Surano, K.A.: Response of plant roots to elevated atmospheric carbon dioxide. *Plant Cell Environ.* 1992. 15. 749–752. Fig. 2. Copyright Wiley-VCH Verlag GmbH & Co. KGaA. Reproduced with permission.)

difference are not well understood. Some suggest that it may be due to nutrient status. For example, Sicher (2005) grew barley (*Hordeum vulgare* L. cv. Brant) seedlings in vermiculite flushed with a complete mineral nutrition solution containing 1.0 mM inorganic phosphate (sufficient phosphorus) or 0.05 mM inorganic phosphate (insufficient phosphorus). Plants grew in controlled environment chambers providing 36 Pa (ambient) or 100 Pa (elevated) CO_2. Elevated CO_2 increased root growth of barley, but the increase was dependent upon the level of phosphorus in the nutrient solution. At harvest (17 days after planting), root dry weight was increased by 12% and 48% by CO_2 enrichment in the phosphorus-sufficient and phosphorus-insufficient treatments, respectively.

Studies have been conducted on below-ground responses of cotton (*G. hirsutum* L. cv. Deltapine 77) plants under FACE (Prior et al. 1994a,b; Prior et al. 1995; Rogers et al. 1997). Plants grew with 360 μmol/mol CO_2 (ambient level) or with 550 μmol/mol CO_2. Dry weights were higher for CO_2-enriched plants than plants grown with ambient CO_2. The distribution of root density per unit volume of soil was investigated both vertically and horizontally. The density of roots increased under CO_2 enrichment at most depths to 90 cm, but the increase was more significant in the upper 45 cm of the soil profile (Figure 6.7). The root length densities also tended to exhibit greater differences between ambient and elevated CO_2 treatments as horizontal distance from the row center increased, which indicated faster and more prolific spread of cotton roots through the soil profile under elevated CO_2. In a later study, Prior et al. (2003) grew cotton in a sandy soil in containers placed in

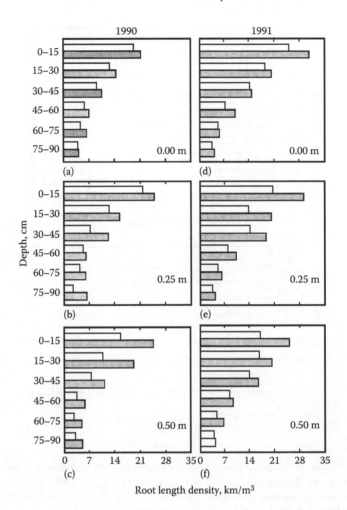

FIGURE 6.7 The effect of CO_2 concentration (open bars = 360 μmol/mol; dark bars = 550 μmol/mol) on cotton root length density at various depth increments and positions (0, 0.25, and 0.50 m) away from the crop row during vegetative growth in 1990 (a, b, and c) and 1991 (d, e, and f). (Reprinted from Rogers, H.H., Runion, G.B., Krupa, S.V., and Prior, S.A. Plant responses to atmospheric carbon dioxide enrichment: Implications in root-soil-microbe interactions, in Allen Jr., L.H., Kirkham, M.B., Olszyk, D.M., and Whitman, C., Eds., *Advances in Carbon Dioxide Effects Research*, American Society of Agronomy, Crop Science Society of America, and Soil Science Society of America, Madison, WI, pp. 1–34, 1997, Fig. 1-3. With permission from the American Society of Agronomy, Crop Science Society of America, and the Soil Science Society of America, Madison, WI.)

open-top field chambers under ambient (360 μmol/mol) and twice ambient (720 μmol/mol) CO_2. As in the earlier study, they found that the elevated level of CO_2 increased root growth.

HORTICULTURAL CROPS

Jasoni et al. (2004) grew onion (*Allium cepa* L. cv. Purplette) in glass tubes for 30 days under two CO_2 treatments: 400 and 1000 μmol/mol. At harvest, root weights were 0.19 and 0.21 g/tube for the 400 and 1000 μmol/mol CO_2 concentrations, respectively, and the difference was not significant. However, the total biomass (shoot plus root growth) was increased under the elevated CO_2 (0.99 and 1.62 g/tube for the 400 and 1000 μmol/mol CO_2 treatments, respectively). Apparently, the 30 day growth period was not long enough for an increase in root growth to become evident under the elevated CO_2 treatment.

Behboudian and Lai (1995) studied the effect of CO_2 enrichment on the distribution of assimilates in tomato plants (*Lycopersicon esculentum* Mill. cv. Virosa) using a ^{14}C-label. They grew the plants in growth chambers maintained at 340 or 1000 µmol/mol CO_2. Forty-eight days after the CO_2 treatments began, they fed ^{14}C to the plants. The roots were a weak sink for ^{14}C-assimilate under both CO_2 treatments, but more ^{14}C was detected in the roots of enriched plants. They pointed out that a complex bi-collateral arrangement of phloem exists in the tomato plant whereby roots are supplied by leaves in the upper part of the plant via an external phloem. A bi-collateral vascular bundle is a bundle having phloem on two sides of the xylem (Esau 1977, p. 503). More studies are needed to understand how carbon moves to tomato roots in relation to its bi-collateral vascular bundles.

Growth of underground storage organs of plants is especially increased under elevated CO_2. Studies by Cummings and Jones (1918) showed that radish (*Raphanus sativus* L.) roots grew faster when the plants were exposed to high CO_2. They were edible sooner and had more carbohydrates and less protein. For potatoes (*Solanum tuberosum* L.), tuber formation did not take place sooner, but, once it started, it was faster and better tubers were produced. Other studies show that under elevated CO_2 numbers, dry weights and diameters of tubers increase (Rogers et al. 1997, p. 13). Parenchyma cell division and expansion and carbohydrate concentrations in tubers also increase (Rogers et al. 1997, p. 13). Under elevated CO_2, the tropical root crop cassava (*Manihot esculenta* Crantz) had a large increase in growth (150%), with a greater partitioning to the roots (Rogers et al. 1997, p. 13). A mean yield ratio of 1.40 for mature and 1.77 for immature root crops under CO_2 enrichment compared with the controls has been reported (Rogers et al. 1997, p. 8). Beneficial effects of increased atmospheric CO_2 may be more significant for the yield of root crops such as carrots (*Daucus carota* L.) and beets (*Beta vulgaris* L.) than for crops grown for their fruit, leaves, or stems (Rogers et al. 1997, p. 6).

Horticulturists have used CO_2 enrichment to increase root formation on plant cuttings (Grant et al. 1992). The improved rooting probably arises as a result of enhanced water relations rather than increased carbohydrate levels (Rogers et al. 1997, p. 7). Increasing the concentration of CO_2 in greenhouse mist systems has increased the percentage of plant cuttings that form roots in many ornamental and floricultural species (Rogers et al. 1997, p. 7). Elevated CO_2 during propagation increased root number in sweet potato (*Ipomoea batatas* Lam.) (Bhattacharya et al. 1985) and peperomia (*Peperomia glabella* A. Dietr.) (Davis and Potter 1983). For leafy pea (*Pisum sativum* L.) cuttings, elevated CO_2 (1800 µmol/mol) increased carbohydrates, water potential, root length, and root dry weight, but it did not increase root number (Davis and Potter 1989). Laforge et al. (1991) rooted raspberry (*Rubus idaeus* L.) in a rooting medium in test tubes under three CO_2 levels in growth chambers. The CO_2 concentrations were 330, 1650, and 3000 µmol/mol CO_2. After 5 weeks, root dry weights were 0.11, 0.30, and 0.38 g/test tube for the three CO_2 concentrations, respectively, and the increase in root weight due to CO_2 treatment was significant. They performed the same experiment on asparagus (*Asparagus officinalis* L.), and fresh root weights for the three treatments were 0.61, 0.56, and 0.75 g/test tube, respectively. The fresh root weight of plants grown at 3000 µmol/mol was significantly more than the fresh weights of plants grown at 330 and 1650 µmol/mol.

PASTURE PLANTS

Newton et al. (1996b) studied plants removed intact from pastures in New Zealand and then grown in growth rooms at 350 µmol/mol CO_2 (ambient; controls) or 700 µmol/mol CO_2. The samples ($1 \times 0.5 \times 0.25$ m) were lifted from the pasture, where the soil was a sand, using a blade attached to a tractor. The major plant species in the pasture were *Lolium perenne* L. (perennial ryegrass), *Trifolium repens* L. (white clover), *Poa pratensis* L. (Kentucky bluegrass), *Cynosurus cristatus* L. (crested dogs-tail), *Agrostis capillaris* L. (cloud-grass), *Anthoxanthum odoratum* L. (sweet vernal-grass), *Paspalum dilatatum* Poir (Dallis grass and the only C_4 species), and *Plantago lanceolata* L. (ribgrass). They found that net CO_2 exchange continued at elevated CO_2

even when the volumetric soil moisture content was less than $0.10\,\text{m}^3/\text{m}^3$. At this moisture content, gas exchange at ambient CO_2 was zero. The additional carbon fixed by plants grown in the elevated CO_2 plots was primarily allocated below ground, as shown by the maintenance of root length density at the same level as in well-watered plots. However, upon re-watering, there was compensatory growth at ambient CO_2, so that the above-ground growth rate exceeded that of plants in plots that had not experienced a moisture deficit.

Before the soil dried out, the plants that were growing at $700\,\mu\text{mol/mol}$ CO_2 had a greater proportion of legumes (mainly white clover) in the harvested herbage than the controls (Newton et al. 1996b). The legume content under elevated CO_2 tended to be reduced under the soil moisture deficit. Even though many studies have demonstrated that under elevated CO_2, legumes have an advantage over non-legumes, as reviewed by Edwards and Newton (2007), the study by Newton et al. (1996b) showed that under water-limited conditions, legumes grow less than grasses, when the CO_2 content is doubled.

Growth and decomposition of roots of pasture plants under ambient and elevated atmospheric CO_2 have been the focus of several studies. Van Ginkel et al. (1996) grew perennial ryegrass (*L. perenne* L. cv. Barlet) for 71 days in two growth chambers, which had continuous $^{14}CO_2$ labeling of the atmospheres at 350 and $700\,\mu\text{mol/mol}$ CO_2. The plants grew in pots containing a loamy sand. After the labeling experiment, the grass roots were dried and ground for use in a subsequent decomposition study, which was done in four wind tunnels. In two wind tunnels, a constant flow of ambient air (mean concentration of $361\,\mu\text{mol/mol}$ CO_2) was maintained, whereas in the other two wind tunnels the ambient air was enriched with $350\,\mu\text{mol/mol}$ CO_2 (mean concentration of $705\,\mu\text{mol/mol}$). The soil was mixed with the ^{14}C-labeled, dried-and-ground grass roots, placed in pots, and perennial ryegrass again was planted, which grew for two seasons. After the second growing season, elevated CO_2 had stimulated shoot and root growth by 13% and 92%, respectively. After the first growing season, the decomposition of roots grown with $700\,\mu\text{mol/mol}$ CO_2 was 19% lower than that of roots grown with $350\,\mu\text{mol/mol}$ CO_2. They concluded that a combination of higher root yields at elevated CO_2 along with a decrease in root decomposition will lead to a longer residence time of carbon in the soil under elevated CO_2.

Using a FACE system near Zürich, Switzerland, Jongen et al. (1995) also found that decomposition decreased under elevated CO_2. They grew *L. perenne* L. (perennial ryegrass) and *T. repens* L. (white clover) at elevated ($600\,\mu\text{mol/mol}$) and ambient ($340\,\mu\text{mol/mol}$) CO_2 concentrations during one growing season. Under elevated CO_2, root growth of both plants increased and root decomposition decreased. A possible explanation for this decrease was the change in tissue composition, such as the increase in the carbon:nitrogen ratio in roots of *L. perenne* at elevated CO_2 concentrations.

Deforestation and logging transform more forests in eastern and southern Amazonia than in any other region of the world (Nepstad et al. 1994). Pastures are the most common type of vegetation grown on the deforested land in Amazonia. They vary in the proportion of grass and woody plants. In managed pastures, woody vegetation is removed by heavy machinery. In the more common degraded pastures, shrubs and trees dominate. Nepstad et al. (1994) studied deep roots in a mature, evergreen forest and in adjacent pastures near the town of Paragominas in the Brazilian state of Pará. The deeply weathered clay soils at the site are common in Amazonia. A degraded pasture had the grasses *Panicum maximum* Jacq. (Guinea grass) and *Brachiaria humidicola* (common name unknown). A managed pasture was planted with *B. brizantha* (common name unknown). Roots were most abundant near the soil surface but were found in deep shafts to depths of 8 m in the managed pasture, 12 m in the degraded pasture, and 18 m in the forest. Fine-root (less than 1 mm in diameter) biomass was lower in the top 10 cm of the degraded pasture soil than in the forest, but no differences occurred at greater depths. In the managed pasture, fine-root biomass was three times greater than in the forest near the soil surface and 15–100 times lower at depth. While Nepstad et al. (1994) did not study pastures under elevated CO_2, their results indicated that as more land in the Amazon is deforested, there will be runoff, because the pastures with shallower root systems cannot withstand the seasonal droughts. Nevertheless, some introduced pastures in South America, such as those on the savannas, will have beneficial effects, because they offset increased anthropogenic CO_2 emissions after burning of tropical rainforests (Fisher et al. 1994).

NATIVE GRASSES

Scientists in Colorado, United States (Pendall et al. 2004a,b; Milchunas et al. 2005a,b; LeCain et al. 2006) studied the effects of elevated CO_2 on root growth in a semiarid short-grass steppe containing C_3 and C_4 grasses. They used six open-top chambers, three with ambient CO_2 (360 µmol/mol) and three with elevated CO_2 (720 µmol/mol CO_2). Three non-chambered plots allowed evaluation of chamber effects. The study was carried out for 5 years (1997–2001). Chambers were placed on the plots before growth started in late March or early April, and they were removed at the end of the growing season in late October. The soil at the site was a fine sandy loam to greater than 1 m depth. Two of the 5 years were wet (first season and third season); two were dry (fourth and fifth seasons); and one had normal precipitation (second season) (Milchunas et al. 2005a). The most abundant species at the study site were the C_4 grass, *Bouteloua gracilis* (H.B.K.) Lag. (blue grama), and the C_3 grasses *Stipa comata* Trin. and Rupr. (needle-and-thread grass) and *Pascopyrum smithii* (Rydb.) A. Love (western wheatgrass) (LeCain et al. 2006).

Estimates of root growth differed among methods of calculation (Milchunas et al. 2005a). Data from large soil cores (20 cm diameter × 60 cm depth) showed that root biomass was increased by elevated CO_2 in 2 of the 5 years (LeCain et al. 2006). The 2 years were the first season (1997–1998, a wet season) and the last season (2000–2001, a dry season) (Pendall et al. 2004a). A root in-growth method (where roots grow into bags that can be pulled out of the soil for root measurements) showed increases in root production of over 60% with elevated CO_2 during years of intermediate levels of precipitation, with smaller effects in wettest year (third season) and no effects in the driest year (fourth season) (Milchunas et al. 2005b). Over the 5 year period, minirhizotron data showed a root-length increase of 52% and a root-length loss of 37%. The overall increase of the total size of the roots was 41% (Milchunas et al. 2005b). The root turnover times were 5.8, 7.0, and 5.3 years for the control (no chamber), ambient CO_2, and elevated CO_2 treatments, respectively (Milchunas et al. 2005a).

The average increase in above-ground productivity in the Colorado studies was 41%, which was primarily attributed to one species, *S. comata* (LeCain et al. 2006). Over the 5 year period, root biomass of this C_3 grass was nearly twice the amount in elevated compared to ambient CO_2 plots in the 0 to 20 cm layer. But elevated CO_2 did not affect the root dry weight of the C_4 grass, *B. gracilis* (LeCain et al. 2006). The increased growth of *S. comata* could alter the ecosystem. The current dominant, high forage quality species, *B. gracilis*, might be replaced by the less desirable *S. comata*. The diameter of roots was larger under elevated than ambient CO_2, but only in the upper soil profile (0–20 cm depth) (Milchunas et al. 2005a). The minirhizotron data, which went to a depth of 59 cm, showed that elevated CO_2 increased branching (Milchunas et al. 2005a). This suggested that the extra CO_2 was affecting the pericycle of the root, where branch roots are initiated. It is not known if the elevated CO_2 from the atmosphere diffused through the soil to the root and through the root to the pericycle, where it affected branching directly, or if the increased branching was due to more water in the soil under elevated CO_2. Berntson and Woodward (1992) found that elevated CO_2 resulted in more branched and longer root systems of *Senecio vulgaris* L. (common groundsel or ragwort). Under low water supply and elevated CO_2, the branching patterns and root lengths of the ragwort plants were similar to those grown under ambient CO_2 with a high water supply. The effect of CO_2 concentration on the pericycle of roots under wet and dry conditions needs to be studied.

Nylon in-growth bags were used to study changes in root biomass during 2 years (1990 and 1991) in a tallgrass prairie in Kansas (Owensby et al. 1993). The dominant plants on the prairie were *Andropogon geradii* Vitman (big bluestem), *A. scoparius* Michx. (little bluestem), and *Sorghastrum nutans* (L.) Nash (Indian grass) (all C_4), and *P. pratensis* L. (Kentucky bluegrass) (C_3).

Open-top chambers were used to supply ambient CO_2 (350 µmol/mol) and doubled CO_2 (700 µmol/mol). In 1990, the root biomass in the in-growth bags was 120 and 270 g/m² for the ambient and doubled CO_2 treatments, respectively. In 1991, the values were 175 and 210 g/m², for the ambient and doubled CO_2 concentrations, respectively. Even though the root growth was increased under elevated atmospheric CO_2, the difference in growth between the C_3 and C_4 roots could not be identified. Research needs to be done on C_3–C_4 communities under elevated CO_2 that will allow distinction of C_3 and C_4 roots.

A question that arises in CO_2 studies is the following. Under elevated atmospheric CO_2, will terrestrial ecosystems accumulate more carbon or will they just process it faster (Norby 1997)? To answer this question, Hungate et al. (1997) exposed two grasslands in central, coastal California, United States, to ambient (360 µmol/mol) and elevated (720 µmol/mol) CO_2 for 3 years (1992–1994). They were on serpentine- or sandstone-derived soils at the Jasper Ridge Biological Preserve of Stanford University, California. In these grasslands, the roots are not long-lived, as in woody, perennial systems (Norby 1997). They determined rates of carbon input to soil by using the ${}^{13}C$-depleted isotopic signature of the CO_2 added to the enriched plots. (The next section discusses carbon isotopes.) They quantified short-term carbon partitioning in below-ground respiration in a separate experiment using a ${}^{13}CO_2$ pulse label in these grassland communities grown in out-door microcosms. By monitoring the rate and amount of $\delta^{13}C$ of CO_2 released from the soil, they could distinguish CO_2 produced from root sources (sometimes called "autotrophic respiration") and heterotrophic or non-root sources (mainly respiration from microorganisms) (Prévost-Bouré et al. 2009). Elevated CO_2 increased below-ground respiration in both the sandstone and serpentine grasslands. The results showed that carbon allocated below ground can contribute to increased root growth of short-lived roots that quickly decompose. This may be one reason why some studies do not always show an increase in root growth, if only biomass is measured at the end of the experiment. The following answers the question at the beginning of the paragraph: At least for the California grasslands, elevated atmospheric CO_2 is processed fast, and they do not accumulate more carbon under elevated CO_2. The results cast doubt on the capacity of the California grasslands to sequester extra carbon (Norby 1997). In a later, 3 year (1998–2001) study at the Jasper Ridge site (Morgan 2002; Shaw et al. 2002) elevated CO_2 (680 µmol/mol) did not increase root biomass.

Lagomarsino et al. (2009) studied CO_2 emissions at a FACE site in Central Italy, where the soil was a Pachic Xerumbrept. [Xerumbrepts are soils of Mediterranean climates, and in the United States, they are close to the Pacific Ocean (Soil Survey Staff 1975, pp. 268–269).] The plants at the site were poplars (*Populus* × *euramericana*). Ambient and elevated CO_2 plots were maintained at about 370 and 550 µmol/mol CO_2, respectively. They found that only a fraction of the extra carbon fixed by photosynthesis in the elevated CO_2 plots contributed to enhanced carbon storage into soil, because of the contemporary stimulation of soil heterotrophic respiration.

CARBON ISOTOPE RATIOS OF PLANTS WITH DIFFERENT PHOTOSYNTHETIC PATHWAYS

Let us now turn to carbon isotopes, which Hungate et al. (1997) and others (Prévost-Bouré et al. 2009) have used in root studies to distinguish CO_2 produced from different sources. The numerical value of carbon isotope ratios is conventionally reported as $\delta^{13}C$. The number represents the difference between the ratio of ${}^{13}C$–${}^{12}C$ found in a given sample and the ratio that exists in a universal standard, expressed as a per mill (‰, or number of one-thousandths) deviation from the standard. The letters PDB refer to the standard, which is based on a fossil mollusk shell from the Pee Dee Formation, a Cretaceous deposit in South Carolina (Knoll 1991). The Pee Dee is a river flowering through North and South Carolina (Friend and Guralnik 1959). The "B" in PDB stands for belemnite, which is the fossil mollusk shell, and it is an extinct kind of cuttlefish (Friend and

Guralnik 1959). The Cretaceous period occurred 130–60 million years ago. (See Figure 9.10 for the geological time scale.) The ratio is as follows:

$$\delta^{13}C = \left\{ \frac{[^{13}C/^{12}C_{sample} - ^{13}C/^{12}C_{standard}]}{[^{13}C/^{12}C_{standard}]} \right\} \times 10^3 \; (\permil, \; PBD)$$

We now need to review the different photosynthetic systems before we proceed with discussing carbon isotope ratios, because the ratios vary according to the photosynthetic system. As noted in Chapter 1, plants are classified into three groups based on their photosynthetic systems: C_3, C_4, or Crassulacean acid metabolism (CAM) plants. The most common photosynthetic cycle is the C_3 or Calvin–Benson pathway. In C_3 plants, CO_2 from the atmosphere is fixed as phosphoglyceric acid, a 3-carbon compound. Ribulose-1,5-bisphosphate carboxylase is the enzyme that catalyzes the combination of CO_2 with ribulose-1,5-diphosphate to form the 3-phosphoglyceric acid (Conn and Stumpf 1963, p. 296). This enzyme is often abbreviated RuP_2 carboxylase and is called Rubisco. In C_4 plants, an additional mechanism is involved, in which atmospheric CO_2 is fixed as oxaloacetic acid, a 4-carbon compound. The reaction, by which CO_2 (actually HCO_3^-) is converted into one of the four carbons of malic and aspartic acids, is through its initial combination with phospho-*enol*pyruvic acid (PEP) to form oxaloacetic acid and $H_2PO_4^-$. Phospho*enol*pyruvate carboxylase, an enzyme widely distributed in plants, is the catalyst involved. This enzyme is often called PEP carboxylase (Salisbury and Ross 1978, p. 140). (We discuss C_3 and C_4 photosynthesis further in Chapters 12 and 13.)

The leaves of C_3 and C_4 plants differ in morphology as well as in the chemical mechanisms of CO_2 fixation. In C_3 plants, chloroplasts are found in the palisade mesophyll and spongy mesophyll cells of leaves. Bundle sheath cells that surround the vascular bundles in C_3 plants are lacking or essentially lacking in chloroplasts (Kirkham 2005, pp. 361–362). In leaves of C_4 plants, however, the chloroplasts that fix CO_2 by the C_3 mechanisms are highly concentrated in the bundle sheath cells. In a C_4 leaf, the chloroplasts that fix CO_2 by the C_4 mechanism are located in the relatively large mesophyll cells that make up the body of the leaf. The 4-carbon malic and aspartic acids (formed from the initial 4-carbon product, oxaloacetic acid) produced by the chloroplasts in these mesophyll cells are transported to the C_3 chloroplasts in the bundle sheaths. The unique leaf structure that is observed in C_4 plants is called "Kranz anatomy" after the term coined by the German botanist Haberlandt (1904), who conducted original anatomical work in the late nineteenth century and found this structure in sugarcane (*Saccharum officinarum* L.) leaves (El-Sharkawy 2009b). Kranz means "garland" or "wreath" in German. The bundle sheath cells form a wreath around the vascular tissue in C_4 plants.

Many succulent tropical species have CAM (Salisbury and Ross 1978, p. 145). Such species open their stomata primarily at night and fix CO_2 into organic acids, especially malic acid. Species with CAM usually lack a well-developed palisade layer of cells in their leaves, and most of the leaf or stem cells are spongy mesophyll. Bundle sheath cells are present but, in contrast to those of C_4 plants, are similar to mesophyll cells. PEP carboxylase is responsible for most of the CO_2 fixation at night, and ribulose bisphosphate carboxylase becomes active during daylight. The role of ribulose bisphosphate carboxylase seems to be identical to its function in bundle sheath cells of C_4 species, that is, refixation of CO_2 lost from organic acids such as malic acid (Salisbury and Ross 1978, p. 145).

C_3–C_4 intermediates also have been studied. Since the elucidation of the C_4 photosynthetic syndrome (Hatch and Slack 1966), researchers have tried to investigate the feasibility and value of interspecific crosses between related C_3 and C_4 species. The objectives were the desire to gain insights concerning the evolutionary development of the C_4 pathway and to enhance photosynthesis and productivity of C_3 crops. Notable in these efforts was the pioneering work done at the Carnegie Institution of Washington at Stanford University in Palo Alto, California, where hybrids were successfully made between C_3 and C_4 plants of saltbush, *Atriplex* sp. (Björkman et al. 1971). But the whole spectrum of components of the C_4 pathway was not detectable in the resulting C_3–C_4 hybrids. Therefore, little or no advantage was found in the breeding materials, in terms of photosynthetic efficiency and productivity. This indicated the genetic complexity of the C_4 systems (M. El-Sharkawy, personal communication, July 12, 2007). Naturally occurring, uncultivated C_3–C_4 intermediate species have been identified belonging to several genera including *Panicum, Moricandia, Flaveria, Mollugo, Neurachne, Alternanthera, Parthenium, Brassica, Cyperus,* and *Diplotaxis* (M. El-Sharkawy, personal communication, July 12, 2007). Although studies on these uncultivated species helped to enhance scientific knowledge about the evolutionary steps that took place in the development of the C_4 syndrome, they were of little or no value from the agronomic point of view. In contrast, research on cassava (*Manihot* sp.) at the international agriculture research center, CIAT (Centro Internacional de Agricultura Tropical), did provide interesting information on C_3–C_4 photosynthetic characteristics and on their implication for crop productivity. Cassava was the first cultivated crop to be considered as a C_3–C_4 intermediate species (El-Sharkawy 2006, 2007, 2009a,b; El-Sharkawy and De Tafur 2010). Because rice (*Oryza sativa* L.) has anatomical characteristics intermediate between those of most C_3 and C_4 grasses, introduction of Kranz anatomy into it may not require radical changes in leaf anatomy (Sage and Sage 2009).

The overall abundance of ^{13}C relative to ^{12}C in plant tissue is commonly less than in the carbon of atmospheric CO_2, indicating that carbon isotope discrimination occurs in the incorporation of CO_2 into plant biomass. Because the isotopes are stable, the information inherent in the ratio of abundances of carbon isotopes, presented by the convention as $^{13}C/^{12}C$, is invariant as long as carbon is not lost (Farquhar et al. 1989). Higher plants with the C_3 pathway of carbon assimilation have a $^{13}C/^{12}C$ ratio about 20‰ less than that in the atmosphere, while plants with the C_4 pathway have a ratio that is lower than the atmosphere by about 10‰ (Farquhar et al. 1982). Plants with CAM exhibit intermediate values between C_3 and C_4 plants, which appear to be related to the relative proportions of C_3 and C_4 fixation by these species (Farquhar et al. 1982). The lower-than-atmosphere ratios show that there is discrimination of the stable isotope of carbon, ^{13}C, relative to the more abundant ^{12}C (Turner 1993). In C_3 plants, the discrimination occurs due to the different diffusivities of ^{13}C and ^{12}C across the stomata and the fractionation by the C_3 enzyme, ribulose bisphosphate carboxylase (Farquhar et al. 1989). This primary carboxylating enzyme in C_3 plants discriminates against ^{13}C because of the intrinsically lower reactivity of ^{13}C (Farquhar et al. 1982).

The ratio in C_3 plants does vary and is not always about 20‰ less than the atmosphere. Turner (1993) showed the carbon isotope discrimination of 15 different C_3 crops, which were wheat (*T. aestivum* L.), barley (*H. vulgare* L.), rice (*O. sativa* L.), sunflower (*Helianthus annuus* L.), peanut (*Arachis hypogaea* L.), common bean (*P. vulgaris* L.), potato (*S. tuberosum* L.), tomato (*L. esculentum* Mill.), crested wheatgrass [*Agropyron desertorum* (Fischer ex Link) Schultes], wild rye [*Leymus angustus* (Trin.) Pilger], orchard grass (*Dactylis glomerata* L.), tall fescue (*Festuca arundinacea* Schreb.), perennial ryegrass (*L. perenne* L.), cotton (*G. hirsutum* L.), and coffee (*Coffea arabica* L.). It was always negative and the discrimination ranged from 14.1‰–15.7‰ for tall fescue to 20.0‰–24.5‰ for wild rye.

Within a specific C_3 species, variation also exists in the ratio, and it has been used successfully to identify plants with high water use efficiency (Farquhar and Richards 1984) or high transpiration efficiency (Turner 1993). Water use efficiency can be defined as the production of dry matter per unit of water consumed in evapotranspiration (Rosenberg 1974, p. 36). (We discuss water use efficiency in detail in Chapter 11.) If soil evaporation is prevented, then water use efficiency is based only on transpiration and it is called transpiration efficiency. Carbon isotope discrimination has been used to identify high transpiration efficiency genotypes of wheat (*T. aestivum* L.), barley (*H. vulgare* L.), peanut (*A. hypogaea* L.), and crested wheatgrass [*A. desertorum* (Fischer ex Link) Schultes] (Turner 1993).

The bases of discrimination against ^{13}C in C_4 plants are more complex (Farquhar 1983), because of the interposition of phospho*enol*pyruvate carboxylase (Farquhar et al. 1982). CO_2 diffuses through stomata, dissolves, is converted into HCO_3^-, and is fixed by PEP carboxylase into oxalacetate (Farquhar 1983). Various transformations then occur that are different for the various C_4 types (see Chapter 13 for these types), but the net result in all cases is that CO_2 is released in the bundle sheath cells and refixed by RuP_2 carboxylase. The bundle sheaths are not gas tight, and some CO_2 and HCO_3^- leak out of these cells into the mesophyll cells where it mixes with other CO_2 molecules that have diffused in through the stomata. The leakage allows some discrimination by the RuP_2 carboxylase (Rubisco) in the bundle sheath cells.

The $^{13}C/^{12}C$ isotope ratio is not a fixed value in C_4 species, and its value depends on how far the leaf Kranz anatomy is developed, on the degree of leakiness of the bundle sheath cells for CO_2, and the separation of the C_4 PEP carboxylase from the C_3 Rubisco.

This is why that in all the C_3–C_4 intermediate plants, known so far, the ratio is similar to values in C_3 species (M. El-Sharkawy, personal communication, April 23, 2008). Hattersley (1982) gives ratios for leaves of 31 different C_4 grasses with different photosynthetic types. The ratios range from −10.6‰ for *P. bulbosum* H.B.K. and −10.6‰ for *P. laevifolium* Hack. to −13.3‰ for *Eragrostis cilianensis* (All.) Lutati. He also reports the ratios for three C_4 dicotyledons, and they range from −12.2‰ for *Amaranthus edulis* Speg. to −14.7‰ for *Atriplex spongiosa* F. Muell. The carbon isotope ratio has not been used successfully to identify C_4 plants with high transpiration efficiency, as it has been used in C_3 plants (Mortlock and Hammer 1999).

The equations governing the carbon isotope ratio of mosses and sedges, as well as those for C_3 plants, are given by White et al. (1994). They presented a new method for reconstructing atmospheric CO_2 concentrations from the past. Instead of using ice cores, they used the $^{13}C/^{12}C$ ratio in mosses and sedges in peat. They exploited the fact that, unlike sedges and most other plants, mosses do not possess stomata and are therefore unable to regulate their uptake of CO_2 and water. The carbon isotope ratio of mosses thus depends on both atmospheric CO_2 concentration and available water, and the carbon isotope ratio of sedges from the same peat can be used to remove the water signal. The method provides a resolution of about a decade, which is much higher than is possible from ice cores. Their results show three sharp increases in atmospheric CO_2 concentration in the past 14,000 years: 12,800 years ago, corresponding to an episode of warming in the North Atlantic region; 10,000 years ago, corresponding to the end of the Younger Dryas cold period; and 4,400 years ago, after which time modern climates were established globally (White et al. 1994). The Younger Dryas is the period from approximately 10,800 to 9,600 BC, when climate in the region around Greenland cooled by 5°–7°C within a few decades and recovered with similar rapidity at the end of the period (Glickman 2000). It is named for the expansion of the geographic range of the arctic herb *Dryas octopetala*. The period since the Younger Dryas is the Recent Epoch, which is also called the Holocene Epoch. It is the last 10,000 years of geologic time (Glickman 2000).

TREES

Matamala et al. (2003) monitored root turnover in two forest plantations growing under FACE: loblolly pine, *Pinus taeda* L., near Durham, North Carolina, planted in 1980, and sweet gum, *Liquidambar styraciflua* L. (a hardwood), in Oak Ridge, Tennessee, planted in 1988. They assessed root turnover with the use of an isotope tracer, which was depleted $^{13}CO_2$. The continuous CO_2 fumigation with FACE technology used the tracer, which altered the $\delta^{13}C$ of plant tissues. The pine and sweet gum forests were enriched with 192 and 150 µmol/mol CO_2, respectively. The $\delta^{13}C$ equaled −44‰ and −55‰ relative to the Pee Dee Belemnite standard for the 192 and 150 µmol/mol CO_2 applied to the pine and sweet gum forests, respectively. Growth at elevated CO_2 did not accelerate root turnover in either the pine or the hardwood forest. Turnover of the carbon in the fine roots (less than 1 mm in diameter) varied from 1.2 to 9 years. The mean residence time of carbon in the fine roots was 4.2 years for the pine forest and 1.25 years for the sweet gum forest. Carbon had a maximum residence (i.e., 95% turnover) in roots less than 1 mm in diameter of about 12 years for pine and 4 years for sweet gum. For thicker pine roots, 2–5 mm in diameter, the mean residence time was 6.3 years, with a maximum residence time greater than 18 years. For sweet gum roots between 1 and 2 mm, the mean residence time was 3 years and maximum residence time was over 9 years. (The same size roots for non-fine roots were not compared for pine and sweet gum.) These long turnover times suggest that root production and turnover in forests have been overestimated and that sequestration of anthropogenic atmospheric carbon in forest soils may be lower than currently estimated.

Berntson and Bazzaz (1996b) determined total root production, total root loss, net root production, and biomass production for seedlings of *Betula papyrifera* Marsh. (canoe or paper birch) and *Acer rubrum* L. (red, scarlet, or swamp maple) grown under two CO_2 concentrations. Half the plants were in glasshouses that had a CO_2 concentration of 400 µmol/mol, which they called the ambient treatment, and the other half grew in glasshouses with 700 µmol/mol CO_2. Elevated CO_2 increased total root production, total root loss, and net root production in seedlings of *Betula* but not *Acer*. However, the effects of CO_2 on root production and loss did not alter the allometric relationship between root production and root loss for either *Acer* or *Betula*. [The term "allometric" means a relationship between two variables of the form $y = ax^b$ where a and b are constants such that $\ln y = b \ln x + \ln a$; b is termed the allometric coefficient. The term is commonly used in morphogenesis studies, where organ dimensions and growth rates frequently can be fit by the relation (Barnes and Beard 1992a).] The results of Berntson and Bazzaz (1996b) showed that even though elevated CO_2 did not affect the relationship between root production and root loss, it may lead to increases in root production and in root loss (turnover).

In another paper, Berntson and Bazzaz (1997) reported fast cycling of fine roots of *B. papyrifera* Marsh. They quantified gross root production and root loss from sequential, in situ images of fine roots of the paper birch at ambient (375 µmol/mol) and elevated (700 µmol/mol) CO_2 atmospheres from 2 weeks following germination through leaf senescence. They found that elevated CO_2 increased standing root length throughout the experiment, but increases were greater early in the season. They concluded that static assessments of below-ground productivity may underestimate fine root productivity. More studies are needed to understand seasonal variations in root architecture to determine growth responses to rising CO_2 (Berntson and Bazzaz 1996a).

Walker et al. (1997) studied the effects of atmospheric CO_2 enrichment on above- and below-ground development of juvenile ponderosa pine (*P. ponderosa* Dougl. ex P. Laws and C. Laws) grown for three seasons in open-top field chambers with one of three CO_2 levels: 350, 525, and 700 µmol/mol. At the first harvest, shoot weight and coarse and fine root weights were increased by elevated CO_2. At the second harvest, shoot and root weights were increased by CO_2 enrichment, but growth stimulation by the 525 µmol/mol CO_2 concentration exceeded that in 700 µmol/mol CO_2. At the third harvest, above- and below-ground growth increases were largely confined to the 700 µmol/mol CO_2 treatment. These results suggested that elevated CO_2 exerts stimulatory effects on shoot and root growth of juvenile ponderosa pine under field conditions, but temporal variation may periodically result in a greater response to a moderate rise in atmospheric CO_2 than to a doubling of the current ambient concentration. Another study with Ponderosa pine grown in open-top, field-exposed chambers showed that elevated CO_2 (525 and 700 µmol/mol; ambient CO_2 was 354 µmol/mol) increased soil exploration by fine roots measured during a 4 year period using minirhizotron tubes (Tingey et al. 2005).

Thomas et al. (1996) studied another pine, Monterey pine (*P. radiata* D. Don), which was grown in large, open-top field chambers at ambient CO_2 (362 µmol/mol) or elevated CO_2 (654 µmol/mol). The increase in number of fine (less than 0.5 mm diameter) roots of 1-year-old trees was estimated using minirhizotron tubes placed horizontally at a depth of 0.3 m. Root growth started in spring and continued until late summer. Roots were observed 4 weeks earlier in the elevated CO_2 treatment. Mean values of CO_2 flux density at the soil surface, F, which was measured with a portable soil respiration system placed on the soil surface, increased from 0.02 to 0.13 g/m²/h during the year, and F was 30% greater for trees grown at the elevated CO_2 concentration than at the ambient CO_2.

Brown et al. (2009) compared root growth using both minirhizotrons and soil cores in a scrub-oak (*Quercus*) ecosystem in central Florida, United States, which was exposed to ambient or twice ambient CO_2 for 10 years starting in 1996. The soil was a sand. The plant community included 75% *Quercus myrtifolia*; 15% *Q. geminata*; and 7% of the following three plants: *Q. chapmannii*, *Serenoa repens* (*Serenoa* is saw palmetto), and *Lyonia ferreginea* (*Lyonia* is a genus with evergreen or deciduous shrubs). Root length and biomass, estimated from minirhizotron data, were comparable to determinations from soil cores. Biomass estimates from minirhizotrons indicated that <0.25 mm diameter roots accounted for nearly 95% of the total root length in 2002. The smallest diameter fine roots (<0.25 mm) were greater under elevated CO_2 during the early years of the study, and the largest (2–10 mm) had greater biomass under elevated CO_2 during the later years of the study.

An experiment was carried out to simulate the effects of severe drought in the rainforest of the Brazilian Amazon (Showstack 2005). The research has implications for climate change, because the amount of carbon removed from the atmosphere during tropical droughts is expected to decrease. While the study did not deal with elevated CO_2, the results are instructive. A total of 1.8 m of rainfall was excluded from an 1 ha plot in the middle of the Tapajos National Forest in Brazil during January 2000 to July 2004. Parts of the rain-excluded forest tolerated the dry condition by absorbing water from deeper in the soil. The study also showed that the sensitivity of large trees (39.6–45.7 m) to drought was greater than expected, and that once the moisture that was stored in deep soil was depleted, they died (Showstack 2005). Hydraulic lift may be occurring in the trees exposed to drought. Hydraulic lift occurs when plant roots extract water from a moist subsoil and release it into a dry topsoil (Song et al. 2000). In arid and semiarid regions, movement of water by plant roots through hydraulic lift appears to be one method that allows plants to survive.

Much research dealing with the effects of CO_2 on below-ground growth of forest tree species has been conducted at Oak Ridge National Laboratory, Oak Ridge, Tennessee (Rogers et al. 1997, p. 11). The species include shortleaf pine (*P. echinata* Mill.), Virginia pine (*P. virginiana* Mill.), white oak (*Q. alba* L.), and yellow poplar (also called tulip-tree; *Liriodendron tulipifera* L.). These tree species have shown increases in root dry weight under elevated CO_2.

C_3 AND C_4 CROPS COMPARED

Rogers et al. (1983) grew soybean [*Glycine max* (L.) Merr.] (C_3) and corn (*Zea mays* L.) (C_4) for 11 weeks in open-top field chambers and exposed the plants to four concentrations of CO_2: 340 (ambient), 520, 718, or 910 μmol/mol CO_2. The plants grew in pots with a mixture of sand, a horticultural substrate (Metro-Mix made by W.R. Grace and Co., Columbia, Maryland), and a sandy clay loam. At harvest, root growth for corn at 340, 520, 718, and 910 μmol/mol CO_2 was 7.4, 15.8, 12.4, and 15.1 g/plant, respectively. For soybeans, these values were 5.1, 10.0, 10.5, and 12.8 g/plant, respectively. The percent increases in root growth were 151% and 104% for the C_3 and C_4 plants, respectively, when the 340 and 910 μmol/mol CO_2 levels were compared. The first increment of CO_2 (going from ambient to 520 μmol/mol CO_2) caused the increase in root growth, and there was no increase in root growth for either species after that first increment. Plants grown at 910 μmol/mol CO_2 produced as many roots as did plants grown at 340 μmol/mol CO_2. Leaf growth of corn increased after the first increment, but not thereafter. Dry weights of corn leaves were 11.1, 12.8, 12.0, and 13.1 g/plant for the four CO_2 levels, respectively. However, leaf growth of soybean continued to increase up to the highest level of CO_2. Dry weights of soybean leaves were 9.6, 13.4, 15.8, and 17.9 g/plant for the four CO_2 levels, respectively.

Root biomass of soybean and sorghum (C_4), grown in a FACE system with ambient and double the ambient levels of CO_2, increased 54% and 73%, respectively (Rogers et al. 1997, p. 11). However, in the study comparing soybean and corn, described in the previous paragraph (Rogers et al. 1983), the C_3 roots (soybean) had a greater percentage increase in growth at elevated CO_2 than did the C_4

roots (corn). No general conclusion can be made about the relative response of C_3 and C_4 roots under elevated CO_2.

CAM PLANTS

A desert plant with the CAM photosynthetic system has been shown to have increased root growth under elevated CO_2. Drennan and Nobel (1996) measured root growth of *Encelia farinosa* Torrey and A. Gray (Asteraceae; C_3), *Pleuraphis rigida* Thurber (Poaceae; C_4), and *Agave deserti* Engelmann (Agavaceae; CAM), which are codominants in the northwestern Sonoran Desert, under current (360 μmol/mol) and doubled (720 μmol/mol) CO_2 concentrations and simulated winter and summer conditions in growth chambers. The doubled CO_2 concentration increased average daily root elongation rates for *A. deserti* under both winter (7%) and summer (12%) conditions, but it had no effect for the other two species. They noted that doubled CO_2 concentrations generally increase root growth for C_3 species and gave no explanation for the lack of response of the roots of *E. farinosa* to doubled CO_2 concentration. They did note the wide range in temperature at the soil surface in the Sonoran Desert, which varies from 0°C to 75°C. Temperatures apparently are more important in controlling the root growth of this C_3 desert species than doubled CO_2.

ROOT-TO-SHOOT RATIOS

Rogers et al. (1994, 1996, 1997) reviewed the literature for variations in R/S ratios of plants under elevated CO_2. In their 1996 review, they identified 264 determinations of R/S of crops under elevated CO_2. In most cases (59.5% of the cases), R/S increased; in a few cases (3.0%), it remained unchanged; and in others (37.5%), it decreased. They concluded that the explanation for these differences probably resided in crop type, resource supply, and other experimental factors. In a study of 27 herbaceous species, R/S decreased in 14 species and increased in 6, and 7 species showed no response (Rogers et al. 1997, p. 7). Newton (1991) reviewed 10 papers in which the R/S of eight C_3 plants and nine C_4 plants (all nonarable) had been studied. [Arable means suitable for plowing (Friend and Guralnik 1959).] In the C_3 species, R/S increased in four of them and decreased in two, and there was no change in two. In the C_4 species, R/S increased in five of them and decreased in two, and there was no change in one. One C_4 plant (*Setaria faberii* Herrm, a type of foxtail) showed both an increase in R/S and a decrease in R/S, depending on the study. Acock and Allen (1985) reviewed the literature and found, in general, an increase in R/S when crops were grown under elevated CO_2. However, BassiriRad et al. (2001) in their review found no consistent pattern in R/S of plants grown under elevated CO_2.

Some have found that drought affects the ratio. Miao et al. (1992) observed more allocation to roots under elevated CO_2 and drought, which suggested that the increased root growth allowed more access to a limited water supply (Rogers et al. 1997, p. 13). A decrease in rainfall favors the carbon partitioning to roots (Lambers et al. 1996). Limited water availability increased the allocation of ^{13}C to roots of wheat grown in columns, so that at booting, 0.38 of shoot carbon was below ground compared to 0.31 in well-watered plants (Gregory et al. 1997). Elevated CO_2 (700 μmol/mol) increased the proportion of wheat roots to total mass by 55% compared with the normal concentration.

Farrar and Williams (1991) reviewed the literature and found that R/S ratios tended to increase for CO_2-enriched herbaceous plants and decrease for CO_2-enriched trees. But R/S for cereals usually remained constant, as Chaudhuri et al. (1986a, 1990) found for sorghum and wheat. The conclusion of Farrar and Williams (1991) for tree species is contradicted by Ceulemans and Mousseau (1994), who presented a table with 84 entries for deciduous tree species under elevated CO_2. Under elevated CO_2, the majority of the experiments that they reviewed showed an increase in R/S, and a small number showed no difference. They concluded that this indicated that CO_2 enrichment preferentially induced extra root storage rather than shoot storage.

ROOT RESTRICTION

Of concern to investigators studying plants in pots is the effect of pot size on root growth. Restricted growth may limit responses of roots to elevated CO_2, and several studies have investigated the topic (Rogers et al. 1997, pp. 3–4). Arp (1991) found that plants grown in small pots had a reduced photosynthetic capacity under elevated CO_2, while plants grown in the field showed no reduction or an increase. Pot volume also determined the effect of elevated CO_2 on the R/S ratio. The R/S ratio increased when root growth was not restricted and decreased in plants grown in small pots. Sionit et al. (1984) compared the response of container- and field-grown soybeans and concluded that the stress imposed on plants by confining the roots may decrease the magnitude of their photosynthetic response to atmospheric CO_2 enrichment. Thomas and Strain (1991) studied photosynthesis of cotton (*G. hirsutum* L.) seedlings under elevated CO_2 with root restriction as a main factor. They found that reduced photosynthetic capacity could be restored by simply repotting to larger containers, thus reducing root restriction. Barrett and Gifford (1995) used three different sizes of pots to study photosynthesis and growth of cotton at three CO_2 concentrations (376, 652, and 935 μmol/mol). Root biomass and R/S ratios decreased with a decrease in rooting volume. Even though these studies showed that restricted root volume affected results of plants grown under elevated CO_2, Coleman and Bazzaz (1992) pointed out that root restriction may be a characteristic of some natural habitats so that unrestricted rooting volume may be as unrealistic as small pots.

Schaffer et al. (1996) found that elevated levels of CO_2 (1000 μmol/mol; ambient was 350 μmol/mol) compensated for the effects of root restriction on growth of banana (*Musa* sp.), because at 1000 μmol/mol CO_2, there was generally no effect of root chamber size on plant dry weight. In another study, Schaffer et al. (1997) determined the effects of atmospheric CO_2 enrichment and root restriction on net CO_2 assimilation and dry mass partitioning in mango (*Mangifera indica* L.). Trees grew in controlled-environment greenhouses at CO_2 concentrations of 350 or 700 μmol/mol. At each CO_2 concentration, trees were grown in 8 L containers, which restricted root growth, or were grown aeroponically in 200 L root mist chambers, which did not restrict root growth. Total plant and organ dry mass was generally higher for plants grown at 700 μmol/mol CO_2 due to increased net CO_2 assimilation. Root restriction reduced net CO_2 assimilation resulting in decreased organ and plant dry mass. In root-restricted plants, reduced net CO_2 assimilation and dry matter accumulation offset the increases in these variables resulting from atmospheric CO_2 enrichment.

Kerstiens and Hawes (1994) studied inadequate rooting volume to see if it reduced growth stimulated by elevated CO_2 in potted tree seedlings. One-year-old cherry (*Prunus* sp.) saplings were grown for one season in naturally lit growth chambers with two concentrations of CO_2 (ambient; ambient +250 μmol/mol) and three different root volumes (4, 10, and 20 L pots). The overall mean effect of high CO_2 on plant dry mass by the end of the season was +24%. The variation of carbon allocation to roots and shoots with pot size was similar in all treatments. They concluded that there is no evidence that inadequate pot volume had a negative impact on the stimulation of growth of this tree species in elevated CO_2.

Berntson et al. (1993) measured growth of C_3 and C_4 plants in pots to provide insight into the effects of CO_2 on their usage of below-ground space. They studied two species: *Abutilon theophrasti* Medic. (velvet-leaf), a C_3 dicot with a deep taproot, and *S. faberii* Herrm. (mentioned above; it is a type of foxtail), a C_4 monocot with a shallow fibrous root system. They responded differently to elevated CO_2. When pot volume increased, *Abutilon* allocated more roots to the tops of pots. This response may have allowed roots to exploit regions of greater nutrient concentration or greater water availability that may occur in the upper soil layers. For *Setaria*, when pots were large, elevated CO_2 resulted in a lower root density at the surface. Root densities for *Abutilon* and *Setaria* were similar to those observed in field conditions for annual dicots and monocots, respectively, and this suggested that studies using pots may successfully mimic natural conditions.

Nobel et al. (1994) studied the effects of current (360 μmol/mol) and elevated (720 μmol/mol) CO_2 concentrations on root and shoot areas and biomass accumulation for *Opuntia ficus-indica*

(L.) Miller, a highly productive CAM species cultivated worldwide. Plants were grown in environmentally controlled rooms for 18 weeks in pots of three soil volumes (2,600, 6,500, and 26,000 cm³), the smallest of which was intended to restrict root growth. For plants in the medium-sized soil volume, basal cladodes tended to be thicker, and areas of main and lateral roots tended to be greater as the CO_2 level was doubled. Increasing the soil volume 10-fold led to a greater stimulation of biomass production than did doubling the CO_2 level. At 18 weeks, root biomass doubled and shoot biomass nearly doubled as the soil volume was increased 10-fold. This study, as well as others cited in this section, shows the need to consider rooting volume in the design of experiments dealing with elevated CO_2.

SUMMARY

Most studies show that root weight is increased with elevated concentrations of CO_2 in the atmosphere. These results have been shown for C_3, C_4, and CAM plants. Growth of root crops, such as potatoes, radishes, carrots, and beets, is especially increased under elevated atmospheric CO_2. Root length is usually increased under elevated CO_2. Consequently, there is often more penetration of roots in the soil profile and spreading under elevated CO_2 compared to ambient levels. Carbon isotope ratios have been used to study the response of roots, as well as shoots, to elevated CO_2. The ratios are conventionally reported as $\delta^{13}C$. The number represents the difference between the ratio of $^{13}C-^{12}C$ found in a given sample and the ratio that exists in a universal standard, expressed as a per mill (‰, or number of one-thousandths) deviation from the standard. The letters PDB refer to the standard, which is based on a fossil mollusk shell (called belemnite) from the Pee Dee Formation, a Cretaceous deposit in South Carolina. Higher plants with the C_3 pathway of carbon assimilation have a $^{13}C/^{12}C$ ratio about 20‰ less than that in the atmosphere, while plants with the C_4 pathway have a ratio that is lower than the atmosphere by about 10‰. Plants with CAM exhibit intermediate values between C_3 and C_4 plants. R/S ratios are variable under elevated CO_2. They can be increased, decreased, or show no change. Root restriction of plants grown in small pots can affect the results of research done under elevated CO_2. Experiments must be designed to allow adequate rooting volume.

In this chapter, we used the data that we obtained in Kansas between 1984 and 1987 for sorghum and winter wheat to calculate how much root weight has increased due to increasing CO_2 concentrations in the atmosphere. The concentration of CO_2 in the atmosphere has increased 55 µmol/mol between 1984 and 2008. With an increase of 55 µmol/mol CO_2 in the atmosphere, root weights of sorghum and winter wheat have increased 4 and 7 g/m², respectively.

7 Elevated Atmospheric Carbon Dioxide: Plant Water Potential, Osmotic Potential, and Turgor Potential

INTRODUCTION

Growth is directly related to plant water potential. The higher the water potential (less negative), the greater is the rate of growth. Figure 7.1 shows this relationship for snap beans (*Phaseolus vulgaris* L.). As the plant approaches the permanent wilting point [often considered to be about −1.5 MPa or −15 bar (Kirkham 2005, pp. 104–107)], the plant dies due to water deficit. Growth of most plants stops before the permanent wilting point is reached. The curve in Figure 7.1 has been drawn assuming that factors such as light, CO_2, and nutrients are constant (Gardner 1973, p. 223).

As we know from our plant physiology classes, water potential is the sum of the osmotic potential and the turgor potential (also called turgor pressure), if we ignore the matric potential and gravitational potential (Kirkham 2005, p. 287). Turgor potential also is directly related to growth. As water potential gets more negative, the turgor potential falls (Figure 7.2; data for snap beans). At the permanent wilting point or about −15 bar, the turgor potential is close to zero (Figure 7.2) and growth stops. One way to maintain the turgor potential is by osmotic adjustment. Osmotic adjustment is the net accumulation of solutes in a cell of a plant in response to a fall in the water potential of the cell's environment (Blum et al. 1996). As a consequence of this net accumulation, the osmotic potential of the cell is lowered, which in turn attracts water into the cell, and this tends to maintain turgor pressure. Bernstein (1961, 1963) at the U.S. Salinity Laboratory in Riverside, California, United States, was one of the first to document osmotic adjustment and show how it can maintain growth under saline conditions.

Turgor potential is also related to stomatal conductance. As shown in Chapter 8, stomatal conductance is the reciprocal of stomatal resistance. Under ambient CO_2, if one plots stomatal conductance versus turgor potential, the values fall on a straight line for both the abaxial (lower) and adaxial (upper) stomata (Figure 7.3; data for snap beans). As the soil water potential decreases below −1 bar, there is a direct relationship between stomatal conductance and soil water potential (Gardner 1973). If stomata are closed, less water is lost from the soil than if stomata are open, and transpiration is decreased. (We consider transpiration in Chapter 10.) Because the turgor potential is influenced by the transpiration rate, there appears to be a direct coupling between stomatal conductance and transpiration. Also plant temperature and transpiration are related, because, if a plant is well watered, the stomata are open, transpirational cooling occurs, and canopy temperature is cool. Conversely, as a plant becomes water stressed, stomata close, transpiration is reduced, and canopy temperature increases (Kirkham 2005, p. 425). The relationship between turgor potential and stomatal conductance (Figure 7.3) is not expected to be unique. It is here that light, CO_2, nutrition, temperature, and the many other factors that affect plant growth should be expected to have an influence (Gardner 1973).

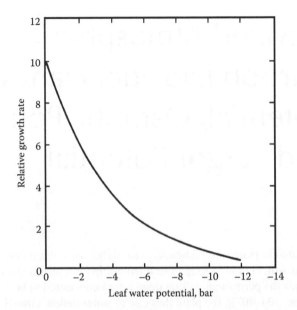

FIGURE 7.1   Relative growth rate as a function of leaf water potential for snap beans (*Phaseolus vulgaris* L.) in a growth chamber. (Reprinted from Gardner, W.R., Internal water status and plant response in relation to the external water régime, in Slatyer, R.O., Ed., *Plant Response to Climatic Factors*, United Nations Educational, Scientific and Cultural Organization, Paris, pp. 221–225, 1973, Fig. 5; Data in the figure come from E.T. Kanemasu and C.B. Tanner, *Plant Physiol.*, 44, 1547, 1969. With permission from the American Society of Plant Biologists and Wilford R. Gardner.)

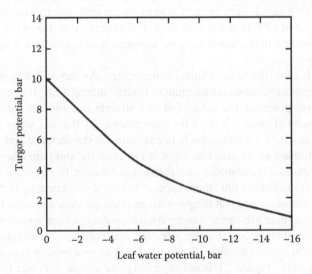

FIGURE 7.2   Relationship between total leaf water potential and turgor potential for snap beans (*P. vulgaris* L.). (Reprinted from Gardner, W.R., Internal water status and plant response in relation to the external water régime, in Slatyer, R.O., Ed., *Plant Response to Climatic Factors*, United Nations Educational, Scientific and Cultural Organization, Paris, pp. 221–225, 1973, Fig. 2; Data in the figure come from E.T. Kanemasu and C.B. Tanner, *Plant Physiol.*, 44, 1547, 1969. With permission from the American Society of Plant Biologists and Wilford R. Gardner.)

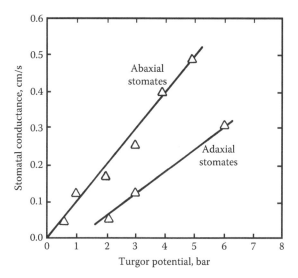

FIGURE 7.3 Stomatal conductance as a function of leaf turgor potential for beans, obtained under field conditions. (Reprinted from Gardner, W.R., Internal water status and plant response in relation to the external water régime, in Slatyer, R.O., Ed., *Plant Response to Climatic Factors*, United Nations Educational, Scientific and Cultural Organization, Paris, pp. 221–225, 1973, Fig. 3; Data in the figure come from E.T. Kanemasu and C.B. Tanner, *Plant Physiol.*, 44, 1547, 1969. With permission from the American Society of Plant Biologists and Wilford R. Gardner.)

Let us now see how elevated CO_2 affects the potentials of plants. We begin with winter wheat and native grasses, the plants that we in the Evapotranspiration Laboratory at Kansas State University studied under field conditions in the 1980s.

WHEAT

In the spring of two growing seasons (1984–1985 and 1985–1986), Chaudhuri et al. (1986, 1987) measured the water potential of winter wheat (*Triticum aestivum* L.) growing in closed-top chambers in a field with four concentrations of CO_2: ambient (330 μmol/mol in 1986; 340 μmol/mol in 1987), 485, 660, and 825 μmol/mol. The soil was a silt loam, which was either well watered (the high water level or field capacity) or kept dry (the low water level or half field capacity). Chaudhuri et al. (1990a,b) give details of the experimental setup, and it is also described in Chapter 6. Water potential was measured with a pressure chamber (Model 3005, Soilmoisture Equipment Corp., Santa Barbara, CA, United States), and adaxial and abaxial stomatal resistances were measured with a steady state porometer (Model LI-1600, Li-Cor, Inc., Lincoln, NE, United States). The total stomatal resistance of the two leaf surfaces combined was determined by using the parallel resistance law (Kirkham 2005, p. 392): $1/R_{total} = (1/R_{adaxial}) + (1/R_{abaxial})$, where R_{total} is the total stomatal resistance and $R_{adaxial}$ and $R_{abaxial}$ are the stomatal resistances of the upper and lower surfaces of the leaf, respectively, and the unit of stomatal resistance is s/cm. Canopy temperature was measured with an infrared thermometer (see Kirkham 2005, pp. 425–435 for a description of the technique).

For both watering regimes, the leaf water potential increased as the CO_2 concentration increased (Table 7.1). As expected, plants that were drought stressed had a lower water potential than plants that were well watered. The water potential of water-stressed plants grown under the highest CO_2 level (–2.10 MPa) was similar to that of the well-watered plants grown under the ambient level of CO_2 (–2.21 MPa), which suggested that elevated CO_2 compensated for reductions in water potential due to a low water supply. The elevated CO_2 also compensated for reductions in yield due to drought

TABLE 7.1

Leaf Water Potential (MPa) of Winter Wheat Supplied with High and Low Water Levels as Affected by CO_2 Concentration during a Two-Year Period (1985 and 1986)

CO_2 μmol/ mol	1985				1986				Aver.
	May 2	May 2	May 3	May 24	May 2	May 6	May 15	May 21	
High water									
340[a]	−1.74[b]	−1.88	−1.89	−2.28	−2.39±0.00[c]	−2.53±0.05	−2.29±0.12	−2.70±0.13	−2.21
485	−1.48	−1.68	−1.69	−2.04	−2.21±0.07	−2.24±0.03	−2.23±0.21	−2.30±0.16	−1.98
660	−1.50	−1.63	−1.68	−1.98	−2.13±0.03	−2.04±0.22	−1.94±0.10	−2.07±0.23	−1.87
825	−1.39	−1.61	−1.56	−1.96	−1.83±0.41	−2.15±0.14	−2.01±0.17	−2.10±0.28	−1.83
Low water									
340	−2.05	−2.27	−2.25	−2.68	−2.62±0.05	−2.86±0.08	−3.07±0.11	−3.04±0.33	−2.61
485	−1.73	−1.93	−1.97	−2.51	−2.36±0.25	−2.81±0.13	−2.94±0.09	−2.88±0.08	−2.39
660	−1.85	−1.85	−1.98	−2.30	−2.30±0.16	−2.64±0.10	−2.72±0.29	−2.52±0.26	−2.27
825	−1.79	−1.73	−1.86	−2.30	−2.14±0.03	−2.36±0.05	−2.24±0.08	−2.38±0.24	−2.10

Sources: Chaudhuri, U.N. et al., Effect of elevated levels of CO_2 on winter wheat under two moisture regimes. In *Response of Vegetation to Carbon Dioxide*, Research Report No. 029. U.S. Department of Energy, Carbon Dioxide Research Division, Office of Energy Research, Washington, DC, 1986, xii+77pp; Chaudhuri, U.N. et al., Effect of elevated levels of CO_2 on winter wheat under two moisture regimes. In *Response of Vegetation to Carbon Dioxide*. Research Report No. 040. U.S. Department of Energy, Carbon Dioxide Research Division, Office of Energy Research, Washington, DC, 1987, xiv+70pp.

Measurements were taken in Manhattan, Kansas, between 11:00 and 13:50 Central Daylight Time, except measurements on 2 May 1985, which were taken between 15:30 and 17:10 Central Daylight Time.

[a] CO_2 concentration was 330 μmol/mol in 1985.

[b] One measurement only.

[c] Mean and standard deviation.

(Chaudhuri et al. 1990). Because water potential and growth are related (Figure 7.1), the water potential and growth data, taken together, indicated that increasing the water potential increased growth.

Under the high water level, stomatal resistance of the wheat increased as the CO_2 concentration increased (Table 7.2 for the 1985 data; Table 7.3 for the 1986 data). However, under the low water level, stomata had similar stomatal resistances, which were higher than those under the well-watered conditions. As shown in Chapter 8 for other studies, elevated CO_2 closes stomata, which conserves water. Therefore, it appeared that the water-stressed plants were able to maintain a higher water potential under elevated CO_2, both under well watered and dry conditions, because CO_2 closed the stomata.

The canopy temperature data paralleled the stomatal resistance data (Table 7.4 for the 1985 data; Table 7.5 for the 1986 data). Canopy temperatures generally increased as the CO_2 concentration increased under the high water level. Under dry conditions, canopy temperatures were usually higher than under the wet conditions, as expected, but they did not increase or decrease in response to CO_2 concentration.

Because elevated CO_2 closed the wheat stomata, the soil water content was probably greater under elevated CO_2 compared to the ambient level. However, we could not confirm this because we did not measure soil water content. Higher water content in soil under elevated CO_2 is documented in Chapter 5. A greater amount of water in the soil under elevated CO_2 may have been a reason for the higher water potentials of wheat grown with elevated CO_2.

TABLE 7.2
Stomatal Resistance (s/cm) of Winter Wheat Supplied with High and Low Water Levels as Affected by CO_2 Concentration during the 1984–1985 Growing Season

CO_2 µmol/mol	April 24	May 2	May 2[a]	May 3	May 3	May 17	May 22	May 22	May 22	May 24	May 24	Aver.
						1985						
High water												
330	0.53	0.20	0.51	0.49	0.47	0.51	0.48	0.57	0.52	0.46	0.55	0.48
485	0.64	0.25	0.69	0.57	0.53	0.64	0.59	0.65	0.60	0.54	0.63	0.58
660	0.67	0.28	0.71	0.62	0.74	0.67	0.62	0.68	0.69	0.52	0.68	0.63
825	0.66	0.32	0.84	0.63	0.82	0.69	0.69	0.73	0.74	0.60	0.70	0.67
Low water												
330	0.77	0.36	0.82	0.79	0.97	0.78	0.81	0.77	0.84	0.76	0.93	0.78
485	0.73	0.35	0.82	0.70	0.84	0.72	0.82	0.78	0.79	0.69	0.86	0.74
660	0.74	0.39	0.91	0.73	0.88	0.74	0.79	0.68	0.82	0.74	0.85	0.75
825	0.78	0.40	0.88	0.76	0.90	0.78	0.80	0.79	0.86	0.82	0.83	0.78
VPD (KPa)	1.54	0.96	1.45	1.48	1.76	1.49	1.60	1.82	1.98	1.43	1.89	

Source: Chaudhuri, U.N. et al., Effect of elevated levels of CO_2 on winter wheat under two moisture regimes, in *Response of Vegetation to Carbon Dioxide*, Research Report No. 029, U.S. Department of Energy, Carbon Dioxide Research Division, Office of Energy Research, Washington, DC, 1986, xii+77pp.

The vapor pressure deficit (VPD) at the time of the measurements is given at the bottom of the table.

[a] Chaudhuri et al. (1986) do not give the time of the measurements of the stomatal resistance. However, they were probably taken at about the same time that the leaf water potential measurements were taken, which were midday and mid-afternoon. On 22 May, when three measurements were taken, the final one was probably taken late in the afternoon. On 24 April and 17 May, when only one measurement was taken, it was probably taken at midday.

TABLE 7.3
Stomatal Resistance (s/cm) of Winter Wheat Supplied with High and Low Water Levels as Affected by CO_2 Concentration during the 1985–1986 Growing Season

CO_2 μmol/mol	1986					
	April 29[a]	May 6	May 15	May 20	May 21	Aver.
High water						
340	0.23	0.24	0.39	0.49	0.46	0.36
485	0.24	0.27	0.45	0.50	0.55	0.40
660	0.23	0.30	0.48	0.51	0.53	0.41
825	0.30	0.32	0.47	0.50	0.60	0.44
Low water						
340	0.29	0.34	0.66	0.70	0.74	0.55
485	0.26	0.36	0.64	0.67	0.78	0.54
660	0.26	0.40	0.65	0.88	0.68	0.57
825	0.38	0.49	0.66	0.67	0.69	0.58

Source: Chaudhuri, U.N. et al., Effect of elevated levels of CO_2 on winter wheat under two moisture regimes, in *Response of Vegetation to Carbon Dioxide*, Research Report No. 040, U.S. Department of Energy, Carbon Dioxide Research Division, Office of Energy Research, Washington, DC, 1987, xiv + 70pp.

Chaudhuri et al. (1987) do not report the vapor pressure deficit in 1986, as they did in 1985.

[a] Chaudhuri et al. (1987) do not give the times of the measurements of the stomatal resistance, but they were probably taken about midday.

GRASSLAND PLANTS

In 1989, He et al. (1992) measured the water potential of two native grasses, big bluestem (*Andropogon gerardii* Vitman; C_4) and Kentucky bluegrass (*Poa pratensis* L.; C_3), growing on a silty clay loam in a tallgrass prairie in Kansas, United States, under two levels of CO_2 (ambient or 337 μmol/mol and two-times ambient or 658 μmol/mol). The plants were exposed to two watering regimes: a high water level or field capacity and a low water level or half field capacity. Measurements were taken midday with a pressure chamber during an entire season from June 22 to October 23, 1989. Under the high water level (Figure 7.4), water potential of both grasses was higher under elevated CO_2 than under ambient CO_2. The average water potentials, plus or minus the standard errors, of big bluestem and Kentucky bluegrass under elevated CO_2 were −0.63 ± 0.04 and −0.70 ± 0.04 MPa, respectively. Under low CO_2, these values were −0.84 ± 0.05 and −0.90 ± 0.05 MPa, respectively. Therefore, elevated CO_2 raised the seasonal water potential by about 0.20 MPa for both grasses with the high water level.

With the low soil-water level (Figure 7.5), the two grasses also had higher water potentials under elevated CO_2 compared to ambient CO_2. The average values, plus and minus the standard error, for big bluestem and Kentucky bluegrass under elevated CO_2 were −0.80 ± 0.05 and −0.98 ± 0.04 MPa, respectively. Under ambient CO_2, these values were −1.10 ± 0.05 and −1.23 ± 0.05 MPa, respectively. Therefore, elevated CO_2 raised the water potentials of big bluestem and Kentucky bluegrass by 0.30 and 0.25 MPa, respectively, under the low water level. The water potentials of the two grasses under the low-water, high-CO_2 conditions were about the same as those under the high-water, ambient-CO_2

TABLE 7.4
Canopy Temperature (°C) of Winter Wheat Supplied with High and Low Water Levels as Affected by CO_2 Concentration during the 1984–1985 Growing Season

CO_2 μmol/mol	1985												
	April 24	May 2	May 2	May 3	May 3	May 8	May 15	May 17	May 22	May 22	May 24	May 24	Aver.
High water													
330	24.9	25.0	24.9	24.5	25.0	26.5	25.0	24.1	26.5	26.0	25.5	30.0	25.7
485	25.5	24.6	24.7	24.4	26.3	27.3	25.3	24.7	27.9	28.1	27.8	32.1	26.6
660	25.2	24.4	25.2	25.0	26.6	27.4	25.8	25.4	27.5	28.5	28.2	32.4	26.8
825	25.8	24.8	25.8	25.2	27.2	28.1	25.6	25.3	28.4	28.9	28.7	32.8	27.2
Low water													
330	25.9	26.2	27.4	28.3	28.3	27.7	26.8	26.0	28.4	28.8	28.5	33.0	27.9
485	25.2	25.9	26.3	27.4	27.4	27.1	26.3	25.4	27.5	28.4	27.8	32.3	27.2
660	24.9	25.8	26.0	27.1	27.3	27.4	25.8	25.8	27.8	28.0	27.5	32.8	27.2
825	25.0	26.3	26.8	27.5	28.0	27.6	26.4	26.1	27.6	28.3	27.9	32.7	27.5
Time of measurements and environmental conditions													
Time, CDT	14:00	11:30	16:00	11:10	13:30	11:30	14:20	13:20	12:35	15:30	11:00	14:20	
Temperature, °C	22.1	20.5	22.3	22.0	24.0	24.8	22.8	21.3	23.9	24.8	25.6	27.7	23.5
RH, %	42	60	46	44	41	52	42	53	46	42	56	49	
PAR, μmol/m²/s	1880	1650	1600	1850	1700	1770	1720	1900	1800	1650	1560	1710	

Source: Chaudhuri, U.N. et al., Effect of elevated levels of CO_2 on winter wheat under two moisture regimes, in *Response of Vegetation to Carbon Dioxide*, Research Report No. 029, U.S. Department of Energy, Carbon Dioxide Research Division, Office of Energy Research, Washington, DC, 1986, xii+77pp.
The time of measurements and the environmental conditions outside of the chambers are given at the bottom of the table.

TABLE 7.5
Canopy Temperature (°C) of Winter Wheat Supplied with High and Low Water Levels as Affected by CO_2 Concentration during the 1985–1986 Growing Season

CO_2 µmol/mol	1986								Aver.
	April 24	April 29	April 30	May 6	May 13	May 15	May 16	May 21	
High water									
340	17.2	26.6	13.7	27.2	25.8	24.4	16.9	24.2	22.0
485	17.1	26.9	13.3	27.3	25.0	26.6	17.5	26.9	22.6
660	16.8	27.6	14.6	28.0	26.1	27.6	17.0	27.3	23.1
825	16.3	28.0	13.8	27.8	26.0	27.4	17.0	28.0	23.0
Low water									
340	17.4	28.4	15.7	30.2	30.1	30.5	19.4	29.9	25.2
485	17.6	28.5	15.6	29.3	27.9	29.7	17.1	30.4	24.5
660	17.1	28.6	14.6	30.1	28.6	27.5	18.5	30.8	24.5
825	16.8	28.7	14.4	30.8	29.1	29.7	18.3	31.0	24.9
Time of measurements and air temperature									
Time, CDT	...[a]	13:50	...	15:00	12:30	14:00	...	15:00	...
Temperature, °C	14.4	28.6	6.7	31.0	25.3	26.5	15.7	28.3	22.1

Source: Chaudhuri, U.N. et al., Effect of elevated levels of CO_2 on winter wheat under two moisture regimes, in *Response of Vegetation to Carbon Dioxide*, Research Report No. 040, U.S. Department of Energy, Carbon Dioxide Research Division, Office of Energy Research, Washington, DC, 1987, xiv+70pp.
The time of measurements and the air temperature outside of the chambers are given at the bottom of the table.
[a] Unknown.

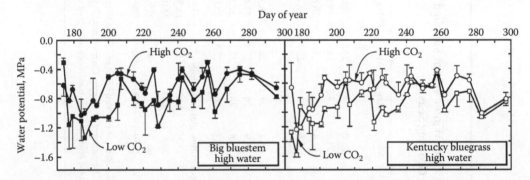

FIGURE 7.4 Water potential of Kentucky bluegrass and big bluestem grown with a high soil-water level (field capacity) and two atmospheric CO_2 concentrations. Vertical bars = ± standard deviation. (Reprinted from He, H. et al., *Trans. Kansas Acad. Sci.*, 95, 139, 1992, Fig. 5, Copyright 1992, Kansas Academy of Science. With permission from the Editor of the Transactions of the Kansas Academy of Science, Fort Hays State University.)

conditions. Therefore, as for winter wheat (Chaudhuri et al. 1986, 1987), CO_2 appeared to compensate for reductions in plant water potential because of drought. For both CO_2 treatments, Kentucky bluegrass had a lower water potential than that of big bluestem. Hicks et al. (1990) also found that a native C_3 perennial bunchgrass, *Stipa leucotricha*, had a lower water potential than did a C_4 bunchgrass, *Bouteloua curtipendula*. They did not study elevated CO_2.

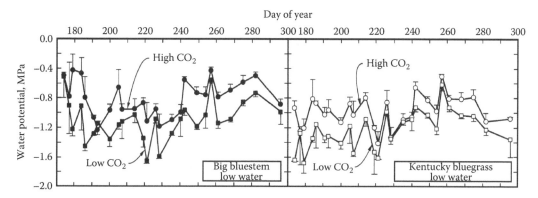

FIGURE 7.5   Water potential of Kentucky bluegrass and big bluestem grown with a low soil-water level (half field capacity) and two atmospheric CO_2 concentrations. Vertical bars = ± standard deviation. (Reprinted from He, H. et al., *Trans. Kansas Acad. Sci.*, 95, 139, 1992, Fig. 5, Copyright 1992, Kansas Academy of Science. With permission from the Editor of the Transactions of the Kansas Academy of Science, Fort Hays State University.)

The plants under the elevated CO_2 had a higher water potential (between 0.2 and 0.3 MPa higher) most likely because there was a higher soil water content under the elevated atmospheric CO_2 conditions compared to the ambient level (see Chapter 5). Thus, we see that in the equation for photosynthesis given in the Preface (CO_2 and H_2O joining to form sugar and oxygen), both CO_2 and H_2O are affected by elevated CO_2, and each factor favors increased growth. Because we have more CO_2 in the atmosphere, the reaction is pushed to the right, which results in more production of sugar and oxygen. With more water in the soil, the water potential of the plants is increased, and this favors growth too (Figure 7.1). Chapter 15 documents the increased growth observed under elevated CO_2.

Research following ours on the tallgrass prairie confirmed our early findings. During two seasons (1991 and 1992), Knapp et al. (1993) measured midday water potential of leaves of *A. gerardii* using a pressure chamber. Plants were exposed to ambient (350 µmol/mol) or elevated (double the ambient or 700 µmol/mol) levels of atmospheric CO_2 in open-top chambers. Plants outside of the chambers and exposed to the ambient level of CO_2 also were measured. Midday xylem pressure potentials of *A. gerardii*, plotted as a function of "Julian date," were higher in plants grown at elevated CO_2 in both years (Figure 7.6). [Xylem pressure potential is another way of expressing the water potential obtained with a pressure chamber (Kirkham 2005, pp. 263–268). Julian date is a former, erroneous term used for "day of year" (Stone 1983).] Xylem pressure potentials were lower in 1991 than 1992 because 1991 was a drier year than 1992. During June and July of 1992, Knapp et al. (1994a) measured the effect of shade on the measurements. The plants were subjected to fluctuations in sunlight similar to that produced by clouds or within canopy shading. Plants were exposed to alternating 8 min periods of full sunlight [more than 1500 µmol/m²/s photon flux density or PFD (light with wavelengths between 0.4 and 0.7 µm or photosynthetically active radiation)] and 5 min periods of shade (350 µmol/m²/s PFD). Even with shade, the plants that grew with elevated CO_2 had a higher water potential than plants that grew with ambient levels of CO_2 (Figure 7.7).

Using open-top chambers, Nelson et al. (2004) and Morgan et al. (2001, 2004) investigated the effects of twice ambient CO_2 (720 µmol/mol) on plant and soil water relations of a semiarid short-grass steppe of northeastern Colorado, United States, from 1997 to 2001. Unchambered plots were designated the controls. Soil at the site was a fine sandy loam. Volumetric soil moisture content was measured below the 15 cm depth using a neutron probe and above this depth using time domain reflectometry. Growing season precipitation ranged from 247 mm in 2000 to 523 mm in 1999, compared to the long-term average of 280 mm. Both 1997 (480 mm) and 1999 (523 mm) were well above

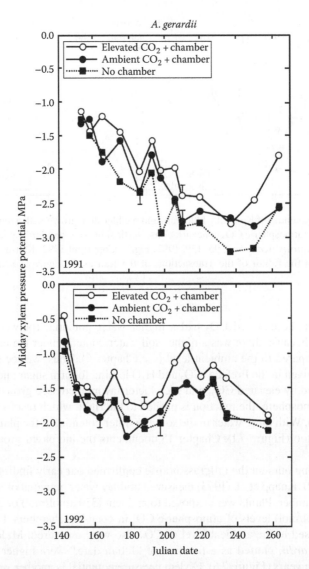

FIGURE 7.6 Seasonal course of midday leaf xylem pressure potential in *A. gerardii* for the 1991 and 1992 growing seasons. Plants were exposed to elevated (double ambient) CO_2 concentrations and ambient CO_2 in open-top chambers. Vertical bars indicate maximum standard errors of the mean for each treatment. (Reprinted from Knapp, A.K. et al., *Int. J. Plant Sci.*, 154, 459, 1993, Fig. 1. With permission. Copyright 1993, University of Chicago Press.)

average rainfall years; 1998 (302 mm) and 2001 (311 mm) were near average; and 2000 (247 mm) was a dry year. Although the total precipitation in 2001 was near the long-term average, spring rains failed to recharge the soil profile following a dry year (2000) and a summer drought decreased soil moisture values to the lowest recorded during the study (Figure 7.8). Despite the differences in seasonal rainfall, total soil water content in the upper 1 m during the season was higher under elevated CO_2 compared to ambient CO_2 (Figure 7.8). When averaged across years, volumetric soil water content in the 15–45 and 75–105 cm depth increments was higher under elevated CO_2 compared to ambient CO_2 over the study period, with a strong trend for higher soil water content at the 45–75 cm depth (Figure 7.9).

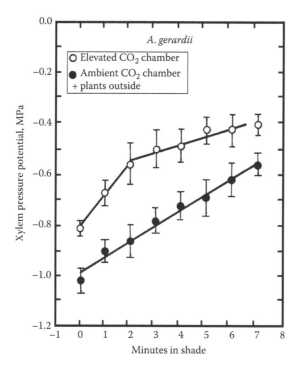

FIGURE 7.7   Effect of atmospheric CO_2 concentration on leaf xylem pressure potential in *A. gerardii* during periods of shade. Measurements were made in the field on plants exposed to ambient or elevated CO_2, and data shown are means ± 1 standard error of the mean. No measurements are reported for the unchambered plots, even though there is a symbol for them. (Knapp, A.K., Fahnestock, J.T., and Owensby, C.E., Elevated atmospheric CO_2 alters stomatal responses to variable sunlight in a C_4 grass. *Plant Cell Environ*. 1994a. 17. 189. Fig. 4. Copyright Wiley-VCH Verlag GmbH & Co. KGaA. Reprinted with permission.)

Averaged over the study period, leaf water potential was enhanced 24%–30% under elevated CO_2 in the major warm- and cool-season grass species of the shortgrass steppe. Specifically, the values were 28.5% for *B. gracilis* H.B.K. (Lag.) or blue grama, C_4; 24.7% for *Pascopyrum smithii* (Rydb.) A. Love or western wheat grass, C_3; and 30.4% for *S. comata* Trin and Rupr. or needle and thread, C_3 (Figure 7.10). The degree of responsiveness in leaf water potential to elevated CO_2 did not differ between C_3 and C_4 plant types. The seasonal data for the water potential of the dominant C_4 grass, *B. gracilis*, in 2000 is shown in Figure 7.11c with the corresponding soil water content (Figure 7.11d) (Morgan et al. 2004). The increase in aboveground biomass due to elevated CO_2 is shown in Figure 7.11a for each of the 5 years as well as the average for the 5 years (Figure 7.11b). Water use efficiency (grams aboveground biomass harvested/kilogram water consumed) was 43% higher under elevated CO_2 (average value: 6.10 g/kg) than under ambient CO_2 (average value: 4.27 g/kg) over the study period (Figure 7.12). (Chapter 11 documents other studies that show an increase in water use efficiency under elevated CO_2.)

The results of Nelson et al. (2004) and Morgan et al. (2004) suggest that a future, elevated CO_2 environment may result not only in increased plant productivity due to improved water use efficiency, but also lead to increased water drainage and deep soil moisture storage in this semi-arid grassland ecosystem. This, along with the ability of the major grass species (blue grama) to maintain a favorable water status under elevated CO_2 (i.e., a higher plant water potential), should result in the short-grass steppe being less susceptible to prolonged periods of drought. However, increases in deep soil water may eventually favor deeper-rooted over shallow-rooted species.

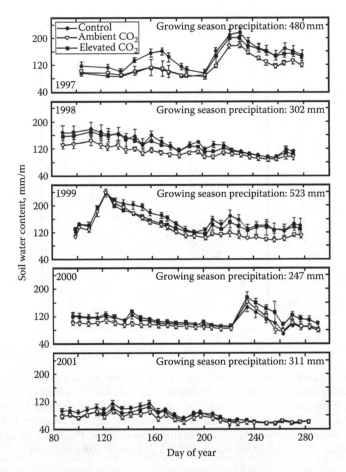

FIGURE 7.8 Total soil water content (mm) in the upper 1 m of the soil profile in unchambered controls and chambered ambient (360 µmol/mol) and elevated (720 µmol/mol) CO_2 plots in five growing seasons (1997–2001) on the shortgrass steppe of Colorado. Total growing season precipitation (mm) is noted for each year of the study. (Reprinted with kind permission from Springer Science+Business Media: *Plant Soil*, Elevated CO_2 increases soil moisture and enhances plant water relations in a long-term field study in semi-arid shortgrass steppe of Colorado, 259, 2004, 169, Nelson, J.A., Morgan, J.A., LeCain, D.R., Mosier, A.R., Milchunas, D.G., and Parton, B.A., Fig. 1. Copyright 2004, Springer.)

The CO_2-induced deep-water drainage beyond the root zone will result in recharge of the ground water (Morgan et al. 2004).

Jackson et al. (1994) studied the effects of elevated CO_2 in an annual grassland in California, United States. The study site was the Jasper Ridge Biological Preserve near Stanford University in Palo Alto, California. (See Chapter 6, which deals with roots, for other studies carried out at this site.) Three field treatments were used: no-chamber controls, open-top chambers with ambient CO_2, and open-top chambers with ambient $CO_2 + 350$ µmol/mol CO_2 (a seasonal average of 723 µmol/mol CO_2). Stomatal conductance, transpiration, and photosynthesis were measured with a closed gas-exchange system. Stomatal conductance and transpiration also were measured using a steady state porometer. Midday leaf water potentials were measured with a pressure chamber. Elevated CO_2 led to lower stomatal conductance and transpiration (approximately 50%) and higher midday leaf water potentials (30%–35%) in the most abundant species of the grassland, *Avena barbata* Brot. (slender wild oat) (Figure 7.13). Higher CO_2 concentrations also resulted in greater midday photosynthetic rates (70% greater on average). The effects of CO_2 on stomatal conductance, transpiration, leaf water potential, and photosynthesis decreased toward the end of the growing season when the plants

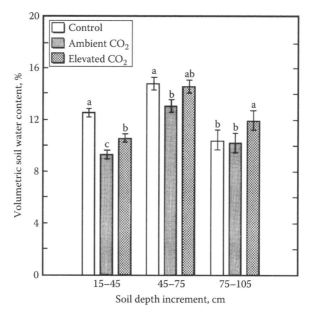

FIGURE 7.9 Percent volumetric soil water content (±1 S.E.; $n=3$) averaged over 5 years of study (1997–2001) of the 15–45, 45–75, and 75–105 cm depth increments in unchambered control and chambered ambient (360 μmol/mol CO_2) and elevated (720 μmol/mol CO_2) plots on the shortgrass steppe of Colorado. Different letters denote significant treatment effects ($P<0.05$; Least Square Means) at each soil depth. (Reprinted with kind permission from Springer Science+Business Media: *Plant Soil*, Elevated CO_2 increases soil moisture and enhances plant water relations in a long-term field study in semi-arid shortgrass steppe of Colorado, 259, 2004, 169, Nelson, J.A., Morgan, J.A., LeCain, D.R., Mosier, A.R., Milchunas, D.G., and Parton, B.A., Fig. 2. Copyright 2004, Springer.)

began to show signs of senescence. Their study confirmed results of others for grasses: elevated CO_2 increases the water potential of leaves.

SOYBEANS

Prior et al. (1991) grew soybean [*Glycine max* (L.) Merr. cv. Bragg] in large containers in open-top field chambers under five atmospheric CO_2 concentrations and two watering regimes. Plants also were grown in pots in unchambered plots under ambient CO_2. The soil was a mixture of a sandy clay loam soil, sand, and commercial potting mix. The CO_2 concentrations were 348 (ambient plots without chambers), 349 (ambient, chambered plots), 421, 496, 645, and 946 μmol/mol. Note that the 421 μmol/mol CO_2 concentration used by Prior et al. (1991) was only 36 μmol/mol more than the globally averaged CO_2 concentration in 2008, which was 385 μmol/mol (Dlugokencky 2009). In 1991, the 421 μmol/mol CO_2 treatment was 72 μmol/mol CO_2 above ambient. Half of the plants were well watered (containers were re-watered when the matric potential, as measured with tensiometers, reached −0.005 to −0.015 MPa), and half of the plants were water stressed (containers were re-watered when the matric potential was between −0.075 and −0.085 MPa). Tensiometer readings started when the plants were at the third-node stage (V3 stage; Fehr and Caviness 1977). Transpiration, stomatal resistance, xylem pressure potential, and water use were measured at two periods during reproductive growth: anthesis (R1 stage) and pod fill (R5 stage). Transpiration and stomatal resistance were measured on the abaxial (lower) leaf surfaces with a steady state diffusion porometer. Water potential was measured with a pressure chamber.

Soybean is more sensitive to water stress during pod and seed development than during vegetative growth. By stage R5 (pod fill), water-stressed plants at the lower CO_2 levels (349, 421, and

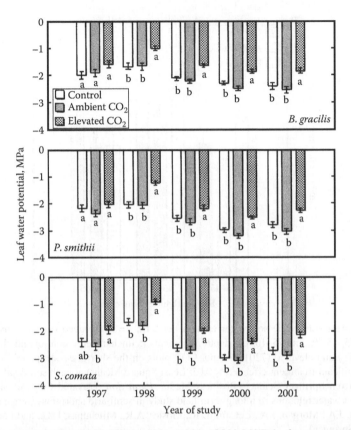

FIGURE 7.10 Seasonal average (1997–2001) midday (1000–1200 h MST) leaf water potential (±1 S.E.; $n=3$) for three common grass species of the shortgrass steppe of Colorado in unchambered control and chambered ambient (360 μmol/mol CO_2) and elevated (720 μmol/mol CO_2) plots. Different letters denote significant treatment effects ($P < 0.05$; Least Square Means) within a year for each grass species. (Reprinted with kind permission from Springer Science+Business Media: *Plant Soil*, Elevated CO_2 increases soil moisture and enhances plant water relations in a long-term field study in semi-arid shortgrass steppe of Colorado, 259, 2004, 169, Nelson, J.A., Morgan, J.A., LeCain, D.R., Mosier, A.R., Milchunas, D.G., and Parton, B.A., Fig. 5. Copyright 2004 Springer.)

496 μmol/mol) depleted soil water faster than the high CO_2 level plants (645 and 946 μmol/mol). During reproductive growth (Figure 7.14), the positive effects of CO_2 enrichment on soil water conservation were observed earlier under the well-watered regime (Figure 7.14a) than under the water-stressed condition (Figure 7.14b). Note in Figure 7.14a, the soil with the plants grown under the ambient CO_2 level (triangles) had the lowest soil matric potentials that were evident 62 days after planting. Although high CO_2 grown plants were larger, water extraction from the soil proceeded at a lower rate. For example, under water-stressed conditions on day 70, the 349 μmol/mol CO_2 treatment (triangles) reached the end of a drying cycle before the 645 (squares) or 946 (diamonds) μmol/mol CO_2 treatments. High CO_2 plants encountered the same low water a few days later. But because the watering cycle was longer, these plants experienced fewer stressful days than low CO_2 plants over the season. Overall, during reproductive growth, water-stressed plants at 349 μmol/mol CO_2 went through 12 drying cycles compared with 10 and 9 at 645 and 946 μmol/mol CO_2, respectively. Under well-watered conditions, plants grown at 349 μmol/mol CO_2 received irrigation water 29 times compared with 24 and 21 times at 645 and 946 μmol/mol CO_2, respectively. Based on these observed values, Prior et al. (1991) inferred that CO_2 enrichment has a strong tendency to alleviate plant water stress when critical periods of water demand occur during reproduction.

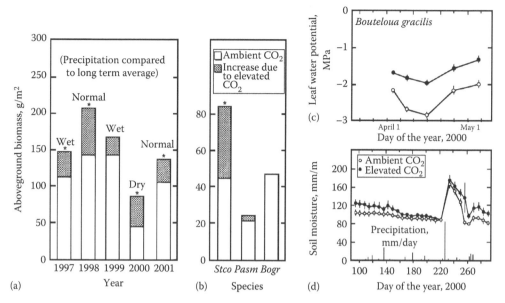

FIGURE 7.11 Yearly aboveground total plant biomass (panel a), species aboveground biomass averaged across 5 years (panel b: species are Stco *S. comata*; Pasm *P. smithii*; and Bogr *B. gracilis*), leaf water potential for the dominant C_4 grass, *B. gracilis* (panel c: year 2000), and soil water content (panel d: year 2000) in a Colorado shortgrass steppe under present ambient and elevated (720 µmol/mol) CO_2 concentrations. In panels (a) and (b), the asterisk (*) indicates significant CO_2 treatment effect for years and species, respectively, at $P < 0.05$ as determined by Tukey's means comparison test. Standard error bars are given with means in panels c and d. Water relations data for 2000 are typical of what occurred in other study years (Morgan et al. 2001; Nelson et al. 2004). (Reprinted with kind permission from Springer Science+Business Media: *Oecologia*, Water relations in grassland and desert ecosystems exposed to elevated atmospheric CO_2, 140, 2004, 11, Morgan, J.A. et al., Fig. 3. Copyright 2004, Springer.)

Leaves of CO_2-enriched soybean plants grown under well-watered conditions had higher stomatal resistances (Figure 7.15a) and lower transpiration rates (Figure 7.15b) relative to well-watered plants grown under ambient CO_2 (circles, Figure 7.15). [Note that Prior et al. (1991) use different symbols for the same CO_2 treatment in different figures. In Figure 7.14, the ambient treatment was represented by triangles. In Figure 7.15, it is represented by circles.] Under water-stressed conditions, the stomatal response at 645 (triangles) and 946 (squares) µmol/mol CO_2 (Figure 7.15c) resulted in transpiration rates (Figure 7.15d) that were similar to those observed under well-watered conditions (Figure 7.15b). Even during early portions of stress periods, lower levels of elevated CO_2 (421 µmol/mol, the diamonds in Figure 7.15, and 496 µmol/mol, the hexagons in Figure 7.15) induced partial stomatal closure that caused lower transpiration rates relative to the 349 µmol/mol control. Thus, even a small rise in CO_2 concentration in the atmosphere (72 µmol/mol above ambient) had a large and measurable effect on stomatal resistance and transpiration rate of soybeans (Figure 7.15). This shows how sensitive stomata are to the CO_2 concentration in the atmosphere.

Midday xylem water potential data for the soybeans from both reproductive periods are shown in Figure 7.16 (top part, anthesis; bottom part, pod fill) as a function of CO_2 concentration. At anthesis (R1), CO_2 enrichment had little effect on reducing water potential under water stress (Figure 7.16a). The line for the well-watered plants (symbols with solid triangles) was only slightly above the line for the water-stressed plants (symbols with solid circles), and the slope of the lines was nearly flat, indicating little response to CO_2. The fact that similar midday leaf water potentials were seen was not surprising because plants at the various CO_2 concentrations had extracted soil water at equivalent rates and had reached about the same soil moisture level by the time of measurement at anthesis (day 60, Figure 7.16a). However, at pod fill (R5), CO_2 had an effect on water potential

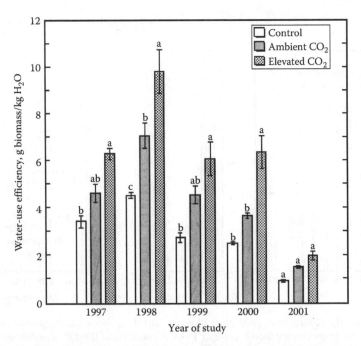

FIGURE 7.12 Yearly (1997–2001) water use efficiency (grams aboveground biomass harvested per kilo-gram water consumed; ±1 S.E.; $n=3$) for unchambered control and chambered ambient (360 μmol/mol) and elevated (720 μmol/mol) CO_2 plots in the shortgrass steppe of Colorado. Different letters denote significant treatment effects ($P<0.05$; Least Square Means) within a year. (Reprinted with kind permission from Springer Science+Business Media: *Plant Soil*, Elevated CO_2 increases soil moisture and enhances plant water relations in a long-term field study in semi-arid shortgrass steppe of Colorado, 259, 2004, 169, Nelson, J.A., Morgan, J.A., LeCain, D.R., Mosier, A.R., Milchunas, D.G., and Parton, B.A., Fig. 6. Copyright 2004, Springer.)

under water-stressed conditions (Figure 7.16b). This was because water-stressed plants grown under elevated CO_2 had lower rates of water use and thus were under a lower level of stress. Plants grown at 349, 421, and 496 μmol/mol CO_2 had water potentials of −1.08, −0.83, and −0.72 MPa, respectively. Because the soil water levels were still relatively high under water-stressed conditions at the 645 and 946 μmol/mol CO_2 levels, their respective water potential values were similar to their well-watered control values (Figure 7.16b). High CO_2 plants showed no signs of wilting, unlike the plants grown with ambient levels of CO_2 (Prior et al. 1991). Thus, soybean plants grown at high CO_2 presumably have an advantage over plants grown at a lower CO_2 level in that they can survive drought better.

PEAS

Davis and Potter (1989) tested three levels of atmospheric CO_2 for their effect on water relations and adventitious root formation in leafy pea (*Pisum sativum* L. cv. Alaska) cuttings. Plants were grown from seeds in a commercial potting mix composed of peat and vermiculite. Ten days after sowing, cuttings were made by excising the seedlings at the first scale leaf and inserting the basal 2–3 cm into the same potting mix. Plants were exposed to three levels of CO_2: ambient (350 μmol/mol) and elevated (675 or 1800 μmol/mol). A thermocouple psychrometer chamber was used to measure water potential of 2 cm long stem segments from the basal ends of the cuttings. Immediately after determination of the water potential, tissue was sealed in a chamber, frozen, thawed, and solute (osmotic) potential was measured. Turgor pressure (turgor potential) was calculated by taking the

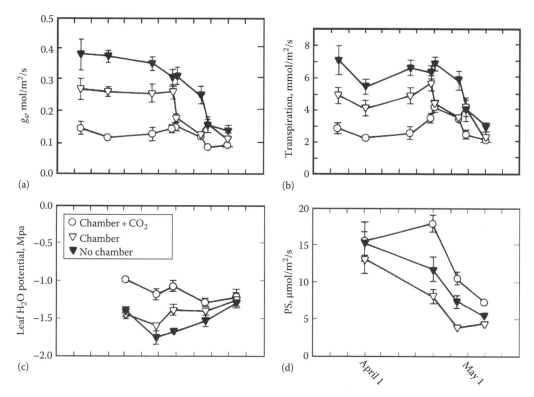

FIGURE 7.13 Midday leaf stomatal conductance (a), transpiration (b), water potential (c), and photosynthesis (d) for *Avena* plants in no-chamber controls (solid triangles), chambered controls (open triangles), and chamber + 350 μmol/mol CO_2 (open circles) (about 700 μmol/mol CO_2). Mean ± standard error are shown ($n = 6$ to 10). Physiological measurements were initiated in mid-to-late March 1993, when the plants were of sufficient size, and continued through plant senescence in mid-May. Measurements were taken on the most recent, fully expanded leaf of random *Avena* plants in each field plot (10 plots per treatment; multiple plants averaged for a given plot). (Reprinted with kind permission from Springer Science+Business Media: *Oecologia*, CO_2 alters water use, carbon gain, and yield for the dominant species in a natural grassland, 98, 1994, 257, Jackson, R.B., Sola, O.E., Field, C.B., and Mooney, H.A., Fig. 1. Copyright 1994, Springer.)

difference between water potential and solute potential. Generally, throughout the first 7 days of the rooting period, turgor potential was 0.1–0.2 MPa higher in the 1800 μmol/mol CO_2 cuttings (triangles, Figure 7.17) than in controls (circles, Figure 7.17). This level of CO_2 raised water potential only after day 4 and did not affect the solute potential. The increase in turgor potential was probably due to reduced transpiration caused by partial stomatal closure in response to high CO_2. Both CO_2 enrichment treatments stimulated root growth, possibly by a combination of high water potential and high turgor potential.

TREES

Eamus et al. (1995) investigated the effect of CO_2 enrichment upon the water relations of *Maranthes corymbosa* Blume and *Eucalyptus tetrodonta* F. Muell, which are species of the wet–dry tropics. (Plants in the genus *Eucalyptus* have the common name of "gum-tree." *Maranthes* apparently has no common name.) Seeds of the two trees were sown under ambient (concentration not given) or enriched (700 μmol/mol) levels of CO_2 in tropical Australia and allowed to grow for 20 months rooted in the ground (type of soil was not stated). For both species, measurements of leaf water

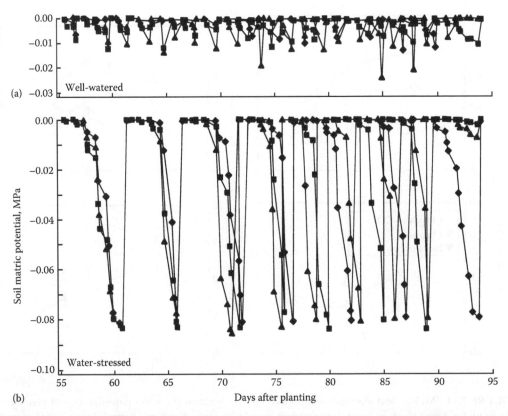

FIGURE 7.14 Changes in soil matric potential (a, b) during reproductive growth (55–94 days after planting) for well-watered and water-stressed soybean grown at 349 (triangles), 645 (squares), and 946 (diamonds) µmol/mol CO_2. For well-watered plants, $n=4$; for water-stressed plants, $n=6$. (Reprinted from *Agric. Ecosyst. Environ.*, 35, Prior, S.A., Rogers, H.H., Sionit, N., and Patterson, R.P., Effects of elevated atmospheric CO_2 on water relations of soya bean, 13, Copyright 1991, Fig. 1, with permission from Elsevier.)

potential, stomatal conductance, and leaf temperature were made during a daytime period. Leaf water potential was measured with a pressure chamber, and stomatal conductance was measured with a porometer. Transpiration rates were calculated using leaf temperature, leaf-to-air vapor pressure difference, and stomatal conductance. Measurements of predawn leaf water potential also were made on *E. tetrodonta*. Pressure–volume curves were developed [see Kirkham (2005, pp. 289–298) for the method] to calculate graphically maximum turgor potential, osmotic potential at zero turgor potential, and relative water content at zero turgor potential.

For *M. corymbosa*, leaf water potential was lower throughout the day for control leaves compared to leaves growing in CO_2-enriched air (Figure 7.18). Mean daily water potentials for *M. corymbosa* under ambient and elevated CO_2 were −0.76 and −0.68 MPa, respectively. Similarly, predawn leaf water potential was lower for control *E. tetrodonta* trees (average value: −0.84 MPa) than for trees grown with CO_2 enrichment (average value: −0.31 MPa). However, mid-afternoon leaf water potentials were lower for *E. tetrodonta* plants growing in CO_2-enriched air (average value: −1.40 MPa) than those growing in ambient air (average value: −1.25 MPa). In both species, transpiration rates and stomatal conductance were lower for trees grown in CO_2-enriched air than for controls (Figure 7.19 for *M. corymbosa*). For both species, maximum turgor potential increased, and osmotic potential at zero turgor decreased for trees grown in CO_2-enriched air. For *M. corymbosa*, maximum turgor potentials under ambient and elevated CO_2 were 1.30 and 1.49 MPa, respectively. For *E. tetrodonta*, these values were 1.30 and 1.72 MPa, respectively. For

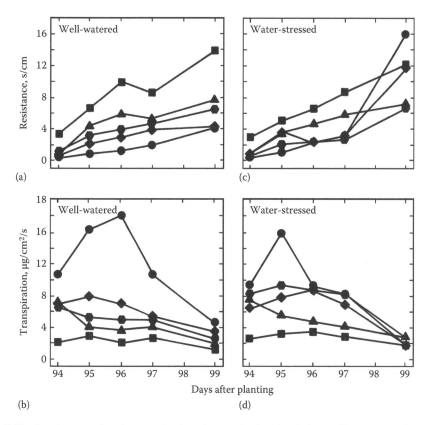

FIGURE 7.15 Leaf stomatal resistance (a, c) and transpiration (c, d) for well-watered (a, b) and water-stressed (c, d) soybean grown in five CO_2 treatments within chambers at pod fill. Circles: 349 μmol/mol; diamonds: 421 μmol/mol; hexagons: 496 μmol/mol; triangles: 645 μmol/mol; and squares: 946 μmol/mol. Measurements were taken during a drying cycle; $n = 8$. (Reprinted from *Agric. Ecosyst. Environ.*, 35, Prior, S.A., Rogers, H.H., Sionit, N., and Patterson, R.P., Effects of elevated atmospheric CO_2 on water relations of soya bean, 13, Copyright 1991, Fig. 5, with permission from Elsevier.)

M. corymbosa, the osmotic potentials at zero turgor potential under ambient and elevated CO_2 were −1.43 and −1.68 MPa, respectively. For *E. tetrodonta*, these values were −1.61 and −2.00 MPa, respectively. There was no difference between CO_2 treatments in the graphically determined relative water content at zero turgor potential, and the relative water content at this point was, on average, 95.25%.

Increased solute accumulation in trees grown in CO_2-enriched air must have occurred in both species because maximum turgor potential increased and the osmotic potential at zero turgor decreased (Eamus et al. 1995). The accumulation of solutes (probably carbohydrates) occurred due to enhanced assimilation rates under CO_2 conditions. These solutes can cause an increased turgor potential, and they may result in an increased ability to withstand low water potentials.

Morse et al. (1993) studied the effect of increasing atmospheric CO_2 concentrations on tissue water relations in *Betula populifolia* Marsh. (gray or white birch), a common pioneer tree species of the northeastern deciduous forests of the United States. Seedlings grew in pots with a mixture of soil, sand, and Turface and under either ambient (350 μmol/mol) or elevated (700 μmol/mol) CO_2 concentrations and with two watering regimes. Mesic plants received 150 mL of water when the soil surface looked dry. Xeric plants received 50 mL of water only when the plants wilted. Components of tissue water relations were estimated from pressure–volume curves of tree seedlings using a pressure chamber.

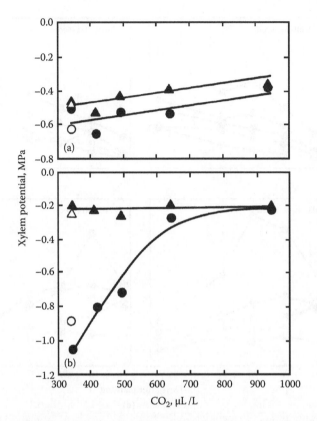

FIGURE 7.16 Leaf xylem potential for well-watered (open and closed triangles) and water-stressed (open and closed circles) soybean grown in open plots (open symbols) and five CO_2 concentrations within chambers (closed symbols) at anthesis (a: day 60) and pod fill (b: day 94). Chamber effect was not significant. Response to CO_2 was not significant in anthesis (a) and for the well-watered plants at pod fill (b); $n = 6$. (Reprinted from *Agric. Ecosyst. Environ.*, 35, Prior, S.A., Rogers, H.H., Sionit, N., and Patterson, R.P., Effects of elevated atmospheric CO_2 on water relations of soya bean, 13, Copyright 1991, Fig. 6, with permission from Elsevier.)

On average, osmotic potentials at full hydration were 0.1 MPa lower with xeric conditions and 0.1 MPa lower under elevated CO_2. These shifts together gave the ambient CO_2, mesic plants the highest osmotic potential at full hydration (−0.92 MPa) and the elevated CO_2, xeric plants the lowest osmotic potential at full hydration (−1.23 MPa). The shift in osmotic potential in response to both CO_2 and drought enabled the plants to withstand lower water potentials. Associated with differences in osmotic potentials were shifts in the relationship between turgor pressure (also called turgor potential) and plant water potential. Typical data obtained from a pressure–volume analysis are shown in Figure 7.20. Under high CO_2, xeric plants maintained turgor to water potentials 0.2–0.7 MPa lower than the mesic plants (Figure 7.20b).

Tschaplinski et al. (1995) studied the effect of elevated CO_2 on osmotic potential and osmotic adjustment of 1-year-old seedlings of American sycamore (*Platanus occidentalis* L.), sweet-gum (*Liquidambar styraciflua* L.), and sugar maple (*Acer saccharum* Marsh.) under drought. They were grown in open-top chambers containing either ambient (343 μmol/mol) or ambient air enriched with 300 μmol/mol CO_2. Plants were well watered daily or subjected to a series of drought cycles. The mean soil water potential at re-watering for drought-stressed seedlings in ambient CO_2 was −0.5, −0.7, and −1.8 MPa for sugar maple, sweet-gum, and sycamore, respectively, compared with −0.2, −0.7, and −1.2 MPa, respectively, under elevated CO_2. Well-watered plants were maintained at soil water potentials greater than −0.1 MPa. Drought tolerance of leaves was assessed by pressure–volume analysis. Drought under ambient CO_2 reduced the osmotic

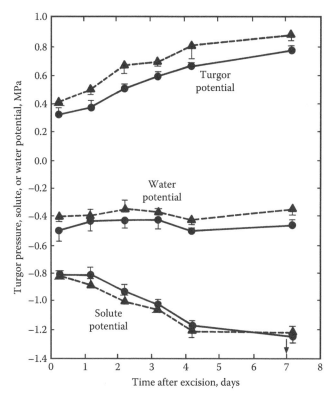

FIGURE 7.17   Water potential, solute potential, and turgor pressure (turgor potential) of basal 2 cm stem segments of leafy pea cuttings rooted in atmospheres containing either 350 (circles) or 1800 (triangles) µmol/mol CO_2 for the first 7 days after excision. Bars indicate standard error of the mean ($n = 8$). Arrow indicates the time of root emergence. (Davis, T.D. and Potter, J.R., Relations between carbohydrate, water status and adventitious root formation in leafy pea cuttings rooted under various levels of atmospheric CO_2 and relative humidity. *Physiol. Plant.*, 1989, 77, 185, Fig. 1. Copyright Wiley-VCH Verlag GmbH & Co. KGaA. Reprinted with permission.)

potential for leaves of sycamore and sweet-gum by 0.30 and 0.61 MPa, respectively, but leaves of sugar maple did not display osmotic adjustment to drought. Elevated CO_2 increased osmotic potential at turgor loss (zero turgor) for leaves of sugar maple by 0.3 MPa under well-watered conditions and 0.48 MPa under drought. Under well-watered conditions, the osmotic potential at zero turgor for sugar maple was −2.22 and −1.89 MPa, respectively, for the ambient and elevated CO_2 conditions. Under drought, the osmotic potential at zero turgor for sugar maple was −2.31 and −1.83 MPa, respectively, for the ambient and elevated CO_2 conditions. Osmotic potential at zero turgor for sycamore and sweet-gum was not affected by CO_2 treatment. Tschaplinski et al. (1995) suggested that the response of sugar maple might be related to its rapid growth causing a depletion of solutes. Even though osmotic potential at zero turgor of the sugar maple shoots was affected by CO_2, elevated CO_2 did not alter the concentration of total solutes in its roots or roots of the other two species.

Tschaplinski et al. (1993) used pressure–volume curves to determine the effects of elevated CO_2 on osmotic adjustment of shoots of loblolly pine (*Pinus taeda* L.) grown in pots with a commercial potting mix. Seedlings were placed in controlled-environment chambers in either 350 or 700 µmol/mol CO_2 with weekly watering for 4 months, after which they were either watered weekly (well-watered treatment) or every 2 weeks (water-stressed treatment) for 59 days. Under well-watered conditions, osmotic potential at full turgor was −1.21 and −1.13 MPa for the ambient and elevated CO_2 treatments, respectively. Under water-stressed conditions, osmotic potential at full turgor was −1.19

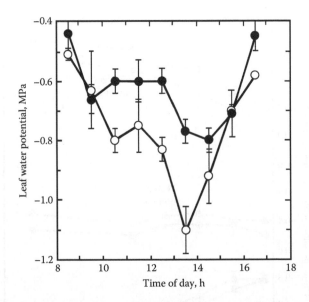

FIGURE 7.18 The time course of changes in leaf water potential of *M. corymbosa* seedlings growing under ambient (open circles) and CO_2-enriched (closed circles) conditions. (Reprinted from Eamus, D. et al., *Aust. J. Bot.*, 43, 273, 1995, Fig. 1. With permission. Copyright 1995, CSIRO Publishing.)

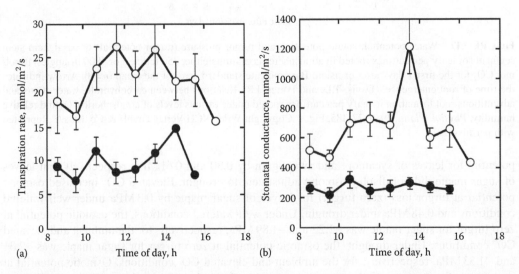

FIGURE 7.19 A comparison of transpiration rates (a) and stomatal conductance (b) of seedlings of *M. corymbosa* growing under ambient CO_2 (open circles) or CO_2-enriched (closed circles) conditions. (Reprinted from Eamus, D. et al., *Aust. J. Bot.*, 43, 273, 1995, Fig. 2. With permission. Copyright 1995, CSIRO Publishing.)

and −1.03 MPa for the ambient and elevated CO_2 treatments, respectively. Differences in osmotic potential at full turgor among the treatments were not different. But differences among treatments were different for osmotic potential at zero turgor. Under well-watered conditions, the osmotic potential at zero turgor was −1.92 and −1.64 MPa for the ambient and elevated CO_2 treatments, respectively. Under water-stressed conditions, the osmotic potential at zero turgor was −2.09 and −1.78 MPa for the ambient and elevated CO_2 treatments, respectively. Thus, elevated CO_2 increased

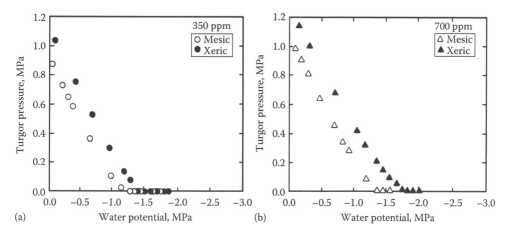

FIGURE 7.20 The relationship between water potential and turgor pressure (or turgor potential) under slow dehydration for a representative branch of gray birch under ambient [(a) 350 ppm] or elevated [(b) 700 ppm] CO_2. (Reprinted with kind permission from Springer Science+Business Media: *Oecologia*, Elevated CO_2 and drought alter tissue water relations of birch (*Betula populifolia* Marsh.) seedlings, 95, 1993, 599, Morse, S.R., Wayne, P., Miao, S.L., and Bazzaz, F.A., Fig. 1. Copyright 1993, Springer.)

osmotic potential of loblolly pine at zero turgor by 0.28 and 0.31 MPa under well-watered and water-stressed conditions, respectively.

Centritto et al. (1999a) grew, from seed, cherry seedlings (*Prunus avium* L.) in soil (type of soil not stated) for two growing seasons in ambient (350 μmol/mol) and elevated CO_2 (two times ambient or 700 μmol/mol) concentrations in open-top chambers. A drying cycle was imposed in both growing seasons to half the seedlings. In the first growing season, the cycle was 69–115 days after emergence, and in the second growing season, it was 212–251 days after emergence. The same seedlings were exposed to drought in the second season that had already experienced drought in the first season. Leaf water potential was determined by using a pressure chamber. Leaf osmotic potential was measured on the same leaves used for determining leaf water potential. The leaves were frozen and sap exuded and the osmotic potential of the sap was measured by using a vapor–pressure osmometer. [See Kirkham (2005, pp. 305–308) for a description of the technique.] Gravimetric soil water content was measured and then converted to volumetric soil water content by multiplying by the bulk density.

As expected, soil water content was strongly affected by water regime, but there were no differences in the rate of soil drying between the CO_2 treatments under either the well-watered or the drought-stressed treatments (Figure 7.21a). Over the course of time, the osmotic potential of plants in elevated CO_2 became lower than that in ambient CO_2 in both well-watered and drought-stressed plants (Figure 7.21b). There were no differences in water potential of well-watered seedlings between elevated and ambient CO_2 (Figure 7.21c). In water-stressed plants, the water potential of plants grown under elevated CO_2 was lower than that of plants grown under ambient CO_2 (Figure 7.21c). However, on about day 110, water potential in ambient CO_2 reached the same negative value as in elevated CO_2, and there were no further differences between the CO_2 treatments. During the second growing season, leaf water potential was affected only by water treatments and not by CO_2 concentration. (The authors do not give data for osmotic potential during the second season.) Centritto et al. (1999a) conclude that, because elevated CO_2 did not improve plant water relations, including leaf water potentials and osmotic potentials, it would not increase the water stress tolerance of cherry seedlings. The reason why the leaf water potential of cherry did not increase under elevated CO_2, as has been observed for other species, needs to be investigated.

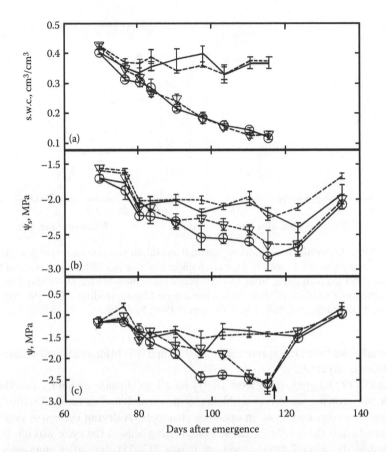

FIGURE 7.21 Time course of well-watered and droughted cherry seedlings grown in ambient and elevated CO_2, during and after the first drought cycle. (a) Soil water content (s.w.c). (b) Osmotic potential (ψ_s). (c) Leaf water potential (ψ). The arrow on the abscissa indicates the end of the drying cycle. Solid lines with no data points show well-watered, elevated CO_2 plants; dashed lines with no data points show well-watered, ambient CO_2 plants; solid lines with open circles for data points show droughted, elevated CO_2 plants; dashed lines with open triangles for data points show droughted, ambient CO_2 plants. Data points are the means of three plants per treatment ± 1 standard error. (Centritto, M., Lee, H.S.J., and Jarvis, P.G., Interactive effects of elevated $[CO_2]$ and drought on cherry (*Prunus avium*) seedlings. II. Photosynthetic capacity and water relations. *New Phytol.*, 1999a. 141, 141, Fig. 1. Copyright Wiley-VCH Verlag GmbH & Co. Reprinted with permission.)

STRESS RELAXATION

Before we can understand the final two papers we are going to discuss in this chapter (Ferris and Taylor 1994a,b), we need to learn about "stress relaxation," which we now consider.

Growth can be defined as irreversible cell enlargement (Hsiao et al. 1976, p. 482). Any reduction in tissue water potential reduces growth (Figure 7.1), sometimes completely stopping when water potential is lowered in a plant by 0.3–0.4 MPa (Hsiao et al. 1976, p. 482). The primary effect of water stress on growth appears to be physical. Turgor pressure is a prerequisite for cell enlargement. In the process of cell expansion, the first step in wall yielding is often viewed as a relaxation of mechanical stress in the wall (Hsiao et al. 1976, p. 484). The increase in cell volume, basic to growth, requires water uptake. In the simplest case, if the osmolality of the cell does not change [see Kirkham (2005, pp. 305–308) for a discussion of osmolality], the only way to create the water potential gradient for uptake (assuming initial equilibrium between cell and surroundings) is through a decrease in the stress borne by the cell wall. This is called "stress relaxation" (Hsiao et al. 1976, p. 484). Turgor stress on the cell wall is prerequisite to stress relaxation. Water uptake, cell wall yielding, and consequently, cell enlargement follow stress relaxation. When cell enlargement is sustained, the processes that produced the initial stress relaxation must be continued. For enlargement to be true growth, the increase in dimension must be irreversible. This requires biochemical modifications of the wall and, commonly, deposition of new wall material. In fact, biochemical events of stress relaxation and wall synthesis are probably not separable from the physical events of water uptake and wall yielding (Hsiao et al. 1976, p. 484).

Like Hsiao et al. (1976), Cosgrove et al. (1984) also explain that cell expansion during plant growth requires simultaneous expansion of the cell wall and absorption of water. Cell enlargement is initiated when biochemical loosening of the cell wall permits the wall to undergo viscoelastic extension in response to hydrostatic pressure of the cell contents. The consequent lowering of the turgor pressure and the water potential of the cell generates a favorable water potential gradient between the growing cell and the apoplastic solution, so as to permit a flux of water into the cell. Absorption of water directly generates the increase in volume, which tends to restore the turgor potential and the water potential toward their original values. Under conditions of steady state growth, stress relaxation of the cell wall, because of loosening, is balanced by an increase in wall stress because of water absorption by the cell, with the result that turgor potential remains constant. The rate of cell extension is thought to be a function of the amount by which turgor potential exceeds the threshold value of turgor pressure required for growth. This is called the "yield threshold." If no water is available for uptake, stress relaxation should result in a drop in turgor pressure until the yield threshold is reached. At this point, further stress relaxation, and hence turgor pressure relaxation, should stop (Cosgrove et al. 1984).

To ascertain the rate and magnitude of stress relaxation of the cell walls in cells of young, rapidly growing stem tissues of pea (*P. sativum* L. cv. Alaska), Cosgrove et al. (1984) used thermocouple psychrometers. They measured water potential of pea internode segments with the psychrometers and osmotic potential by exuding sap from frozen tissue and measuring its osmolality, which can be converted to osmotic potential [see Kirkham (2005, p. 308) for this conversion]. They could calculate turgor potential by subtracting osmotic potential from water potential. If continual measurements of water potential are made in the psychrometric chamber and no water is available for uptake, then the water potential will fall over time. If the osmotic potential of the cell does not change during this time, then when the water potential reaches a minimum value after several hours in the thermocouple psychrometer chamber, the turgor pressure (or turgor potential) should be equal to this minimum water potential. This minimum water potential value attained when relaxation stops is equal to the turgor pressure at the yield threshold (Cosgrove et al. 1984, p. 51). Cosgrove found that the osmolality of the pea segments did not change over time, and, hence, the osmotic pressure did not change during the time that the pea segments were in the thermocouple psychrometers. So the minimum value of water potential was the yield threshold for turgor pressure. They found this value for turgor pressure to be about 2.8 bar (0.28 MPa) for the pea internode segments.

NATIVE HERBS

Using our knowledge about stress relaxation, let us turn to the work by Ferris and Taylor (1994a), who used the thermocouple psychrometer technique of Cosgrove et al. (1984) to see the effects of elevated levels of CO_2 on the water relations and cell wall extensibility of young growing leaves of four chalk downland (i.e., southern) herbs in England. The plants were *Sanguisorba minor* Scop. (common name: burnet, a plant of the rose family); *Lotus corniculatus* L. (birds-foot trefoil, in the legume family); *Anthyllis vulneraria* L. (kidney vetch or woundwort from its ancient use in healing wounds, also in the legume family); and *Plantago media* L. (hoary plantain in the plantain family). Elevated CO_2 has contrasting effects on leaf growth when the four plants are grown together. The leaves of herbs such as *S. minor* show large, positive responses to CO_2 and may increase at the expense of some legumes such as *A. vulneraria*, a species for which they found only small effects

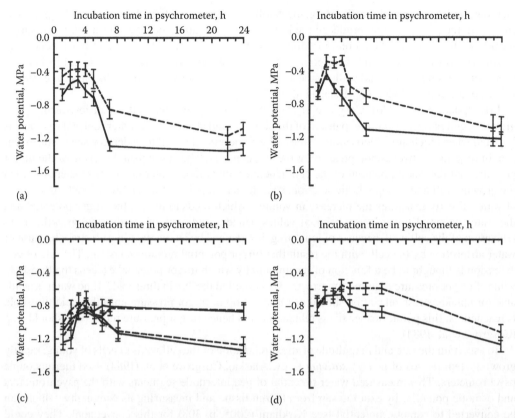

FIGURE 7.22 Variation in water potential with time for leaf discs incubated in psychrometers at 25°C. The data show young leaves of *P. media* samples at (a) midday and (b) predawn and both young leaves (bottom lines) and mature leaves (top two lines) of *A. vulneraria* sampled at (c) midday and (d) predawn (young leaves only) following exposure to either $350\,\mu mol/mol$ CO_2 (dashed lines) or $590\,\mu mol/mol$ CO_2 (solid lines) for 20 days. Each point represents the mean value of 10 measurements ± one standard error. (Ferris, R. and G. Taylor, Elevated CO_2, water relations and biophysics of leaf extension in four chalk grassland herbs. *New Phytol.*, 1994, 127, 297, Fig. 2. Copyright Wiley-VCH Verlag GmbH & Co. KGaA. Reprinted with permission.)

of elevated CO_2 on leaf growth. During June 1992, 10 plants of each species were potted in plastic tubes containing vermiculite and put in controlled environmental cabinets exposed to ambient ($345–355\,\mu mol/mol$) and elevated ($590\,\mu mol/mol$) CO_2 concentrations. After 20 days of exposure to the CO_2 treatments, measurements of water potential were made using thermocouple psychrometers over a 24h period. Yield threshold was measured by stress relaxation of excised leaf discs within the psychrometers, as described above (Cosgrove et al. 1984). After 24h, solute potential was determined on the same leaf discs by freezing them and putting them back in the thermocouple psychrometers to measure solute potential.

Figure 7.22 illustrates the stress relaxation of young, growing leaves of two species, *P. media* (Figure 7.22a and b) and *A. vulneraria* (Figure 7.22c and d), sampled at night (predawn) and at midday. Similar measurements were made on leaves of *S. minor* and *L. corniculatus* (data not shown). Plants grown with elevated CO_2 had lower water potentials than plants grown with ambient CO_2. Figure 7.22c shows both mature and young leaves of *A. vulneraria*. The upper two curves (less negative values) are the data for the mature leaves. Fully expanded mature leaves of *A. vulneraria* reached vapor equilibrium after approximately 4h and, unlike the stress relaxation of growing, young leaves, water potential remained stable over the 24h and there was no difference between CO_2 treatments.

Values of turgor pressure were calculated from Figure 7.22, using the data after 2–4 h (highest or least negative values) and the measurements of osmotic potential. For three species, *S. minor, P. media*, and *A. vulneraria*, solute potential decreased in elevated CO_2 compared to ambient CO_2. Under ambient CO_2, the midday solute potential for *S. minor, P. media*, and *A. vulneraria* were, on average, −1.391, −1.291, and −1.473 MPa, respectively. Under elevated CO_2, these values were −1.510, −1.427, and −1.737 MPa, respectively. The osmotic potential of *L. corniculatus* was not affected by CO_2 concentration. For three species, *S. minor, P. media*, and *L. corniculatus*, there was no effect of elevated CO_2 on leaf cell turgor pressure. But for *A. vulneraria*, turgor pressure increased in elevated CO_2. Under ambient CO_2, the midday turgor pressure of *A. vulneraria* was 0.665 MPa, and under elevated CO_2 it was 0.952 MPa. Yield turgors were calculated after 24 h of incubation, and data showed that there was no effect of elevated CO_2 on yield turgor of any of the four species.

Ferris and Taylor (1994b) followed up their previous work with the leaves of the four chalk downland herbs (Ferris and Taylor 1994a) by studying their roots under ambient and elevated CO_2. During March 1993, 10 plants (10–15 days old) of each species were placed in plastic tubes as described previously (Ferris and Taylor 1994a) with one modification. Clear plastic tubes were used to allow measurement of root extension rate without opening the tubes. Root cells measured in the zone of elongation were longer in all species exposed to elevated CO_2. Water potential, solute potential, turgor pressure, and yield turgor were measured by stress relaxation of excised root tips placed in psychrometers. Figure 7.23 shows the stress relaxation of young growing roots of *P. media* sampled predawn and at midday (graph inset). Similar measurements were made on the roots of the other three species, *S. minor, L. corniculatus*, and *A. vulneraria* (data not shown). No differences were detected in values of solute potential made after either 4 or 8 h incubation within the psychrometers for the four species, so the data in Figure 7.23 can be used to determine yield threshold.

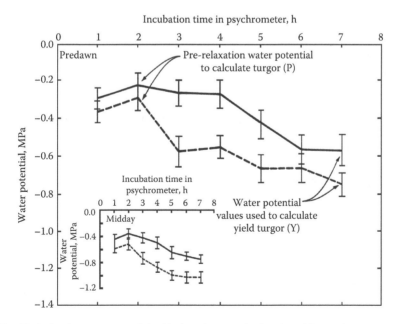

FIGURE 7.23 Variation in water potential with time for root tips incubated in psychrometers at 25 °C. The data show young growing roots of *P. media* sampled predawn and at midday (graph inset) following exposure to either 350 μmol/mol CO_2 (solid lines) or 590 μmol/mol CO_2 (broken lines) for 37 days. Each point represents the mean value of 10 measurements ± one standard error. (Reprinted from Ferris, R. and G. Taylor, *J. Exp. Bot.*, 1994, Fig. 3 by permission of Oxford University Press.)

The water potential of roots of plants exposed to elevated CO_2 was lower than that of roots of plants exposed to ambient CO_2 (Figure 7.23). For *P. media*, highest values (least negative) of water potential of −0.35 (ambient) and −0.52 MPa (elevated CO_2) were observed at 2 h in the midday samples (Figure 7.23, inset), while highest values of water potential of −0.23 (ambient) and −0.29 MPa (elevated CO_2) were observed at night (Figure 7.23, main part). Figure 7.23 shows that water potential declines quickly in the young roots exposed to elevated CO_2, suggesting more rapid stress relaxation of the growing tissue. The solute potentials decreased for all species following exposure to elevated CO_2. For plants exposed to ambient CO_2, the average daytime solute potential for *S. minor*, *L. corniculatus*, *A. vulneraria*, and *P. media* were −0.685, −0.912, −0.992, and −1.036 MPa, respectively. For plants exposed to elevated CO_2, these values were −1.090, −1.138, −1.184, and −1.415 MPa, respectively. Because the roots of all four species had lower solute potential under elevated CO_2 compared to ambient, this suggested that the roots may be accumulating vacuolar solutes (Ferris and Taylor 1994b). Turgor pressure of *S. minor* was higher under elevated CO_2 (0.727 MPa) compared to ambient CO_2 (0.529 MPa). Turgor pressure of the other three species was not affected by CO_2 treatment. The authors conclude that the data in Figure 7.23 show that, when cells are excised from their water supply, a decrease in water potential of the roots occurred with time and this appeared to be more rapid for plants exposed to elevated CO_2. That is, elevated CO_2 increased the speed of cell wall loosening necessary for growth.

SUMMARY

Most studies show that elevated CO_2 increases the water potential of plants (makes it less negative). Stomatal closure under elevated CO_2, along with increased soil water content as a result of stomatal closure (less water is used when stomata are closed), is probably the main reason for the increased water potential under augmented CO_2. More growth occurs at higher water potentials, so increased water potential may be a reason that plants grow better under elevated CO_2 compared to ambient CO_2. Thus, we see that in the equation for photosynthesis given in the Preface (CO_2 and H_2O joining to form sugar and oxygen), both CO_2 and H_2O are affected by elevated CO_2 and each factor favors increased growth. Because we have more CO_2 in the atmosphere, the reaction is pushed to the right, which results in more production of sugar and oxygen. With more water in the soil, the water potential of the plants is increased and this favors growth too.

There are few reports concerning the effects of elevated CO_2 on osmotic potential, and results for osmotic potential are not consistent. Some studies show that elevated CO_2 lowers osmotic potential, perhaps because of increased carbohydrates produced under elevated CO_2. Other works show that osmotic potential may be increased under elevated CO_2 because of depletion of solutes due to increased growth under elevated CO_2. There are even fewer studies concerning the effect of elevated CO_2 on turgor potential, and, again, results are not consistent. Turgor potential can be raised, lowered, or not changed by elevated CO_2.

8 Elevated Atmospheric Carbon Dioxide: Stomatal Conductance

INTRODUCTION

Because lack of water is potentially a problem for higher plants, survival and optimal growth require that stomatal conductance be sensitive to all factors influencing plant growth (Gardner 1973), including CO_2. As we mentioned in Chapters 1 and 5, elevated CO_2 closes stomata and, therefore, is an antitranspirant. In this chapter, we consider specific studies that have documented stomatal conductance (or its reciprocal, stomatal resistance) under elevated CO_2. Before we turn to elevated CO_2, we first consider leaf resistances, the units of stomatal conductance, and, briefly, the physiology of stomatal movements.

LEAF RESISTANCES

Four resistances control CO_2 transport (photosynthesis): the boundary layer resistance, the cuticular resistance, the stomatal resistance, and the mesophyll resistance. [See Kirkham (2005, pp. 388–392) for a discussion of leaf resistances.] We shall discuss the cuticular resistance and the stomatal resistance in the last paragraph of this section. The *boundary layer resistance* is due to the thin layer of still air above all leaves, and its thickness (which determines its resistance) depends on the roughness of the leaf, the wind speed, and the shape of the leaf. The boundary layer resistance is usually small. A typical value of 0.15 s/cm is the default value used in porometry (Li-Cor, Inc., 1985). We shall not consider the boundary layer resistance further. The *mesophyll resistance* is a collection of resistances and includes the solubility of CO_2, the diffusion of CO_2 in the liquid phase from the mesophyll cell wall to the chloroplast, and all the metabolic factors that affect the photosynthetic rate, including the primary carboxylation reaction of photosynthesis (Troughton and Slatyer 1969). Troughton (1969) showed that the mesophyll resistance is less affected by water stress than stomatal resistance. He found that in cotton (*Gossypium hirsutum* L. var. Deltapine smooth leaf), the mesophyll resistance did not vary with the relative leaf water content down to 75%, but increased progressively as relative water content dropped from 75% to 56%.

Diffusion of CO_2 in the intercellular airspaces of the leaf mesophyll is one of the many processes that can limit photosynthetic carbon assimilation. The limitation can be substantial, reducing CO_2 assimilation rates by 25% or more in some leaves (Parkhurst 1994). But in many agricultural plants, intercellular gaseous diffusion probably limits CO_2 assimilation by at most a few percent, because they usually have thin, amphistomatous leaves. Substantial limitations occur in the thicker, often hypostomatous, leaves of wild plants that occupy large areas of the earth (Parkhurst 1994). However, Syvertsen et al. (1995) found that hypostomatous peach [*Prunus persica* (L.) Batsch] leaves had a high internal conductance to CO_2 transfer from the stomatal cavity to sites of carboxylation, probably because the leaves are thin.

Let us consider the internal conductance to CO_2. The reciprocal of the mesophyll resistance can be considered to be the *internal CO_2 conductance* in leaves or the *mesophyll conductance*. Singsaas et al. (2003) define mesophyll conductance as "a measure of the transfer capacity of

147

CO_2 between the leaf internal airspaces and the site of carboxylation in the chloroplast and is a fundamental property of leaves that may influence photosynthetic capacity." Bunce (2009) defines mesophyll conductance as "the conductance of the pathway for CO_2 movement from the intercellular airspace to the site of fixation inside the chloroplast during photosynthetic CO_2 fixation." Singsaas et al. (2003) investigated the effects of a high CO_2 and low CO_2 atmospheric concentration on five different C_3 species [bean (*Phaseolus vulgaris* L.), cucumber (*Cucumis sativus* L.), spinach (*Spinacia oleracea* L.), aspen (*Populus tremuloides* L.), and sweet gum (*Liquidambar styraciflua* L.)]. The high and low CO_2 concentrations for bean, cucumber, and spinach were 700 and 500 µmol/mol, respectively. The high and low CO_2 concentrations for aspen and sweet gum were 560 and 360 µmol/mol, respectively. They measured both sun and shade leaves of sweet gum. The mesophyll conductance, which they abbreviated to g_m, at the high CO_2 concentration was the same as the g_m at the low CO_2 concentration for all species, except the sun leaves of sweet gum. For the sun leaves of sweet gum, the g_m at 560 µmol/mol CO_2 was 0.32 mol/m²/s/bar, and the g_m at 360 µmol/mol CO_2 was 0.17 mol/m²/s/bar. Interestingly, the internal CO_2 concentration (C_i) was not affected by the atmospheric CO_2 concentration, and it was the same under the two CO_2 treatments. For the five species under the two CO_2 treatments, it averaged 412 µmol/mol CO_2. This is in contrast to our results that showed for both C_3 and C_4 grasses (Kentucky bluegrass and big bluestem, respectively); the internal CO_2 concentration increased when the atmospheric CO_2 concentration increased (Kirkham et al. 1991; He et al. 1992).

Stomatal resistance is one of the two resistances making up *epidermal resistance*. The other epidermal resistance is the *cuticular resistance*. Both the stomatal resistance and the cuticular resistance control transpiration and photosynthesis, and the two resistances act in parallel. [Kirkham (2005, pp. 388–389) discusses resistors in series and parallel.] Water or CO_2 diffuses either through the cuticle or a stoma. Usually, the cuticular resistance is ignored, because it is so much greater than the stomatal resistance. The boundary layer resistance and the stomatal resistance act in series. Water or CO_2 has to diffuse both through a stoma and through the air over or under the leaf.

UNITS

The reciprocal of stomatal resistance is *stomatal conductance*. Two types of units are given for stomatal conductance. The one most commonly used now defines a molar conductance, and the dimensions are mole per length squared per unit time (e.g., mol/m²/s) (Jones 1992, p. 55). The other unit, stomatal resistance, is based on considering a stoma as a true resistor such as one has in an electrical circuit. A resistor is a device used in a circuit primarily to provide resistance (Friend and Guralnik 1959). In developing the equations used for stomatal resistance, the stomatal pore is considered to be a tube, and water diffusing through this tube meets a resistance depending upon how large the stomatal pore is (Hsiao 1975).

Because stomatal resistance is a resistance, the early workers who designed the first porometers preferred using units of stomatal resistance rather than stomatal conductance (C.B. Tanner, University of Wisconsin, personal communication, c. 1973). And all the early literature reports stomatal conductance as either a conductance (e.g., cm/s) or a resistance (s/cm). Measurements made with early porometers were based on calibrations to true resistances (van Bavel et al. 1965; Kanemasu et al. 1969). For example, the Kanemasu et al. (1969) porometer was calibrated using a thin metal plate with holes drilled in it to simulate the resistance of pores on a leaf. One placed the metal plate over a piece of wet filter paper and secured the plate to a solid plastic bottom. The wet filter paper simulated a leaf. One timed with a stop watch the time it took for a humidity sensor, which was connected to a meter that read the sensor's output, to wet up between two pre-designated points on the meter. Using the measurements taken with wet filter paper, one could plot time versus a known resistance and make a calibration chart. Then, one placed the leaf on the humidity sensor and read the time for the humidity sensor to change between the two designated points. The time then was related to a resistance. Because the early porometers were calibrated, they gave accurate measurements of stomatal resistance (Cornelius van Bavel, Professor Emeritus, Texas A&M University, personal communications, May 23, 2006). The units reported were s/cm.

Today, most people use portable photosynthetic systems that report stomatal conductance using the molar conductance (e.g., mmol/m^2/s). The reason that it is now common to report units as a mole flux density (mol/m^2/s) is that biochemical reactions concern the number of molecules (Jones 1992, p. 55) and resistance is not considered. Units for photosynthetic rate are based on number of molecules of CO_2 (e.g., μmol/m^2/s). The units reported by portable photosynthetic systems are so ubiquitous in the literature that people no longer even know the term "stomatal resistance." However, people who model evapotranspiration still use resistance units, when they consider stomatal control in their equations (Allen et al. 1998, p. 191).

In 2006, a new leaf porometer was put on the market (Model SC-1, Steady State Diffusion Porometer; Decagon Devices, Pullman, Washington). It reports units in both stomatal resistance units (s/m or m^2 s/mol) and stomatal conductance units (mmol/m^2/s) (Decagon Devices 2006). Jones (1992, p. 56) gives the equations for the conversion between the two types of units (stomatal conductance and stomatal resistance). If one assumes that a leaf is at sea level and the temperature is 25°C, the equations are as follows:

$$r^m \, (\text{m}^2 \, \text{s/mol}) = 2.5r \, (\text{s/cm}) = 0.025r \, (\text{s/m}) \tag{8.1}$$

$$g^m \, (\text{mol/m}^2/\text{s}) = 0.04g \, (\text{mm/s}) \tag{8.2}$$

where

g^m defines a molar conductance

r^m defines the corresponding molar resistance (Jones 1992, p. 55)

r is stomatal resistance and g is stomatal conductance (reciprocal of stomatal resistance), when one considers the stomata acting as resistors

Let us work a sample problem using these equations. Let us assume that the stomatal resistance is 2 s/cm, a reasonable value for a well-watered, non-stressed plant. This is $r = 200$ s/m. Using Equation 8.1,

$$r^m = 0.025 \, (200 \, \text{s/m}) = 5 \, \text{m}^2\text{s/mol}$$

We can take its reciprocal to get 0.2 mol/m^2/s, which is 200 mmol/m^2/s. This value falls into reasonable ranges, which we shall consider in the paragraph after next.

Using Equation 8.2, The reciprocal of 2 s/cm is 0.5 cm/s or 5 mm/s.

$$g^m = 0.04 \, (5) = 0.2 \, \text{mol/m}^2/\text{s or } 200 \, \text{mmol/m}^2/\text{s}$$

The two values check. Each equation gives a conductance of 200 mmol/m^2/s for a resistance of 2 s/cm.

One should always compare values of stomatal conductance or stomatal resistance that one measures with values in the literature to make sure that they are reasonable. Jones (1992, p. 145) gives typical values for maximum conductance for different types of plants. The values range from about 40 to 590 mmol/m^2/s or 1 to 15 mm/s. Note that these are maximum values. The maximum conductance of 15 mm/s, or 1.5 cm/s, corresponds to a minimum resistance of 0.67 s/cm. C.H.M. van Bavel (personal communication, April 21, 2006) states that a minimum stomatal resistance is 0.02 s/cm. Under stressed conditions, resistances can get very high (e.g., greater than 40 s/cm; Kirkham 1983a,b,

2008). Hufstetler et al. (2007) give values of minimum epidermal conductance for 23 soybean genotypes. The values range from 8.5 to 18.5 mmol/m²/s.

FACTORS THAT CONTROL STOMATAL MOVEMENTS

Stomata are constantly opening and closing in response to many environmental factors, including
humidity, light, temperature, and CO_2 concentration, to maintain a viable balance between CO_2
gained in photosynthesis and water lost in transpiration (Heath and Mansfield 1969; Hetherington
and Woodward 2003). Figure 8.1 is a summary of these factors. A positive sign in the figure shows

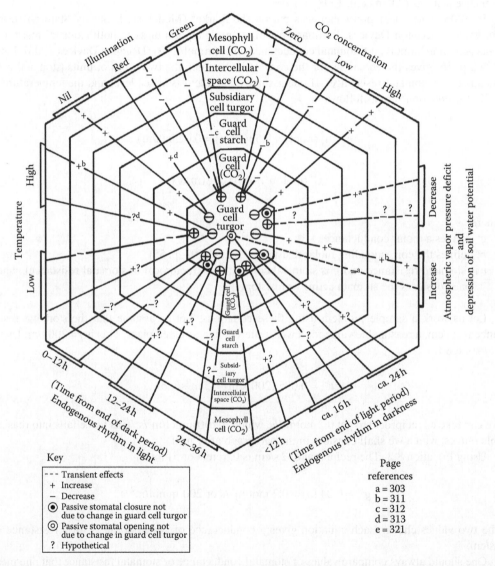

FIGURE 8.1   An attempt to depict the effects of the various factors known to bring about stomatal movements.
The letters in the figure refer to the pages, cited in the lower right of the figure, that discuss the various factors
controlling stomatal movements. These pages are in the article by Heath and Mansfield from which this figure
has been taken. (Reprinted from Heath, O.V.S. and Mansfield, T.A, The movements of stomata, in Wilkins,
M.B., Ed., *The Physiology of Plant Growth and Development*, McGraw-Hill, New York, 1969, pp. 301–332,
Fig. 9.9, Copyright 1969, McGraw-Hill, with permission from T.A. Mansfield, Lancaster University, Lancaster,
United Kingdom, and Malcolm B. Wilkins, Regius Professor Botany, Glasgow, United Kingdom.)

that guard cell turgor is increased and stomata open. A negative shows that guard cell turgor is decreased and stomata close.

Under non-stressed conditions, stomata of plants with the C_3 and C_4 type of photosynthesis are open during the day and closed at night (de Willigen et al. 2005). Stomata of plants with the CAM (Crassulacean acid metabolism) type of photosynthesis are open at night and are closed during the day. They open in the dark in response to a reduction in CO_2 concentration, and, thus, in CAM species, chloroplast photosynthesis is not required for stomatal opening (Von Caemmerer and Griffiths 2009). The one commercially important plant with the CAM type of photosynthesis is pineapple (*Ananas comosus* Merr.). Some stomata of C_3 and C_4 plants are "leaky" at night, and these plants have nocturnal water loss, like CAM plants (Green et al. 2006). But we shall not consider this topic here.

Now, let us look in more detail at the factors that control stomatal movements, not yet considering elevated CO_2.

STOMATAL OPENING

For over a 100 years, it has been known that guard cells take up potassium (Macallum 1905). In the mid part of the twentieth century, Japanese workers studied potassium in guard cells. Imamura (1943) found an abundance of potassium in guard cells in open stomata, but little in closed ones. Yamashita (1952) also found that potassium content in guard cells was correlated with stomatal movements. In the 1950s, Fujino (in Japanese publications) showed that guard cells of open stomata contained large concentrations of potassium, but little in closed ones in the dark. The Japanese work was unknown in the western world until 1967, when Fujino published a paper in English showing that potassium uptake is necessary for stomatal opening (Fujino 1967). At about the same time, an independent study in the western world found the same results (Fischer 1968).

During stomatal opening, guard cells take up K^+ and Cl^-, extrude H^+, and produce malate^{2-} (Assmann 1993, p. 348; Nilson and Assmann 2007). Data indicate that channels specific for each ion operate in moving the ions across the guard cell membrane (Assmann 1993, p. 346). Passive uptake of K^+ down an electrical gradient, created by H^+ extrusion, is mediated by K^+-selective ion channels (Assmann 1993, p. 350). These inward K^+ channels show selectivity for K^+ over Na^+ and other monovalent cations. The conductance and density of these channels are sufficient to account for the K^+ uptake observed during stomatal opening. During stomatal opening, production of malate^{2-} and $2H^+$ replenishes H^+ lost via H^+ extrusion and provides a counterion for K^+ uptake, thus reducing energetically costly Cl^- uptake (Assmann 1993, p. 350).

Solute increases attributable to K^+ salts do not always account completely for the guard cell osmotic potential observed in open stomata, which suggests a requirement for production of other solutes (Assmann 1993, p. 351). Organic solutes such as sucrose might be imported, produced from starch breakdown, or synthesized from products of the photosynthetic carbon reduction pathway, also called the Calvin–Benson cycle (Assmann 1993, p. 351). As noted before (see the section "Carbon Isotope Ratios of Plants with Different Photosynthetic Pathways" in Chapter 6), C_3 plants have only this photosynthetic pathway, while C_4 plants have the C_3 pathway along with another pathway called the Hatch-Slack pathway.

STOMATAL CLOSURE

Stomatal closure does not simply involve the inhibition or reversal of ion transport processes necessary for opening, but rather requires unique ion transport systems (Assmann 1993, p. 351). Given the theory that K^+ moves passively across the guard cell membrane, the K^+ loss associated with stomatal closure requires membrane depolarization. Because the electrochemical gradient favors passive anion efflux, opening of plasma membrane anion channels is a mechanism for rapid membrane depolarization that could drive K^+ efflux. Several different types of anion channels have been described in guard cells. One is called an R-type (rapid-type) anion channel, which activates rapidly in response to depolarization and then inactivates. R-type channels have the permeability sequence: $NO_3^- > Cl^- > malate^{2-}$. Thus, they are termed anion channels rather than Cl^- channels. The second type of anion channel is the S-type (slow-type), which activates and deactivates slowly in response to voltage. Like the R-type, S-type anion channels are not completely selective for Cl^-. The third type of anion channel described in guard cells is an SA (stretch-activated) anion channel, which is activated by stretch. It exhibits decreased closed time as membrane stretch is increased. This channel is highly selective for Cl^- over K^+ (Assmann 1993, pp. 351–352).

The genes encoding anion efflux channels in plant plasma membranes remain unknown (Negi et al. 2008; Vahisalu et al. 2008). An *Arabidopsis* gene, *SLAC1* (*SLOW ANION CHANNEL-ASSOCIATED 1*, At1g2480), which mediates CO_2 sensitivity in regulation of plant gas exchange, has been isolated (Negi et al. 2008; Vahisalu et al. 2008). The SLAC1 protein is a distant homologue of bacterial and fungal C4-dicarboxylate/malic acid transport proteins and is localized specifically to the plasma membrane of guard cells (Negi et al. 2008). *SLAC1* is preferentially expressed in guard cells (Vahisalu et al. 2008). The plasma membrane protein SLAC1 is essential for stomatal closure in response to CO_2, abscisic acid, ozone, light/dark transitions, humidity change, calcium ions, hydrogen peroxide, and nitric oxide. Mutations in *SLAC1* impair S-type anion channel currents that are activated by cytosolic Ca^{2+} and abscisic acid but do not affect R-type anion channel currents or Ca^{2+} channel function (Vahisalu et al. 2008). These results suggest that SLAC1-family proteins have an evolutionarily conserved function that is required for the maintenance of anion homeostasis at the cellular level (Negi et al. 2008).

Removal of organic osmotica during stomatal closure occurs via glycolysis and subsequent oxidative phosphorylation, by reincorporation into starch and by loss to the apoplast (Assmann 1993, p. 353). [The apoplast is the part of the plant considered to be nonliving, such as cell walls; see Kirkham (2005, p. 218) for a further description of the apoplast.] Guard cells have been observed to lose as much as 60% of their $malate^{2-}$ to the external medium upon stomatal closure, presumably through $malate^{2-}$-permeable anion channels (Assmann 1993, p. 353).

SECOND MESSENGERS

As we have seen before, mechanisms involved in stomatal opening include extrusion of H^+, K^+ channels mediating K^+ influx, a Cl^- carrier, and production of organic solutes (Assmann 1993, p. 353). Stomatal closure requires membrane depolarization, which can be driven by anion efflux, followed by loss of K^+ through outward K^+ channels, and regulated decreases in organic osmotica (Assmann 1993, p. 353). Alterations of any of these processes can affect stomatal aperture; such changes are triggered by alterations in the level of activity of two so-called second messengers, Ca^{2+} and H^+ (pH) (Assmann 1993, pp. 353–354).

Calcium is an important second messenger for stomatal closure (Chasan 1995). Exogenously applied Ca^{2+} is potent in reducing stomatal aperture (Assmann 1993, p. 357). However, there are also data suggesting that Ca^{2+} may play a role in stomatal opening. It is known that second messengers can have both activating and inactivating effects on a process (Assmann 1993, p. 357). Assmann and Wang (2001) review how Ca^{2+} affects guard cell oscillations. Calcium affects both K^+ channels and anion channels. Calcium has been shown to increase and decrease according to atmospheric CO_2 levels. Studies of the guard cells of *Arabidopsis* show a high CO_2-induced stomatal closing and a low CO_2-induced stomatal opening (Young et al. 2006).

pH may act as an extracellular signal (Assmann 1993, p. 358). As extracellular pH becomes more alkaline, water stress increases, which results in stomatal closure (Assmann 1993, p. 358). A more alkaline extracellular pH decreases inward K^+ movement in guard cells. Conversely, acidic extracellular pH activates inward K^+ channels (Assmann 1993, p. 359). Thus, H^+ extrusion during stomatal opening, which can decrease extracellular pH from 7.2 to 5.1, may additionally activate the pathway for K^+ influx (Assmann 1993, p. 359).

ENDOGENOUS SIGNALS

Messengers that can be produced within the leaf or translocated from a distant site of synthesis include plant hormones (Assmann 1993, p. 359). (For the definition of a hormone, see the section "Ethylene" in Chapter 10.) For over 40 years, it has been known that hormones produced in the roots affect shoot water relations (Vaadia and Waisel 1967). During the last 15 years, the number of known plant hormones has grown from 5 (cytokinins, ethylene, gibberellins, auxins, and abscisic acid) to at least 10 (Santner and Estelle 2009). The five new ones are brassinosteroids, jasmonic acid, salicyclic acid, nitric oxide, and strigolactones. Here, we shall consider only the first five ones. However, we note in passing that strigolactones are in a group of terpenoid lactones thought to be derived from carotenoids (Gomez-Roldan et al. 2008). They control different aspects of plant growth, especially branching of shoots (Klee 2008). The presence of strigolactones is reduced in mutants of *Arabidopsis*, pea (*Pisum sativum* L.), and rice (*Oryza sativa* L.) that are highly branched (Klee 2008). This hormone suppresses formation of lateral shoots.

Cytokinins are produced in roots and travel to leaves where they promote stomatal opening. Cytokinins can be sprayed on plant leaves, and stomata open more widely than they are under normal conditions (Kirkham et al. 1974). The rate of transpiration has been used as a bioassay for cytokinin activity (Luke and Freeman 1967). Reduced cytokinin production in stressed roots results in increased stomatal resistance.

Ethylene, the only gaseous hormone, is the simplest organic compound that affects plants. It is active in trace amounts, which makes its detection under non-stressed conditions difficult. Ethylene production in plants increases following damage, wounding, or stress from a variety of sources. This increase has been called wound ethylene. Wound ethylene is produced when a plant is drought stressed. The effects of ethylene on stomata are contradictory. Some experiments show no response of stomata to ethylene, while others have found that ethylene increases stomatal resistance (Kirkham 1983a). Ethylene usually does not decrease stomatal resistance. In Chapter 10, we compare ethylene and CO_2 and explain why CO_2 is not a hormone, even though it is a small-molecular-weight gas like ethylene.

Little information exists concerning the effect of gibberellins on transpiration, and reviews on plant water relations often do not include this hormone (Davies and Zhang 1991; Assmann 1993). Information does indicate that gibberellic acid stimulates transpiration and decreases stomatal resistance (Livnè and Vaadia 1965; O'Leary and Tarquinio Prisco 1970).

The first group of plant hormones to be studied was the auxins, and indole-3-acetic acid (IAA) was one of the first auxins to be isolated. The fundamental concept that the growth of plants, and the interrelation between their parts, is controlled by hormones stems from the classical work of Charles Darwin and his son, Francis Darwin (C. Darwin 1897). Francis Darwin, a botanist, is known for his work on stomata (Darwin 1897, 1898; Darwin and Acton 1909). In Charles Darwin's last book, *The Power of Movement in Plants*, first published in 1880, many experiments are described concerning the response of plants to light and gravity. Among these was the finding that the tip of a grass seedling controls the ability of the whole seedling to curve toward light (Thimann 1969). It is now known that auxins are involved in this curvature response. Auxins are required for cell elongation, as well as for cell proliferation, but they have a multitude of additional effects. Horticulturists use auxins to promote root formation on cuttings. Depending on the type of auxin and the auxin concentration, either stomatal opening or stomatal closure may occur (Assmann 1993, p. 362). IAA is a natural auxin, and it stimulates stomatal opening at concentrations from $10 \mu M$ to $1 mM$ (Assmann 1993, p. 362). With the synthetic auxin 1-NAA (1-napthylacetic acid), maximal stimulation of opening occurs at $5 \mu M$ and inhibition of opening occurs at $0.5 mM$ (Assmann 1993, p. 362).

The hormone that has been most widely studied in relation to stomata is absciscic acid (ABA), because it strongly promotes stomatal closure (Assmann 1993, p. 359). ABA is produced in response to stress, particularly drought stress. Because exogenous Ca^{2+} application also inhibits stomatal opening, an ABA–Ca^{2+} link has been sought (Assmann 1993, pp. 359–360). A link between ABA application and Ca^{2+} uptake has not yet been demonstrated (Assmann 1993, p. 360). Studies have been done to determine whether the effects of ABA application on ion transport mechanisms are the same as effects resulting from Ca^{2+} application. ABA inhibits inward K^+ currents in guard cells, an effect similar to that of elevated intracellular Ca^{2+} and one that could contribute to ABA inhibition of stomatal opening (Assmann 1993, pp. 360–361). Outward K^+ current is enhanced by ABA, consistent with ABA promotion of stomatal closure (Assmann 1993, p. 361). With water stress, ABA levels in leaves rise, preventing influx and/or causing efflux of K^+ and anions from guard cells. Because of the consequent reduction in guard cell turgor pressure, stomatal apertures are reduced. Ca^{2+} appears to be a second messenger for an inhibitory stimulus such as ABA. Recent research with guard cells suggests that CO_2 and ABA enhance sensitivity to calcium, and ABA receptors and CO_2-binding response proteins are being identified (Kim et al. 2010).

Even though ABA is important in controlling stomatal aperture, it is probably the balance of hormones (Tal et al. 1970) and second messengers such as calcium (Assmann 1993, pp. 357–358) that determine the size of the stomata. Cytokinins decrease stomatal resistance and increase the resistance of the root to absorption of water. The general effect of cytokinins, therefore, is to reduce plant turgor and their concentration decreases in stressed plants. In contrast, ABA increases stomatal resistance and decreases the resistance of the root to water flow. This hormone increases the plant turgor, and its concentration increases in stressed plants. Rapid changes in concentrations of both cytokinins and ABA have been demonstrated in water-stressed plants. The plant undergoes alternating periods of low turgidity during the day, when absorption lags behind water loss by transpiration, and high turgidity at night, when transpiration lags behind water absorption. The periodic changes of water status may induce periodic fluctuations of hormonal concentrations, which influence stomatal and root resistances (de Willigen et al. 2005). The balance of hormones in water-stressed plants may cause the "aftereffect" that is observed in stomata (Fischer et al. 1970). Stomata commonly fail to resume regular opening after a period of wilting of leaves despite rapid recovery of leaf water content. The aftereffect may be due to an accumulation of an inhibitor like ABA or a deficiency of a substance that promotes stomatal opening like a cytokinin.

Aquaporins

The hydrophobic nature of the lipid bilayer of a membrane presents a considerable barrier to the free movement of water into the cell and between intracellular compartments. However, plasma membranes (around the cell) and tonoplasts (around vacuoles) can be rendered more permeable to water by proteinaceous transmembrane water channels, called aquaporins (de Willigen et al. 2005). Aquaporins are a class of water-channel proteins that have been found in nearly all living organisms. They are highly expressed in plant membranes. They are members of a family of transmembrane channels known as the *major intrinsic protein* (MIP). Water movement through aquaporins can be modulated rapidly. Evidence suggests that these channels may facilitate water movement in drought-stressed tissues and promote the rapid recovery of turgor on watering.

The factors that determine internal CO_2 conductance (defined in the first part of this chapter) are only partly understood. Aquaporin is a candidate for study, because CO_2 can permeate through some plant aquaporins (Hanba et al. 2004). Hanba et al. (2004) measured the internal conductance of CO_2 diffusion and stomatal conductance in transgenic rice leaves that overexpressed a barley aquaporin. They found that increasing the level of aquaporin increased the internal CO_2 conductance. The increase in the internal CO_2 conductance was accompanied by a 14% increase in CO_2 assimilation rate and a 27% increase in stomatal conductance. Uehlein et al. (2003) found that a tobacco plasma membrane aquaporin NtAQP1 facilitates CO_2 membrane transport and has a function in photosynthesis and stomatal opening. The overexpression of NtAQP1 increases membrane permeability for CO_2 and water and increases leaf growth. Their results indicated that NtAQP1-related CO_2 permeability is of physiological importance under conditions where the CO_2 gradient across a membrane is small, as in the case between the atmosphere and the inside of a plant cell.

Uehlein et al. (2008) showed that the aquaporin NtAQP1 is localized on the inner chloroplast membrane and the plasma membrane in tobacco, and they presented evidence that it facilitates transport of CO_2 into the chloroplast, which impacts photosynthesis (Eckardt 2008a).

ENVIRONMENTAL SIGNALS

In addition to water availability, environmental signals with effects on stomatal aperture are light, humidity, and CO_2. Red and blue light stimulate stomatal opening (Assmann 1993, p. 363; Mott 2009). Chlorophyll absorbs blue as well as red light and, thus, mediates some of the stomatal opening stimulated by blue wavelengths. However, there is much evidence for the presence in guard cells of a blue light photoreceptor distinct from chlorophyll (Assmann 1993, pp. 363–364). The most definitive evidence is that the action spectrum for light-stimulated stomatal opening does not parallel the absorption efficiency of chlorophyll in the blue region of the spectrum (Assmann 1993, p. 364). Light may also contribute to solute buildup via production of sugars from products of the Calvin–Benson photosynthetic cycle (Assmann 1993, p. 366). The stomata of the fern *Adiantum capillus-veneris* are an exception to the general observation that red and blue lights stimulate stomatal opening. The stomata of this fern lack a blue-light-specific opening response but open in response to red light (Doi and Shimazaki 2008).

Reduction in ambient humidity or an increase in vapor pressure deficit (Katul et al. 2009) causes stomatal closure, but the reason for the closure is poorly understood (Assmann 1993, p. 368). Experiments suggest that transpirational water loss is the important signal. Guard cells respond metabolically to humidity, for example, by losing K^+. Guard cells may also be responding to regulatory metabolites such as ABA that are carried in the transpiration stream (Assmann, 1993, p. 369). Talbott et al. (2003) identified air relative humidity as a key environmental factor that mediated changes in stomatal response to CO_2. They grew broad bean (*Vicia faba* L. cv. Windsor Long Pod) in a greenhouse under a normal (40%–70%) and a high (95%) relative humidity. The high relative humidity was obtained with atomizer nozzles, which supplied a fine water mist to the air moving through the plant canopy. The aperture of stomata of *V. faba* grown at the normal relative humidity was the same size at 400, 650, and 900 μmol/mol CO_2 (about 10 μm). But when the relative humidity was increased by misting to 95%, the stomatal aperture decreased as the CO_2 concentration increased. Average apertures of stomata grown at 400, 650, and 900 μmol/mol CO_2 were 12, 6, and 3 μm, respectively. They could not ascertain the specific parameter sensed by the stomata, in response to changes in air relative humidity. However, their data showed that air humidity regulated stomatal response of *V. faba* to CO_2, and they concluded that such a response might allow stomatal opening inside dense foliage canopies with low light and low CO_2 conditions (Talbott et al. 2003).

Elevated CO_2 typically stimulates stomatal closure, while lowered CO_2 concentrations have the opposite effect (Assmann 1993, p. 367). It is the intercellular concentration of CO_2, and not the external concentration, or the concentration within the pore, to which the guard cells are responding (Assmann 1993, p. 368). The mechanism of the CO_2 response is unknown (Assmann 1993, p. 368).

In an atmosphere free from CO_2 and in the light, Darwin (1898) found that stomata remain open. He used a horn hygroscope, made from the awn of *Stipa* (spear grass). He placed the awn on the stomatal side of a leaf. The amount that the awn twisted away from the leaf gave an indication (an "index") of whether the stomata were open or closed. When placed on the astomatal side of a leaf, the index remained at zero. On the stomatal side, the index rose in 5 or 10 s to a position of approximate rest (no twisting), a position that varies roughly with the degree of transpiration (Darwin 1898, pp. 533–534). Schellenberger [1896; cited by Darwin (1898)] stated that in the absence of CO_2, the stomata shut, and they do so for the same reason that darkness is supposed to close them. That is, assimilation is checked, which causes a loss of assimilative material, and, therefore, loss of turgor (Darwin 1898, p. 608). Schellenberger placed garden plants under a large bell jar with CO_2-free air in it. (The CO_2 was removed by passing air through 10% KHO, as he wrote the formula. It is also

KOH, or potassium hydroxide.) After 2 days, he found all the plants with shut stomata, while control plants in a CO_2-containing atmosphere were open. Darwin (1898) attributes the closure to the dryness of the air in Schellenberger's experiments. He placed two beakers containing cut leaves of *Tropæolum majus* L. (garden nasturtium) under two bell jars—one with KHO and a control jar that also included a vessel of water. Without the vessel of water, the stomata closed. He states, "I have no doubt that the difference between the stomata in the two jars depended on the dryness of the air" in the jar with KHO. He did further studies with *Narcissus* sp. (depending on the species, the common names include narcissus, daffodil, jonquil), *Campanula pyramidalis* L. (chimney bellflower), *T. majus* L., and *Teedia lucida* (common name unknown) and got similar results as with his first experiments with *Tropæolum*. He concluded that when illuminated, stomata remain open in the absence of CO_2 (Darwin 1898, p. 610). Because most researchers are concerned with elevated CO_2 in the atmosphere, not CO_2-free air, little recent research has been done to study stomata when CO_2 is not present.

Bunce (2007) studied a low, but not zero, atmospheric CO_2 concentration (100 μmol/mol) on stomatal conductance of four species exposed to a drying root medium. They were velvetleaf (*Abutilon theophrasti* L.), soybean [*Glycine max* (L.) Merr.], cotton (*G. hirsutum* L.), and a type of cocklebur (*Xanthium strumarium* L.). Control plants were grown at an ambient CO_2 concentration of 380 μmol/mol. The control rooting medium was vermiculite flushed daily with a complete nutrient solution. He dried the rooting medium by withholding nutrient solution for 2–5 days before he made measurements. As expected, the stomatal conductance of the dry plants grown at 380 μmol/mol CO_2 was lower than that of plants grown under the wet (control) treatment at 380 μmol/mol. The stomatal conductance of plants grown under the dry treatment and with 100 μmol/mol CO_2 in the air was intermediate between the stomatal conductances of plants grown under the wet 380 μmol/mol CO_2 treatment and the dry 380 μmol/mol CO_2 treatment. For example, the stomatal conductances of soybean grown under the wet 380 μmol/mol CO_2, dry 380 μmol/mol CO_2, and dry 100 μmol/mol CO_2 treatments were 342, 164, and 285 mmol/m²/s, respectively. (He did not report measurements of plants grown at 100 μmol/mol CO_2 under the wet conditions.)

In experiments separate from the root-medium-drying ones, Bunce (2007) dried the air around sunflower (*Helianthus annuus* L.) and Sitka spruce [*Picea sitchensis* (Bong.) Carriere] by exposing leaves to leaf-to-air vapor pressure differences (*D*) between 0.5 and 3.2 kPa. He again exposed the plants to the two CO_2 concentrations: 100 and 380 μmol/mol CO_2. For sunflower at 380 μmol/mol CO_2, stomatal conductance decreased from about 1000 to 200 mmol/m²/s as *D* increased from 0.6 to 2.6 kPA. But at 100 μmol/mol CO_2, stomatal conductance of sunflower stayed the same (about 1000 mmol/m²/s) as *D* decreased from 0.5 to 2.6 kPa. For spruce, the stomatal conductance of plants grown at 380 and 100 μmol/mol CO_2 decreased as *D* increased from 0.6 to 3.1 kPa. At the higher *D*'s, stomatal conductances were similar, but below about 1.5 kPa, stomatal conductance of spruce at 100 mmol/m²/s was higher than that at 380 mmol/m²/s. That is, low CO_2 concentration opened the stomata. Bunce (2007) concluded that drying of the rooting medium or drying of the air did not prevent a low CO_2 concentration in the air (100 μmol/mol) from increasing stomatal conductance. However, he noted that the mechanism by which stomatal conductance responds to CO_2 concentration is unknown, as did Assmann (1993).

Kamakura and Furukawa (2008) found that 350 μmol/mol CO_2 was a key concentration for causing stomatal closure. They studied the responses of individual stomata of ipomoea (also called morning glory) (*Ipomoea pes-caprae* L.) to CO_2 concentrations ranging from 0 to 900 μmol/mol. Air supplied from an air compressor was passed through a cylinder containing soda lime [soda lime is a white powdery mixture of sodium hydroxide, NaOH, and calcium oxide or lime, CaO (Friend and Guralnik 1959)] to remove CO_2. The CO_2-free air then was mixed with CO_2 from a CO_2 cylinder to obtain the desired CO_2 concentration. Each leaf was exposed to CO_2-free air for 60 min in the light to obtain a steady-state stomatal aperture. Then the leaves were exposed to CO_2 concentrations of 50, 200, 350, 700, or 900 μmol/mol for 40 min, and microscopic images of the stomatal pores were observed. A CO_2 concentration of 350 μmol/mol or higher induced stomatal closure, whereas concentrations below $350\,\mu mol/mol$ did not. The time it took stomata to close decreased with increasing CO_2 concentration. Stomatal pore width at the $350\,\mu mol/mol$ CO_2 concentration started out at $4.5\,\mu m$ (after the CO_2-free air exposure) and ended up with a minimum diameter of $2\,\mu m$ after $40\,min$. Stomatal pore width at $900\,\mu mol/mol$ CO_2 again started out at $4.5\,\mu m$ (after the CO_2-free air exposure) and ended up with a minimum diameter of $0.5\,\mu m$ after only $20\,min$.

We now turn to experiments that describe the effects of elevated levels of CO_2 on stomatal conductance.

STOMATAL CONDUCTANCE AND ELEVATED CO_2

ANNUAL CROPS

Tuba et al. (1994) grew winter wheat (*Triticum aestivum* L. cv. MV 16) in a calcareous sandy soil in open-top chambers in Hungary at two concentrations of CO_2: ambient ($350\,\mu mol/mol$) and elevated ($700\,\mu mol/mol$). Anatomical and physiological measurements were taken on flag leaves. Stomatal conductance was measured with a mass flow porometer [Delta-T Devices, Cambridge, U.K.; see Kirkham (2005, p. 396) for a picture of this instrument.] Water use efficiency was determined as the ratio of net CO_2 assimilation rate to transpiration rate, and both rates were determined with an infrared gas analyzer. The method allowed ambient CO_2 concentrations during measurements to be changed in the sensor where the flag leaf was measured, and these ambient concentrations were 60, 100, 180, 350, 500, 700, and $1000\,\mu mol/mol$. The instrument also determined the internal CO_2 concentration when the leaves were exposed to these different ambient CO_2 concentrations. Note that the "ambient CO_2" concentration in the leaf chamber during measurement was different from the ambient CO_2 concentrations in the air that the plants were exposed to during their growth cycle (350 and $700\,\mu mol/mol$).

Stomatal density did not differ between the high CO_2-treated and control plants. The adaxial stomatal densities (number per mm^{-2}) at 350 and $700\,\mu mol/mol$ were 33.38 ± 3.34 and 33.75 ± 2.31, respectively. The abaxial stomatal densities (number mm^{-2}) for the 350 and $700\,\mu mol/mol$ CO_2 treatments were 29.63 ± 2.97 and 26.63 ± 2.62, respectively. The data confirm that wheat has a higher stomatal density on the adaxial surface than on the abaxial surface, as found by many researchers (e.g., see Meidner and Mansfield 1968; the data from Meidner and Mansfield are reproduced in Table 9.1). The values of stomatal conductance in the high CO_2 plants were lower than in the control at all ambient CO_2 concentrations used during measurement (Figure 8.2a). They were also lower at the corresponding internal CO_2 concentrations (Figure 8.2b).

The lower stomatal conductance in the plants grown under high CO_2 compared to the control plants resulted in a lower transpiration rate. The net assimilation rate of plants grown under the high CO_2 concentration was higher at 700 and $1000\,\mu mol/mol$ than leaves grown at ambient CO_2. The ratio of net assimilation to transpiration rate (the water use efficiency) was higher for plants grown at high CO_2 at these two values (Figure 8.3). Increased water use efficiency is a characteristic change in plants exposed to high CO_2, as further documented in Chapter 11. This is of importance for wheat and other plants in arid and semiarid conditions (Tuba et al. 1994).

Stanghellini and Bunce (1993) did an experiment similar to that of Tuba et al. (1994), except they also included the effect of humidity and temperature. They grew tomato (*Lycopersicon esculentum* L.) in fertilized, vermiculite-filled pots, at 350 and $700\,\mu mol/mol$ (or, using their units, 350 and $700\,cm^3/m^3$), presumably in growth chambers (the growth environment is not defined). At the beginning of the measurement period (27–31 days after sowing), a leaflet of either the fourth or fifth leaf was enclosed in a gas exchange chamber where the CO_2 concentration, air temperature, and humidity could be controlled. Gas exchange was measured using an infrared gas analyzer at five CO_2 concentrations, which they called the "external CO_2 concentration": 100, 200, 350, 700, and $1000\,\mu mol/mol$. They used three air temperatures (18°C, 25°C, and 32°C), measured with a thermocouple pressed against the lower side of the leaf. Vapor-pressure difference (the leaf was assumed

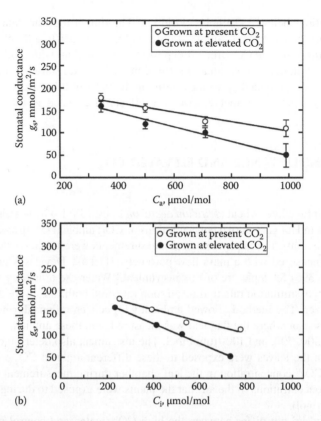

FIGURE 8.2 (a) Stomatal conductance (g_s) as a function of ambient CO_2 concentration (C_a) and (b) of intercellular CO_2 concentration (C_i) in the flag leaves of winter wheat (*T. aestivum* "MV 16") grown from germination at present (350 μmol/mol) and elevated (700 μmol/mol) CO_2 concentration. Measurements were made at the beginning of flowering. (Reprinted from *J. Plant Physiol.*, 144, Tuba, Z., Szente, K., and Koch, J., Response of photosynthesis, stomatal conductance, water use efficiency and production to long-term elevated CO_2 in winter wheat, 661, Fig. 2, Copyright 1994, with permission from Elsevier.)

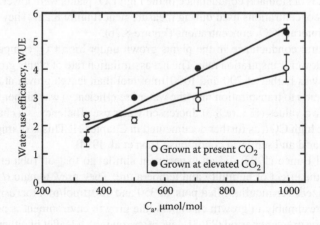

FIGURE 8.3 Water use efficiency as a function of ambient CO_2 concentration (C_a) in the flag leaves of winter wheat (*T. aestvium* "MV 16") grown from germination at present (350 μmol/mol) and elevated 700 μmol/mol CO_2 concentration. Measurements were made at the beginning of flowering. (Reprinted from *J. Plant Physiol.*, 144, Tuba, Z., Szente, K., and Koch, J., Response of photosynthesis, stomatal conductance, water use efficiency and production to long-term elevated CO_2 in winter wheat, 661–668, Fig. 6, Copyright 1994, with permission from Elsevier.)

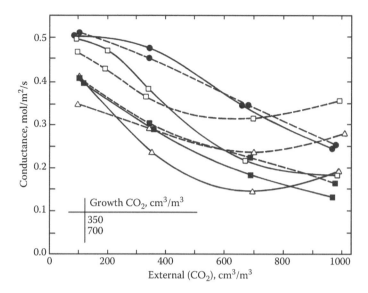

FIGURE 8.4 Mean leaf conductance of tomato to water vapor with respect to prevailing CO_2 in saturating irradiance [2000 μmol (photon) m^{-2} s^{-1}], at a leaf temperature of 18 (triangles), 25 (squares), and 32 (circles) °C, and a vapor pressure difference of 1.1 (open symbols) and 2.2 (closed symbols) kPa, respectively. Solid lines refer to plants grown in 350 cm^3 (CO_2) m^{-3} and dashes to 700 cm^3 (CO_2) m^{-3}. (Reprinted with kind permission from Springer Science+Business Media: *Photosynthetica*, Response of photosynthesis and conductance to light, CO_2, temperature and humidity in tomato plants acclimated to ambient and elevated CO_2, 29, 1993, 487, Stanghellini, C. and Bunce, J.A., Fig. 7. Copyright 1993, Springer.)

to be saturated at its actual temperature) was controlled at either 1.1 or 2.2 kPa. While external CO_2 concentration, temperature, and humidity were varied on the leaf being measured, the rest of the plant was kept at the same conditions during the growth period (temperature of 25°C, relative humidity of 70%, and CO_2 concentration of either 350 or 700 μmol/mol). [The vapor pressure deficit, different from the vapor pressure difference, for 70% relative humidity can be calculated to be 9.50 mb or 95 kPa. See Kirkham (2005, p. 247) for the calculations.]

Figure 8.4 shows the short-term responses of the stomata to the various environmental conditions imposed during the measurement period. In some of these combinations, stomatal conductance of plants grown full-time at 700 μmol/mol CO_2 was higher than that of plants grown full-time at 350 μmol/mol, opposite to what is normally found. For example, for the "external CO_2 concentration" of 350 μmol/mol (see this value on the abscissa of Figure 8.4), a temperature of 18°C, and a vapor pressure difference of 1.1 kPa during the measurement, plants grown full-time at 700 μmol/mol CO_2, 25°C, and 70% relative humidity (open triangles connected by dashed lines in Figure 8.4) had a higher stomatal conductance than plants grown full time at 350 μmol/mol CO_2, 25°C, and 70% relative humidity (open triangles connected by solid lines in Figure 8.4). Stangellini and Bunce (1993) point out that their data, which show that stomatal conductance of plants grown under elevated CO_2 is higher than that of plants grown under ambient CO_2, are contrary to what has been observed in many agricultural crops (Cure and Acock, 1986). They do not give an explanation for this discrepancy.

Bunce (1993) did further experiments concerning the interaction of elevated CO_2 and humidity. The water content of air surrounding leaves is known to affect stomatal aperture, but the mechanism of the response is not known. Of all the stomatal responses, the humidity response is the most mysterious (Assmann 1993, p. 368). He determined the difference in the partial pressure of water vapor between the surface of the leaf and the air, which he abbreviated "surface Δw." He made comparisons between plants grown and measured at 350 and 700 μPa/Pa partial pressures of CO_2

for amaranth (*Amaranthus hypochondriacus* L.; a C_4 plant), soybean [*Glycine max* (L.) Merr.; a C_3 plant), and sunflower (*H. annuus* L.; a C_3 plant). He grew the plants in pots with fertilized vermiculite and placed them in controlled environmental chambers. Gas exchange measurements were made using a photosynthetic system (Bunce 1992). Whole, attached leaves (terminal leaflets in soybeans) were enclosed in a cuvette for these measurements. Dew point temperatures of the air entering and leaving the cuvette were measured using an optical-condensation sensor. Leaves were exposed to Δw of 8–13 mPa/Pa for about 1 h, and then the Δw was increased in steps of 4–5 mPa/Pa. Steps were increased after the leaf gas exchange became constant (usually 30 min or less). Maximum Δw was about 30 mPa/Pa, which is a common midday summer value at Beltsville, Maryland, where Bunce (1993) carried out his experiments.

Increasing the Δw decreased leaf conductance to water vapor in all species at both growth CO_2 partial pressures (Figure 8.5). In amaranth (Figure 8.5a) and soybean (Figure 8.5b), stomatal conductance was higher at 350 μPa/Pa CO_2 (open circles; Figure 8.5a and b) than at 700 μPa/Pa CO_2 (open squares; Figure 8.5a and b) for all values of Δw. At 5 mPa/Pa Δw, the leaf conductances of amaranth and soybean were reduced by 60% and 57%, respectively, when the CO_2 concentration

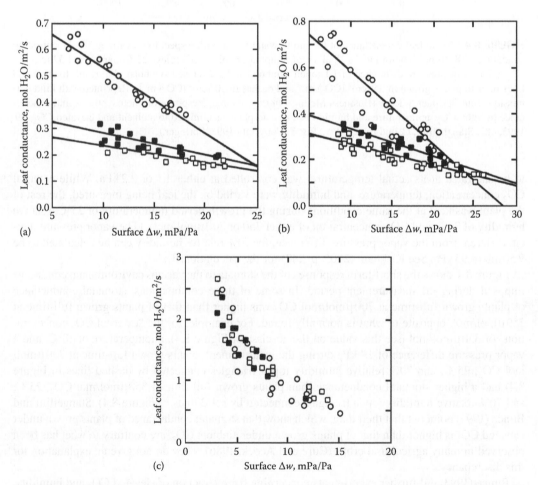

FIGURE 8.5 The responses of leaf conductance to the Δw at the leaf surface for leaves of (a) amaranth, (b) soybean, and (c) sunflower grown at 350 and 700 μPa/Pa CO_2 partial pressures. Lines are linear regressions: open circles: grown and measured at 350 μPa/Pa; open squares: grown and measured at 700 μPa/Pa; closed squares: grown at 350 and measured at 700 μPa/Pa. (Bunce, J.A.: Effects of doubled atmospheric carbon dioxide concentration on the responses of assimilation and conductance to humidity, *Plant Cell Environ.*, 1993, 16, 189, Fig. 1. Copyright Wiley-VCH Verlag GmbH & Co. KGaA. Reprinted with permission.)

was doubled (Figure 8.5a and b), which is more than the 34% decrease observed as an average value for many C_3 plants (Cure and Acock 1986). In contrast, there was no effect of CO_2 concentration on response of conductance to Δw in sunflower (Figure 8.5c). Bunce (1993) stated that the lack of response of leaf conductance to CO_2 concentration, observed here in sunflower, could be related to the cultivar, growth, or measurement conditions. Previous studies have also found that the leaf conductance of sunflower is not affected as the CO_2 concentration increases (Goudriaan and van Laar 1978; Graan and Boyer 1990).

Bunce (2004) reviewed the effects of elevated CO_2 on crop plants undergoing drying. In experiments with maize (*Zea mays* L.) and soybeans, plants were grown in pots filled with soil in open-top chambers in the field and then were moved into controlled-environment chambers to impose soil water deficits. The results showed that the stomatal conductance for both maize and soybean was higher under ambient CO_2 than under elevated CO_2, and the stomatal conductance decreased for both species as the leaf water potential became more negative (Figure 8.6a and b; top part). The lower part of Figure 8.6 shows stomatal conductance relative to the mean values of well-watered plants under the two CO_2 concentrations. When relative stomatal conductance is plotted, differences in conductance between elevated and ambient CO_2 are not evident, because the values are relative to well-watered conditions, not to a common CO_2 concentration (like ambient). The relative conductance varies between about 1.0 under well-watered conditions and almost 0 under dry conditions, values that one would expect.

Frechilla et al. (2002) showed that the response of stomata to elevated CO_2 varies depending upon where the plant is grown (growth chamber or greenhouse). They grew broad bean (*V. faba* L. cv. Windsor Long Pod) in pots with a commercial potting mix and kept them well watered.

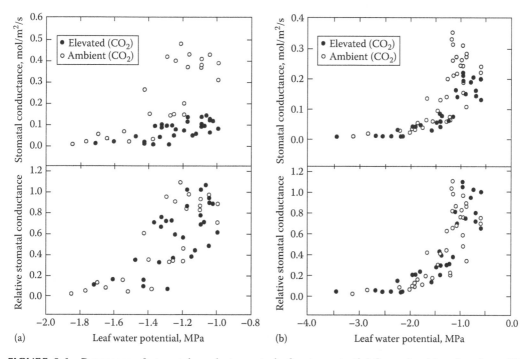

FIGURE 8.6 Responses of stomatal conductance to leaf water potential for maize (a) and soybean (b) grown and measured at the current ambient [CO_2] or at twice the current ambient [CO_2] under field conditions. For each species, both absolute values of conductance (i.e., the measured values) and conductances relative to the mean values of well-watered plants are given. (Reprinted with kind permission from Springer Science+Business Media: Carbon dioxide effects on stomatal responses to the environment and water use by crops under field conditions, *Oecologia*, 2004, 140, 1, Bunce, J.A., Fig. 2. Copyright 2004, Springer.)

Plants were grown either in a growth chamber or a greenhouse. The growth chamber environment was constant (85%±3% relative humidity, 12 h light at 550 μmol/m²/s, and 25/15°C day/night air temperatures), while the greenhouse environment was variable (50%–75% relative humidity, natural light at the University of Los Angeles, California, where the experiments were done, and 25°C–30°C/15°C–20°C day/night temperatures). In the greenhouse, a plastic enclosure was constructed that allowed elevation of the CO_2 concentration. Plants (either from the greenhouse or growth chamber) were transferred to the enclosure in the morning and allowed to equilibrate for 2 h before an initial measurement of stomatal apertures was taken. Plants were kept for 1 h at each CO_2 concentration (400, 650, and 900 cm³/m³ or 400, 650, and 900 μmol/mol CO_2) before measurement of stomatal apertures. Stomatal apertures were determined from digitized video images of stomata in epidermal peels using a microscope connected to a digital imaging camera. Average apertures of stomata from greenhouse-grown plants were 11.9±0.4 μm under the ambient CO_2 concentration prevailing at the onset of the experiments. A 1 h exposure to 650 and 900 cm³/m³ resulted in stomatal apertures of 11.9±0.4 (same values as at beginning of the experiment) and 11.5±0.5 μm, respectively. The data showed that stomata from greenhouse-grown leaves were insensitive to changes in the CO_2 concentration in the 400–900 cm³/m³ range. In contrast, average stomatal apertures of growth-chamber-grown leaves were 12.7±0.5 μm under ambient CO_2, and 9.8±0.5 and 7.5±0.4 μm after a 1 h exposure to 650 and 900 cm³/m³ CO_2, respectively. Thus, apertures decreased to about 60% of their initial value when CO_2 was increased to 900 cm³/m³ (Figure 8.7). They could not identify a specific reason for the difference in response of the stomata to the two environments and said that further research was needed to determine it.

Chen et al. (1995) also found differences in stomatal conductance when they switched soybean plants between high and low CO_2 atmospheres. They grew soybean [*Glycine max* (L.) Merr. cv. Jack] plants in a controlled environment during the winter and in the field during the summer. In the controlled environment, plants grew in pots that contained a mixture of soil, peat, and perlite placed in open-top chambers with CO_2 concentrations of 400 or 800 μmol/mol. The seemingly high concentration for ambient (400 μmol/mol) was the concentration of the outside air during the winter in Urbana, Illinois, where the authors were located. In the field, the plants grew in a silt loam soil. Open-top chambers were used to maintain CO_2 concentrations of 350 or 700 μmol/mol in the field.

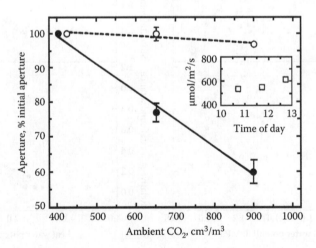

FIGURE 8.7 The CO_2 response of stomata from *V. faba* grown under greenhouse (open circles) and growth chamber (closed circles) conditions. Aperture change in response to a 1 h exposure to 650 or 900 cm³/m³ ambient CO_2 is shown as a percentage of the initial aperture value. Data are the average of three experiments ± standard error. Error bars are contained within the symbol where not shown. Inset shows greenhouse light levels over the course of the experiment. (Reprinted from Frechilla, S., Talbott, L.D., and Zeiger, E., *J. Exp. Bot.*, 2002, Fig. 1 by permission of Oxford University Press.)

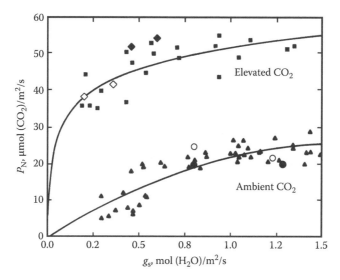

FIGURE 8.8 Comparison of leaf net photosynthesis (P_N) and stomatal conductance (g_s) for soybean grown at enriched (700 μmol/mol) and ambient CO_2 levels. Both field and controlled environment P_N and g_s values are shown in the figure. The data obtained by Chen et al. (1995) (the large symbols) are superimposed on a plot of P_N versus g_s and fitted curves from their previous study (the small solid squares and small solid triangles) (Chen et al. 1993). The large symbols are as follows: open diamonds: ambient-grown leaves measured at enriched CO_2 levels; closed diamonds: leaves grown and measured at enriched CO_2 levels; closed circles: leaves grown at enriched CO_2 levels but measured at ambient CO_2 levels; open circles: ambient-grown leaves measured at ambient CO_2 levels. (Reprinted with kind permission from Springer Science+Business Media: *Photosynthetica*, Soybean stomatal acclimation to long-term exposure to CO_2-enriched atmospheres, 31, 1995, 51, Chen, X.M., Begonia, G.B., and Hesketh, J.D., Fig. 1. Copyright 1995, Springer.)

In both environments, they switched the plants between the two CO_2 concentrations, and, 24 h after the switch, they determined leaf gas exchange rates on the youngest, most fully expanded terminal leaflet using a portable photosynthetic system (Model LI-6200, Li-Cor, Inc., Lincoln, Nebraska).

Their results are shown in Figure 8.8. In this figure, they have included data from a previous experiment (Chen et al. 1993), in which they grew soybean outside in open-top chambers with either ambient or elevated (700 μmol/mol) CO_2. In the figure, the small squares represent data from the elevated CO_2 treatment and the small triangles represent data from the ambient treatment in this 1993 experiment (Chen et al. 1993). The large symbols in Figure 8.8 show the data from the 1995 experiment (Chen et al. 1995). Note that most of the data are from the 1993 experiment. For plants that grew in the controlled environment at 800 μmol/mol CO_2 and then were switched to the 800 μmol/mol CO_2 (i.e., no change in concentration), the stomatal conductance was 0.58 μmol/m²/s (value on abscissa in Figure 8.8; the closed diamond); and for plants that grew at 400 μmol/mol before and then were switched to 400 μmol/mol CO_2, the stomatal conductance was 1.27 μmol/m²/s (open circle, Figure 8.8; I got the exact numbers for stomatal conductance given in a table in their paper, and this symbol in the figure does not line up exactly with the number given in the table). Similar results were observed for the soybeans in the field grown under the same CO_2 concentration before and after the switch. That is, the stomatal conductance of plants grown at the elevated level of CO_2 was two times smaller compared to the stomatal conductance of plants grown at the ambient level of CO_2 (0.44 μmol/m²/s; closed diamond, Figure 8.8, versus 0.82 μmol/m²/s, open circle, Figure 8.8). When plants were switched from a high level of CO_2 (800 μmol/mol CO_2 in the controlled environment; 700 μmol/mol in the field) to the ambient level (400 μmol/mol in the controlled environment or 350 μmol/mol in the field), the stomatal conductance increased in response to the current conditions. Plants grown at 800 μmol/mol CO_2 and then 400 μmol/mol CO_2 (800/400) in the controlled environment had a stomatal conductance of 1.29 μmol/m²/s (closed circle, Figure 8.8), and plants grown at 700/350 μmol/mol in the field had a stomatal conductance of 0.81 μmol/m²/s (closed circle, Figure 8.8). When plants were switched from the ambient level of CO_2 to the elevated level of CO_2, the stomatal conductance reflected that of the elevated level of CO_2. It was low. Plants grown at 400/800 μmol/mol CO_2 in the controlled environment had a stomatal conductance of 0.36 μmol/ m²/s (open diamond, Figure 8.8), and plants grown at 350/700 μmol/mol in the field had a stomatal conductance of 0.20 μmol/m²/s (open diamond, Figure 8.8). The results indicated that stomatal conductance reflects the current CO_2 atmospheric concentration, not previous exposures. The effects of elevated CO_2 on stomatal conductance were negated 24 h after exposure to the ambient level. These results are consistent with the observation that stomata respond quickly to a change of their environment. When Ishibashi and Terashima (1995) sprinkled artificial rain on green bean leaves (*P. vulgaris* L.), stomata closed completely within 2 min. (Why rainfall is a "stress" to the bean leaves and closes their stomata is not understood.)

In a FACE (free-air CO_2 enrichment) study in Japan, Shimono et al. (2010) grew three different cultivars of rice (*O. sativa* L.) under flooded conditions for two growing seasons. Elevated-CO_2 plots received ambient + 200 μmol/mol CO_2. Stomatal conductance of all cultivars was generally higher in the morning and around noon than in the afternoon, and elevated CO_2 decreased stomatal conductance by up to 64% over the 2 years, with a mean decrease in stomatal conductance of 23%.

GRASSLAND SPECIES

Maherali et al. (2002) measured stomatal conductance of C_3 and C_4 grassland species exposed to CO_2 concentrations ranging from sub-ambient to super-ambient. Measurements were taken on three dominant species in a grassland near Lubbock, Texas: *Solanum dimidiatum* Raf., a C_3 perennial forb; *Bromus japonicus* L., a C_3 annual grass; and *Bothriochloa ischaemum* (L.) Keng, a C_4 perennial grass. On the grassland, the C_3 species are mostly active early in the growing season (spring) and the C_4 species dominate by mid-summer. Elongated chambers, 60 m in length, 1 m wide, and 1 m tall, were placed on the grassland. The soil had 35%–55% clay in the top 40 cm, and it was kept well watered during the experiment by irrigation. The CO_2 concentration gradients in the sub-ambient and super-ambient chambers ranged from 360 to 200 μmol/mol and from 550 to 360 μmol/ mol, respectively. Treatments began in May 1997 and continued in each growing season through 2000. Measurements were taken with an open gas exchange system (LI-6400, Li-Cor, Inc., Lincoln, Nebraska) when each species was at its peak abundance, which was June 2000 for *S. dimidiatum*, April 2000 for *B. japonicus*, and August 1999 for *B. ischaemum*. Stomatal conductance (g_s) declined with increasing CO_2 concentration for all species (Figure 8.9). *S. dimidiatum* (C_3) had the highest overall stomatal conductance among the three species, followed by *B. japonicus* (C_3) and *B. ischaemum* (C_4), which is in agreement with other studies in which C_3 and C_4 plants have been grown under the same conditions. Under these situations, C_3 plants have a higher stomatal conductance, or lower stomatal resistance, than C_4 plants [e.g., Kentucky bluegrass, C_3, versus big bluestem, C_4 (He et al. 1992); sunflower, C_3, versus sorghum, C_4 (Rachidi et al. 1993a)]. The stomatal conductance of the C_4 plant (*B. ischaemum*) was least affected by the changing CO_2 concentration. The reduction in stomatal conductance with increasing CO_2 concentration, even from sub-ambient levels, suggests that changes in evapotranspiration and soil drainage in grassland ecosystems may have already occurred (Maherali et al. 2002).

Maherali et al. (2003) looked at the effect of humidity on the stomata in the grassland under different CO_2 concentrations. Methods were similar to those described earlier (Maherali et al. 2002), except they studied only two plants: *B. japonicus*, C_3, and *B. ischaemum*, C_4. They measured stomatal conductance under different D. The value of D was manipulated within the leaf cuvette of the portable photosynthetic system by diverting a greater percentage of air through the desiccant (Drierite). Data were analyzed statistically to determine probability (P) values. Stomatal

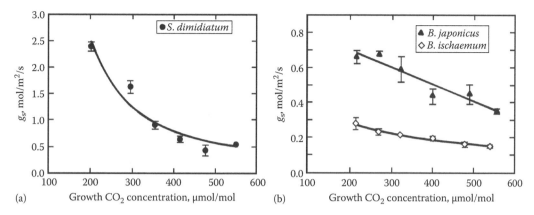

FIGURE 8.9 Mean (±1 standard error) values of steady-state stomatal conductance (g_s) for *S. dimidiatum* (a), *B. japonicus*, and *B. ischaemum* (b) plants grown across a range of atmospheric CO_2. Note the difference in scales on the *Y*-axis between (a) and (b). (Maherali, H., Johnson, H.B., and Jackson, R.B.: Stomatal acclimation over a subambient to elevated CO_2 gradient in a C_3/C_4 grassland. *Plant Cell Environ.*, 2002, 25, 557. Fig. 1. Copyright Wiley-VCH Verlag GmbH & Co. KGaA. Reprinted with permission.)

conductance (g_s) and photosynthesis (*A*) responded linearly to increases in CO_2 from sub-ambient to elevated levels of CO_2 (Figure 8.10). In *B. japonicus*, g_s declined with increasing CO_2 at all levels of *D*. The response of g_s to CO_2 was strongest (i.e., the steepest slope) at the lowest *D* (1 kPa, Figure 8.10a). Stomatal conductance of *B. ischaemum* also declined with rising CO_2, but the response was similar across levels of *D* (Figure 8.10c). Photosynthesis (*A*) was stimulated by rising CO_2 at all levels of *D* in *B. japonicus*. Photosynthesis of *B. ischaemum* increased with increasing CO_2, but the response was little changed by *D*. They suggested that the relative insensitivity of stomata to *D* in *B. ischaemum* may be associated with drought tolerance or be a general feature of species with low maximum stomatal conductance (i.e., C_4 species).

Kawamitsu et al. (1993) also found that stomata of a C_4 species were insensitive to humidity, as the atmospheric CO_2 concentration increased, but a C_3 species did respond both to humidity and CO_2 concentration. They grew rice (*O. sativa* L. cv. Nagakurabozu), a C_3 species, and green panic (*Panicum maximum* L. cv. Trichoglume) [Bailey (1974) classifies this as *P. maximum* Jacq. and gives its common name as guinea-grass; it is a perennial grass], a C_4 species, for 4 weeks in two growth chambers: one set at 35% and one set at 85% relative humidity. The plants grew in pots with a fertilized gravel and vermiculite mixture. Fully expanded, young leaves were placed in a chamber, where gas exchange measurements could be made and where CO_2 partial pressures could be varied between 0 and 100 Pa (called the "ambient" CO_2 concentration). Carbon-dioxide assimilation rates increased with increasing ambient CO_2 partial pressure in both species (Figure 8.11a; squares for rice; circles for panic grass). In *P. maximum*, humidity pretreatments did not alter the assimilation rate. At high CO_2 partial pressure, the CO_2 assimilation rate for *O. sativa* grown at low humidity (Figure 8.11a, open squares) was lower than that for *O. sativa* grown at high humidity (Figure 8.11a, closed squares). The leaf conductance of *P. maximum* (Figure 8.11b, circles) was lower than that of *O. sativa* (Figure 8.11b, squares) at all CO_2 partial pressures, as we have noted before when leaf conductance of C_3 and C_4 plants is compared in the same experiment. In addition, leaf conductance of *O. sativa* was reduced by low humidity at all ambient CO_2 partial pressures, but in *P. maximum*, humidity had no influence on leaf conductance (Figure 8.11b). Kawamitsu et al. (1993) said that they did not know why the leaf conductance of green panic, a C_4 species, did not respond to humidity. They noted that at ambient CO_2 concentration, the internal CO_2 concentration is often saturating in C_4 species. Thus, photosynthesis of C_4 plants may be insensitive to partial stomatal closure caused

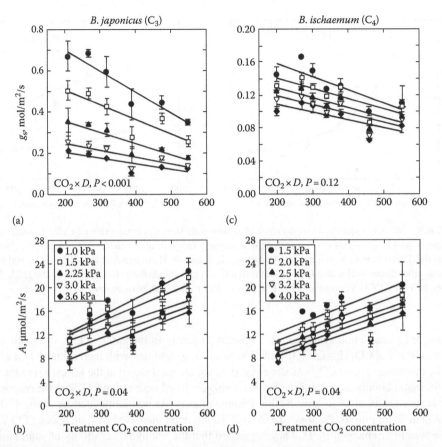

FIGURE 8.10 The response of mean (±1 standard error) stomatal conductance, g_s (a,c), and photosynthesis, A (b,d), to growth CO_2 concentration (µmol/mol) at contrasting D in *B. japonicus* and *B. ischaemum*. Each data point represents the mean of three to five individuals. Response to growth CO_2 concentration was fitted with linear least squares regression models. The P of the interaction, $CO_2 \times D$, is given in each part of the figure. (Maherali, H., Johnson, H.B., and Jackson, R.B.: Stomatal sensitivity to vapour pressure difference over a subambient to elevated CO_2 gradient in a C_3/C_4 grassland, *Plant Cell Environ.* 2003, 26, 1297, Fig. 5. Copyright Wiley-VCH Verlag GmbH & Co. KGaA. Reprinted with permission.)

by a large vapor pressure deficit, and this may represent an advantage of C_4 photosynthesis in warm, dry environments.

WEEDS

He et al. (2005) studied the effect of elevated levels of CO_2 on pokeweed (*Phytolacca americana* L.), which is a tall and stout perennial herb with the C_3 photosynthetic pathway. The genus occurs in tropical and warm regions, mostly in America, but it also exists in Africa and Asia. In North America, the weed ranges from Quebec and Ontario to northern Mexico. *P. americana* has an unpleasant odor and a large poisonous root (Fernald 1950). It grows in open disturbed areas, particularly forest edges and canopy gaps. They grew the plants in 12 chambers under two different CO_2 treatments (350 and 700 µmol/mol CO_2) and three temperature regimes, which were 26°C/20°C (day/night temperatures) (control), 30°C/24°C, and 28°C/24°C. Plants grew under well-watered conditions in plastic pots with a mixture of fertilized garden soil and commercial potting medium. Gas exchange was measured with a portable photosynthetic system (Model LI-6400, Li-Cor, Inc., Lincoln, Nebraska) at the seedling, flowering, and mature-seed stages (65, 91, and 119 days after

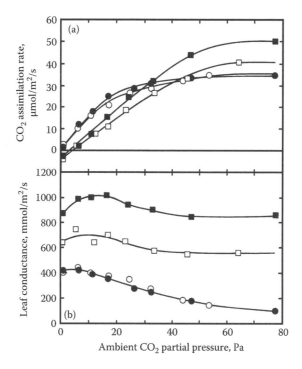

FIGURE 8.11 The response of CO_2 assimilation rate (a) and leaf conductance (b) to ambient CO_2 partial pressure for leaves of *O. sativa* (a C_3 plant) (squares) and *P. maximum* (a C_4 plant) (circles). Plants were grown in high humidity (RH 85%, closed symbols) or in low humidity (35%, open symbols) at 25°C air temperature, 35 ± 2 Pa CO_2, and 1,700 μmol/m²/s PFD (photon flux density). Measurements were performed at 30°C leaf temperature, 1,700 μmol/m²/s PFD, 21 kPa oxygen partial pressure, and 0.7 kPa vapor pressure difference. CO_2 partial pressure was altered stepwise from high to low values. (Reprinted from Kawamitsu, Y., Yoda, S., and Agata, W., *Plant Cell Physiol.*, 1993, Fig. 2, by permission of Oxford University Press. Copyright 1993, Japanese Society of Plant Physiologists.)

planting, respectively). Elevated CO_2 led to a reduction in stomatal conductance at both the flowering and mature-seed stages, while elevated temperature did not affect stomatal conductance (Figure 8.12b). Photosynthesis increased under elevated CO_2 91 days after planting, but was not affected by elevated temperature (Figure 8.12a). The effect of elevated CO_2 on photosynthesis was less pronounced by the time the seeds were mature (119 days after planting; Figure 8.12a). He et al. (2005) pointed out that models predict CO_2 and temperature interactions, which they did not observe in this study. They concluded that it is important to incorporate experimental measurements into predictive, climate-change models for proper understanding of plant responses to changing atmospheric CO_2 and air temperatures.

TREES

Berryman et al. (1994) studied stomatal conductance of leaves of seedlings of the tropical tree eucalyptus (*Eucalyptus tetrodonta* F. Muell) exposed to different CO_2 concentrations and leaf temperatures. Plants grew in pots with a mixture of vermiculite, sand, and peat that were placed in ventilated, plastic tents maintained at ambient (355 μmol/mol) and elevated (700 μmol/mol) levels of CO_2. Plants were taken out of the tents and placed in a gas exchange laboratory 1 day prior to measurements. Ambient CO_2 concentrations around the leaf were varied by mixing pure CO_2 and CO_2-free air. When plants from the CO_2-enriched tents were used, the CO_2 concentration within the leaf chambers and around the plants in the laboratory was maintained at

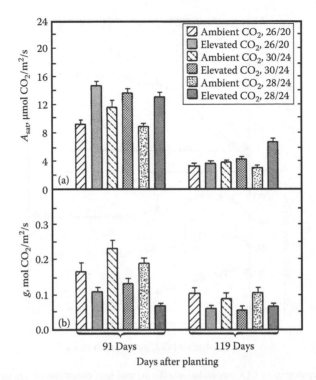

FIGURE 8.12 Leaf-level net photosynthesis at saturating light (A_{sat}) and stomatal conductance (g) of *P. americana*. Measurements were taken at the following three stages of development: seedling (August 4, 2001, 65 days after planting), flowering (August 30, 2001, 91 days), and mature seed (September 27, 2001, 119 days). Results only from the flowering and mature-seed stages are shown. The numbers in the upper right-hand part of the figure show the day/night temperatures (°C) of the different treatments. Mean ±1 standard error is shown for each treatment (*n* = 8). (Reprinted from He, J.-S. et al., *Int. J. Plant Sci.* 166, 615, 2005, Fig. 4. With permission. Copyright 2005, University of Chicago Press.)

700 μmol/mol. The fourth leaf down from the apex of each plant was sealed into a leaf chamber. Two controlled temperature water baths were used to regulate leaf temperature. A thermocouple inside each leaf chamber measured leaf temperature. The internal CO_2 concentration in the leaf (C_i) and the stomatal conductance (g_s) were calculated according to equations in the literature. Stomatal conductance was greater for the ambient-grown plants than for plants grown under elevated CO_2, but the differences between the two CO_2 treatments decreased as the leaf temperature increased between 20°C and 40°C (Figure 8.13a). Leaf internal CO_2 concentration (C_i) was higher in CO_2-enriched plants (Figure 8.13b), as has been observed in other experiments (Kirkham et al. 1991; He et al. 1992). Stomatal conductance declined linearly with increasing C_i for both CO_2 treatments (Figure 8.13c). The data of Berryman et al. (1994) confirm that increasing leaf temperatures decrease stomatal conductance under ambient CO_2 and that stomatal conductance decreases under elevated CO_2.

During a 4 year period (1998–2001), Herrick et al. (2004) measured stomatal conductance of sweet gum (*L. styraciflua* L.) grown at the Duke (in North Carolina, United States) FACE facility. Sweet gum is a tall tree with shining, maple-like leaves, spiny balls of fruit, and fragrant juice (Friend and Guralnik, 1959). The FACE site contained six 30 m diameter experimental rings, three with elevated CO_2 (average concentration during the experiment of 572 μmol/mol) and three with ambient CO_2 (average: 376 μmol/mol). Gas exchange measurements were measured twice a year around June 25 (about 68 days after leaf initiation) and September 1 (about 50 days before leaves

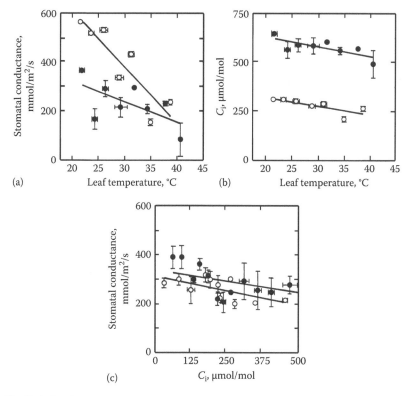

FIGURE 8.13 Relationships between (a) stomatal conductance (mmol/m²/s) and leaf temperature (°C); (b) intercellular CO_2 concentration (μmol/mol) and leaf temperature (°C); and (c) stomatal conductance (mmol/m²/s) and intercellular CO_2 concentration (μmol/mol) for *E. tetrodonta*. Open circles are for plants grown under ambient CO_2. Filled circles are for plants grown under enriched CO_2. Error bars are ± standard error. Each data point is the mean of three replicate leaves. (Reprinted from Berryman, C.A., Eamus, D., and Duff, G.A., *J. Exp. Bot.*, 1994, Fig. 4 by permission of Oxford University Press.)

began to senesce). They measured soil moisture content using time–domain reflectometry in the upper 30 cm of the soil profile. The experimental forest occurs on a nutrient-poor, clay-rich loam soil that is typical of many upland areas in the southeastern United States. The experimental sweet gum trees were 7–11 m high in 1997. The leaves at the top of the crown were exposed to full sunlight and leaves at the bottom of the crown were in deep shade. Sun leaf irradiance was typically saturated and varied between 1100 and 1400 μmol/m²/s during midday on sunny days. Diffuse irradiance for the shade leaves was about 50 μmol/m²/s. [For definitions of direct and diffuse radiation, see Kirkham (2005, pp. 416–418).] Gas exchange measurements were made using a portable photosynthetic system (Model LI-6400, Li-Cor, Inc., Lincoln, Nebraska). Over the 4 year period, CO_2 enrichment decreased stomatal conductance by an average of 31% in sun leaves (Figure 8.14, top) and 25% in shade leaves (Figure 8.14, middle). Stomatal conductance was more than twice as high for sun leaves than for shade leaves. Low soil moisture (Figure 8.14, bottom) resulted in low stomatal conductance (Figure 8.14, top and middle). The decrease in stomatal conductance of sweet gum leaves under elevated CO_2 was sustained over 4 years, which indicated that stomatal sensitivity to CO_2 did not subside with time.

Bunce (1992) grew apple (*Malus domestica* Borkh.), basket oak (*Quercus prinus* L.), and English oak (*Q. robur* L.), which are woody angiosperms, in controlled environment chambers at either 350 (ambient) or 700 μmol/mol CO_2 and found that the stomatal conductance of plants grown at elevated CO_2 concentration did not differ from the conductance of plants at ambient CO_2 concentration, but

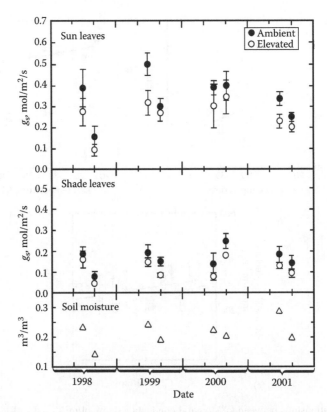

FIGURE 8.14 Steady-state stomatal conductance (g_s) of sun and shade leaves of overstory *L. styraciflua* trees growing at the Duke Forest FACE experiment in ambient (closed symbols) and elevated (open symbols) CO_2. Gas exchange was measured around June 25 and September 1 of each year. Measurements were made at saturating irradiance and were restricted to the hours between 10:00 and 15:00 h on sunny days. Each point is the mean of three rings (±standard error) for each CO_2 treatment. Soil moisture during the measurement period is added for reference. (Herrick, J.D., Maherali, H., and Thomas, R.B.: Reduced stomatal conductance in sweetgum (*Liquidambar styraciflua*) sustained over long-term CO_2 enrichment. *New Phytol.*, 2004, 162, 387, Fig. 1. Copyright Wiley-VCH Verlag GmbH & Co. KGaA. Reprinted with permission.)

gave no reason for this lack of response. Ellsworth (1999) states that the stomatal responses of trees to elevated CO_2 are generally smaller than those documented for crops. Although rare, there are reports of insensitivity to CO_2 concentration in annual angiosperms. As an example, Bunce cites the work by Bierhuizen and Slatyer (1964), who studied the effects of CO_2 concentrations ranging between 200 and 2000 ppm on stomatal resistance of cotton (*G. hirsutum* L.). However, upon close examination of their paper, one cannot see the actual data to substantiate this observation. Bierhuizen and Slatyer (1964) report in their abstract that "there was an almost linear response to CO_2 concentrations up to values of 600–800 ppm." That is, stomatal resistance decreased (or stomatal conductance increased) between 200 and about 800 ppm CO_2. But they present no data in their paper to back up this statement.

Results for broad-leafed tree species like eucalyptus (Berryman et al. 1994) and sweet gum (Herrick et al. 2004), which show that stomata close at elevated CO_2, sometimes differ from those of an evergreen species, loblolly pine (*Pinus taeda* L.). Ellsworth (1999) grew the pine at the Duke FACE facility during the 1996 growing season, which was dry during the main portion of the growing season. Rainfall between May and September in 1996 was 30% lower than the 30 year normal for this period. He found that photosynthesis was stimulated 50%–60% by elevated CO_2 (560 μmol/mol CO_2) compared with ambient controls (360 μmol/mol CO_2). But stomatal conductance of loblolly pine grown under ambient CO_2 was the same at that grown under elevated CO_2. The nonresponsiveness of the pine stomata to elevated CO_2 might have been related to the low soil water content. In a later study at the Duke FACE site, Domec et al. (2009) found that elevated CO_2 (200 μmol/mol above the ambient level) reduced stomatal conductance of loblolly pine, but only when the plants were not under water stress. They took measurements in 2007, when there was a severe summer drought in the Atlantic Southeast. Because of the limited rainfall in 2007, extractable soil water in the upper 30 cm of the soil profile was nearly exhausted by early August and remained so until sizable rains came in October. The rooting zone was shallow (about 35 cm). The stomatal conductance of loblolly pine grown under the ambient CO_2 was higher than that of pine grown under the elevated CO_2 in May 2007. But differences in stomatal conductance between the ambient and elevated CO_2 plots were not evident in July or August, when plants were water stressed.

Teskey (1995) also found that elevated CO_2 (660 μmol/mol) had no effect on stomatal conductance of loblolly pine compared to plants grown at the ambient level of CO_2 (330 μmol/mol). He placed chambers over branches in the upper canopy of a 21-year-old loblolly pine plantation in Georgia (United States) and applied different concentrations of CO_2 to the chambers. The soil was an eroded clay with low nutrient availability. The site was irrigated to maintain soil water potentials above −0.05 MPa, as indicated by tensiometers placed at depths of 0.15 and 0.30 m near each study tree. Teskey (1995) had no explanation as to why the loblolly pine stomata in his study did not respond to CO_2. He said that there is little consistency among studies of trees concerning stomatal conductance in elevated CO_2. In a 2 year study in the United Kingdom, Heath et al. (2005) found that stomatal conductance of pine (*Pinus*; species not given), along with five other tree species (the names of the other five trees that they studied are not given by them), declined as the CO_2 concentration increased. They grew the tree species under four CO_2 concentrations: ambient (376 μmol/mol) and ambient plus 100, 200, or 300 μmol/mol.

We need to consider why stomatal conductance of pine, like loblolly pine in the study of Teskey (1995), may not be affected by elevated CO_2, while other plants (crops, grasses, broad-leafed trees) generally have a lower stomatal conductance under elevated CO_2. The guard cells of stomata of crops and grasses usually respond to intercellular CO_2 concentration, and the lack of such a response in pines and other C_3 plants may be considered exceptional (Ellsworth 1999). [Pine is a C_3 species. See Morison (1987a, pp. 230–231) for a list of C_3 and C_4 plants.] Ellsworth (1999) did not discuss the anatomy of conifer leaves and compare them to broad-leafed woody plants, which might explain the lack of response of conifer stomata to elevated CO_2. The needle-like leaves of conifers, including pine [conifers are cone-bearing plants, mostly evergreens, like pine, spruce, fir, cedar, and yew (Friend and Guralnik 1959)], have a low ratio of surface to volume, a typical xeromorphic character (Esau 1965, p. 446). As in other conifers, the leaf epidermis in pine is heavily cuticularized and has such thick walls that the cell lumina are almost obliterated. The fiber-like cells of the hypodermis also have thick walls and form a compact layer interrupted only beneath the stomata. [See Kirkham (2005, pp. 209, 329) for a discussion of the hypodermis, a layer or layers of cells beneath the epidermis distinct from the underlying ground tissue cells (Esau 1977, p. 512).] Thick cell walls impose considerable resistance on CO_2 diffusion inside leaves (Hanba et al. 2004). In many genera, including *Pinus*, the front cavity of the stoma is typically filled with a whitish or dark and granular occluding material. Because this material is porous and the pores are filled with air, the stomata appear white superficially, a circumstance that facilitates their recognition from the surface. The guard cells are sunken and are overtopped by the subsidiary cells (Esau, 1965, p. 446). This xeromorphic anatomy of the stomata of pine may allow CO_2 to enter for photosynthesis, but inhibit the exit of water. Pines and other evergreen gymnosperms adapt to cold northern regions and must conserve water during the cold months of the year when the water in the soil is frozen (Kirkham 2005, p. 327). The anatomy of pine stomata may be a partial reason for their lack of response to elevated CO_2. There is a paucity of field data for forest stands, including conifers, under elevated CO_2 compared to grasslands (Ellsworth 1999) or agricultural crops, which points out the need for more research.

C$_3$ AND C$_4$ PLANTS COMPARED

We already have considered C$_3$ and C$_4$ plants in the previous section. However, here, we specifically compare the stomatal conductance (or its reciprocal, stomatal resistance) of C$_3$ and C$_4$ plants under elevated CO$_2$. We remember that the three resistances that control photosynthesis (ignoring cuticular resistance) are the boundary layer resistance, the stomatal resistance, and the mesophyll resistance. In this section, we are concerned with stomatal resistance and how CO$_2$ affects it differently in C$_3$ and C$_4$ plants. Chapter 12 also compares C$_3$ and C$_4$ plants under elevated CO$_2$, but not specifically this resistance. We note that plants with the C$_4$ photosynthetic system have a lower internal CO$_2$ concentration (C_i) than plants with the C$_3$ photosynthetic system. For example, Dai et al. (1993) showed that at 330 μmol/mol atmospheric CO$_2$ concentration, the internal CO$_2$ concentrations (C_i) in maize (C$_4$) mesophyll cells, maize bundle sheath cells, and wheat (C$_3$) mesophyll cells were 200, 290, and 850 μmol/mol CO$_2$, respectively (see Dai et al. 1993, their Figure 7). Because of the lower C_i of C$_4$ plants, they have a steeper CO$_2$ gradient between the air and the leaf interior and, therefore, a greater photosynthetic rate than C$_3$ plants (He et al. 1992). He et al. (1992) found that at ambient CO$_2$ concentration (337 μmol/mol), the internal CO$_2$ concentration of the C$_3$ grass, Kentucky bluegrass (*Poa pratensis* L.), was 267 μmol/mol and the internal CO$_2$ concentration of the C$_4$ grass, big bluestem (*Andropogon gerardii* Vitman), was 92 μmol/mol. When atmospheric CO$_2$ concentration was about doubled (658 μmol/mol), the internal CO$_2$ concentrations of Kentucky bluegrass and big bluestem were 448 and 258 μmol/mol CO$_2$, respectively. Under elevated CO$_2$, the internal CO$_2$ concentration of the C$_4$ grass (258 μmol/mol) was about the same as the internal CO$_2$ concentration of the C$_3$ grass under the ambient CO$_2$ concentration (267 μmol/mol). These data

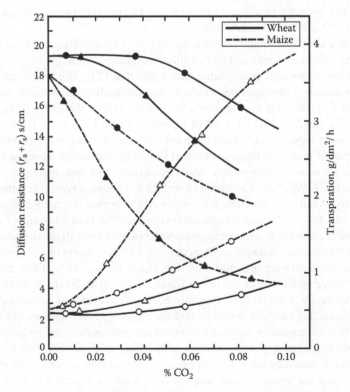

FIGURE 8.15 Transpiration (solid symbols) and diffusive resistance of air and stomata (open symbols) of a maize and a wheat leaf as affected by the atmospheric CO$_2$ concentration at a light intensity of 0.4 (circles) and 0.05 (triangles) cal/cm^2/min (400–700 nm). (Reprinted from Akita, S. and Moss, D.N., *Crop Sci.*, 13, 234, 1973, Fig. 1. With permission. Copyright 1973, Crop Science Society of America.)

indicated that in future atmospheres with elevated CO_2, the internal CO_2 concentration of C_4 plants might be equal to the internal CO_2 concentration that C_3 plants now have.

He et al. (1992) and Rachidi et al. (1993a) reported that under ambient CO_2 concentrations, the stomatal resistance of C_4 plants is higher than that of C_3 plants. He et al. (1992) found that stomatal resistance was increased in both the C_3 and C_4 grasses under doubled CO_2, and the increase was greater for the C_4 grass (60.4%) than for the C_3 grass (8.4%). Under ambient CO_2 concentration, Knapp (1993) studied three C_4 grasses and three C_3 grasses on a tallgrass prairie in Kansas. The C_4 grasses were *Sorghastrum nutans* (L.) Nash (Indian grass), *A. gerardii* Vitman (big bluestem), and *P. virgatum* L. (switch grass). The C_3 grasses were *B. inermis* Leyss (awnless or Hungarian brome), *Agropyron smithii* Rydb. (western wheatgrass), and *Elymus canadensis* L. (*Elymus* is the genus for wild rye). He found that the maximal stomatal conductance of the C_4 grasses was about two times lower than that of the C_3 grasses. The average stomatal conductance of the C_4 grasses was 283.2 mmol/m²/s, and the average stomatal conductance of the C_3 grasses was 537.4 mmol/m²/s. He also found that the stomatal density was higher on the C_4 grasses (average: 263.0/mm²) than the C_3 grasses (average: 190.1/mm²). The guard cells were longer in the C_3 species than the C_4 species. The average guard cell lengths in the C_3 and C_4 species were 41.9 and 32.7 μm, respectively.

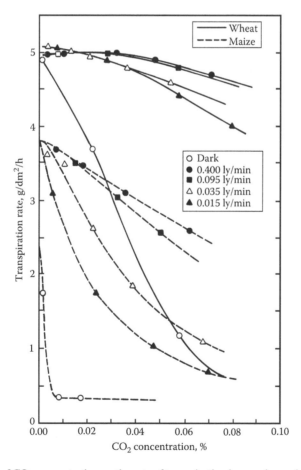

FIGURE 8.16 Effect of CO_2 concentration on the rate of transpiration by a maize and a wheat leaf in various light intensities in Langleys/min (cal/cm²/min, 400–700 nm) at a leaf temperature of 30.5°C. (Reprinted from Akita, S. and Moss, D.N., *Crop Sci.*, 12, 789, 1972, Fig. 2. With permission. Copyright 1972, Crop Science Society of America.)

Akita and Moss (1973) grew wheat (*T. aestivum* L., C_3) and maize (*Z. mays* L., C_4) in pots in a greenhouse and made measurements of gas exchange on the youngest, fully expanded, attached leaves in a laboratory. They made the measurements under two different light intensities (0.4 and 0.05 cal/cm^2/min for light in the 400–700 nm range). From equations they presented, they were able to calculate a diffusion resistance as the sum of r_a (boundary layer resistance) and r_s (stomatal resistance). Figure 8.15 shows the transpiration rate of leaves (solid symbols) and diffusion resistances of air and stomata (open symbols) for a wheat leaf and a maize leaf at various CO_2 concentrations. Transpiration decreased, and the diffusion resistance increased (i.e., stomatal resistance increased), as the CO_2 concentration increased. At both light intensities, maize had a lower transpiration rate, and a higher diffusion resistance, than wheat. The difference between C_3 and C_4 plants in diffusion resistance and transpiration rate became more pronounced as the CO_2 concentration increased. The curves for diffusion resistance of maize and wheat kept increasing and showed no leveling off as the CO_2 concentration increased. Earlier work by Akita and Moss (1972) showed that stomata of the C_3 species, wheat, were less prone to closure than were stomata of the C_4 species, maize, as the CO_2 concentration was increased. The more-open stomata of wheat resulted in higher transpiration rates for the wheat compared to maize (Figure 8.16).

SUMMARY

Most data show that elevated CO_2 decreases stomatal conductance in plants, including annual crops, grassland species, weeds, and deciduous and evergreen trees with broad leaves. When stomatal resistance of C_3 and C_4 plants has been compared, elevated CO_2 increases stomatal resistance of both, and the resistance is higher in C_4 plants compared to C_3 plants. The stomatal conductance of conifers, like pines with needles for leaves, appears to be little affected by elevated CO_2 especially under dry conditions. More research is needed to understand the physiological and anatomical reasons for the response of conifer stomata to elevated CO_2.

9 Elevated Atmospheric Carbon Dioxide: Stomatal Density

INTRODUCTION

As we noted in the previous chapter, CO_2 controls stomatal opening and closing. In addition, it may also control stomatal number (Lake et al. 2001). As we know from books on stomatal physiology (Meidner and Mansfield 1968), a plant species is characterized by the number of stomata on the upper and lower surfaces of its leaves (Table 9.1). However, this number may not have been always constant due to changing CO_2 concentrations in the atmosphere. In this chapter, using data from modern experiments and historical records, we look at the effect of elevated CO_2 on stomatal density. Stomatal density is defined as the number of stomata per unit of interveinal leaf area (Salisbury 1927). The densities are different on the upper and lower surfaces of leaves, as shown in Table 9.1.

WOODWARD'S 1987 DISCOVERY: STOMATAL DENSITY DECREASES WITH INCREASING CO_2 CONCENTRATION

In 1987, Woodward (1987) at the University of Cambridge in England reported the "startling result" (Morison 1987b) that the leaves of tree species collected from the midlands of England two centuries ago differed in stomatal density from leaves collected in the 1980s. He observed a decrease in the density of stomata in specimens of eight different temperate, arboreal species stored in the herbarium in the Department of Botany at the University of Cambridge. They were *Acer pseudoplatanus* L. (sycamore maple), *Carpinus betulus* L. (European hornbeam), *Fagus sylvatica* L. (European beech), *Populus nigra* L. (black poplar), *Quercus petraea* Lieblein (Durmast oak), *Q. robur* L. (English oak), *Rhamnus catharticus* L. [listed by Fernald (1950) and Bailey (1974) as *R. cathartica*] (common buckthorn), and *Tilia cordata* Mill. (small-leaved linden). Although the absolute stomatal densities differed between species, a similar relative decrease in density was found in all cases (Figure 9.1) with a mean reduction of 40% between 1787 and 1981. Over the same period, the ice-core record suggested that the CO_2 in the atmosphere increased by 60 µmol/mol or 21%.

This decline was confirmed by experiments under controlled environmental conditions (Woodward 1987). An increase in the mole fraction of CO_2 from 280 to 340 µmol/mol caused a decrease in stomatal density of 67% (Figure 9.2).

Stomatal density says nothing about the size of the guard cells or the size of the epidermal cells surrounding the guard cells. Cells on the surfaces of leaves can be small (e.g., under drought or salinity) or large (e.g., under low light intensities). A drought-stressed leaf, for example, may have a higher stomatal density than a well-watered leaf, and differences in stomatal density between these two leaves would be due just to cell expansion, not to changes in atmospheric CO_2. Therefore, an index, called the stomatal index, has been developed. The stomatal index is defined as follows (Woodward 1987):

$$\text{stomatal index} = \frac{(\text{stomatal density})}{(\text{stomatal density} + \text{epidermal cell density})} \times 100$$

TABLE 9.1
Stomatal Frequencies, Dimensions, and Pore Areas as Percentages of Total Leaf Areas

Scientific Name	Common Name	Frequencies per mm^2		Dimensions of Stomatal Apparatus, μm		Length of Pore, μm		Pore Area as Average Percentage of Total Leaf Area; Pore Width Taken as 6 μm
		Upper	Lower	Upper	Lower	Upper	Lower	
Osmunda regalis	Royal fern	...[a]	67	...	56×38	...	30	0.5
Phyllitis scolopendrium	Hart's tongue fern	...	59	...	77×42	...	38	0.55
Pteridium aquilinum	Brake fern	...	85	...	46×28	...	21	0.45
Abies nordmanniana	Nordmann fir	...	110	...	49×28	...	17	0.45
Cedrus deodara	Deodar cedar	84	85	38×31	38×31	24	24	1.0
Pinus sylvestris	Scotch pine	120	120	28×28	28×28	20	20	1.2
Picea pungens	Colorado spruce	39	39	49×45	49×45	12	12	0.25
Larex decidua	Larch	14	16	42×26	42×26	20	20	0.15
Allium cepa	Onion	175	175	42×38	42×38	24	24	2.0
Avena sativa	Oat	50	45	52×31	56×26	20	19	0.5
Hordeum vulgare	Barley	70	85	42×21	38×21	17	17	0.65
Triticum aestivum	Wheat	50	40	56×31	53×28	28	28	0.63
Zea mays	Corn	98	108	38×19	43×24	12	16	0.7
Aesculus hippocastanum	Horse chestnut	...	210	...	28×14	...	13	0.7
Carpinus betulus	European hornbeam	...	170	...	32×24	...	13	0.6
Eucalyptus globulus	Tasmanian blue gum	...	370	...	31×23	...	10	0.9
Quercus robur	English oak	...	340	...	28×18	...	10	0.8
Tilia europaea	Linden	...	370	...	25×18	...	10	0.9
Helianthus annuus	Sunflower	120	175	35×25	32×29	15	17	1.1
Medicago sativa	Alfalfa	169	188	26×17	25×17	9	13	0.8
Nicotiana tabacum	Tobacco	50	190	31×25	31×25	14	14	0.8
Pelargonium zonale	Zonal geranium	29	179	48×40	44×38	24	23	1.2
Ricinus communis	Castor bean	182	270	31×21	38×24	12	24	2.1
Vicia faba	Broad bean	65	75	46×25	46×25	28	28	1.0
Xanthium pennsylvanicum	Cocklebur	173	177	39×25	39×25	17	17	1.5
Sedum spectabilis	Stonecrop	28	35	32×32	33×31	21	20	0.32
Tradescantia virginia	Spiderwort	7	23	67×38	70×42	49	52	0.35

Source: Reprinted from Meidner, H. and Mansfield, T.A., *Physiology of Stomata*, McGraw-Hill, New York, 1968, 179pp, Table 1.1. Copyright 1968, McGraw-Hill, with permission from T.A. Mansfield, Lancaster University, United Kingdom.

[a] No stomata on upper surface.

All dimensions are in μm.

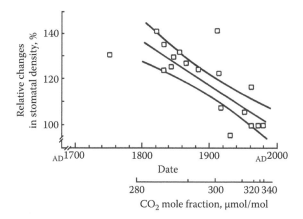

FIGURE 9.1 Abaxial stomatal densities of herbarium-stored leaves of *A. pseudoplatanus* (sycamore maple), *C. betulus* (European hornbeam), *F. sylvatica* (European beech), *P. nigra* (black poplar), *Q. petraea* (Durmast oak), *Q. robur* (English oak), *Rhamus catharticus* (common buckthorn), and *T. cordata* (small-leaved linden). Leaves had been stored in the herbarium in the Department of Botany, University of Cambridge in the United Kingdom. Only leaves on reproductive shoots were sampled, with the assumption that these leaves had developed in full irradiance. Five leaves of each species were sampled from different dates, back to AD 1750, and from collections made in the midlands of England. Stomatal densities varied between species by a factor of about two; however, the changes in stomatal densities relative to recent collections (1970–1981) were similar for all species. Reconstructed changes in atmospheric CO_2 based on ice-core studies are also included. The linear regression line, with 95% confidence limits, shows a 40% reduction in the ratio of stomatal densities over a period of 200 years, $r = -0.828$. (Reprinted by permission from Macmillan Publishers Ltd. *Nature*, Woodward, F.I., Stomatal numbers are sensitive to increases in CO_2 from pre-industrial levels, 327, 617, 1987, Fig. 1, Copyright 1987.)

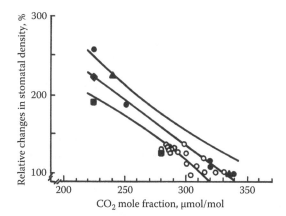

FIGURE 9.2 Comparison of the experimental effects of a change in the CO_2 mole fraction on stomatal density with the putative effects of CO_2 shown in herbarium material (open circles). Experimental observations were made on *A. pseudoplatanus* (closed circles), *Q. robur* (closed diamonds), *R. catharticus* (closed triangles), and *Rumex crispus* (yellow dock) (closed squares). Plants were grown in small airtight enclosures for a period of 3 weeks in a 16-h photoperiod at 18°C, a mean leaf water vapor pressure deficit of 1.2 kPa, and an irradiance of 373 μmol/m²/s. The night temperature was 7°C. The CO_2 mole fractions supplied to the enclosures were controlled with a gas diluter and measured with an infrared gas analyzer. The linear regression for the experimental observations, on leaves which had been initiated and developed in the treatments, had a slope of $-1.12\% \pm 1.5\%$ μmol/mol CO_2 (95% confidence limit) ($r = -0.940$). The linear regression coefficient for the herbarium material (open circles) was $-0.61\% \pm 0.8\%$ μmol/mol CO_2 ($r = -0.858$). (Reprinted by permission from Macmillan Publishers Ltd. *Nature*, Woodward, F.I., Stomatal numbers are sensitive to increases in CO_2 from pre-industrial levels, 327, 617, 1987, Fig. 2, Copyright 1987.)

TABLE 9.2
Response of Stomatal Index to Changes in CO_2 Mole Fraction[a]

| Species | Common Name | CO_2 Mole Fraction, $\mu mol/mol$ | Stomatal Index, % | |
			Adaxial Surface	Abaxial Surface
A. pseudoplatanus	Sycamore maple	225	0	14.9 ± 0.8
		340	0	6.7 ± 1.1
Q. robur	English oak	225	0	17.4 ± 1.1
		340	0	9.6 ± 1.0
R. crispus	Yellow dock	225	3.5 ± 0.8	15.5 ± 0.7
		340	0.5 ± 0.3	11.8 ± 0.9
Vaccinium	Velvet leaf bilberry	250	13.1 ± 1.9	17.5 ± 2.4
myrtillus		340	2.5 ± 0.8	16.9 ± 2.2
		450	1.2 ± 1.6	14.2 ± 2.7
L. esculentum	Tomato	350	26.4	31.0
		650	25.6	29.5
		1000	26.6	27.6
Liquidambar	Bilsted (sweet gum)	340	0	25 ± 6.8
styraciflua		520	0	28 ± 4.5
		718	0	29 ± 6.8
		910	0	26 ± 4.5

Source: Reprinted by permission from Macmillan Publishers Ltd. *Nature*, Woodward, F.I., Stomatal numbers are sensitive to increases in CO_2 from pre-industrial levels, 327, 617, 1987, Table 1, Copyright 1987.

[a] Values are means ± standard error of the mean. Stomatal index = [(stomatal density)/(stomatal density + epidermal cell density)] × 100.

Woodward (1987) found that changes in stomatal density were similar to changes in stomatal index (Table 9.2), which indicated that the effect of CO_2 was on stomatal development, not simply a response involving the control of epidermal cell expansion. Woodward found that neither stomatal density nor stomatal index was affected by CO_2 levels exceeding the ambient level (340 $\mu mol/mol$).

Why would stomatal density and stomatal index have decreased as CO_2 concentration in the atmosphere has increased? Woodward (1987) suggested that the decrease is to increase water use efficiency (WUE). On a leaf level, WUE can be defined as the photosynthetic rate divided by the transpiration rate. In Chapter 11, we consider in detail the effect of elevated CO_2 on WUE. Here we mention it in connection with stomatal density. As we noted in the previous chapter, a leaf must control gas exchange during photosynthesis and transpiration to maintain its viability. Woodward (1987) determined stomatal conductance and WUE of maple leaves grown under a high (340 $\mu mol/$ mol) and a low (225 $\mu mol/mol$) CO_2 concentration (Figure 9.3). Leaves grown at the higher level of CO_2 had a lower stomatal conductance and a higher WUE than maple leaves grown at a low level of CO_2. At 340 $\mu mol/mol$ CO_2 and a vapor pressure deficit of 0.7 kPa, WUE was 2.2 times greater than at 225 $\mu mol/mol$ CO_2. Woodward's experiment suggests that the photosynthetic rate has been maintained at a relatively constant level throughout the centuries as CO_2 has increased, but water lost has become less due to increased WUE. The higher density of stomata in earlier times resulted in more water being lost from the leaves, and, consequently, WUE was lower than today. Woodward (1987) concludes that "in the lower CO_2 levels of the pre-industrial era the uptake of CO_2 should have been more expensive in terms of water loss than the present day." On an evolutionarily scale, if a plant species is trying to survive droughts and maintain dry-matter production (speaking teleologically),

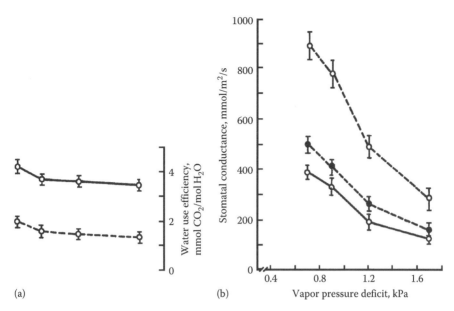

FIGURE 9.3 The responses of (a), water use efficiency or WUE (the ratio of instantaneous rates of photosynthesis to transpiration) and (b), stomatal conductance in *A. pseudoplatanus* to variations in the water vapor pressure deficit of the air. Open circles and dashed lines: measurements on leaves grown at a CO_2 mole fraction of 225 µmol/mol; open circles, solid lines: measurements on leaves grown at a CO_2 mole fraction of 340 µmol/mol; closed circles, dashed lines: measurements on leaves grown at a CO_2 mole fraction of 340 µmol/mol and then transferred to a mole fraction of 225 µmol/mol for 1 week before measurements (the stomatal density of measured leaves was not significantly different from leaves maintained continually at 340 µmol/mol). Measurements of transpiration, stomatal conductance, and photosynthetic rate were made with a portable gas analyzer at different water vapor pressures in the environments for growth. (Reprinted by permission from Macmillan Publishers Ltd. *Nature*, Woodward, F.I., Stomatal numbers are sensitive to increases in CO_2 from pre-industrial levels, 327, 617, 1987, Fig. 3, Copyright 1987.)

a lower stomatal density would be valuable for survival. However, Woodward does not mention if conditions have become drier in the midlands of England over the last two centuries, which would have made this adaptation (lower stomatal density) desirable.

To back up his conclusion (Woodward 1987), Woodward and Bazzaz (1988) studied the stomatal density and stomatal index of plants along an altitudinal gradient in Scotland. (See Appendix for a biography of Fakhri Bazzaz.) One population of *Nardus stricta* L. (Poaceae family; common name: matgrass; a densely matted wiry perennial that lives in poor grasslands) (Fernald 1950, p. 132) growing at high altitudes (860 m) was exposed to a lower partial pressure of CO_2 than another population growing at lower levels (366 m). They simulated outdoor conditions experimentally by keeping the mole fraction of CO_2 constant, as it is with altitude, but changed the partial pressure of CO_2, which declines as one ascends a mountain. The stomatal densities of the two *Nardus* populations under the two partial pressures (100 kPa = about 1 atm and 78 kPa, the pressure equivalent to an altitude of 2000 m) are shown in Figure 9.4. Plants grew at partial pressures of CO_2 between the low level of 22.5 and 34 Pa (current level of CO_2 in 1988). Both populations of *Nardus* showed an increase in abaxial (lower leaf surface) and adaxial (upper leaf surface) density, and significantly so at $P = 0.05$ for the population from 860 m, when the partial pressure of CO_2 was reduced. They concluded that the population of *Nardus* originating from the lower CO_2 environment at 860 m showed a greater response to the change in CO_2 partial pressure than the population from the higher CO_2 environment at 366 m. This response suggested to them that the currently increasing levels of CO_2 will cause greater reductions in stomatal density the greater the altitude.

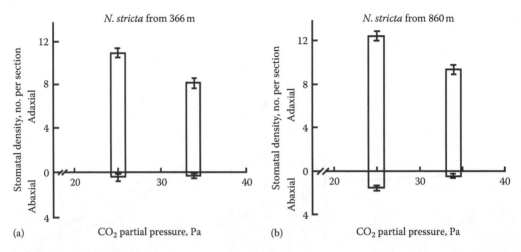

FIGURE 9.4 Stomatal responses of two populations of *N. stricta* (matgrass) to changes in the partial pressure of CO_2 at two atmospheric pressures, 100 and 78 kPa. The pressure of 78 kPa is equivalent to an altitude of 2000 m. Width of the histograms indicates the mean daily variation in CO_2 partial pressure. Stomatal densities measured as the number of stomata per microscope section, with 12 sections per leaf and 5 leaves per treatment: (a) Population from 366 m and (b) population from 860 m. (Reprinted from Woodward, F.I. and Bazzaz, F.A., *J. Exp. Bot.*, Fig. 1, 1988 by permission of Oxford University Press.)

IMPORTANCE OF HERBARIUMS TO STUDY THE HISTORICAL RECORD OF STOMATA

As might be expected, the surprising finding by Woodward (1987), i.e., that stomatal density decreases as atmospheric CO_2 concentration increases, has sparked many subsequent studies. Researchers have scurried to their herbariums to study the historical record. As an aside, this discovery is a reason to maintain herbariums. Since 1990, many articles have been written about whether or not herbarium collections should be maintained (Chalmers et al. 1990; Clifford et al. 1990; Gyllenhaal et al. 1990; Harley 1990; Jarzembowski 1990; Stevens 1990; Stirton et al. 1990; West and Conn 1990; Ingram et al. 1994; Knapp et al. 2002; Gropp 2003, 2004; Lilien 2003; Raven 2003; Ebach and Holdrege 2005). They are expensive to keep; and current tough economic times dictate that research monies go to molecular biology rather than to taxonomy, which takes a back seat to experimentation and quantification (Gould 1994). Even the Academy of Natural Sciences in Philadelphia, North America's first natural history museum founded in 1812, might not be saved due to lack of money to support it. Its vast natural history collections come from some of America's founding pioneers: Benjamin Franklin, Thomas Jefferson, Meriwether Lewis, and William Clark (Dalton 2007).

However, natural history museums and their collections are essential to see what plants now extinct looked like (Vogel 2004) and to document the planet's current biological diversity (Agosti 2003). They also serve vital roles in the identification of invasive and disease species, which enables biological control (Causey et al. 2004; Miller et al. 2004). Taxonomists and botanists have lost their jobs to molecular biologists. At Kansas State University, general botany was no longer taught after 2006. The job cuts at the Academy of Natural Sciences in Philadelphia exemplify a disturbing trend that threatens our understanding of biodiversity (LaPolla 2005). Training of taxonomists is essential so that conservation biologists and decision-makers can make scientifically informed policies in efforts to preserve life (Knapp et al. 2002; Wheeler et al. 2004). In particular, taxonomists are critically needed now to study the impact of climate change on flora (Dalton 1999; Causey et al. 2004). Students who have a passion to understand biological diversity should not be denied training because universities no longer give courses in systematics. The great scientist, Ernst Mayr

(1904–2005), the Agassiz Professor of Zoology at Harvard University and director of its Museum of Comparative Zoology (Diamond 2005), mastered ornithology by studying in natural history museums (Meyer 2004; Pennisi 2004). Mayr said, "Great scientists need more than genius. They also need to be in the right place at the right time" (Pennisi 2004). If Mayr were a student today at one of America's universities, where natural history collections are being eliminated, he might not have had the chance to make his monumental contributions in evolutionary biology.

CONFIRMATION OF THE 1987 DISCOVERY

In agreement with Woodward's (1987) data, herbarium specimens from the University of Barcelona, Spain (Catalonia), a Mediterranean climate, showed an average decrease between 1750 and 1990 in stomatal density of 14 different species, representing trees, shrubs, and herbs. The CO_2 concentration during this time increased from 278 µmol/mol CO_2 to 348 µmol/mol (Peñuelas and Matamala 1990). The decrease was steeper during the initial slower increases in CO_2 atmospheric levels as compared with the relatively faster CO_2 increases in recent years.

Also, in agreement with Woodward's (1987) findings, Beerling and Kelly (1997) found a decrease in stomatal density over a nearly 70-year period (1927–1995). They studied a wide range of tree, shrub, and herb species (60 species) in the United Kingdom and compared them to data collected by Salisbury (1927). Ice core studies indicated that the atmospheric CO_2 concentration in 1927 was ~55 µmol/mol lower than the then present day (in 1997) value of 355 µmol/mol. They concluded that firmer predictions for future changes in stomatal density across many species, expected as a possible result of anthropogenically related CO_2 increases, are now possible.

CONTRADICTIONS TO THE 1987 DISCOVERY: NO EFFECT OF ELEVATED CO_2 ON STOMATAL DENSITY

Körner (1988) tested the hypothesis that global increases of atmospheric CO_2 reduce stomatal density. He examined species from low and high altitudes of alpine plants from central Europe, especially near Innsbruck, Austria (600 m above sea level) and the Ötzta Alps (3000 m above sea level) collected between 1864 and 1918 and compared them with collections obtained between 1976 and 1986 (largely 1985). When all data were pooled (185 historical and 104 recent determinations), no significant differences were seen at either altitude. In fact, he was surprised that the means of such different collections were so similar. He concluded that changes of stomatal density due to the global increase in CO_2 appear to be smaller than experimental noise, and that "very large data sets" will be required "to obtain adequate resolution with respect to structural responses to altered atmospheric CO_2-levels."

Why did Körner not see differences in stomatal density with CO_2 concentration, while Woodward (1987) and Woodward and Bazzaz (1988) did? Körner suggested that intraspecific variation is large and defies a valid comparison, if only a few locations are sampled. He pointed out that differences in stomatal density are biased by microsite variation. He states, "[I]n these elevations climatic differences other than atmospheric pressure are attributable to greater extent to microsite differences than altitude." He does not state what these differences are, but wind might be one factor. Although I know of no studies relating wind speed to stomatal density, wind has a profound effect on water loss (Kirkham 1978). Light is another factor. Leaves that originate in the shade tend to be less thick, and have a thinner epidermis, fewer stomata, and a lower density of hairs, all characters associated with the enhancement of light trapping (Moore 1992). Values of stomatal density and stomatal index in shaded leaves of *Alnus glutinosa* (L.) Gaertn. (black alder) were lower (71% and 93%, respectively) than those for leaves in the sun (Poole et al. 1996). Poole et al. (1996) discuss the implications of their findings for paleoclimatic interpretations and emphasize the need for caution when drawing conclusions based solely on stomatal characters.

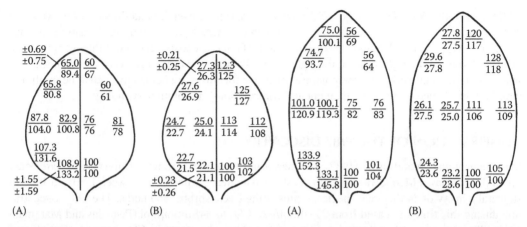

(A)   (B)   (A)   (B)

FIGURE 9.5 (Left half) Mean stomatal densities (A) and stomatal pore lengths (B) on various points of adult leaves of medium insertion level of *Nicotiana sanderae* (garden tobacco). On left half of the leaf is shown at A: number of stomata per 1 mm^2, upper side (n = 450) and lower side; at B in the same way the pore length in μm (n = 220). On right half of leaf is shown the same values in per cent of values at blade base. Mean errors of indicated averages in apical and basal part (the numbers are also shown in the figure) are: 65.0 ± 0.69; 89.4 ± 0.75; 108.9 ± 1.55; 133.2 ± 1.59; 27.3 ± 0.21; 26.3 ± 0.25; 22.1 ± 0.23; 21.1 ± 0.26. (Reprinted with kind permission from Springer Science+Business Media: *Biol. Plant.*, The distribution pattern of transpiration rate, water saturation deficit, stoma number and size, photosynthetic and respiration rate in the area of the tobacco leaf blade, 5, 1963, 143, Slavík, B., Fig. 3.) (Right half) The same values as in left half of the figure with adult leaves of medium insertion level from leaf rosettes of sugar beet, *Beta vulgaris* variety "Dobrovická A" (n = 750) (density) and n = 375 (length). (Reprinted with kind permission from Springer Science+Business Media: *Biol. Plant.*, The distribution pattern of transpiration rate, water saturation deficit, stoma number and size, photosynthetic and respiration rate in the area of the tobacco leaf blade, 5, 1963, 143, Slavík, B., Fig. 4.)

Körner (1988) also suggested that differences in his data and those of others might be due to sampling. Woodward (1987) studied stomata only on leaves from reproductive shoots, with the assumption that these leaves had developed in full irradiance. Woodward does not mention where along the leaf he took his observations, even though, undoubtedly, he must have made comparisons using the same portion of every leaf. Care must be taken to sample the same point on a leaf, because stomatal number varies from tip to base and from midvein to edge (Slavík 1963) (Figure 9.5). When O'Leary and Knecht (1981) studied stomatal density of green beans (*Phaseolus vulgaris* L.) as affected by two levels of CO_2 (400 or 1100 μmol/mol), they made impressions at a point midway between the midvein and leaf margin and halfway between the apex and base of the leaflets. When they did this, they found no difference in stomatal density on the adaxial surface, but there was a lower stomatal density on the abaxial surface of leaves in the higher CO_2 concentrations.

Körner (1988) also might not have observed a relationship between stomatal density and CO_2 concentration, because his records only went back to 1864, early industrial times. Woodward's data extend to 1750, before industrial times. The Industrial Revolution, according to Webster's Dictionary (Friend and Guralnik 1959), is "the change in social and economic organization resulting from the replacement of hand tools by machine and power tools and the development of large-scale industrial production: applied to this development in England from about 1760." As noted in Chapter 1, the increase in CO_2 in the atmosphere is thought to be due to the CO_2 produced due to the industrial revolution.

Malone et al. (1993) also pointed out the importance of time in observing changes in stomatal density due to CO_2 concentration. They measured stomatal density on *Avena sativa* L. (oat; an annual with the C_3 photosynthetic pathway), *Triticum aestivum* L. (wheat; annual; C_3), *Prosopis glandulosa* Torr. var. *glandulosa* (honey mesquite, a woody, perennial legume; C_3), and *Schizachyurium scoparium* (Michx.) Nash. (little bluestem, a perennial prairie grass; C_4). The plants grew in a

TABLE 9.3
Stomatal Density and Stomatal Size of Leaves of Winter Wheat Grown in the Field under Two Moisture Regimes and Two CO$_2$ Concentrations

| Treatment | Stomatal Density (mm^{-2}) | | Stomatal Size | | | |
| | | | Length (μm) | | Width (μm) | |
	Adaxial	Abaxial	Adaxial	Abaxial	Adaxial	Abaxial
Well watered, 340 μmol/mol CO$_2$	64.6 ± 3.9	45.0 ± 3.8	53.3 ± 1.5	49.8 ± 0.2	29.5 ± 4.2	31.4 ± 0.2
Water stressed, 340 μmol/mol CO$_2$	79.2 ± 5.8	46.7 ± 1.1	48.9 ± 3.3	47.0 ± 3.8	23.9 ± 2.4	25.0 ± 0.4
Well watered, 825 μmol/mol CO$_2$	66.1 ± 7.1	52.0 ± 6.4	58.8 ± 1.8	52.5 ± 0.9	30.8 ± 1.1	29.7 ± 1.3
Water stressed, 825 μmol/mol CO$_2$	64.6 ± 3.4	49.3 ± 3.2	51.9 ± 1.3	48.6 ± 2.9	26.6 ± 0.9	28.6 ± 1.5

Source: Chaudhuri, U.N. et al., Effect of elevated levels of CO$_2$ on winter wheat under two moisture regimes, *Response of Vegetation to Carbon Dioxide*, Research Report No. 050, U.S. Department of Energy, Carbon Dioxide Research Division, Office of Energy Research, Washington, DC, 1989, ix + 49pp. Unpublished data.

Mean and standard error are shown. Stomatal density is the mean of four values. Stomatal size is the mean of 16 values.

growth chamber across a sub-ambient CO$_2$ gradient from about 200–350 μmol/mol. Only on the adaxial surface of wheat did they see a decreased stomatal density as CO$_2$ concentration increased. They concluded that the other three species lacked the ability to alter stomatal density in response to increasing atmospheric CO$_2$ concentration in a single generation (annuals) or growing season (perennials).

Similarly, Chaudhuri et al. (1989) saw no effect of elevated carbon dioxide (825 μmol/mol) on stomatal density in winter wheat (*T. aestivum* L. cv. Newton) grown under two moisture regimes (field capacity and about one-half field capacity) in closed-top chambers in a field with a silt loam soil in Kansas (Table 9.3). Water stress increased the stomatal density under ambient CO$_2$ (340 μmol/mol), but not under elevated CO$_2$. Consequently, under dry conditions, stomatal density was less at elevated CO$_2$ compared to ambient CO$_2$. As noted above, drought decreases stomatal size, and Chaudhuri et al. (1989) also found this. Under both CO$_2$ concentrations, drought decreased the length and width of the stomata (Table 9.3). Note that Chaudhuri et al. (1989) measured the length and width of the entire stomatal complex and not the size of the stomatal pore (aperture). Water lost through stomatal pores is not related to stomatal length and width, but to the size of the stomatal aperture. Stomata may be present but not functioning (opening and closing) (Esau 1965, p. 158).

BROWN AND ESCOMBE'S DIAMETER LAW

Because stomata are dynamic and quickly respond to changes in environmental variables, it is difficult to measure stomatal aperture. However, theoretical considerations show that the shape of the aperture might affect stomatal density. Important investigations on the diffusion of gases and liquids through small openings were carried out by Brown and Escombe (1900). These studies showed that the rates of diffusion through small, single apertures are proportional to the diameters and not to the areas of the openings. This agreed with results previously established by Josef Stefan (1835–1893) for the converse case of evaporation from circular surfaces of water (Maximov 1929, p. 172). Brown and Escombe (1900) found that their "diameter law" holds true also for the case of

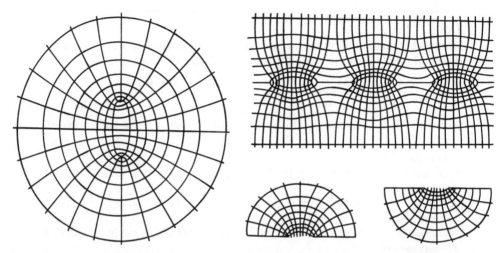

FIGURE 9.6 Diagrammatic representation of the diffusion of water vapor through small openings. Left: Diffusion through a single opening in a vertical septum; the fan-like, diverging lines show the courses of the diffusing particles; perpendicular to them are concentric lines of equal vapor density. Right (above): Diffusion through a horizontal multi-perforate septum (three openings are represented). Right (below): Diffusion through single openings of the same size. (From Maximov, N.A., *The Plant in Relation to Water*, George Allen and Unwin, Ltd., London, 1929, p. 173, Fig. 24. Copyright 1929, George Allen & Unwin, Ltd.)

diffusion through a number of small openings (a "multi-perforate septum"). From this, it follows that more water vapor will diffuse in unit time through several small apertures than through a single larger opening with an area equal to the combined areas of the smaller ones (Kirkham 2005, p. 383). If gases of different composition (e.g., dry and moist air) are diffusing through neighboring holes, then the flux from one hole will affect the flux from another hole. The "lines of flow" of the diffusing molecules, which normally tend to diverge fan-wise as they issue from the apertures (Figure 9.6), show interference with one another, thus slowing down the rate of diffusion. Brown and Escombe found experimentally that such interference begins when the distance between the apertures is somewhat less than 10 times the diameter of the holes (Maximov 1929, p. 172). The fact that the rate of diffusion through small openings is proportional not to the area, but to the diameter of the opening, greatly increases the possible amount of diffusion that can take place through a multi-perforate septum (a leaf surface).

The results of Brown and Escombe (1900) suggest that maybe we should not be looking just at stomatal density, but we also should be measuring stomatal diameters and how close they are. Indeed, in some studies with elevated CO_2, stomatal diameter has been measured and the results are variable. Radoglou and Jarvis (1990a) found no effect of CO_2 enrichment on the length of the stomatal pore of four different clones of poplar (*Populus* sp.). However, Malone et al. (1993) found that aperture length of two wheat (*T. aestivum* L.) cultivars (Yaqui 54 and Seri M82) increased slightly as the CO_2 concentration increased from about 200–350 μmol/mol, but a statistically significant increase occurred only on the adaxial surface of Seri M82 (Figure 9.7). We need more studies to see how stomatal diameter is affected by CO_2 concentration. Mathematicians who understand transport phenomena should be able to help in these studies. They could quantify and predict CO_2 fluxes across a multi-perforate septum under varying CO_2 levels. Their predictions could be confirmed by experiments.

AMPHISTOMATOUS AND HYPOSTOMATOUS LEAVES

Five of the species that Woodward (1987) studied (*Acer, Carpinus, Tilia*; the two *Quercus*) are hypostomatous (Meidner and Mansfield 1968, Table 9.1; Woodward 1987). That is, leaves of these

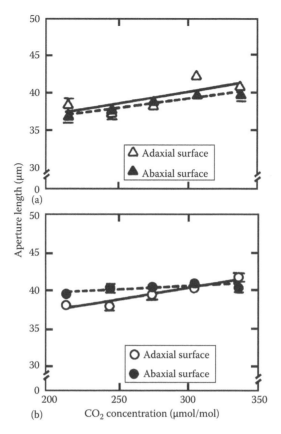

FIGURE 9.7   Linear regressions of length of stomatal aperture on adaxial and abaxial surfaces of flag leaves of Yaqui 54 wheat (a) and Seri M82 wheat (b) against $[CO_2]$, 50 days after planting in a sub-ambient CO_2 chamber. Data points are the mean of 60 observations per CO_2 concentration. Vertical bars represent two standard errors of the mean. Aperture length on the adaxial surface of Seri M82 increased significantly ($r^2 = 0.93$, $P < 0.008$). There was no statistically significant response of aperture length on either flag leaf surface of Yaqui 54 (adaxial, $r^2 = 0.43$; abaxial, $r^2 = 0.64$) or on the abaxial surface of Seri M82 flag leaves ($r^2 = 0.54$). (Reprinted from Malone, S.R. et al., *Am. J. Bot.*, 80, 1413, 1993, Fig. 2. With permission. Copyright 1993, Botanical Society of America.)

trees have stomata only on the abaxial (lower) surface. *Populus* is amphistomatous. That is, it has stomata on both the adaxial and abaxial surfaces of its leaves. I do not know if the other two species that Woodward studied (*Fagus, Rhamnus*) are hypostomatous or not. Many tree species are hypostomatous (Table 9.1). Woodward might have gotten different results if he had studied only amphistomatous leaves, which typify most plants other than trees (Table 9.1). The stomatal resistance (or its reciprocal, stomatal conductance), as well as the water potential and transpiration rate, of the upper and lower surface of amphistomatous leaves varies. In corn (*Zea mays* L.), the upper surface has a lower stomatal density and a higher stomatal resistance and lower transpiration rate than the lower surface (Kirkham 1986). One would think that plants with a complete lack of stomata on one surface (i.e., the adaxial surface in many tree species) would respond differently to elevated levels of CO_2 than plants that have stomata on both surfaces. It would be interesting to see if amphistomatous tree leaves had a different slope in Figure 9.1 (from Woodward 1987) as compared to hypostomatous tree leaves.

Woodward and Kelly (1995) surveyed 100 species and made 122 observations and found an average reduction in stomatal density of 14.3% with CO_2 enrichment, with 74% of the cases exhibiting a reduction in stomatal density. Forty-three of the observations were made in controlled environments and under a typical range in CO_2 enrichment of 350–700 µmol/mol. For these cases, the average stomatal density declined by 9%, and 60% of the cases showed reductions in stomatal density. When analyses were restricted to these 43 observations, amphistomatous samples more frequently had greater changes in stomatal density than did hypostomatous samples.

SENSITIVITY OF STOMATA TO CO_2

Woodward (1988) points out that some stomata are "sensitive" and some stomata are "insensitive" to changes in CO_2 partial pressure. *A. pseudoplatanus* L. (sycamore maple), one of the plants that he studied in his 1987 paper, is a woody plant with stomata that respond to CO_2. He found that the stomatal conductance of *A. pseudoplatanus* decreased as the CO_2 partial pressure increased from 28 Pa (the level of CO_2 in the atmosphere in 1700) to 34 Pa (the level of CO_2 in the atmosphere in 1988) (Woodward 1988). But a minority of plants, including species of *Pinus* (pine) and *Picea* (spruce), possessed stomata with a limited response of stomatal conductance to CO_2. *Acer* is an angiosperm and *Pinus* and *Picea* are gymnosperms. (In Chapter 1, we discuss the differential response of angiosperms and gymnosperms to elevated CO_2 under drought. Elevated CO_2 appears to compensate for reductions in growth due to drought more in angiosperms than gymnosperms.) Stomata of some plants are also not responsive after watering. For example, genotypes of soybean [*Glycine max* (L.) Merr.] with low stomatal frequently have been found to be unresponsive to irrigation and more tolerant of water stress, with overall low yield (Buttery et al. 1993). The absence of a stomatal response to CO_2 has limited consequences for photosynthesis and WUE at current levels of CO_2. But in atmospheres enriched in CO_2, such plants have slightly higher photosynthetic rates but lower water use efficiencies than CO_2 sensitive species (Woodward 1988). Below CO_2 partial pressures of 34 Pa, WUE tends to be higher in the CO_2-insensitive species. Therefore, in analyzing the relationship between stomatal density and CO_2 concentration, one must consider stomatal sensitivity to CO_2.

STUDIES SINCE WOODWARD'S 1987 DISCOVERY

In addition to the previously mentioned studies, other research, in follow-up of Woodward's work (1987), has monitored stomatal density in response to CO_2 concentration. The experiments have been done under controlled environmental conditions, and herbarium specimens have not been part of them. The results are not consistent. They show that stomatal density is either not affected or increased by elevated CO_2. For example, results from the previously mentioned study by Radoglou and Jarvis (1990a), who grew four poplar clones in an open-top growth chamber inside a glasshouse in Scotland, agreed with Körner's herbarium data, which showed no change in stomatal density with CO_2 enrichment. Both stomatal density and the stomatal index of the clones were the same under ambient CO_2 (350 µmol/mol) and doubled CO_2 (700 µmol/mol) (Figure 9.8). Lack of differences in stomatal density indicated that CO_2 enrichment had no effect on the initiation of the number of stomata during ontogenesis.

Apel (1989) studied eight different species (*Hordeum vulgare* L., barley; *T. aestivum* L., wheat; *Secale cereale* L., rye; *A. sativa* L., oat; *Z. mays* L., corn; *Vicia faba* L., bean; *Lycopersicon esculentum* Mill., tomato; *A. pseudoplatanus* L., maple) and found that four responded to elevated CO_2 (1500 µmol/mol) with an increase in stomatal density, compared to plants grown under ambient CO_2 (345 µmol/mol). The four that responded were barley, corn, tomato, and maple. The stomatal density of the other species was not affected by CO_2 level. Apel (1989) also studied stomatal density of 12 French (green) bean (*P. vulgaris* L.) cultivars under the same two CO_2 concentrations. He found a range of responses, but all 12 cultivars had an increased stomatal density under the elevated level of CO_2. He defined the stomatal density under the ambient level as 100%; stomatal density was increased between 114% and 154% under the elevated level of CO_2, depending upon the cultivar.

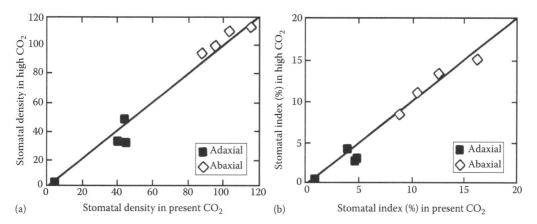

FIGURE 9.8 Relationship between stomatal density (in number per mm^2) (a) and stomatal index (in %) (b) of plants in high and present CO_2 concentration. Poplar clones measured were Columbia River, Beaupre, Robusta, and Raspalje. Each point is the mean of three plants. Three fields were measured on each surface of four leaves per plant. (Reprinted from Radoglou, K.M. and Jarvis, P.G., 1990a. *Ann. Bot.*, Fig. 2, 1990 by permission of the Oxford University Press.)

Rowland-Bamford et al. (1990) grew rice (*Oryza sativa* L.) under varying levels of CO_2 (160–900 μmol/mol) and observed an increase in stomatal density as CO_2 concentration increased from 160 to 330 μmol/mol (Figure 9.9). Enrichment above this level resulted in no further increase in stomatal density. Abaxial stomatal density was more influenced by increases in CO_2 than that on the adaxial surface.

During a 2-year study, Kimball et al. (1986, 1987) grew cotton (*Gossypium hirsutum* L.) in open-top chambers under ambient CO_2 (340 μmol/mol) and elevated CO_2 (640 μmol/mol). Stomatal density was determined for the adaxial and abaxial sides of fully expanded cotton leaves. In 1986, stomatal density on both leaf surfaces was increased by elevated CO_2. Overall in 1986, an increase of 10% was observed in stomatal density. In 1987, trends were similar, but the increase (6%) in stomatal density under CO_2 enrichment was not significant. Sator et al. (1995) grew white clover

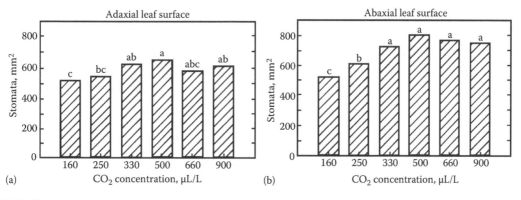

FIGURE 9.9 Stomatal densities of flag leaves of rice grown under a range of CO_2 concentrations, on (a) adaxial surface and (b) abaxial surface. Different letters denote a significant difference at the $P < 0.01$ level; six leaves were sampled and each surface of each leaf was measured in five random locations. Measurements were made 14 weeks after the imposition of the CO_2 treatments. The unit of μL/L is equivalent to the unit μmol/mol. (Reprinted from *Environ. Exp. Bot.*, 30, Rowland-Bamford, A.J., Nordenbrock, C., Baker, J.T., Bowes, G., and Allen Jr., L.H., Changes in stomatal density in rice grown under various CO_2 regimes with natural solar irradiance, 175, Fig. 1, Copyright 1990, with permission from Elsevier.)

(*Trifolium repens* L.) in open-top chambers under two CO_2 concentrations (380 and 670 μmol/mol). They found no difference in stomatal densities of both leaf sides with respect to CO_2 concentrations. Poole et al. (2000) grew *A. glutinosa* Gaertn. (black alder) in growth chambers under the current ambient (360 μmol/mol CO_2 and doubled ambient (720 μmol/mol CO_2) levels of CO_2 and found no change in stomatal density.

Miller-Rushing et al. (2009) measured stomatal density in 27 individual trees growing in the Arnold Arboretum in Boston, Massachusetts, over the last 100 years. They examined leaves from 74 herbarium specimens collected from three genera—*Acer* (maples), *Quercus* (oaks), and *Carpinus* (hornbeams), which were sampled between 1893 and 2006. During this period, global average atmospheric CO_2 concentrations increased by 86 ppm. Although stomatal density declined during the lifetimes of individual trees, the changes were not dependent on the changes in CO_2 concentrations.

Estiarte et al. (1994) found no significant differences in stomatal densities or stomatal indexes of a hard red spring wheat (*T. aestivum* L. cv. Yecora rojo) or sour orange (*Citrus aurantium* L.) trees grown at high CO_2 concentrations in two different CO_2 enrichment systems (free-air CO_2 enrichment for wheat and open-top chambers for orange trees). Their results were in accordance with most of the previous results obtained in short-term experimental studies, which suggest that plants do not acclimate to increasing CO_2 concentration by changing stomatal density within a single generation.

However, Mao et al. (2005) found striking differences in the distribution of stomata on the upper and lower surfaces of birch (*Betula platyphylla* Suk.), when plants were exposed to different levels of CO_2 and nitrogen. They grew 2-year-old birch seedlings in sand in either a greenhouse (control plants with 300 μmol/mol CO_2) or in a growth chamber (700 μmol/mol CO_2). Plants were watered with Hoagland's nutrient solution containing low (2 mM) or high (16 mM) amounts of NH_4NO_3. Table 9.4 shows stomatal densities after 2 months of growth at the end of the study.

The complete lack of stomata on the upper leaf surface under the elevated CO_2 and low nitrogen is an outstanding result and deserves further study. The results for the upper leaf surface of plants grown under low nitrogen confirm Woodward's findings—i.e., that elevated CO_2 reduces stomatal density. However, under low nitrogen, stomatal density on the lower leaf surface under elevated CO_2 was greater than that under the ambient level of CO_2. The authors concluded, "[W]hen the balance between CO_2 concentration and mineral nutrition is destroyed, CO_2 operates not only as the substrate of carbon nutrition but also as the factor affecting the morphology of the plant assimilation

---

TABLE 9.4
Stomatal Density on the Upper and Lower Surfaces of Leaves of 2-Year-Old Birch Seedlings Grown for 2 Months in Sand and Exposed to Two Atmospheric CO_2 Concentrations and Watered with Two Different Levels of Nitrogen in the Nutrient Solution

	350 μmol/mol CO_2	700 μmol/mol CO_2
Upper leaf surface (mm^{-2})		
2 mM NH_4NO_3	267 ± 2	0
16 mM NH_4NO_3	Not determined	361 ± 3
Lower leaf surface (mm^{-2})		
2 mM NH_4NO_3	240 ± 1	447 ± 5
16 mM NH_4NO_3	Not determined	248 ± 4

Source: Reprinted from Mao, Z.-J. et al., *Russ. J. Plant Physiol.*, 52, 171, 2005, Table 2. With permission from P. Yu. Voronin, Timiriyazev Institute of Plant Physiology, Copyright 2005.

apparatus." Their data suggest that the lack of stomata on the upper leaf surfaces of many tree leaves (Table 9.1) might be related, in part, to the balance of carbon and nitrogen in the plants during leaf development.

In contrast to results described above, in which stomatal density has not been affected or increased by elevated CO_2, the study by Paoletti and Gellini (1993) confirmed Woodword's (1987) observations. They studied leaf samples of *F. sylvatica* L. (beech) and *Q. ilex* L. (holm oak) collected over the last 200 years. They used material in three herbaria in Italy and also collected fresh samples. Both herbarium and fresh samples came from two regions of central Italy, Tuscany and Lazio. The leaves were examined using a scanning electron microscope. They observed a reduction in the abaxial stomatal density in beech and oak of 43% and 28%, respectively, between 1810 and 1990. They noted that the scanning electron microscope provides greater accuracy and reliability, which is especially important with herbarium samples because they have usually suffered from harsh preservation treatments and frequent mechanical manipulations. Thus high magnifications are often required to recognize a stoma. This was particularly true for the holm oak, where abrasion was necessary to remove the abaxial pubescence and altered the appearance of the epistomatal area. They concluded that the increase in atmospheric CO_2 has already had adaptive effects on stomatal density (i.e., a decrease in stomatal density) in these two tree species grown in temperate and temperate-warm zones.

Knapp et al. (1994b) found that, under field conditions near Manhattan, Kansas, United States, stomatal density in leaves of *Andropogon gerardii* Vitman (big bluestem, a C_4 grass) was higher ($190 \pm 7\,mm^{-2}$) in plants grown at ambient CO_2 compared to plants grown in chambers with twice the ambient level of CO_2 ($121 \pm 5\,mm^{-2}$) for one growing season (1993). They do not give the CO_2 concentration in the atmosphere, but in 1993, the year of their study, the concentration in Manhattan, Kansas, was about $340\,\mu mol/mol$ (Kirkham et al. 1991). Most stomata were on abaxial leaf surfaces in this grass, but the ratio of adaxial to abaxial stomatal density was greater at elevated levels of CO_2. However, for a forb, *Salvia Pitcheri* Torr. (blue sage), grown under the same conditions, stomatal density was greater at elevated CO_2 ($218 \pm 12\,mm^{-2}$) compared to plants grown at the ambient level of CO_2 ($140 \pm 6\,mm^{-2}$). The ratio of stomata on adaxial and abaxial surfaces did not vary in this herb. Guard cell lengths were not affected by elevated CO_2 for either species. These results indicated that stomatal responses to elevated CO_2 are species specific and that changes could occur in just one season's exposure to elevated CO_2.

STUDIES OF FOSSIL PLANTS

Van de Water et al. (1994) measured stomatal density of limber pine (*Pinus flexilis* James) needles (leaves) preserved in pack rat middens from the Great Basin in the United States. The Great Basin is in the states of Idaho, Nevada, Utah, and Arizona. Data revealed shifts during the last 30,000 years. Stomatal density decreased about 17% during deglaciation from 15,000 to 12,000 years ago, concomitant with a 30% increase in atmospheric CO_2. The stomatal density at the time of their study (0 years ago) was $99.7\,mm^{-2}$ and 28,020 years ago it was $130.7\,mm^{-2}$. Stomatal density declined abruptly between 15 and 12 ka (1000 years ago). Evidence from ice cores indicates that ambient CO_2 levels varied from 190 to 200 parts per million by volume (ppmv) (or $190–200\,\mu mol/mol$) during glacial maxima to 270–280 ppmv (or $270–280\,\mu mol/mol$) during interglacials. Relatively rapid increases have accompanied deglaciation. They divided the last 30,000 years into three periods on the basis of the ice core record of atmospheric CO_2 levels: glacial (30–15 ka), transition (15–12 ka), and Holocene (also called the Recent Epoch, the last part of the Quaternary Period) (since 12 ka). Mean stomatal densities in the glacial, transition, and Holocene were 117.0, 111.1, and $97.2\,mm^{-2}$ respectively. They noted that sensitivity of stomatal density to changing CO_2 levels is less for gymnosperms than for angiosperms. Stomatal densities in limber pine, a gymnosperm, decreased 17% during the increase in atmospheric CO_2 concentration from glacial to Holocene times; but in willow (*Salix* sp.), an angiosperm, they decreased by about 30%.

Robinson (1994a) hypothesized that angiosperms were better able to adjust to the declining CO_2 concentration in the atmosphere than were gymnosperms, because they contained taxa that evolved quickly (e.g., the grasses). During Late Tertiary evolution about 20 million years ago, ambient atmospheric CO_2 concentration declined from values above $500 \mu mol/mol$ to values below $200 \mu mol/mol$ (Robinson 1994a). Figure 9.10 shows the geological timescale. The Cenozoic Era, which began 60 million years ago, is divided into the Tertiary and Quaternary, with the Quaternary more recent (Spencer 1962, pp. 16–17, 357). The Tertiary is also called the Paleogene Period, and the Quaternary is also called the Neogene Period (Spencer 1962; Figure 9.10). Since industrialization, stomatal densities have dropped 20%–40% in various angiosperms but have not varied in limber pine. Thus stomatal densities in conifers may be sensitive to the lower CO_2 of glacial atmospheres but not to future CO_2 enrichment (Robinson 1994a).

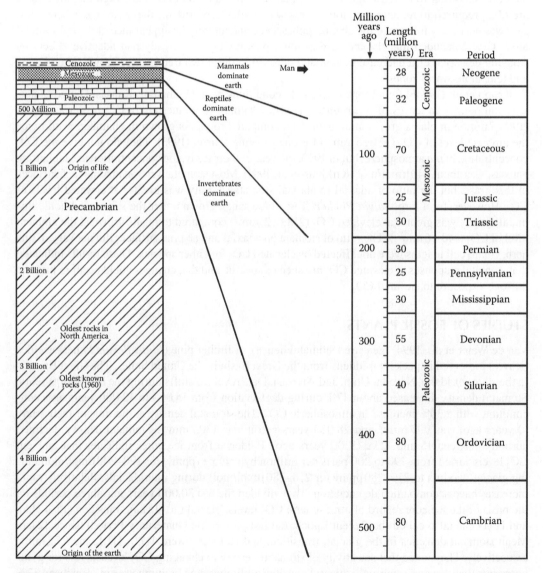

FIGURE 9.10 Diagrammatic representation of the geological timescale. (Reprinted from Spencer, E.W. *Basic Concepts of Historial Geology*, Thomas Y. Crowell Company, New York, 1962, Figs. 1–8. With permission from Wolters Kluwer/Lippincott Williams & Wilkins.)

Woodward in his 1987 paper said that stomatal density did not respond to CO_2 levels exceeding a concentration of about 300 µmol/mol. This effect has been termed the "CO_2 ceiling" (Roth-Nebelsick 2005). For the lower CO_2 concentrations of 12,000–30,000 years ago, the results of Van de Water et al. (1994) were consistent with those of Woodward (1987)—i.e., that stomatal density decreases with an increase in CO_2 in the atmosphere. However, some studies do show a response above 300 µmol/mol, For example, Wagner et al. (2005) studied leaf remains of five species {*Q. nigra* L., black oak; *A. rubrum* L., red maple; *Myrica cerifera* L., wax-myrtle or candleberry; *Ilex cassine* L. [also named *I. vomitoria* Ait., because it causes vomiting; it has the common name of yaupon, a plant in the holly family whose leaves are used for tea (Fernald 1950)]; and *Osmunda regalis* L., royal fern} that were preserved in precisely dated peat deposits in the subtropical area of north-central Florida. They pointed out that stomatal density data come mainly from temperate and subarctic environments, and there is only sparse information from plants from the warm temperate to subtropical and tropical realms. The stomatal index decreased as atmospheric CO_2 increased from 310 to 370 µmol/mol between 1944 and 1998. They found that leaf remains of the widely occurring species *M. cerifera* and *I. cassine* were especially suitable for CO_2 reconstructions and that they could be used to extend the stomatal frequency proxy into the warm-temperate to subtropical realms of North America. Roth-Nebelsick (2005) stated that not only do stomatal parameters (e.g., stomatal density) respond to the CO_2 concentration, but stomatal reaction of plants itself might influence the climate because evapotranspiration of vegetation is a significant factor for climate. (We talk about elevated CO_2 and evapotranspiration in Chapter 10.) Because tree leaves (and their stomata) are often well preserved in fossil records, they could be used as sources of data to study changes in climate over geological time periods.

Hovenden and Schimanski (2000) found that stomatal density of *Nothofagus cunninghamii* (Hook.) Oerst., a beech of the South Pacific with a well preserved fossil record stretching back 50 million years, was dependent upon CO_2 concentration. Their results supported those of Van de Water et al. (1994), which suggested that stomatal density may be more sensitive at low levels of CO_2 in the atmosphere than current levels. They grew four different genotypes from Tasmania at current (~370 µmol/mol) and depleted (~170 µmol/mol) levels of atmospheric CO_2. Growth was less in the lower CO_2 concentration for all genotypes. Mean stomatal length at low CO_2 concentration was 97% of that at current CO_2 concentration. The magnitude of the effect on the length of the stomatal pore (~3%) was trivial compared to changes in stomatal density (up to 20%). For example, in one clone, stomatal density at 370 µmol/mol was 510 mm^{-2} and at 170 µmol/mol it was 400 mm^{-2}. Well preserved fossil leaves of *N. cunninghamii* date back approximately 1 million years, and the oldest fossil leaves of *N. cunninghamii* have a greater stomatal density than leaves of extant plants. Therefore, because *N. cunninghamii* existed before the last glacial maximum, it may serve as a useful indicator of past climate. This is important because the species has survived large changes in atmospheric CO_2 concentrations, as follows: a significant lowering of atmospheric CO_2 concentration, to a minimum of approximately 160 µmol/mol 15,000–20,000 years ago; a sharp increase of atmospheric CO_2 concentration at the end of the last glaciation; and a sustained increase occurring over at least the last 9000 years (Hovenden and Schimanski 2000).

McElwain and Chaloner (1995) measured stomatal density and stomatal index on leaves/axes of four species of fossil plants that predated Tertiary leaves and extended observations back into the Paleozoic Era. The Paleozoic Era is divided into the following epochs (or periods): Cambrian (oldest) (520–440 million years ago or Mya), Ordovician (440–360 Mya), Silurian (360–320 Mya), Devonian (320–265 Mya), Carboniferous [made up of the Mississippian (265–235 Mya) and Pennsylvania (235–210 Mya)], and Permian (youngest; 210–185 Mya) (Spencer 1962, pp. 16–17 and 262; Figure 9.10). They studied the stomatal density of Early Devonian, Carboniferous, and Early Permian plants and compared them to the nearest living equivalents. Observed changes in the stomatal data gave support to the suggestion from physical evidence that atmospheric CO_2 concentrations fell from an Early Devonian high of 10–12 times its present value to one comparable to that of the present day by the end of the Carboniferous. Devonian atmospheric CO_2 concentrations of 3000–3600 and 4200–6000 µmol/mol have been suggested from modeling of the global carbon budget and from carbon isotopic ratios, respectively. Assuming a pre-industrial atmospheric CO_2 concentration of 300 µmol/mol, this implies that Devonian ambient CO_2 was 10–20 times higher than the present. A Siluro-Devonian peak is believed to have been succeeded by a Carbo-Permian low (about 300 µmol/mol). Extremely low stomatal densities of Devonian plants (4.5 and 4.3 mm^{-2}) contrasted with the high values of their two nearest living equivalents (309.1 and 40.1 mm^{-2}). Similarly, there was a difference between the low stomatal densities of the Devonian species and those of the Carboniferous, which were between one and two orders of magnitude higher than in the Devonian plants. They concluded that stomatal density of fossil leaves has potential value for assessing changes in atmospheric CO_2 concentration throughout geological time.

Retallack (2001) confirmed that stomatal density of fossil leaves can indicate the CO_2 concentration in ancient atmospheres. He used fossil leaves of four genera of plants that are closely related to the present-day *Ginkgo* tree (common name: ginkgo or gingko) to reconstruct past atmospheric CO_2 concentrations. The genus *Ginkgo* has a fossil record of well-preserved leaf cuticles at least back to the late Triassic Period (229 million years ago, Mya). The genera were *Rhachiphyllum* (early Permian, 295–265 Mya); *Lepidopteris* (mid Permian to late Triassic, 258–200 Mya); *Tatarina* (late Permian, 252–250 Mya); and *Ginkgo* (late Triassic to recent, 229–0 Mya). In an inverse method, he used the stomata in the fossil leaves as biological sensors of past levels of atmospheric CO_2 (Kürschner 2001). He used stomatal index instead of stomatal density to minimize the effects of stomatal position within leaves, water stress, and differences in insolation that can affect the density. Retallack's (2001) data for Permian times were questioned because preservation of cuticles from Permian times is generally too fragmentary for the stomatal index to be reliably determined (Kerp 2002). However, his estimates are the best available until more fossils can be analyzed (Retallack 2002).

Chen et al. (2001) found that the stomatal index of four fossil species of *Ginkgo* (*G. coriacea, G. huttoni, G. yimaensis,* and *G. obrutschewii*) from the Jurassic and Cretaceous Periods in the Mesozoic Era were 60%–40% lower than the modern values recorded for extant *G. biloba*. They calculated stomatal ratios, which they defined as the stomatal index of the modern value divided by that of fossil leaves. The stomatal ratios for *G. coriacea* (Early Cretaceous), *G. huttoni* (Middle Jurassic), *G. yimanensis* (Early to Middle Jurassic), and *G. obrutschewii* (Early Jurassic) were 2.8, 1.7, 3.6, and 1.4, respectively. For example, the stomatal index of *Ginkgo biloba* (modern plant) had a mean value of 9.3 and that of *G. huttoni* had a mean value of 5.5. So the stomatal ratio was 1.7. These ratios agreed fairly well with RCO_2 values defined as the ratio of atmospheric CO_2 in the past relative to the pre-industrial value of 300 µmol/mol. They concluded that leaves of plants of *Ginkgo* can provide a reasonable measure of ancient atmospheric CO_2 change.

McElwain et al. (1995) investigated the impact of different concentrations of ambient CO_2 on stomatal density and index of *Salix cinerea* L. (gray willow). They compared fossil leaves (9190 years Before Present, BP) and present-day material from the same site in southwest Ireland with nineteenth-century herbarium material from Ireland. Stomatal density and index were both found to be inversely related to CO_2 concentration. The fossil plants (9190 BP; 259 µmol/mol CO_2 concentration) had a mean stomatal density of 947.2 mm^{-2}; herbarium plants (AD 1856–1875; CO_2 of 285 µmol/mol) had a mean stomatal density of 773.3 mm^{-2}; and present-day plants (AD 1992; CO_2 of 353 µmol/mol) had a mean stomatal density of 603.3 mm^{-2}.

Because the studies showing the inverse relation between atmospheric CO_2 concentration and stomatal frequency on fossil leaves had not considered needles, Kouwenberg et al. (2003) studied the responsiveness of four native North American conifer species [*Tsuga heterophyll* Sarg., a type of hemlock; *Picea glauca* Voss, white spruce; *P. mariana* BSP (Britton, Sterns, Poggenburg), black spruce; and *Larix laricina* Koch, American larch] to a range of historical CO_2 concentrations

(290–370 µmol/mol). Due to the stomatal patterning in conifer needles, the stomatal index of these species was not affected by CO_2. So Kouwenberg et al. (2003) developed a new measure of stomatal frequency based on the number of stomata per millimeter of needle length. It decreased with increasing CO_2. Because of their sensitive response to CO_2, combined with a high preservation capacity, they concluded that fossil needles of the four conifer species have a potential for quantifying past atmospheric CO_2 fluctuations. Studies probably will have to consider long periods of time, because Luomala et al. (2005) found no effect of elevated CO_2 (693 µmol/mol) on the stomatal density of Scots pine (*P. sylvestris* L.) during 3 years of growth in closed-top chambers.

The results of Kouwenberg et al. (2003) also contrast with those of Van de Water et al. (1994), who said that the sensitivity of stomatal density to changing CO_2 is less for gymnosperms than for angiosperms. The correct methods must be used, however, to study gymnosperm stomata. If they are, perhaps stomatal densities in conifers may be shown to be not only sensitive to the lower CO_2 of glacial atmospheres but also to future CO_2 enrichment.

Understanding the way that plants responded to changes in atmospheric CO_2 concentration in the past will increase our ability to predict the responses of vegetation to future changes in atmospheric CO_2 concentration. As we have noted in this chapter, many herbarium and paleoecological studies have shown that increases in atmospheric CO_2 concentration have been accompanied by a decrease in the stomatal density of leaves (Beerling and Kelly 1997). Long and short timescales have revealed that the natural CO_2 fluctuations affect stomatal density. However, more studies of plants in herbariums, not yet investigated, need to be carried out to predict the relationship between stomatal density and CO_2 concentration.

STOMATAL ANATOMY AND ELEVATED CO_2

Most of the studies relating atmospheric CO_2 concentration to stomatal density have considered stomatal numbers and not the anatomy of the stomata themselves. Boetsch et al. (1996) studied the influence of elevated CO_2 concentration (670 µmol/mol) on the structure, distribution, and patterning of stomata and their subsidiary cells. Extra subsidiary cells, beyond the normal complement of four per stoma, were associated with nearly half the stomatal complexes on leaves grown in elevated CO_2. Substomatal cavities of complexes with extra subsidiary cells appeared larger than those found in control leaves. The extra subsidiary cells developed from the epidermal cell population during development. Because enriched CO_2 had no effect on the structure or patterning of guard cells, but resulted in the formation of additional subsidiary cells, the authors concluded that separate and independent events pattern the two cell types.

Maherali et al. (2002) studied C_3 and C_4 plants in a grassland grown across a range of CO_2 concentrations (20–550 µmol/mol). They found larger stomatal sizes in *Solanum dimidiatum* (a C_3 perennial forb) at sub-ambient CO_2 concentrations compared to higher concentrations. The authors said that the increased stomatal size would facilitate CO_2 diffusion into the leaf under CO_2 starvation. However, in contrast to Woodward's (1987) finding, they saw that stomatal density increased in this species with increasing CO_2 concentration. A C_4 plant, *Bothriochloa ischaemum* (a perennial grass) did show a decrease in stomatal density with increasing CO_2.

HOW DOES STOMATAL DENSITY CHANGE WITH CO_2 CONCENTRATION?

It seems reasonable to me that stomatal density should increase with increased CO_2 concentrations. It is well known that stomata close when atmospheric CO_2 concentration increases (see Chapter 8). A good illustration of the decrease in stomatal conductance with increase in CO_2 concentration is shown in Figure 9.11 for a grassland plant, little bluestem [*Schizachyrium scoparium* (Michx.) Nash] (Polley et al. 1997). And when stomata close, CO_2 cannot be taken up for photosynthesis and growth. There is a direct relationship between stomatal conductance and growth (Figure 9.12)

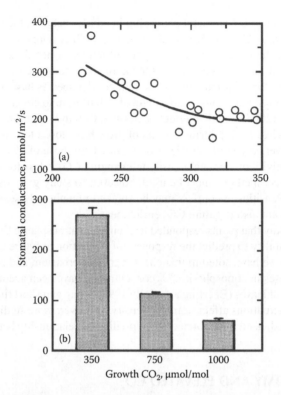

FIGURE 9.11 Stomatal conductance of sunlit leaf blades of the C_4 grass *S. scoparium* (little bluestem) as a function of the CO_2 concentration at which plants were grown. (a) The line is a regression through single measurements per plant with an infrared gas analyzer. (b) Vertical bars denote 1 standard error of the mean of 18 daily averages of leaf (blade) conductance. Note that the scale of the *y*-axis differs in (a) and (b). (Reprinted from Polley, H.W. et al., *J. Range Manage.*, 50, 278, 1997. Fig. 1. With permission. Copyright 1997, Society for Range Management.)

(Gardner 1973). I observed this relationship for wheat. Under well watered conditions, a drought-sensitive wheat cultivar had a higher stomatal conductance and grew more than a drought-resistant cultivar (Kirkham 1989). If stomata are closed in a higher CO_2 world, plants are going to need more of them to get the same amount of CO_2 that they get under current levels of CO_2.

Additional studies are needed to answer the question, "How does stomatal density change with CO_2 concentration?" New methods make determination of stomatal density less labor intensive and allow archivable specimens for revisiting samples (Gitz and Baker 2009). In these studies, we may need to focus on plants in the wild rather than crops. The great variety of wild plant species warrants further study (Doley 2004). Boyer (1982) showed that during a 33-year period (1935–1968) average yield of soybean [*Glycine max* (L.) Merr.] cultivars increased from about 3400 to 4000 kg/ha. The increasing yield suggests that plant breeders inadvertently are breeding for cultivars that keep their stomata open. To understand effects of CO_2 on stomatal density, we may have to avoid crops whose evolutionary history is determined by man. Man may also alter normal stomatal development through genetic modification. The number and distribution of stomata are ultimately under genetic control. Mutations in the genes TOO MANY MOUTHS (TMM) (Nadeau and Sack 2002) and YODA (a mitogen-activated protein kinase kinase kinase [sic] gene) (Sack 2004; Bergmann et al. 2004) disrupt normal patterning of stomata in *Arabidopsis*. A gene named STOMAGEN increases stomatal density when overexpressed in *Arabidopsis* (Kondo et al. 2010). Thus, native plants, whose response is not confounded by breeding or genetic modification, might be used to understand how CO_2 concentration affects stomatal density.

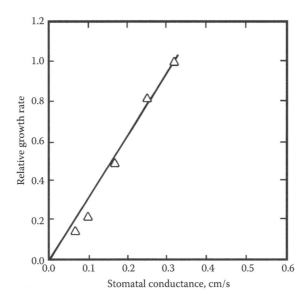

FIGURE 9.12 Relative growth rate as a function of stomatal conductance for green beans in a growth chamber. (Reprinted from Gardner, W.R., Internal water status and plant response in relation to the external water régime, in Slatyer, R.O., Ed., *Plant Response to Climatic Factors*, United Nations Educational, Scientific and Cultural Organization, Paris, 1973, pp. 221–225, Fig. 4. With permission from the American Society of Plant Biologists and Wilford R. Gardner; Data in the figure come from E.T. Kanemasu and Tanner, C.B., *Plant Physiol.*, 44, 1547, 1969.)

SUMMARY

The data are contradictory and show that stomatal density decreases, increases, or stays the same under elevated atmospheric CO_2. Results apparently depend on several factors, including species, CO_2 concentration, and time of exposure to the elevated CO_2 concentration. Several authors have demonstrated, both experimentally and from fossil and herbarium material, an inverse relationship between stomatal density and increasing ambient CO_2 levels. These investigations suggest that reductions in stomatal density are a physiological response by plants that increases WUE and may, therefore, influence their growth and survival under moisture-limited conditions. However, other authors have not confirmed these observations and have suggested that variations in stomatal density may be more strongly influenced by factors other than CO_2, such as altitude and microclimatic conditions including light and moisture. We need more studies showing the interaction of water (wet and dry periods) and CO_2 concentration on stomatal density during long times and over different geological periods.

APPENDIX

BIOGRAPHY OF FAKHRI A. BAZZAZ

Fakhri Bazzaz, one of the first to study the physiology of plants under elevated CO_2, was born in Baghdad, Iraq, on June 16, 1933. He was the son of Abdul-Latif and Munifa Bazzaz. He came to the United States in 1958 and became a U.S. citizen. He married Maarib D.L. Bakri on August 25, 1958. They had two children: a daughter, Sahar, and a son, Ammar. He received his bachelor of science degree at the University of Baghdad in 1953 and his master of science degree and PhD at the University of Illinois in 1960 and 1963, respectively (Marquis Who's Who 2007). He was a professor at the University of Illinois, Urbana, between 1977 and 1984. He was head of the Department of Plant Biology (1980–1984) and acting director of the School of Life Sciences (1983–1984) at the University of Illinois. He joined the Department of Organismic and Evolutionary Biology at Harvard University in Cambridge, Massachusetts in 1984. Between 1988 and 1997, he was the H.H. Timken Professor of Science, and from 1997 until his death in 2008, he was the Mallinckrodt Professor of Biology. In 2000, he was named Leverhulme Professor in British Universities (Bangor, Sheffield, Edinburgh, Cambridge, Oxford, and Imperial College).

He received many honors, including being named fellow of Clare Hall, Cambridge University, United Kingdom (1981); fellow of the American Association for the Advancement of Science (1987); fellow of the American Academy of Arts and Sciences (1989); and fellow for the Japan Society for the Promotion of Science (1993). He was the recipient of a Guggenheim fellowship (1988); the German Humboldt Research Prize (1997); and the Nevada Medal (2004). He received the Illinois Achievement Award in 2003 and became an honorary member of Phi Beta Kappa (2006) and of the World Innovation Foundation (2004). He was a member of the Ecological Society of America and the British Ecological Society (Marquis Who's Who 2007).

He taught many classes including introduction biology, introductory botany, plant ecology, vegetation of North America, ecological methods and instrumentation, population biology of plants, plant–animal co-evolution, tropical ecology, experimental plant ecology, and plant population biology. He served on the editorial boards of *Oecologia, Tree Physiology, Plants Today*, the *Journal of Vegetation Science*, and *Global Change Biology*. He was a prolific publisher and has authored or coauthored over 330 scientific articles. He coedited four books and wrote one book as sole author, *Plants in Changing Environments: Linking Physiological, Population, and Community Ecology* (Cambridge University Press, Cambridge, United Kingdom, 1996, 320pp).

His research concentrated on understanding how species respond to disturbance in nature and how they adapt to it. It was strongly experimental, and he carried out his studies in Harvard Forest and in the tropical forests of Mexico and Malaysia. A key part of his research was the potential impact of elevated CO_2 on ecosystems. In these studies, he used annual plants from a wide range of ecosystems, tropical trees, and temperate forest trees.

He guided 56 students through to their MS degree or PhD. Thirty-three postdoctoral fellows trained under him. Nineteen visiting scientists stayed in his laboratory for extended periods of time. I was one of them. Dr. Bazzaz accepted me into his laboratory to study during my sabbatical leave in 1990. I had contacted two other laboratories to ask if I could spend my sabbatical leave there, but with no success. So I was thrilled that Dr. Bazzaz invited me to come to Harvard as a visiting scholar, because he was the most important person doing research concerning the effects of elevated levels of CO_2 on plants. He supplied me with my own office, invited me to attend the weekly laboratory meetings, introduced me to other famous professors in the Department of Organismic and Evolutionary Biology and in other departments on campus, and took me to visit Harvard Forest. He and his wife invited me to their home in Lexington, Massachusetts, to meet his students. Kristen Norweg, his long-time administrative assistant, gave me invaluable help. The discussions that I had with Dr. Bazzaz and his students provided the most intellectually stimulating environment that I had had since my graduate student days at the University of Wisconsin. His students have gone on to distinguished careers. After I left Harvard, Dr. Bazzaz wrote letters to support me in my various endeavors.

About a year after my sabbatical leave, he told me that he had had a heart attack, but had recovered from it. He died in Massachusetts on February 6, 2008, at the age of 74.

10 Elevated Atmospheric Carbon Dioxide: Transpiration and Evapotranspiration

INTRODUCTION

We have seen that elevated CO_2 reduces stomatal conductance (Chapter 8). We now investigate the effects that this reduction has on transpiration and evapotranspiration.

TRANSPIRATION UNDER GREENHOUSE CONDITIONS

An increase in stomatal conductance causes an increase in transpiration. This is illustrated in Figure 10.1, where there is a positive sign between leaf conductance and transpiration. However, conductance and transpiration are decoupled to some extent, particularly in greenhouses, where insufficient cooling of the leaves through transpiration causes heat damage that occurs at high radiation (Nederhoff et al. 1992). This is explained in Figure 10.1, which depicts a course of events following an increase in CO_2 in a greenhouse (Nederhoff 1994, p. 61). High CO_2 causes partial stomatal closure (reduced leaf conductance—note the negative sign between CO_2 and leaf conductance in Figure 10.1). This initially reduces transpiration of leaves, which slightly increases leaf temperature. There is a negative sign between transpiration and leaf temperature in Figure 10.1 because the greater the transpiration rate, the cooler is the leaf temperature. Reduction of transpiration also decreases absolute humidity. [Absolute humidity is the mass of water vapor present in unit volume of the atmosphere, usually measured as grams per cubic meter. It may also be expressed in terms of the actual pressure of the water vapor present (Weast 1964, p. F-45).] If the absolute humidity is decreased, the vapor pressure deficit (VPD) increases, first in the boundary layer and later in the surrounding air. Increased leaf temperature and reduced absolute humidity imply an increase of the VPD between the stomatal cavities and the air (VPD in leaf–VPD in air; Figure 10.1, bottom). An increase in VPD reduces the stomatal conductance, thus enforcing the effect of the CO_2 concentration on stomatal conductance. [See also Figure 8.1. This figure shows that when atmospheric VPD is increased, guard-cell turgor is decreased and stomata close.] However, increased VPD increases transpiration through a larger driving force, thus counteracting the effect of stomatal closure through the CO_2 concentration on transpiration (Allen et al. 1985, p. 13; Nederhoff 1994, pp. 60–61).

Leaf area also affects these processes. Plants tend to produce more leaf area when enriched with CO_2, which provides a greater transpirational surface area. Both the increased leaf area and the increased leaf temperature tend to offset reductions in crop transpiration that might be expected from elevated CO_2 (Allen et al. 1985, p. 13). Consequently, no effect of CO_2 on transpiration is sometimes observed in greenhouses (Nederhoff et al. 1992). Under outside conditions, there is usually some airflow, even if the air appears calm, and this wind speed is enough to avoid the heat damage seen in greenhouses and to dissipate quickly any CO_2 that can be contained in a greenhouse (Allen 1973).

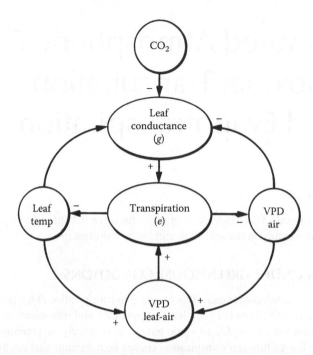

FIGURE 10.1 Schematic representation of the thermal and hydrologic feedback loops influencing leaf conductance and canopy transpiration of plants in a greenhouse. Symbols: + synergistic; − antagonistic effect. (Reprinted from PhD dissertation, Nederhoff, E.M., Effects of CO_2 concentration on photosynthesis, transpiration and production of greenhouse fruit vegetable crops, p. 61, Fig. 3.1, Copyright 1994, with permission from Elsevier; also reproduced from *Sci. Hort.*, 52, Nederhoff, E.M., Rijsdijk, A.A., and de Graaf, R., Leaf conductance and rate of crop transpiration of greenhouse grown sweet pepper (*Capsicum annuum* L.) as affected by carbon dioxide, 283, Fig. 1, Copyright 1992, with permission from Elsevier.)

CARBON DIOXIDE ENRICHMENT IN GREENHOUSES

We are not going to consider greenhouses, because our main concern is the increasing CO_2 in the open air, which is occurring due to man's activities. However, in this section we briefly cover elevated CO_2 in greenhouses, where CO_2 enrichment has long been used to promote growth (Northen and Northen 1973, p. 40). Enriching the greenhouse air with additional CO_2 improves the growth of flowers, increases their quality and numbers, and often hastens plants into bloom. The amount that gives best result is about 1200 ppm. Many commercial growers in the United States use CO_2 enrichment for such flowers as carnations (*Dianthus caryophyllus* L.), geraniums (*Geranium* sp.), roses (*Rosa* sp.), chrysanthemums (*Chrysanthemum* sp.), snapdragons (*Antirrhinum majus* L.), Easter lilies (*Lilium longiflorum* Thunb.), calla [improperly called calla lily (Laurie and Ries 1950, p. 432)] (*Zantedeschia* sp.), gloxinias (*Sinningia speciosa* Benth. & Hook.), African violets (*Saintpaulia ionantha* Wendl.), hyacinths (*Hyacinthus* sp.), bedding plants, and greenhouse vegetables such as tomatoes (*Lycopersicon esculentum* Mill.) and cucumbers (*Cucumis sativus* L.).

Various methods of adding CO_2 have been tried, and the best are special propane or natural gas burners (Northen and Northen 1973, p. 41). Enrichment is most effective during the winter or during cool weather when the vents can remain closed. Opening them even 1 or 2 in. (2.5–5 cm) allows the CO_2 to escape. Fans are operated to circulate the enriched air evenly. When crop improvement can mean the difference between profit and loss for a commercial grower, the extra expense and care necessary with CO_2 enrichment pays off.

ETHYLENE

We now spend a section discussing ethylene, because it is also a low-molecular-weight gas like CO_2 and, the two can compete in biological reactions, which is discussed below. Ethylene ($CH_2=CH_2$) is the simplest organic compound that affects plants. The recognition that ethylene is a plant hormone followed primarily from studies of its effects on the ripening of fruit. The basic observation was that harvested lemons (*Citrus limon* Burm.) were being ripened during shipment in a heated railroad car, and when the kerosene stove was replaced by more modern convection heaters in storage rooms, the lemons failed to ripen (Thimann 1972a, p. 213). Thus, the ripening was not just a temperature effect, but the fumes from the incompletely burnt oil were clearly implicated. Only much later was the ripening agent in such fumes shown to be ethylene.

That illuminating gas (gas used in burners to light homes before electricity) had powerfully injurious effects on plants has been known since 1879 when Fahnestock painstakingly listed all the plants and the damage that they suffered when gas leaked into his greenhouse in Philadelphia (Thimann 1977, p. 386). Abscission was the most common response, and it usually was severe. The same responses have been seen in street trees, where a gas main under the street develops a leak. The trees abscise most of their mature leaves. As early as 1864, observations were recorded on the effect of illuminating gas on vegetation in the streets of European cities (Rouen, Berlin, Hamburg, and Hanover) (Crocker and Knight 1908). Ethylene promotes abscission strongly, sometimes at concentrations less than one-hundredth part per million (Thimann 1977, p. 386). In 1901, the responses elicited in pea (*Pisum sativum* L.) plants by illuminating gas were traced to ethylene, this being the only active compound present in sufficient quantity (Thimann 1972a, p. 213).

Ethylene is produced naturally when a plant is subjected to stress, such as drought. It is difficult to measure ethylene produced by plants, especially under field conditions, because it is a gas and readily escapes. Rose (1985) was probably the first to measure ethylene production under cropping conditions in a field. She found that leaves of 10-day-old seedlings of 'Kanking,' a cultivar of winter wheat (*Triticum aestivum* L.) known to be drought-resistant, when excised in the field and allowed to wilt, evolved more ethylene (150 nL/g/h after 4 h of excision) than did leaves of 'Ponca,' a cultivar known to be drought-sensitive (50 nL/g/h). Without stress, ethylene production was small and the same for the two cultivars (<10 nL/g/h).

Carbon dioxide at relatively high concentrations (5%–20%) competes with ethylene in its effects on plants (Thimann 1972b, p. 230). Elevating the CO_2 concentration in the atmosphere in which fruit [e.g., apples (*Malus sylvestris* Mill)] is stored inhibits fruit ripening (Janick 1972, pp. 422–424). This is because of the competition between ethylene and CO_2 (Thimann 1972b, p. 230). High CO_2 inhibits ethylene production. (Elevating the CO_2 reduces the amount of oxygen in the storage atmosphere, so respiration is reduced and this also reduces the rate of ripening.) In addition to preventing fruit from ripening, which high CO_2 causes, high CO_2 opens the *Phaseolus* (bean) plumular hook, which ethylene or auxin closes; prevents epinasty caused by ethylene; delays leaf abscission; and hastens the sprouting of potato (*Solanum tuberosum* L.) tubers, which ethylene and auxin inhibit (Thimann 1972b, p. 230). The balance between CO_2 and ethylene during storage of potatoes controls their color, which is important to the potato industry (Daniels-Lake and Prange 2009). They must have a light (not darkened) color to be acceptable for commercial use.

The heaters used to elevate CO_2 in greenhouses for enhanced growth also can produce ethylene. We discuss this topic no further, except to point out that CO_2 is not a hormone, even though it is a small molecule, like ethylene; and both CO_2 and ethylene have effects on plant growth. Unlike ethylene, CO_2 is an inorganic compound. For a substance to be a hormone, it needs to satisfy three requirements (Folke Skoog, Lecture Notes, Botany 870, University of Wisconsin, Madison, United States, October 25, 1967): (1) it is a naturally occurring substance; (2) it has effects on the plant in low concentrations (otherwise nutrients, sugars, and enzymes would be hormones); and (3) it is produced in one part of the plant and moves to another part of the plant where it has an effect (otherwise vitamin B_1 would be considered a hormone, but this vitamin can have effects on the organ where it is produced). Carbon dioxide is produced by a plant, but it does not need to move to have effects on plant growth. The threshold for its effects to be noted or measurable (e.g., stomatal closure) have yet to be determined. As we have seen in our experiments, an increase in CO_2 concentration from 330 to 485 ppm (330–485 μmol/mol, a difference of 155 μmol/mol CO_2) caused measurable stomatal closure (Chaudhuri et al. 1986a). In contrast, the level of ethylene that must be reached in a tissue of a fruit to cause ripening is nearly the same in all fruit—around 0.1–1 ppm (Thimann 1972a, p. 217).

As we have pointed out previously, CO_2 concentration has to be high for it to have deleterious effects on plant growth. For example, in Chapter 3, we saw that cotton (*Gossypium hirsutum* L.) roots elongated when the CO_2 concentration in the soil was 24% (240,000 ppm), if the soil was not compacted (see Figure 3.6). Nevertheless, the fact that stomata are so sensitive to CO_2 in the atmosphere, which is just a trace gas in the air [0.03849% CO_2 (Duglokencky 2009) compared to 21% O_2 and 78% N_2], is remarkable. Raising the CO_2 concentration just 0.0155% (155 ppm) starts to close stomata (Chaudhuri et al. 1986a). Stomata have evolved to be highly responsive to trace amounts of CO_2 in the atmosphere. A plant can detect a small change in CO_2 in the air. This is why blowing on stomata closes them. Xu and Kirkham (2003) used the fact that leaves sense a change in CO_2 in the air to obtain a relationship between photosynthetic rate and internal CO_2 concentration of sorghum [*Sorghum bicolor* (L.) Moench]. Dr. Xu placed a leaf in the sensor chamber of a closed gas exchange system (LI-6200 Portable Photosynthetic System, Li-Cor, Inc., Lincoln, Nebraska, United States) and then blew into it. The CO_2 concentration in the chamber was recorded on the readout, and, when a high concentration was reached in the chamber, the chamber was closed and then readings were taken until the CO_2 concentration had fallen to about zero. Photosynthetic rate versus internal CO_2 concentration was then plotted. One could not do such a test in an open exchange system (e.g., the LI-6400, Li-Cor, Inc., Lincoln, Nebraska, United States).

We now turn from CO_2 enrichment in greenhouses and ethylene to experiments that have been done to see the effects of elevated levels of CO_2 on transpiration of plants.

TRANSPIRATION UNDER ELEVATED CO₂

Rabbinge et al. (1993, p. 68) say that, with the doubling of atmospheric CO_2, transpiration in C_3 plants will be reduced by 10%–20%, and assimilation will be stimulated by 40%; and in C_4 plants only transpiration will be reduced by up to about 25%. Lockwood (1995) points out that the CO_2 concentration in the atmosphere between the years 1000 and 1800 was relatively constant and was 270–290 μmol/mol. It started to rise around 1800 because of the Industrial Revolution. When he wrote his article (1995) it was 353 μmol/mol. Thus, the 25% increase in atmospheric CO_2 above preindustrial levels could already have produced (assuming a linear relationship) a transpiration reduction of 2.5%–5% for common C_3 plants and up to 6% for C_4 plants (Lockwood 1995).

Let us compare this average reduction in transpiration by doubling of atmospheric CO_2 to the change in stomatal conductance. Morison (1987a) reported the stomatal conductance of 25 species under ambient (330 μmol/mol) and twice ambient (660 μmol/mol) CO_2 concentrations. His list included nine C_4 species and 16 C_3 species. The stomatal conductance at 660 μmol/mol CO_2 was linearly related to the stomatal conductance at 330 μmol/mol. The relationships for C_3 and C_4 species were indistinguishable, and a simple linear relation sufficed for predicting stomatal conductance in future atmospheres. He found that, for a doubling of CO_2 concentration, stomatal conductance was reduced by 40%. A number of reviews suggest that doubling of CO_2 will increase stomatal resistance (reciprocal of stomatal conductancy) by 57% ± 14% (Rosenberg et al. 1989).

Now let us look at some specific experimental data.

GRASSES

Kirkham et al. (1993) reported stomatal resistance and transpiration rate of big bluestem (*Andropogon gerardii* Vitman, a C_4 plant) during two seasons grown under ambient and doubled CO_2 concentrations on a tallgrass prairie in Kansas. The figures for stomatal resistance and transpiration rate for the big bluestem are given in Figures 5.9 and 5.10, respectively. As explained in that chapter, the reduced transpiration under elevated CO_2 resulted in increased soil water content. Under well-watered conditions, the average transpiration rate for big bluestem during two seasons (1989 and 1990) under ambient CO_2 was 155 mg/m²/s and under doubled CO_2 it was 116 mg/m²/s. The average stomatal conductance (taking the reciprocal of the values for stomatal resistance as shown by Kirkham et al., 1993, in their Table 3) during the two seasons under ambient CO_2 was 0.454 cm/s and under doubled CO_2 it was 0.279 cm/s. We can calculate that doubled CO_2 under well-watered conditions reduced the stomatal conductance by about 40%, as was given as the general rule by Morison (1987a). That is, (0. 454−0.279 cm/s)/0.454 cm/s=0.385% or 38.5%. Under wet conditions, average transpiration rate of the big bluestem during the two seasons under doubled CO_2 was reduced by 25.2%. That is, (155−116 mg/m²/s)/155 mg/m²/s=0.252% or 25.2%. This number agrees almost exactly with the general prediction of Rabbinge et al. (1993) for C_4 plants, as described above. The data of Kirkham et al. (1993) confirm that transpiration is reduced less by doubled CO_2 than stomatal conductance. Under dry conditions, the average transpiration rate for big bluestem during the two seasons under ambient CO_2 was 151 mg/m²/s and under doubled CO_2 it was 106 mg/m²/s; the average stomatal conductance during the two seasons under ambient CO_2 was 0.363 cm/s and under doubled CO_2 it was 0.256 cm/s. We calculate that under dry conditions, doubled CO_2 reduced the stomatal conductance by 29.4%, and transpiration was reduced by 29.8%. So with drought, the stomatal conductance and transpiration rate were reduced by about the same for this C_4 plant.

C_3 AND C_4 PLANTS COMPARED

It is difficult to compare physiological measurements taken on C_3 and C_4 plants in the tallgrass prairie because they are not active at the same time. The warm-season C_4 plants dominate during the hot summer months and the cool-season C_3 plants dominate during the cooler months of spring and fall. The leaves of big bluestem are too small for the measurements in the spring and fall, and the leaves of Kentucky bluestem are too small for the measurements during the summer. However, He et al. (1992) were able to compare a C_3 plant (Kentucky bluegrass; *Poa pratensis* L.) and a C_4 plant (big bluestem) grown on the tallgrass prairie in the fall of 1989, when leaves of both plants were large enough to measure with a portable photosynthetic system (Model LI-6200, Li-Cor, Inc., Lincoln, Nebraska, United States). Temperature of the air also was measured with the LI-6200 and subtracted from the temperature of the leaf to determine how much cooler or warmer than air the leaves were. Transpiration and canopy temperature are related, because as the transpiration increases the leaves get cooler. All the plants grew under well-watered conditions, so, only the effect of doubled CO_2, and not of drought, could be examined. Doubled CO_2 reduced the transpiration rate of big bluestem (C_4) by an average of 35.0% (from 66.2 to 43 mg/m²/s) and reduced the transpiration rate of Kentucky bluegrass (C_3) by 6.5% (from 93.3 to 87.2 mg/m²/s) (Figure 10.2). Because transpiration was reduced under elevated CO_2, the leaves tended to be warmer than under ambient CO_2, although there was great variability in the measurements (Figure 10.3). I converted the values for stomatal resistance, obtained by He et al. (1992) (Figure 10.4) into stomatal conductance. Doubled CO_2 reduced the stomatal conductance of big bluestem by an average of 37.9% (from 0.203 to 0.126 cm/s) and reduced the stomatal conductance of Kentucky bluegrass by 7.6% (from 0.262 to 0.242 cm/s). The small reduction in the stomatal conductance of Kentucky bluegrass under doubled CO_2 is an exception to the general observation that doubled CO_2 decreases stomatal conductance by 40% (Morison 1987a).

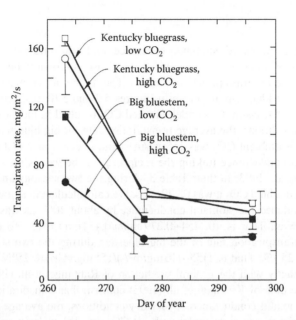

FIGURE 10.2 Transpiration rate of Kentucky bluegrass (open symbols) and big bluestem (closed symbols) with a high (twice ambient) (circles) and a low (ambient) (squares) atmospheric CO_2 concentration. Vertical bars = ±standard deviation. Only half the bar is shown for clarity. (Reprinted from He, H. et al., *Trans. Kansas Acad. Sci.*, 95, 139, 1992, Fig. 3, Copyright 1992, Kansas Academy of Science. With permission from the Editor, Transactions of the Kansas Academy of Science, Fort Hays State University.)

Reports of elevated CO_2 having little or no effect on stomatal conductance, such as occurred with Kentucky bluegrass in the experiment of He et al. (1992), are rare, but they do exist in the literature. Morison (1987a) excluded data from two publications when he determined the relationship between stomatal conductance at 660 and at 330 µmol/mol, because they showed no influence of CO_2 on stomatal conductance. The publications were by Beadle et al. (1979), who found no effect on stomatal conductance of *Picea sitchensis* (Bong.) Carr. (sitka spruce) of atmospheric CO_2 concentrations between 20 and 600 µmol/mol, and Bierhuizen and Slatyer (1964), whose atypical results for cotton (*G. hirsutum* L.) grown at atmospheric CO_2 concentrations between 200 and 2000 µmol/mol CO_2 were discussed under the section "Trees" in Chapter 8.

Dugas et al. (1997) compared the transpiration rate of C_3 and C_4 plants under elevated CO_2 in the same experiment. Their objective was to measure hourly and daily whole-plant transpiration rates from the C_4 plant grain sorghum [*S. bicolor* (L.) Moench] and the C_3 plant soybean [*Glycine max* (L.) Merr.], grown under ambient (359 µmol/mol) and elevated (705 µmol/mol) CO_2 in the air. Plants were well-watered throughout the experiment. Transpiration measurements were made for 22 days in August, 1994, at Auburn, Alabama, United States, using stem flow gauges (Dynamax, Inc., Houston, Texas, United States) on plants growing in open-top chambers (OTC). The soil was a loamy sand. Transpiration rates were consistently greater from plants grown under the ambient CO_2 concentration for both sorghum and soybean (Figure 10.5). Figure 10.5 shows the daily totals (g/m²/day) of transpiration. Sorghum under ambient CO_2 had the highest transpiration rate (e.g., in Figure 10.5, 1082 g/m²/day on Day of Year (DOY) 221), followed by soybean under ambient CO_2, followed by sorghum under the elevated CO_2. The lowest transpiration rate was that of soybean under the elevated CO_2 concentration. For all days of the experiment, the average daily sorghum transpiration was 1128 and 772 g/m²/day for plants growing under ambient and elevated CO_2 concentration, respectively. Equivalent rates per unit ground area were 2.9 and 2.0 mm/day. This is a 32% reduction in transpiration due to the increased CO_2 concentration. Corresponding averages for soybean were 731 and 416 g/m²/day—equivalent to a 43% reduction. Equivalent rates per unit ground area

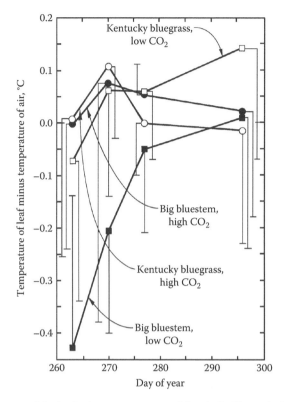

FIGURE 10.3   Temperature of the leaf minus temperature of the air for Kentucky bluegrass (open symbols) and big bluestem (closed symbols) with a high (twice ambient) (circles) and a low (ambient) (squares) atmospheric CO_2 concentration. Vertical bars = ±standard deviation. Only half the bar is shown for clarity. (From He, H. et al., *Trans. Kansas Acad. Sci.*, 95, 139, 1992. Unpublished data.)

for soybean were 4.4 and 2.5 mm/day. Note that the C_3 plant lost more water per unit ground area than the C_4 plant. Thus, the reduction of transpiration under elevated CO_2 concentration was greater for the C_3 plant than the C_4 plant. Dugas et al. (1997) conclude that part of the hypothesized larger increase of water use efficiency for C_3 plants versus C_4 plants with increasing CO_2 may be due to a larger decrease of transpiration for C_3 plants, as well as increased biomass production. Water use efficiency is discussed in Chapter 11.

Bremer et al. (1996) used sap-flow gauges to measure transpiration of C_3 and C_4 plants in a tall-grass prairie in Kansas. The C_3 plant was the forb ironweed [*Vernonia baldwini* var. *interior* (Small) Schub.] and the C_4 plants were the grasses big bluestem (*A. gerardii* Vitman) and Indiangrass [*Sorghastrum nutans* (L.) Nash]. Ten OTC were used—five were exposed to ambient CO_2 concentration or 350 μmol/mol and five were exposed to twice the ambient CO_2 concentration or 700 μmol/mol. Five unchambered plots served as controls. The soil at the site was a silty clay loam. Because of frequent rain during the experimental period (summer of 1993), all data were collected under well-watered conditions. Abaxial stomatal conductance was measured on big bluestem and Indiangrass (no measurements were taken on ironweed) using a diffusion porometer (AP3, Delta T Devices, Cambridge, U.K.).

When the chambered plots were compared, transpiration of ironweed under elevated CO_2 was 33% less than that of ironweed under ambient CO_2 in the chambers (Figure 10.6). The chamber affected transpiration, and transpiration was 19% higher in the chambered plots with ambient CO_2 than in the controls. The increase in transpiration in the chamber was explained by the differences in the heights of the ironweed plants. Control plots (no chamber over them) were shorter

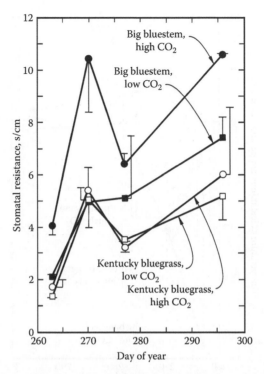

FIGURE 10.4 Stomatal resistance of Kentucky bluegrass (open symbols) and big bluestem (closed symbols) with a high (twice ambient) (circles) and a low (ambient) (squares) atmospheric CO_2 concentration. Vertical bars = ±standard deviation. Only half the bar is shown for clarity. (Reprinted from He, H. et al., *Trans. Kansas Acad. Sci.* 95, 139, 1992, Fig. 4, Copyright 1992. With permission from the Editor, Transactions of the Kansas Academy of Science, Fort Hays State University.)

(mean height = 0.74 m) than those in the chambered, ambient (mean height = 0.88 m). Elevated CO_2 reduced transpiration by 18% in big bluestem (Figure 10.7) and 22% in Indiangrass (Figure 10.8). The environmental effect of the chamber reduced transpiration by 21% in big bluestem and 24% in Indiangrass. Decreased radiation, decreased wind speed, and increased humidity (lower VPD) inside the chamber probably caused this reduction. The reduction in transpiration due to elevated CO_2 was related to an increase in stomatal resistance [Figure 10.9, where Bremer et al. (1996) call it "plant resistance"]. Bremer et al. (1996) converted stomatal resistance (Figure 10.9) to stomatal conductance. Stomatal conductance was 54% and 39% less in the chambered plots with elevated CO_2 than the chambered plots with ambient CO_2, for big bluestem and Indiangrass, respectively. Stomatal conductance decreased from 7.9 to 3.6 mm/s in big bluestem and from 5.3 to 3.2 mm/s in Indiangrass under CO_2 enrichment. Soil water was consistently highest under elevated CO_2, reflecting the large reductions in transpiration. A CO_2-induced decrease in stomatal conductance, causing a reduction in transpiration, ultimately might reduce plant water use, allowing prairies to be better able to withstand periods of drought than under current CO_2 levels (Bremer et al. 1996).

WHEAT

Senock et al. (1996) measured transpiration rate using sap-flow gauges. They did their measurements on winter wheat (*T. aestivum* L. cv. Yecora Rojo) grown in the free-air CO_2 enrichment (FACE) system of the U.S. Department of Agriculture at the Maricopa Agricultural Center, 40 km south of Phoenix, Arizona, United States. At the site, circular rings (25 m diameter) were either

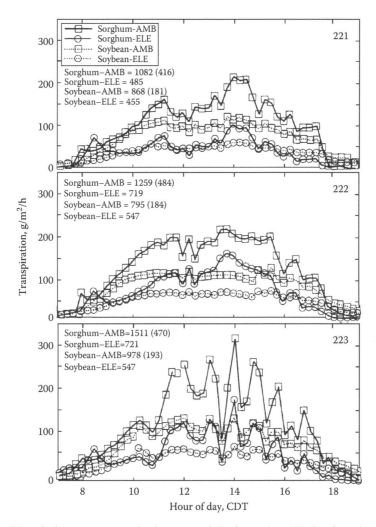

FIGURE 10.5 The 15 min average transpiration, per unit leaf area (one sided), of sorghum and soybean growing under ambient and elevated CO_2 concentrations (359 and 705 μmol/mol, respectively) on days 221, 222, and 223 (August 9–11, 1994). Daily total (g/m²/day) of transpiration and least significant difference between transpiration from plants growing under ambient and elevated concentration ($n=8$) are shown for each species and day. CDT, Central Daylight Time. (Reprinted from *Agric. Forest Meteorol.*, 83, Dugas, W.A., Prior, S.A., and Rogers, H.H., Transpiration from sorghum and soybean growing under ambient and elevated CO_2 concentrations, 37, Copyright (1997), Fig. 2, with permission from Elsevier.)

fumigated with CO_2 (550 μmol/mol) or left exposed to ambient CO_2 levels (about 370 μmol/mol). Under irrigated conditions, sap flows were different between the two CO_2 treatments (Figure 10.10 showing data for DOY 88 through 91). Daily water use in the elevated treatment (18–28 g/day) ranged from 19% to 26% less than that in the control treatment (24–38 g/day). Stomatal conductance was measured on day 89 and, in support of the differences between treatments on this day, the low sap flow in the FACE treatments coincided with a nearly 50% lower leaf stomatal conductance than in the control treatment through most of the day. The range in reduction in transpiration of the C_3 plant winter wheat in the FACE system (19%–26%) is similar to the reductions in transpiration observed by Bremer et al. (1996) for the C_3 plant ironweed (33%) and the C_4 plants big bluestem and Indiangrass (18% and 22%, respectively). Nevertheless, the studies show, for both C_3 and C_4 plants, that transpiration is reduced under elevated CO_2.

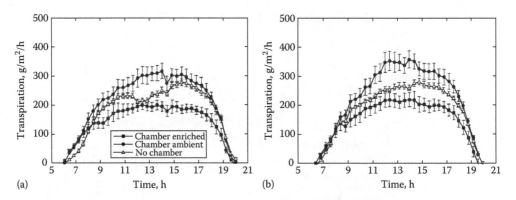

FIGURE 10.6 Transpiration (sap flow) of ironweed (*V. baldwini*) under CO_2-enriched chamber, ambient-CO_2 chamber, and no-chamber treatments on (a) DOY 178 (June 27) and (b) DOY 179, 1993. Data represent the mean sap flow per unit leaf area as measured by six sap-flow gauges. Vertical bars represent ± one standard error of the mean. Time is Central Standard Time. (Reprinted from Bremer, D.J. et al., *J. Environ. Qual.*, 25, 691, 1996, Fig. 1, Copyright 1996. With permission from the American Society of Agronomy.)

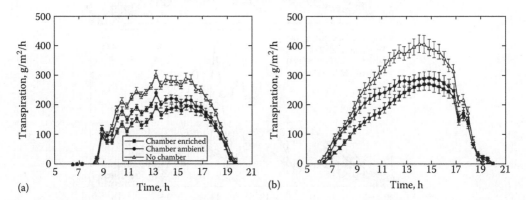

FIGURE 10.7 Transpiration of big bluestem (*A. gerardii*) under CO_2-enriched chamber, ambient-CO_2 chamber, and no-chamber treatments on (a) DOY 221 (August 9) and (b) DOY 234 (August 22), 1993. Data represent the mean sap flow per unit leaf area as measured by 10 sap-flow gauges. Vertical bars represent ± one standard error of the mean. Time is Central Standard Time. (Reprinted from Bremer, D.J. et al., *J. Environ. Qual.*, 25, 691, 1996, Fig. 3, Copyright 1996. With permission from the American Society of Agronomy.)

Samarakoon et al. (1995) grew two cultivars of wheat (*T. aestivum* L. cv. Matong and cv. Quarrion) in two greenhouses in Canberra, Australia, to see the effect of elevated CO_2 and soil water content on transpiration. Quarrion had a 40% lower stomatal conductance than Matong genetically. One greenhouse had the ambient CO_2 concentration (350–360 µmol/mol) and the other was elevated by 350 µmol/mol (CO_2 concentration in the air was about doubled). Plants grew in a fertilized potting soil consisting of loam, sand, and peat moss. The soil water content was brought to 30% (weight basis), which was just below field capacity. Half of the pots were given no more water (dry treatment). In the other half, the soil was kept near 30% water content by weighing the pots. Soil water content was calculated from the initial water content less the cumulative water loss per pot. I assume that "Day 1" was the day when all pots were watered. Some plants were harvested on days 41 and 49, and the last harvest was on day 64.

Transpiration per unit leaf area was reduced by doubled CO_2 in both cultivars under drying soil (Figure 10.11a and b) and in wet soil (Figure 10.12). In drying soil, the decline in transpiration accelerated after day 30 (Figure 10.11a and b). Plotting the data against soil water content (Figure 10.11c and d) showed that high CO_2 reduced the transpiration rate by about 30% until the soil had about

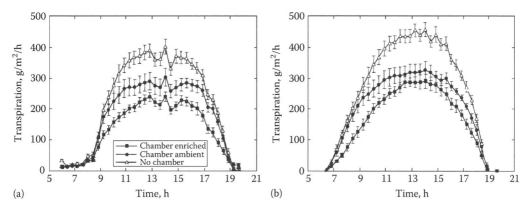

FIGURE 10.8 Transpiration of Indiangrass (*S. nutans*) under CO_2-enriched chamber, ambient-CO_2 chamber, and no-chamber treatments on (a) DOY 237 (August 25) and (b) DOY 238, 1993. Data represent the mean sap flow per unit leaf area as measured by 10 sap-flow gauges. Vertical bars represent ± one standard error of the mean. Time is Central Standard Time. (Reprinted from Bremer, D.J. et al., *J. Environ. Qual.*, 25, 691, 1996, Fig. 4, Copyright 1996. With permission from the American Society of Agronomy.)

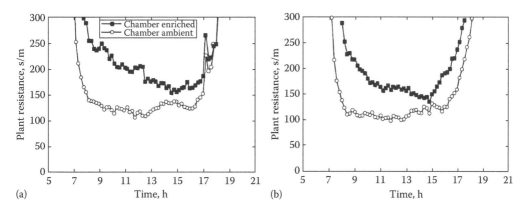

FIGURE 10.9 Plant resistance of (a) big bluestem on DOY 234 (August 22) and (b) Indiangrass on DOY 238 (August 26), 1993, under CO_2-enriched chamber and ambient-CO_2 chamber treatments. (Reprinted from Bremer, D.J. et al., *J. Environ. Qual.*, 25, 691, 1996, Fig. 7, Copyright 1996. With permission from the American Society of Agronomy.)

15% water content. In drying soil, both cultivars maintained a greater leaf area under high CO_2, but in wet soil, leaf area was not enhanced much by elevated CO_2. Water use per plant was determined by knowing the transpiration per unit leaf area and leaf area per plant. In drying soil (Figure 10.13a and b), during the first part of the experiment (up to about day 30), CO_2-enriched plants used less water and the leaf area increase during this period was insufficient to offset the reduction in transpiration rate. From day 30 on, the difference in water use by the low and high CO_2 plants was not significant for either cultivar. In wet soil, both cultivars under elevated CO_2 used less water after about day 30 (Figure 10.13c and d). But Matong used more water than Quarrion, which reflects its higher stomatal conductance.

EVAPOTRANSPIRATION—GENERAL PRINCIPLES

The studies so far have dealt only with transpiration. We now look at those that consider evapotranspiration, which is defined by the Soil Science Society of America (1997) as the "combined loss of water from a given area, and during a specified period of time, by evaporation from the soil surface

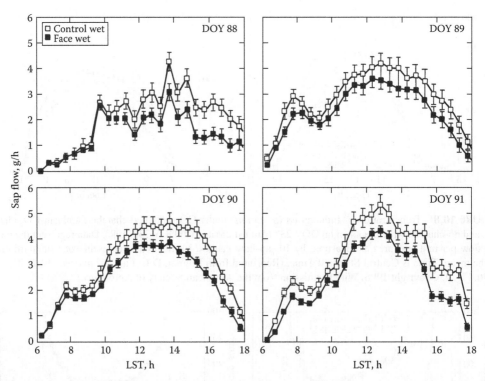

FIGURE 10.10 Diurnal sap flow by treatment on 4 days (DOY 88, 89, 90, and 91 or March 29, 30, 31, and April 1, 1993) at the free-air CO_2 exchange (FACE) site in Maricopa, Arizona. Values plotted on DOY 88, 90, and 91 are means of 20 sap-flow gauges ± standard errors. Values plotted on DOY 89 are means of 10 gauges ± standard errors. LST is local standard time. Environmental conditions on DOY 88 were partly cloudy with daily global irradiance of $20\,MJ/m^2$ and maximum air temperature of 18°C. DOY 89–91 were mostly clear with daily global irradiances of 23–$25\,MJ/m^2$ and maximum air temperatures of 23°C–25°C. (Senock, R.S., Ham, J.M., Loughin, T.M., Kimball, B.A., Hunsaker, D.J., Pinter, P.J., Wall, G.W., Garcia, R.L., and LaMorte, R.L., Sap flow in wheat under free-air CO_2 enrichment, *Plant Cell Environ.*, 1996, 19, 147, Fig. 5, Copyright Wiley-VCH Verlag GmbH & Co. KGaA. Reprinted with permission.)

and by transpiration from plants." Transpiration is fairly easy to measure. Transpiration of an individual leaf is a routine readout of portable photosynthetic systems, such as those made by Li-Cor, Inc. in Lincoln, Nebraska, United States. It also is determined in a greenhouse by weighing pots. The amount of water lost between two readings divided by the time gives the transpiration rate. In these measurements, water lost from the soil is usually ignored, because plants lose so much more water compared to water lost from the soil. However, one can always have a bare-soil pot to see the rate of evaporation. For precise measurements, the soil can be covered with plastic so that only the water lost from the plants is measured (Xin et al. 2008). Covering the top of soil in pots in a greenhouse with gravel or sand will also reduce evaporation.

Measurements of evapotranspiration are more difficult. It is not easy to measure (Allen et al. 1998, p. 9). Measurements of evapotranspiration are divided into three classes: first, the water balance or hydrologic methods; second, micrometeorological methods; and third, empirical methods (Tanner 1967, p. 535). We shall discuss the water balance equation below. Micrometeorological methods require measurements of flux density of water vapor in the boundary layer of the atmosphere. We first turn to an empirical method.

The simplest method to estimate evapotranspiration is to measure pan evaporation—an empirical method (Tanner 1967, pp. 567–568). In all greenhouse experiments, even if one has a bare pot to get soil evaporation, one still should determine pan evaporation, which shows the evaporative

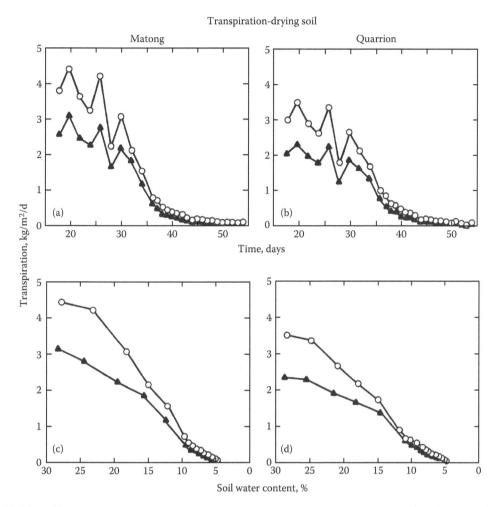

FIGURE 10.11   Transpiration per unit leaf area against time (a and b) and against soil water content (c and d) for wheat cultivars Matong (a and c) and Quarrion (b and d) growing in drying soil under low CO_2 (open circles) and high CO_2 (solid triangles) concentrations. In (a) and (b), the high and low CO_2 curves were significantly different on almost all days ($P < 0.001$ for most days and 0.01 and 0.05 for a few days). No statistical analysis was done on (c) and (d). (Reprinted from Samarakoon, A.B. et al., *Aust. J. Plant Physiol.*, 22, 33, 1995, Fig. 2, Copyright 1995. With permission from CSIRO Publishing.)

demand of the atmosphere. Pan evaporation is a measurement at official weather stations (LeGrand and Myers 1976). The pan at these places has flat sides. In field studies, one can get pan evaporation rates from the nearest official weather station. In a greenhouse, one puts out a pie pan, and ignores the sloping edges. One weighs the pan daily and the difference between weight one day and weight the next day divided by 24 h gives the pan evaporation rate in grams per day. The amount of water lost in one day in grams is converted to cm per day by dividing through by the area of top of the pan, assuming 1 g = 1 cm³.

Under field conditions, these open pans are interpreted to measure evapotranspiration rates (Rosenberg 1974, p. 198). In regions where advection is not important [i.e., non-windy regions; see Kirkham (2005, p. 464) for a discussion of advection], free water evaporation from pans may indeed give realistic estimates of potential evapotranspiration. We need to distinguish evapotranspiration and potential evapotranspiration. A simple definition of potential evapotranspiration is "the amount of water evaporated (both as transpiration and evaporation from the soil) from an area of

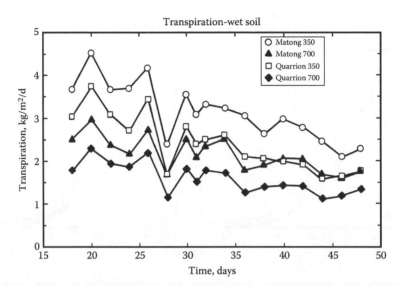

FIGURE 10.12 Time course of transpiration per unit leaf area in continuously wet soil for wheat cultivars Matong and Quarrion under low (350 μmol/mol) and high (700 μmol/mol) CO_2 conditions. Transpiration at high CO_2 was significantly lower (P less than 0.001 for most days and 0.01 or 0.05 for a few days) than at low CO_2 for both cultivars. High CO_2 Matong and low CO_2 Quarrion curves were significantly different (P less than 0.05) up to day 26 but not for most days thereafter. (Reprinted from Samarakoon, A.B. et al., *Aust. J. Plant Physiol.*, 22, 33, 1995, Fig. 3, Copyright 1995. With permission from CSIRO Publishing.)

continuous, uniform vegetation that covers the whole ground and that is well supplied with water" (Glickman 2000). Evapotranspiration is usually less than potential evapotranspiration. For a fuller discussion of potential evapotranspiration, see Kirkham (2005, pp. 455–468). Usually, the crop water use can be taken as 60%–90% of pan evaporation in climates where one does not need to consider advection (Rosenberg 1974, p. 198). In more arid localities and where advection is considerable (i.e., places like Kansas), the pan may give unrealistically low values (Rosenberg 1974, p. 198). So other methods have to be used. Methods to determine potential evapotranspiration are included under empirical methods (Tanner 1967, pp. 555–559).

Another term used in evapotranspiration studies is "reference evapotranspiration." When one determines potential evapotranspiration, the uniform vegetation used to get the potential evapotranspiration needs to be defined. It is often alfalfa (*Medicago sativa* L.) or a short grass. Then, one can determine the actual evapotranspiration of a crop under consideration and "reference" it to the one used to get potential evapotranspiration (e.g., alfalfa, if that is the crop that has been used to get potential evapotranspiration) (Loyd R. Stone, Kansas State University, United States, personal communication, November 5, 2008). One can then determine the ratios of the actual evapotranspiration determined for the crop (see below for different methods) in comparison to the supposedly maximum (potential or reference evapotranspiration). These ratios can be plotted versus growing season, such as Rachidi et al. (1993b) did for sorghum [*S. bicolor* (L.) Moench] and sunflower (*Helianthus annuus* L.). [They found that the ratios of (measured evapotranspiraton)/(reference evapotranspiration) during the growing season were always lower for sorghum, a drought-resistant crop, than for sunflower, a crop that uses much water.]

Scientists who need accurate measurements of evapotranspiration do not use pans. Much has been written concerning the different methods to determine evapotranspiration. They require environmental measurements, such as solar radiation, temperature, humidity, and wind speed, which then are put into (often difficult) equations to solve for evapotranspiration (Rose 1966, pp. 25–26, 78–87, and 178–185; Tanner 1967; Monteith 1973, pp. 176–179; Kanemasu et al. 1979;

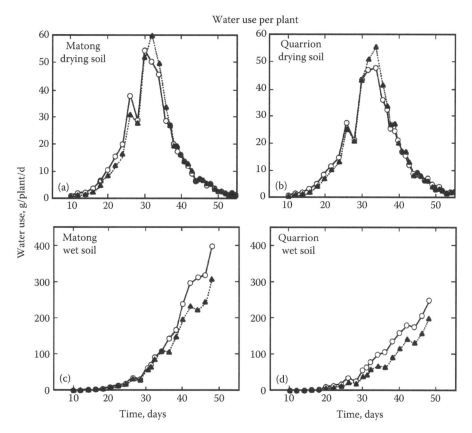

FIGURE 10.13 Time course of water use per plant under drying soil (a and b) and wet soil (c and d) for wheat cultivars Matong (a and c) and Quarrion (b and d) under low CO_2 (open circles) and high CO_2 (closed triangles) concentrations. In drying soil (a and b), the daily plant water use curves were significantly different (P less than 0.05) for Matong up to day 22 and for Quarrion up to day 16. In wet soil (c and d), water use per plant was significantly lower ($P<0.05$) at high CO_2 in Matong on most days except days 18–34 and for Quarrion on almost all days. (Reprinted from Samarakoon, A.B. et al., *Aust. J. Plant Physiol.*, 22, 33, 1995, Fig. 6, Copyright 1995. With permission from CSIRO Publishing.)

American Society of Agricultural Engineers 1985; Jones 1992, pp. 112–130). If interested readers look at these references, they should note that not all scientists accept the term "evapotranspiration" and only use the term "evaporation." For example, Monteith (1973) never uses the term "evapotranspiration" in his book. Such scientists consider the equations developed for water loss from a soil surface to be similar to that for a leaf surface—all termed "evaporation." The evaporation term is part of the "heat budget" at the surface of the earth. Kirkham (2005, p. 415) gives the equation for the heat budget (also called the heat exchange at the surface of the earth). Heat normally is lost when water evaporates from the soil or plant leaves, but heat can be gained when dew forms on leaves.

As noted, the simplest method to estimate evapotranspiration is the measurement of pan evaporation. The next simplest method is the water balance method. In this method, all the water that comes into an area and is lost from an area is accounted for. Tanner (1967, p. 535) gives the water balance equation, as follows:

$$ET = \frac{P - (V_r + V_L + V_i + \Delta V_w + \Delta V_s)}{A} \qquad (10.1)$$

where

ET is evapotranspiration (cm or cm/time)

P is precipitation (cm or cm/time)

V_r is volume of surface and subsurface runoff from the catchment (cm^3)

V_L is leakage from the catchment and not included in runoff (cm^3)

V_i is the volume of intercepted water (cm^3)

ΔV_w is the volume change in ground water storage (cm^3)

ΔV_s is the volume change in water stored above the water table (cm^3)

A is the area of the catchment (cm^2)

The different terms in the soil water balance equation can be determined with accuracy in a lysimeter, because it isolates the crop root zone from its environment and allows measurement of the terms that are difficult to measure in Equation 10.1 (Allen et al. 1998, p. 12). A *lysimeter* is a device in which a volume of soil, which may be planted to vegetation, is located in a container to isolate it hydrologically from the surrounding soil (Tanner 1967, p. 537). The lysimeters are weighed. Chaudhuri et al. (1990) describe such a lysimeter system. The underground boxes that they used were isolated hydrologically and weighed to determine water lost by a crop. Evapotranspiration can be obtained with an accuracy of a few hundredths of a millimeter with lysimeters (Allen et al. 1998, pp. 12–13). Lysimetry is the only hydrological method in which the experimenter has complete knowledge of all the terms in Equation 10.1. It is an independent check on the suitability of micrometeorological methods to determine evapotranspiration and for calibrating empirical formulas (Tanner 1967, p. 536). It is the standard by which all methods should be compared (C.B. Tanner, University of Wisconsin, Madison, United States, personal communication, c. 1973).

When Equation 10.1 is applied to large natural catchments and to drainage and irrigation districts, it is hoped that $V_L = 0$ and that P, V_r, and A are measured with the required precision. The ΔV_w, ΔV_s, and V_i terms can be neglected in comparison to other terms, if the average ET is computed over a sufficiently long period. Leakage is frequently encountered in drainage and irrigation districts and in their associated storage and distribution facilities (Tanner 1967, p. 535). If the ground is flat, runoff can be neglected. If the area is irrigated, then the amount of irrigation water must be added to the precipitation in the water balance equation (Rose 1966, p. 178).

One hydrological method to get evapotranspiration is the soil water depletion method. However, to use the depletion method, drainage errors must be acceptable (Tanner 1967, pp. 535–536). Soil water depletion measurements, for example obtained by using a neutron probe, are based on the water balance equation (Equation 10.1), where ΔV_s is measured and V_r (runoff) is controlled (e.g., by the use of berms). In humid regions, measurement of soil water depletion rarely provides reliable evapotranspiration measurements, because of frequent rains and resulting drainage (Tanner 1967, p. 536). But in arid and semiarid areas, like Kansas, soil water depletion is a good way to get evapotranspiration. One still needs to know if drainage is negligible. If it is not, then it needs to be included in the water balance. Water can be draining from a soil even when the soil is at field capacity. For example, Stone et al. (2008) found that at 100% available water [remember available water is the difference between field capacity and permanent wilting point; see Kirkham (2005, p. 107)], there was not 100% water stored in fallow soil in the west-central Great Plains of the United States, which meant that some of this water had drained out of the soil profile. Their data showed the limitation of the definition of field capacity, which assumes that drainage is negligible at field capacity. Stone et al. (2008) studied storage of soil water only under ambient CO_2, not elevated CO_2. But we note that under elevated CO_2, soil would store more water (Bremer et al. 1996; Ferretti et al. 2003; see also Chapter 5). The wetter soils under elevated CO_2 are the result of improved soil water conservation due to reduced stomatal conductance under elevated CO_2.

Those who ignore drainage, when it is significant, are overestimating the values of evapotranspiration. In dry regions, one can make estimates about drainage to know if it needs to be included in the water balance equation (Equation 10.1). If a season is extremely dry, then one can be fairly sure that drainage is negligible. Also, one can install tensiometers at depth. These readings have to be below the root zone. In Kansas, sunflower can deplete water between the 2.7-m depth (Rachidi et al. 1993a) and 3.1-m depth (Stone et al. 2002). Because time domain reflectometry can measure only up to about the 1-m depth (Kirkham 2005, pp. 197–198), this method cannot be used to determine evapotranspiration. Neutron probes have to be used. Two tensiometers at different depths will tell the direction of water movement [see Kirkham (2005, pp. 51–52 and 63–64) to determine the direction of movement of water in the soil using tensiometers]. If a tensiometer at depth reads −500 to −600 cm [remember about −1000 cm = −1 bar; see Kirkham (2005, pp. 41, 44, 57) for conversion of units], then one can be fairly sure that the soil is dry and drainage is negligible. But if the tensiometer reads −300 cm (the so-called field capacity value, often considered to be −1/3 bar or −300 cm), then one needs to measure drainage for exact measurements of evapotranspiration (Loyd R. Stone, personal communication, November 3, 2008). If drainage is negligible, then evapotranspiration rates are calculated as the total of profile water depletion, rainfall, and irrigation in a given time period divided by the number of days (Stone et al. 1982).

The maximum depth that Loyd R. Stone has put his neutron probes in Kansas is 3.2 m (10.5 ft), even though they could be put at deeper depths (L.R. Stone, personal communication, November 5, 2008). The 3.2-m depth ensures that one can observe changes in water content at deep depths. Even though a root might not be extracting water at deep depths, water at depth can move up in a soil profile to replace water extracted by deep roots. Under irrigated conditions, this does not occur because roots are not at deep depths under irrigation. If one can observe no changes in water content at deep depths under dryland conditions over time, then one can be sure that no deep drainage is occurring. Shallow rains that might occur during a season will not be influencing water at deep depths and draining out of the system. So the depletion method (using neutron probes to get changes in water content) can be used to determine evapotranspiration under dryland conditions. However, under irrigated conditions in semiarid regions, if one measures depletion only to, say, the 2-m depth, drainage may be occurring and one needs to account for it in the water balance equation (Equation 10.1). In this respect, it is easier to get evapotranspiration under dryland conditions than irrigated conditions, because one does not need to account for drainage under dryland conditions, if the neutron probe readings are deep enough to make sure no drainage is occurring.

Many other, more complicated equations than Equation 10.1 have been developed to measure evapotranspiration or potential evapotranspiration. Stone and Horton (1974) compared methods to estimate evapotranspiration. They studied the methods by van Bavel (1966) (micrometeorological method), Penman (1948) (empirical method), and the energy-budget method of Bowen (1926) (micrometeorological method). Penman's (1948) method for estimating potential evapotranspiration uses both energy balance and aerodynamic principles. His method includes a term for wind speed. The Priestley–Taylor (1972) method to measure potential evapotranspiration (empirical method), which is commonly used because it only requires measurements of solar radiation and temperature, does not have a term for wind (Kanemasu et al. 1976; Rosenthal et al. 1977). (For a biography of C.H.B. Priestley, see Appendix.) A simplified form of the Priestley–Taylor equation is given by Campbell (1977, p. 140):

$$\lambda E_p = a(T_a + b)S_t \qquad (10.2)$$

where
E_p is the potential evapotranspiration (g m/s)
λ is the latent heat of vaporatization (cal/g)
a is a constant and a typical value of $a = 0.025°C^{-1}$
T_a is the ambient air temperature (°C)
b is a constant and a typical value for b is 3°C
S_t is the total short-wave flux density on a horizontal surface (W/m²)
The constants a and b are empirically determined

Campbell (1977, pp. 140–141) gives a sample problem, which we now present. We remember that $1\,W = 1\,J/s$ and $1\,cal = 4.186\,J$. The latent heat of vaporization is obtained from a table showing its values at different temperatures [e.g., Kirkham (2005, Table 3.1, p. 31)]. At 20°C it is 586.0 cal/g or 2450 J/g, rounding to three significant figures.

PROBLEM

What is the E_p for a forest with average $S_t = 300\,W/m^2$ and $T_a = 20$°C?

Answer

$$\lambda E_p = 0.025°C^{-1}(20 + 3)C \times 300 \ W/m^2 = 173\,W/m^2$$

$$E_p = \frac{173 \ J/s/m^2}{2,450 \ J/g} = 0.071\,g/m^2/s$$

or $0.071 \ g/m^2/s \times 86,400\,s/day \times 10^{-6}\,m^3/g = 6.1mm/day$

The value obtained in this sample problem can be compared to one of the highest values ever measured for evapotranspiration (Stone et al. 1982). The value was obtained for irrigated alfalfa (cultivar Cody) during one of the hottest and driest summers in Kansas (1980). This high rate was 13 mm/day and it occurred between June 21 to July 1, when the maximum temperature was 43°C (109°F) on two dates during that interval. Rosenberg (1974, p. 181) reports a higher value of 14.66 mm/day for Sudan grass [*S. sudanense* (Piper) Stapf.] growing in a field near Phoenix, Arizona, United States. The value comes from a paper by van Bavel et al. (1962), who determined evapotranspiration using lysimetry. We see that the value for potential evapotranspiration calculated for the forest is about half the value of maximum evapotranspiration measured.

The effect of wind in areas like Kansas should be considered in evapotranspiration equations. For accurate procedures to determine crop evapotranspiration, the Food and Agriculture Organization (FAO) recommends a modification of the Penman method, called the "Penman–Monteith" method (Allen et al. 1998, pp. 17–38). Penman and Monteith published no paper together with this modified equation. It is so-called because ideas from both Penman (1948) and Monteith (1965) were incorporated into the equation. For Monteith's description of the use of the Penman formula and how it needs to be modified under different environmental conditions (e.g., arid regions), see Monteith (1973, pp. 178–183). The error in the Penman formula under arid conditions can be removed by introduction of a physiological resistance, depending on the diffusion of water vapor through the stomata, as shown by Monteith (1973, pp. 180–183). Forms of the Penman–Monteith equation, as now used by researchers, are given by Allen et al. (1998, p. 19) and by Howell and Evett (2004). Nederhoff et al. (1992) and Nederhoff and de Graaf (1993) used the Penman–Monteith equation to study the effect of elevated levels of CO_2 on transpiration of greenhouse-grown crops.

Rosenberg et al. (1989) used the Penman–Monteith model of evapotranspiration to estimate the possible changes in evapotranspiration in a greenhouse-warmed world. They point out that evaporation must increase in a warmer world because the capacity of the air for water vapor increases 5%–6% per °C. They give 46 different simulations including different leaf areas, different air temperatures, different solar radiations, and four different ecosystems (irrigated alfalfa, a tallgrass prairie, a forest, and a wheat field), but they do not consider elevated atmospheric CO_2. Ten of their simulations showed an average decrease in evapotranspiration with standard deviation of 6.3% ± 3.3%, and 35 of the simulations showed an average increase in evapotranspiration with standard deviation of 14.7% ± 9.5%, and one shows no change. As we shall see in the data we report below in experiments where CO_2 has been elevated, measured evapotranspiration in most studies has decreased in response to elevated CO_2. Rosenberg et al. (1989) point out that even though the

Penman–Monteith model is one of the most complete and mechanistically based of the commonly used methods for estimating evapotranspiration, it has limitations. In particular, it does not consider the effects of soil moisture on evapotranspiration, so changes in precipitation cannot be simulated directly but only by varying the resistance of diffusion of water vapor through the stomata, which is called the "surface resistance" (Allen et al. 1998, p. 19) or the "canopy resistance" (Rosenberg et al. 1989). In the simulations of Rosenberg et al. (1989), they estimated canopy resistance by using measurements of stomatal resistance.

In a simulation study where resistance to water movement in the soil surface was considered, as well as elevated CO_2 (Jagtap and Jones 1989), the model predicted diurnal water use patterns of irrigated soybean [*Glycine max* (L.) Merr.] at 330 (ambient), 660, and 990 µmol/mol CO_2. It also predicted a reduction in canopy evapotranspiration under the two higher CO_2 levels compared to the ambient level. However, even in the model of Jagtap and Jones (1989), simulated evapotranspiration at the three CO_2 levels was usually higher than the measured evapotranspiration. Their data show the importance of actual measurements of evapotranspiration under elevated CO_2.

Increased evaporation from the land surface, due to the enhanced temperatures associated with global warming, as predicted by models, may be negligible, because of suppressed transpiration due to elevated levels of CO_2 (Lockwood 1995). After a crop canopy covers the ground, transpiration is the most important factor controlling evapotranspiration—not evaporation from the soil. When a canopy covers the ground, after planting, it depends upon the crop and the time that it attains a leaf area index greater than one. Remember leaf area index is defined as the area of leaves above a unit area of ground taking only one side of each leaf into account (Kirkham 2005, p. 457). Brun et al. (1972) showed that for soybean and sorghum fields the proportion of water lost as transpiration is approximately 50% of the total evapotranspiration at a leaf area index of 2. This proportion increases to 95% of the total evapotranspiration at a leaf area index of 4. If one does not know the leaf area index, then one might estimate visually the percentage cover of the soil with plants. This is done when using the infrared thermometer to make measurements of canopy temperature (Kirkham 2005, pp. 425–435). Measurements should be made on plants in plots with at least 75% of the ground covered, so the instrument records canopy temperature, not soil temperature (Kirkham et al. 1984).

EVAPOTRANSPIRATION UNDER ELEVATED CO_2

Now let us turn to experiments in which evapotranspiration has been measured when plants have been exposed to elevated levels of CO_2. In each of the studies described below, the methods to get evapotranspiration require the solving of equations with several variables, and I shall not describe these equations. The purpose of this section is to see the effect of elevated CO_2 on evapotranspiration, not to study methods to get evapotranspiration. The interested reader can get the papers to see the equations.

Overdieck and Forstreuter (1994) determined the evapotranspiration of young European beech (*Fagus sylvatica* L.) trees growing in greenhouses at ambient (360 µmol/mol) and elevated (698 µmol/mol) CO_2 concentrations. Saplings grew in a sandy loam soil under well-watered conditions under the two CO_2 regimes from November, 1991 until August, 1992, when evapotranspiration measurements were made. They determined transpiration using a cuvette system and an infrared gas analyzer, and they calculated leaf conductance from their measurements. When measurements began, the leaf area index was 4.2 and 6.1 in the ambient and elevated CO_2 treatments, respectively. Leaf transpiration rates were 18% lower in the elevated CO_2 treatment than in the ambient CO_2 treatment. Leaf conductance was decreased by 32% under elevated CO_2. The elevated CO_2 treatment resulted in a 14% reduction in stand evapotranspiration. Figure 10.14 shows 24 h evapotranspiration rates of the beech saplings over the period August 1–21, 1992, plotted against daily means of air temperature and VPD. In both CO_2 treatments, evapotranspiration increased linearly at a rate of 0.2 kg H_2O m^{-2} day^{-1} for each 1°C rise in air temperature between 14°C and 25°C. Because evapotranspiration was reduced by elevated CO_2, they concluded that water losses from deciduous forest stands will be reduced by a doubling of CO_2 concentration.

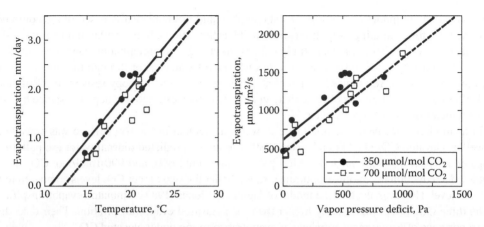

FIGURE 10.14 Relationship between evapotranspiration and both air temperature and VPD during August (1992) in 4-year-old beech stands ($n = 36$) exposed to ambient or elevated CO_2 in greenhouses. (Reprinted from Overdieck, D. and M. Forstreuter, *Tree Physiol.*, Fig. 3, 1994 by permission of Oxford University Press.)

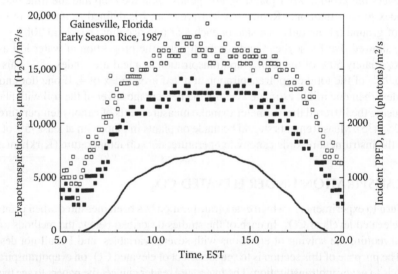

FIGURE 10.15 Diurnal trends in photosynthetic photon flux density (PPFD, solid line) and evapotranspiration rates for the 160 (open symbol) and 900 (closed symbol) μmol/mol [CO_2] treatments at 74 days after planting (DAP) in an experiment with rice. (Reprinted from Baker, J.T. et al., *Agron. J.*, 82, 834, 1990a, Fig. 5, Copyright 1990. With permission from the American Society of Agronomy.)

Baker et al. (1990a) considered the effects of subambient and elevated levels of CO_2 on the evapotranspiration rate of rice (*Oryza sativa* L. cv. IR-30). They grew the plants in controlled-environment chambers exposed to natural sunlight. Daytime CO_2 concentration was maintained at either 160 or 250 μmol/mol (subambient CO_2 treatments), 330 μmol/mol (ambient CO_2 treatment), or 500, 660, or 900 (superambient CO_2 treatments). The soil was a fertilized fine sand, and the soil was flooded for rice growth. Evapotranspiration decreased with increasing CO_2 concentration. Figure 10.15 shows the reduction in the evapotranspiration rate 74 days after planting under the elevated CO_2 treatment (900 μmol/mol) compared to the subambient (160 μmol/mol). On this day, the total daily water loss for the 160, 250, 330, 500, 660, and 900 μmol/mol CO_2 concentrations were 768.9, 793.7, 628.5, 639.5, 632.6, and 567.3 mol (H_2O) m^{-2}, respectively. They do not give evapotranspiration rates for the individual CO_2 treatments, so we cannot calculate the percentage reduction in evapotranspiration by doubling the CO_2.

Using OTC, Ham et al. (1995) measured the effect of elevated CO_2 on evapotranspiration of a tallgrass prairie during a 34-day period in July and August, 1993. The vegetation at the site was dominated by two C_4 grass species, big bluestem (*A. gerardii* Vitman) and Indiangrass (*S. nutans* (L.) Nash). The soil was a silty clay loam, and frequent rainfall kept soil water near field capacity. The CO_2 concentration was either ambient or twice ambient (the concentrations are not given). They measured net CO_2 exchange (NCE) (NCE=canopy photosynthesis – soil and canopy respiration) and evapotranspiration by using methods they described in a previous paper (Ham et al. 1993). Elevated CO_2 did not strongly influence NCE in the prairie when water was not limiting (Figure 10.16a). The minimal effect of CO_2 on photosynthesis was consistent with previous leaf-level measurements at the prairie site (Kirkham et al. 1991). Doubling the CO_2 concentration in the air reduced evapotranspiration by 22% (Figure 10.16b). The greater NCE and lower evapotranspiration resulted in higher water use efficiency under CO_2 enrichment compared to ambient CO_2 (9.84 vs. 7.26 g CO_2 kg^{-1} H_2O). Ham et al. (1995) conclude that the most noteworthy effect of CO_2 enrichment was a reduction in ecosystem evapotranspiration.

Diemer (1994) measured NCE and evapotranspiration of an alpine grassland dominated by *Carex curvula* (authority not given; it is a sedge and sedges are C_3 monocots) grown in OTC and treated with doubled the ambient (528 μbar/bar CO_2 ≈ 680 ppm) and ambient (257 μbar/bar) partial pressure

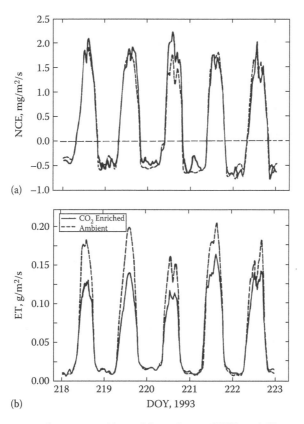

FIGURE 10.16 Diel patterns of ecosystem (a) net CO_2 exchange (NCE) and (b) evapotranspiration (ET) in CO_2-enriched and ambient chambers between DOY 218 and 222 (August 6–10, 1993). (Reprinted from *Agric. Forest Meteorol.*, 77, Ham, J.M., Owensby, C.E., Coyne, P.J., and Bremer, D.J., Fluxes of CO_2 and water vapor from a prairie ecosystem exposed to ambient and elevated atmospheric CO_2, 73, Copyright (1995), Fig. 5, with permission from Elsevier.)

of CO_2 in the Swiss Central Alps (2470 m above sea level). (We remember from Chapter 9, where we studied stomatal density of plants at different elevations, that the ambient CO_2 partial pressure in the air decreases as the distance above sea level increases.) A third treatment consisted of plants that grew with no chambers. The soil was a well-watered, podsolized luvisol. [A podzol is a type of relatively infertile soil found typically in forests and consisting of a thin, ash-colored layer overlaying a brown, acidic humus. Podzol comes from the Russian word "ashlike" (Friend and Guralnik 1959). A luvisol (from the Latin, *luo*, to wash) is a soil with distinct clay accumulation (Nachtergaele 2005).] The three treatments were divided, and half received nitrogen fertilizer (50 kg/N/ha) and the other half received no nitrogen fertilizer. Plants were exposed to the different CO_2 treatments in 1992, and measurements were carried out from August 13 to 26, 1993. In addition, on August 30–31, 1993, between 22:00 and 02:00 h (4 h), the treatments were reversed. That is, high CO_2 plots received ambient CO_2 and low CO_2 plots received high CO_2. Measurements that were taken when the plots were not reversed were labeled the "G" measurements, i.e., the measurements were determined at the "growth" CO_2 partial pressure.

In chambered, unfertilized plots, NCE was 98% higher under elevated CO_2 compared to ambient CO_2. In the fertilized plots, NCE did not differ between the elevated and ambient CO_2 concentrations. In the unfertilized plots (Figure 10.17a), evapotranspiration was on average 21% lower at the elevated CO_2 concentration compared to the ambient CO_2 concentration. In Figure 10.17, focus is on the bars with G, because they give the data for the long-term exposure to the different CO_2 concentrations. The bars without G are the measurements taken when the CO_2 was reversed only for 4 h during one night. In the fertilized plots (Figure 10.17b), the evapotranspiration was similar at both CO_2 levels. Diemer (1994) concludes that for these alpine plant communities adding nutrients to the soil may induce the same effects on gas exchange as increased CO_2 does.

So far, the studies have not considered drought and its effects on evapotranspiration. Burkart et al. (2004) investigated the effects of both elevated atmospheric CO_2 and drought on evapotranspiration of spring wheat (*T. aestivum* L. cv. Minaret) grown in OTC in Braunschweig, Germany. The soil was a loamy sand. The ambient and elevated CO_2 concentrations were 408 and 684 μmol/mol, respectively. Plant-available water was calculated for a 50-cm depth of soil. For the well-watered plants and water-stressed plants, plant-available water was >40 and 10–30 mm, respectively. Figure 10.18 shows an example of diurnal courses of evapotranspiration of all treatments between days 78 and 82 after emergence. Evapotranspiration of the control canopy (well-watered; ambient CO_2) reached a maximum value of 14 mmol/m²/s. Under well-watered conditions, elevated CO_2 reduced evapotranspiration by up to 40% in this period. During the season, elevated CO_2 under well-watered conditions decreased evapotranspiration between 15% and 18%. Drought stress decreased evapotranspiration to 20%–40% of the well-watered canopy. Under drought, no consistent effect of CO_2 concentration on evapotranspiration was observed. The authors conclude that, under drought stress, elevated CO_2 will not reduce evapotranspiration and hence the water use.

Li et al. (2004) also considered the effect of drought on evapotranspiration under elevated CO_2. They grew spring wheat (cultivar Dingxi No. 8654) in pots with loess soil placed in growth chambers at two atmospheric CO_2 concentrations (ambient or 350 μmol/mol and elevated or 700 μmol/mol), two levels of soil moisture (soil kept near field capacity or watered when water content fell to 45%–60% of field capacity), and five nitrogen treatments (ranging between 0 and 450 kg/h/m²). Evapotranspiration was reduced by CO_2 enrichment (Figure 10.19). For the two water levels, the mean decreases (the average of five nitrogen levels) in total evapotranspiration with the doubling of CO_2 in the well-watered treatments and the drought-stressed treatments were 4% and 8.6%, respectively. The mean decreases in stomatal conductance (averaged over the nitrogen levels) due to elevated CO_2 were 41% and 51% for the well-watered treatments and drought-stressed treatments, respectively. The 41% decrease agrees with the observation of Morison (1987a), cited above, that doubling of the CO_2 concentration in the air reduces stomatal conductance 40%. Li et al. (2004) found that, under elevated CO_2, reduction of evapotranspiration was greater in drought-stressed treatments than in well-watered treatments. Their results showed that the CO_2-enrichment-induced

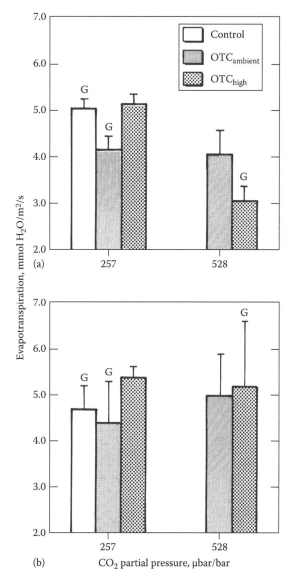

FIGURE 10.17   Evapotranspiration of a sedge-dominated grassland within OTC during gas exchange measurements at treatment CO_2 partial pressures [the CO_2 partial pressure that the plants grew at long-term or "growth" (G) CO_2 concentration] and short-term reversed CO_2 partial pressures. Control plots had no chamber over them. Means and standard deviations of 4–12 OTCs per treatment are plotted. (a) Responses of unfertilized plots; (b) fertilized plots. Bars without G indicate measurements taken when treatments were reversed for a short-term (4 h). (Reprinted with kind permission from Springer Science+Business Media: *Oecologia*, Mid-season gas exchange of an alpine grassland under elevated CO_2, 98, 1994, 429, Diemer, M.W. modified from Fig. 2. Copyright 1994, Springer.)

decrease in evapotranspiration almost compensated for the increase in evapotranspiration brought about by the higher leaf area under adequate nitrogen and water supply. Kimball et al. (1999) in Arizona also studied evapotranspiration of wheat under different CO_2 and nitrogen levels. They grew spring wheat (cultivar Yecora Rojo) under well-watered conditions in a FACE facility exposed to 550 µmol/mol CO_2 (approximately 200 µmol/mol above current ambient levels of about 360 µmol/mol). Evapotranspiration was lower in the FACE plots, by about 6.7% and 19.5% for high and low nitrogen plots, respectively.

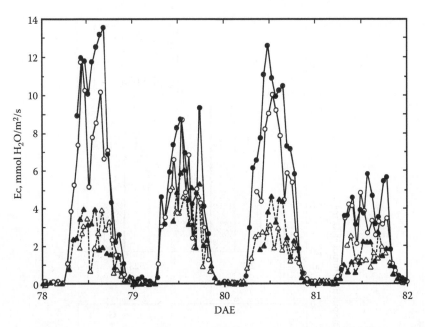

FIGURE 10.18 Time course of canopy evapotranspiration (Ec) from days after emergence (DAE) 78–81 of the 1999 season. Well-watered: circles; drought-stressed: triangles; ambient CO_2 concentration: closed symbols; elevated CO_2: open symbols. Data points represent 10 min averages measured hourly. (Reprinted from *Eur. J. Agron.*, 21, Burkart, S., Manderscheid, R., and Weigel, H.J., Interactive effects of elevated atmospheric CO_2 concentrations and plant available soil water content on canopy evapotranspiration and conductance of spring wheat, 401, Fig. 3, Copyright 2004, with permission from Elsevier.)

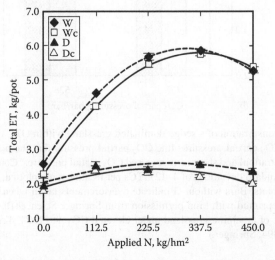

FIGURE 10.19 Effects of CO_2 enrichment on total evapotranspiration of spring wheat (kg per pot) under different soil nitrogen and water levels. W, Wc, D, Dc represent the well-watered, well-watered+CO_2, drought-stressed, and drought-stressed+CO_2 treatments, respectively. Data points are mean ± the standard error ($n=3$). (Reprinted from *Agric. Water Manage.*, 67, Li, F., Kang, S., and Zhang, J., Interactive effects of elevated CO_2, nitrogen and drought on leaf area, stomatal conductance, and evapotranspiration of wheat, 221, Copyright (2004), Fig. 5, with permission from Elsevier.)

The results of Li et al. (2004) for the evapotranspiration of wheat under drought do not agree with those of Burkart et al. (2004). Under drought, Burkart et al. (2004) found no consistent effect of CO_2 on evapotranspiration, while Li et al. (2004) found that evapotranspiration was reduced by elevated CO_2. The difference may be due to leaf area. Burkart et al. (2004) saw no increase in leaf area index due to elevated CO_2 in either the well-watered or drought-stressed treatments. In the experiment of Li et al. (2004), leaf area was increased by elevated CO_2 under both watering regimes. The data indicated that whether or not evapotranspiration is reduced under elevated CO_2 and drought will depend upon the leaf area. No savings in water can be expected in canopies where elevated CO_2 stimulates an increase in leaf area relatively more than it closes stomata (Drake et al. 1997; Engel et al. 2004). This occurred in the experiment of Chaudhuri et al. (1990) with wheat. Under both well-watered and drought-stressed conditions, water use was greater in the treatments with elevated CO_2, even though stomata closed with elevated CO_2 (Chaudhuri et al. 1986b, 1987). Water was not saved under elevated CO_2, probably because leaf area was increased under elevated CO_2.

SUMMARY

Most studies show that under well-watered conditions doubled CO_2 reduces transpiration by an average of about 20%, stomatal conductance by an average of about 40%, and evapotranspiration by an average of about 10%–14%. Whether elevated CO_2 reduces evapotranspiration under drought depends on leaf area. If elevated CO_2 increases leaf area more than it reduces water loss due to transpiration, evapotranspiration will not be reduced. In this chapter, we also considered ethylene, a low-molecular-weight gas like CO_2. However, unlike CO_2, it is a hormone and has effects on plants at low concentrations (0.1–1 ppm). The threshold concentration for effects of CO_2 on plants is not known, but our experiments showed that elevating the atmospheric CO_2 concentration by only 155 ppm above ambient closed stomata.

APPENDIX

BIOGRAPHY OF CHARLES H.B. PRIESTLEY

Charles H.B. Priestley, a pioneer of modern meteorology, was born in Highgate, London, England, on July 8, 1915, the third of four children (Garratt 1999). He graduated with first-class honors in applied mathematics from Cambridge University in 1937 and specialized in hydrodynamics and thermodynamics. He remained at Cambridge for another year and earned a degree in economics. He entered the British Meteorological Office in 1939, 6 months after the outbreak of World War II. He was put to work studying turbulent diffusion in the lower atmosphere. Early in the war, the Meteorological Office sent him to Canada for 2 years to carry out work on the dispersion of heavy gases. In 1943, after his return to England, he was transferred to the upper-air forecasting unit to support the bombing offensive against Germany. Together with a number of other British and American meteorologists, he was involved in the successful D-Day (June 6, 1944) weather forecast for General Eisenhower (Garratt 1999).

In late 1945, the Executive Committee of Australia's Council for Scientific and Industrial Research (CSIR) wrote to Sir David Brunt, head of the Department of Meteorology at Imperial College, London, asking him to recommend someone to head a new Australian research group in the atmospheric sciences. Brunt approached Priestley, and, as a result, he was appointed to CSIR in September 1946. His main task was to establish a new section of meteorological physics. [In 1949, under the Science and Industry Research Act, the council was reconstituted as the Commonwealth Scientific and Industrial Research Organization (Garratt 1999).]

In less than two decades, the group that Priestley formed became one of the leading atmospheric research teams in the world. He served as officer-in-charge of his section in Melbourne, Australia.

In 1955, he became the first chief of the Division of Atmospheric Research, a position he held for 26 years (Garratt 1999).

Priestley's research in meteorology was broad. He made original contributions in many areas, including the dynamics of the atmospheric general circulation, free and forced convection in the atmosphere near the ground, the handover of vertical fluxes from small-scale turbulence close to the earth's surface into the larger-scale motions above, and the assessment of surface heat flux and evaporation using large-scale parameters (Garratt 1999). In 1972, he proposed the establishment of an Australian observatory to measure the background composition of the atmosphere. This resulted in the Cape Grim Baseline Air Pollution Station in northwestern Tasmania, one of only a handful of baseline monitoring stations in the world. In the 1960s, he was one of a number of eminent scientists entrusted with formulating the Global Atmospheric Research Programme (GARP) of the World Meteorological Organization (WMO) (Garratt 1999).

Priestley visited the United States. In 1951, he spent 2 months conferring with a number of American university groups. In 1957, he had a 7 month sabbatical leave in the Department of Meteorology at the University of Chicago. During his time in Chicago, he gave lectures and the University of Chicago Press later published them in a monograph *Turbulent Transport in the Lower Atmosphere* (Garratt 1999).

He won many honors. He was elected fellow of the Australian Academy of Science in 1955 and fellow of the Royal Society in 1966. In 1973, he received the WMO's highest honor, the IMO (International Meteorological Organization) Prize. In 1974, he received the Rossby Research Medal, the American Meteorological Society's highest honor. The citation read, "for his fundamental contributions to the understanding of turbulent processes and the links between small-scale and large-scale dynamics in the atmosphere." In 1976, he received the Order of Australia and the Flinders Medal, which is the highest award given by the Australian Academy of Science to a physical scientist (Garratt 1999). In 1978, he was elected as an Honorary Member of the American Meteorological Society. In 1983, the Australian branch of the Royal Meteorological Society established an award known as the Priestley Medal in recognition of his contributions to science of meteorology. It is awarded every 2 years for excellence in meteorological or oceanographic research. The inaugural Priestley award with lecture took place in 1995.

Priestley retired as chief of the CSIRO Division of Meteorological Physics (now Atmospheric Research) in 1972. He was chairman of the CSIRO Environmental Physics Research Laboratories from 1973 to 1977 and, after that, part-time professor in the Department of Mathematics at Monash University. He retired from active service in the late 1980s. His hobbies were golfing and listening to classical music. He died on May 18, 1998, after several years of ill health. He is survived by his wife, Connie, and their two daughters and son and six grandchildren (Garratt 1999).

In Kansas, we know Priestley for the Priestley–Taylor equation (Priestley and Taylor 1972). It is an equation of choice to determine potential evapotranspiration, because of its simplicity. One can determine potential evaporation only by knowing the solar radiation and temperature. Kirkham (2005, pp. 464–466) gives examples of the use of the Priestley–Taylor equation to determine potential evapotranspiration of winter wheat, maize, and sorghum.

The Australian soil physicist, John Philip (1927–1999) [for a biography, see Kirkham (2005, pp. 168–170)], gave the Priestley lecture in 1997 (Philip 1998). In his lecture, Philip (1998) reminiscences about Priestley, whom he knew personally. He calls him "Bill." He said, "December 1946 was an important month for Bill Priestley and also for me. Both of us were in Melbourne. On December 23 Bill got off the ship and was having his first experience of Australia. Two days earlier I had been in the old Wilson Hall [at the University of Melbourne], having conferred on me the degree of Bachelor of Civil Engineering. In the aftermath of World War 2, the Suez Canal was closed and Bill and his bride Connie had suffered the six weeks rigor of the voyage of the *Dominion Monarch* from Southampton around the Cape. Bill was packed in a crowded cabin for 10 men; and his bride in a crowded cabin for 10 women. Bill had arrived into David Rivett's CSIR. Rivett's principle of scientific management was simple and effective: find the best man to head up the task; then give him the maximum freedom and help to get on with it. (Rivett recognized that, in uncorrupted English, words such as 'man' and 'him' serve for both genders when neither is specified.) Though its brief was based in science, the British Meteorological Office sat fair and square in the Civil Service, with its attendant rigidities. One can envisage how attractive it was to the 31-year-old-Priestley to be plucked from the ranks of the Met. Office and given carte blanche to set up a unit to pursue fundamental research into meteorology, even if it was at the other end of the world." Philip relates how Priestley, in 1953, was instrumental in setting up a Section of Hydrology in CSIRO, despite opposition to its establishment. Philip (1998) says, "It is ironic that it is only now, in 1997, that CSIRO has an entity, Land and Water, with essentially the brief [Priestley] proposed 43 years ago. Opposition to these ideas came from many soil and water people darkly suspicious of an approach with a strong flavor of physics and mathematics. The art of abstracting physically well-based entities from the real world that are amenable to mathematics was not much understood." Philip (1998) concludes his lecture by saying, "...in the spirit of Bill Priestley, we can ensure that our environmental research sits squarely and seriously in geophysics."

11 Elevated Atmospheric Carbon Dioxide: Water Use Efficiency

INTRODUCTION

In Kansas and other regions of the world, the proportion of water returned to the atmosphere by evaporation may be 90% or more (Rosenberg 1974, p. 160). Many schemes have been proposed to obtain water for these dry regions: artificial rainmaking, sea-water distillation, and towing icebergs to regions where the melted water can be used for irrigation. However, manipulations of evapotranspiration (ET) itself may be the best way to conserve water, because so much water is lost by ET. As we shall see in this chapter, one way to conserve water may be to grow plants under elevated levels of CO_2. Before we look at the effects of elevated levels of CO_2 on water conservation, we first define efficient water use and look at its historical aspects.

DEFINITIONS AND HISTORICAL ASPECTS OF EFFICIENT WATER USE

For centuries, man has been concerned with the efficient use of water in the production of crops. Indeed, the ability to grow crops and manage their needs for water probably was a requisite for civilization (Tanner and Sinclair 1983, p. 1). In this section, we want to give a general historical perspective of "efficient water use." We begin first by defining the term. Water use efficiency (WUE) is defined as the "biomass production per unit cropped area for a unit of water evaporated and transpired from the same area" (Tanner and Sinclair 1983, p. 2), or

$$WUE = \frac{\text{biomass produced}}{\text{water consumed}} \tag{11.1}$$

where biomass produced and water consumed to produce that biomass can each be expressed in grams (e.g., g dry matter/g H_2O).

Water use efficiency can be based either on the evapotranspiration (called the "ET efficiency") or the crop transpiration (called the "T efficiency"). The difference is important, because suppression of soil evaporation and prevention of weed transpiration can improve the ET efficiency. However, it need not improve the T efficiency, which is a measure of crop performance. These two water efficiencies also may be based on either the total dry matter production or the marketable yield, and the basis should always be stated. The reciprocal of transpiration efficiency is called the "water requirement" (WR) of plants or

$$WR = \frac{\text{water consumed}}{\text{biomass produced}} \tag{11.2}$$

and the units are g H_2O/g dry matter or kg H_2O/kg dry matter. The units cancel, so WR is sometimes expressed with no units.

The WR is also called the "transpiration ratio" (Briggs and Shantz 1913a, p. 7). The transpiration ratio is defined by meteorologists as "the ratio of weight of water transpired by a plant during its growing season to the weight of dry matter produced (usually exclusive of roots)" (Glickman 2000).

Briggs and Shantz (1913a,b), based in Akron, Colorado, were two of the pioneers in determining the WR of plants. They defined "WR" as "the ratio of the weight of water absorbed by a plant during its growth to the weight of dry matter produced" (Briggs and Shantz 1913a, p. 7). They grew plants in "[s]ubstantial cans of heavily galvanized corrugated iron" (40 cm in diameter and 66 cm tall) (Briggs and Shantz 1913a, p. 9). Containers of this size provided a soil mass adequate for the normal development of plants. Each can was fitted with a heavy, galvanized-iron cover through which holes were punched, suitable in size and number to accommodate the crop under consideration. The openings around the stems of the plants were sealed with wax. Therefore, the water lost from the cans was due to transpiration and not evaporation from the soil. The soil was a rich, dark loam (Briggs and Shantz 1913a, p. 18). At the top of the center of each container, they inserted a 13 cm (5 in.) diameter, unglazed, empty flowerpot with a drainage hole. The soil immediately around the lower part of the flowerpot was replaced by fine gravel. After the lid of the can was in place, a hole was made in the center of the lid and a cork was placed in it. The cork was in the center of the flowerpot. Watering was done by removing the cork, and water was added by means of 2 L flasks. A filled flask was inverted in a funnel placed in the opening in the center of the cover. This arrangement acted as a Mariotte system, keeping the water at a constant level in the receiving flowerpot until the flask was emptied (Briggs and Shantz 1913a, p. 11). Daily weighing was made of each can using a spring balance having a capacity of 150 kg.

In dry regions like Kansas, we want to increase WUE. As we can see from Equation 11.1, it can be increased by increasing the amount of biomass produced per unit of water used or, if the biomass produced remains the same, by decreasing the amount of water used to produce that biomass.

The increase in WUE in dry regions has been a goal of agriculturists for over a hundred years. Let us now return to the pioneering experiments in WUE. At the opening of the twentieth century, considerable work was in progress measuring WR, which is, as noted, the inverse of T efficiency. King (1899, 1901, 1905, 1914) at the University of Wisconsin in Madison was the first in the United States to research the water required to produce field crops (Tanner and Sinclair 1983, p. 3). [King (1911) was also the author of a classic work entitled *Farmers of Forty Centuries*, which described his trip to China in the early 1900s. He made careful notes to see how people could farm the same fields for 4000 years without destroying their productivity. The book has been reprinted by Rodale Press, Emmaus, Pennsylvania, and is available for reading.] King used small lysimeters in a greenhouse and in fields to determine the WR of crops. Widtsoe at the University of Utah in Salt Lake City and Kiesselbach at the University of Nebraska in Lincoln initiated research in 1902 [(Tanner and Sinclair 1983, p. 3); for references, see the experiment station reports for these states]. [Widtsoe (1911) is also well known for his studies on dry farming. The methods that he described for agricultural production in areas with low rainfall apply equally well today as they did at the beginning of the twentieth century.]

Briggs and Shantz (1913a) compiled much of the early WR data from their container experiments beginning in 1910 (Tanner and Sinclair 1983, p. 3). The WR work of Kiesselbach and of Briggs and Shantz was concerned with total dry matter and transpiration on a plant basis rather than on a land-area basis. Briggs and Shantz (1913b) and Kiesselbach (1916) reviewed the work that had been done up to that time. They omitted water-culture experiments and those on seedlings. They found that 14 researchers had worked on WUE prior to 1900. Since that time, there has been an increasing amount of work on water use and WUE, all done with more elaborate equipment and experiments than were done in the 1800s. However, these early experiments still remain the foundations of our modern understanding of crop production. One of the best summaries of the WR of many different plants is that compiled by Shantz and Piemeisel (1927).

The early workers considered the effect of *soil water content*, *climate*, *plant nutrition*, and *pests* on the WR. Let us see what their conclusions were for each of these.

SOIL WATER CONTENT

Briggs and Shantz (1913b) felt that they could make no firm conclusion about the effect of soil water content on the WR. In some experiments, however, the WR usually decreased a little when growth was limited by water deficits. The same result was obtained by Kiesselbach (1916), who found that, with a deficient water regime, the dry matter of corn (*Zea mays* L.) stalks, ears, and leaves was reduced 37%, 28%, and 10%, respectively. The WR based on ear weight decreased 4.3% and the WR based on the total dry matter decreased 10%. Later, Briggs and Shantz (1917) showed that the WR of wheat (*Triticum aestivum* L.) was essentially the same under deficient soil water (dry matter decreased 40%) as when water was adequate. King (1914) reached the same conclusion from experiments with corn, oats (*Avena sativa* L.), and potato (*Solanum tuberosum* L.). Several decades later, de Wit (1958) analyzed experimental data that also showed similar T efficiencies regardless of water shortages.

CLIMATE

The seasonal changes in both T and ET efficiencies found with exposures to different radiation, temperature, and humidity regimes indicated to many early researchers that climate exerted a major influence on water use efficiencies (Tanner and Sinclair 1983, p. 4). However, no systematic relations between T and ET efficiencies and the climate were developed until 1910. Kiesselbach (1910), Kiesselbach and Montgomery (1911), and Montgomery and Kiesselbach (1912) showed that daily and weekly transpiration from corn was correlated with evaporation from 1 gal (3.8 L) glazed, stone jars filled with water. Although most of their experiments were out of doors, one attempt to control environmental conditions is of interest. Montgomery and Kiesselbach (1912) grew corn in two greenhouses with one kept dry and ventilated and the other kept humid by wetting floors and atomizing water into the air. These treatments extended through June and July (summer months). The relation of WR to saturation deficit deduced from Montgomery and Kiesselbach's data by Briggs and Shantz (1913b) was not recognized by Montgomery and Kiesselbach, who sought a relation with relative humidity. However, Briggs and Shantz recognized early that, with well-watered plants, transpiration variation with climate caused changes in WUE as the season progressed between years and between locations and that saturation deficit or free-water evaporation (pan evaporation) (Tables 11.1 and 11.2) could be used to normalize the transpiration component.

TABLE 11.1
Data Recalculated by Briggs and Shantz (1913b)
from Montgomery and Kiesselbach (1912)

Parameter	Humid	Dry	Humid/Dry
Water requirement	214	340	0.63
"Free-water" evaporation, cm	9.4	16.8	0.56
Mean saturation deficit, mbar	19.0	31.3	0.61

Source: Reprinted from Tanner, C.B. and Sinclair, T.R., Efficient water use in crop production: Research or re-search? In Taylor, H.M., Jordan, W.R., and Sinclair, T.R., Eds., *Limitations to Efficient Water Use in Crop Production*, American Society of Agronomy, Crop Science Society of America, and Soil Science Society of America, Madison, WI, 1983, pp. 1–27, Table 1. With permission. Copyright 1983, American Society of Agronomy.

Note: WR can be normalized by free-water evaporation or saturation deficit.

TABLE 11.2
Water Requirement of Grimm Alfalfa at Different Stations in the Great Plains

Location	Growth Period (1912)	WR	Pan Average (mm/day)	WR/Pan
Williston, North Dakota	July 29–September 24	518 ± 12	4.04	128
Newell, South Dakota	August 9–September 6	630 ± 8	4.75	133
Akron, Colorado	July 26–September 6	853 ± 13	5.74	149
Dalhart, Texas	July 26–August 31	$1,005 \pm 8$	7.77	129

Source: Reprinted from Tanner, C.B. and Sinclair, T.R., Efficient water use in crop production: Research or re-search? In Taylor, H.M., Jordan, W.R., and Sinclair, T.R., Eds., *Limitations to Efficient Water Use in Crop Production*, American Society of Agronomy, Crop Science Society of America, and Soil Science Society of America, Madison, WI, 1983, pp. 1–27, Table 2. With permission. Copyright 1983, American Society of Agronomy.

Note: Data are from Briggs and Shantz (1917).

PLANT NUTRITION

Briggs and Shantz (1913b) reviewed data from about 20 experiments showing the effect of varying fertility on WR and concluded that with poor soils, the WR may be reduced one-half to two-thirds by increasing fertility (Tanner and Sinclair 1983, p. 5). In all but five experiments, evaporation from the soil was included in the water use and could bias the WR for smaller plants with less transpiration. In experiments where evaporation was prevented, the WR did not increase significantly as the fertility and yield decreased, until the dry weight of the plants had decreased approximately 50% because of malnutrition. It appears from these data that, unless malnutrition is fairly severe, T efficiency based on total dry matter is not greatly affected. However, with severe nutrient deficiencies, T efficiency decreased markedly. Kiesselbach (1916) provided extensive data from corn grown in large, sealed containers filled with well-fertilized to severely deficient soils showing the same trend. The T efficiency changed little, if at all, until nutrient deficiency reduced yield to about half that on the well-fertilized soil. de Wit (1958) arrived at a similar conclusion from other data.

PESTS

Tanner and Sinclair (1983) found few data on the effects of disease and insect pests on WUE. Rusts decrease transpiration in the early stage of infection but, on sporulation, cause epidermal leakage of water vapor and increase water loss in light and dark. (Water loss from stomata occurs most often in light when stomata usually are open.) The powdery mildews cause a similar decrease in transpiration and then an increase of "transpiration." These "transpiration" trends have been found to decrease the T efficiency. For example, Johnston and Miller (1934) found that the T efficiency (total dry matter base) of wheat was halved with early rust infection and when based on grain yield was 25-fold smaller. Infection at later stages of development produced correspondingly smaller decreases. WUE data for vascular wilts and *Fusarium* root rot are lacking, but the limited data suggest that T efficiency may be changed little. Yield is decreased with little change in T efficiency. Tanner and Sinclair (1983, p. 6) found no reports on how WUE was affected by insect pests.

In sum, the early workers showed that T efficiency (or WR, the inverse of T efficiency) is apparently the same for drought-stressed and well-watered crops, varies with climate, changes little with fertility unless plants are severely malnourished, and appears to be little affected by pests, even though data are lacking to confirm this observation.

These pioneering experiments showed that it is hard to change WUE. This is not surprising, because growth is directly related to stomatal opening (see Figure 9.12). The more open that the stomata are, the more CO_2 can be taken up for growth. But, at the same time, more water will be lost through transpiration with open stomata. If the stomata are closed, less CO_2 is taken up, less water is lost, but photosynthesis is reduced too. So biomass and transpiration appear tightly linked, and it would seem hard to change the ratio as defined in Equation 11.1 (WUE) or Equation 11.2 (WR). However, the type of photosynthetic system that plant has (C_3 or C_4) is a strong determiner of WUE, and we now turn to the WR of C_3 and C_4 plants.

WATER REQUIREMENT OF C_3 AND C_4 PLANTS

Table 11.3 shows the WR of various crops, as determined by Briggs and Shantz (1913b). I have added to the table which plants have the C_3 photosynthetic system and which plants have the C_4 photosynthetic system. (See Section "Carbon Isotope Ratios of Plants with Different Photosynthetic Pathways" in Chapter 6 for the difference in these photosynthetic systems. Compare also Chapters 12 and 13). The C_4 photosynthetic system was not discovered until the mid-1960s (El-Sharkawy and Hesketh 1965; Kortschak et al. 1965; Hatch and Slack 1966), decades after Briggs and Shantz did their work on WR. However, from Briggs and Shantz's early work, we can see that plants can be divided into two groups: those with WRs greater than 350 g H_2O/g dry matter and those with WRs less than 350 g H_2O/g dry matter. The C_3 plants are the ones with WRs greater than 350 g H_2O/g dry matter, and the C_4 plants are the ones with WRs less than 350 g H_2O/g dry matter. The one exception

TABLE 11.3
Water Requirement for Various Crops in Akron, Colorado

Crop (Common Name)	Crop (Scientific Name)	C_3 or C_4	Water Requirement
Wheat	*T. aestivum* L. (common), *T. dicoccum* Schrank (emmer), and *T. durum* Desf. (durum)	C_3	507
Oats	*A. sativa* L.	C_3	614
Barley	*Hordeum vulgare* L.	C_3	539
Rye	*Secale cereale* L.	C_3	724
Corn	*Z. mays* L.	C_4	369
Sorghum	*S. bicolor* (L.) Moench	C_4	306
Millet, foxtail	*S. italica* Beauv. (formerly, *Chaetochloa italica* Scribn.)	C_4	275
Peas	*Pisum sativum* L.	C_3	800
Sweet clover	*Melilotus* sp.	C_3	709
Alfalfa	*Medicago sativa* L.	C_3	1068
Buckwheat	*Fagopyrum esculentum* Moench	C_3	578
Rape	*B. napus* L.	C_3	441
Potatoes	*S. tuberosum* L.	C_3	448
Sugar beets	*Beta vulgaris* L.	C_3	377
Russian thistle (a weed)	*S. kali* L. var. *tenuifolia* Tausch (formerly, *S. pestifer* Nels.)	C_4	336
Amaranth (tumbleweed or pigweed)	*A. graecizans* L.(tumbleweed) or *A. retroflexus* L. (pigweed)	C_4	303
Artemisia	*Artemisia frigida* Willd.	C_3	765

Source: Data in last column are from Briggs, L.J. and Shantz, H.L., *The Water Requirement of Plants. II. A Review of the Literature*, U.S. Department of Agriculture Bureau of Plant Industry Bulletin No. 285, U.S. Department of Agriculture, Washington, DC, 1913b, Table LXXVI, last column, p. 90.

Note: Values are based on dry matter, not grain, yield (g H_2O/g dry matter).

is corn (Table 11.3), which has a slightly higher WR than 350 g H_2O/g dry matter (369 g H_2O/g dry matter). This value (369 g H_2O/g dry matter) is an average value, and some of the corn varieties that Briggs and Shantz studied did have WRs less than 350 g H_2O/g dry matter (Briggs and Shantz 1913a, p. 47). Briggs and Shantz (1913a) noted the unusually low WR of corn, sorghum [*Sorghum bicolor* (L.) Moench], millet (*Setaria italica* Beauv.), the weed Russian thistle (*Salsola pestifer* Nels.), and the amaranthine weeds (*Amaranthus retroflexus* L. or pigweed and *A. graecizans* L. or tumbleweed) (see Briggs and Shantz 1913a, pp. 25, 27, 28, and 34, respectively, for the three crops and weeds) compared to other crops that they studied. Corn, sorghum, millet, Russian thistle, and the amaranthine weeds are all C_4 crops. Briggs and Shantz say concerning the weeds (1913a, p. 34) as follows: "Probably more water is needlessly lost through the growth of these and similar weeds than from any other cause. When soil moisture is available, they make a rapid and luxuriant growth, and the water consumed by them is a complete loss, except insofar as they contribute organic matter to the soil. Sorghum, corn, or millet could take the place of these weeds with no greater consumption of soil moisture." Thus, even Briggs and Shantz (1913a, p. 34) unknowingly placed these C_4 plants into one group, the one with the highest water economy.

Briggs and Shantz (1913a) did not consider the effect of atmospheric CO_2 concentration on WR. But in their review of the literature, they noted (Briggs and Shantz 1913b, pp. 65–66) that Sorauer (1880) grew rape (*Brassica napus* L.) in free air and under two bell jars. In one of the bell jars, he placed a solution of "potassium hydrate" (probably potassium hydroxide) to reduce the CO_2 of the air. His results showed that the WR was increased by reducing the CO_2 content of the atmosphere. The WRs of rape in free air, under the bell jar, and under the bell jar with potassium hydrate were 282 ± 11, 243 ± 5, and 355 ± 5 g H_2O/g dry matter, respectively. When the CO_2 was reduced in the atmosphere, the WR increased. No experiments with elevated CO_2 were reported by Briggs and Shantz (1913b).

Since the work of Briggs and Shantz, the WR of C_3 and C_4 plants has been compared. In general, the WR for C_3 plants ranges between 450 and 950 g H_2O/g dry matter, and for C_4 plants, it ranges between 250 and 350 g H_2O/g dry matter (Salisbury and Ross 1978, p. 165).

The reasons that C_4 plants have a lower WR than C_3 plants are probably due to two facts: (1) the photorespiration (respiration in the light) of C_3 plants, which is generally lacking in C_4 plants, and (2) the higher stomatal resistance of C_4 plants compared to C_3 plants. Let us consider each of these.

PHOTORESPIRATION

As noted previously (Chapter 6), in C_4 plants, C_3 photosynthesis is carried out in the bundle sheath cells and C_4 photosynthesis is carried out in the mesophyll cells of the leaves. High rates of photorespiration are restricted to C_3 plants (Siedow and Day 2000, p. 718). Photorespiration is lacking in C_4 plants, except in the bundle sheath cells, where the C_3 photosynthesis takes place (Salisbury and Ross 1978, p. 165). Let us now look at photorespiration. In C_3 plants, when they are in the light, regular respiration is suspended and it is replaced by a different process called photorespiration (Zelitch 1967). When C_3 leaves are darkened, respiration is preceded by outbursts of CO_2 evolution. Because these outbursts occur only after a period of illumination, it is thought that they represent an overshoot of the greater CO_2 evolution that occurs in light. The effect of oxygen on photorespiration also suggests that light respiration and dark respiration are two different processes. Oxygen has no effect on dark respiration, but it stimulates photorespiration. Plants with C_4 photosynthesis, such as corn, do not have a CO_2 burst in darkness and do not produce CO_2 in light at any oxygen concentration. In most C_3 plants, photosynthetic rates decline as much as 50% when the oxygen concentration is doubled from the ambient value of 21% (Siedow and Day 2000, p. 716). Conversely, photosynthesis is stimulated as much as twofold when the oxygen concentration is decreased to less than 2%. In C_3 plants, the CO_2 compensation point increases with increasing oxygen concentration. [The CO_2 compensation point is the concentration of external CO_2 at which net photosynthetic CO_2 exchange equals zero (Siedow and Day 2000, pp. 716–718).] These results suggest a competition between oxygen and CO_2 during photosynthesis. The CO_2 compensation point for C_3 plants ranges between 20 and 100 μmol/mol, and for C_4 plants it ranges between 0 and 5 μmol/mol (Siedow and Day 2000, p. 718).

The origin of photorespiration is found in the kinetic properties of Rubisco (ribulose 1,5 bisphosphate carboxylase/oxygenase), the enzyme that catalyzes the uptake of CO_2 in the initial reaction of photosynthetic carbon metabolism in C_3 plants (see Chapters 6, 12, and 13). The first step in photorespiration is associated with the oxygenase activity of Rubisco (Siedow and Day 2000, p. 727). Phosphoglycolate formed during the oxygenase reaction is metabolized through a photorespiratory carbon cycle to save 75% of the carbon in the form of phosphoglycerate. The remaining 25% is lost as CO_2 (Siedow and Day 2000, p. 727). The breakdown of glycolate does not occur in C_4 plants (Zelitch 1967, p. 239). The CO_2 in C_4 plants is recycled within the leaf and is not lost as CO_2. Differences in photorespiration among species may account for the high photosynthetic efficiency of plants that do not photorespire [C_4 plants like corn, sugar cane (*Saccharum officinarum* L.), and sorghum] compared with plants that do photorespire [C_3 plants like tobacco (*Nicotiana tabacum* L.), soybean (*Glycine max* Merr.), and wheat]. The dry matter production for C_3 plants averages 22 ± 0.3 t/ha/year, and for C_4 plants it averages 39 ± 17 t/ha/year (Salisbury and Ross 1978, p. 165).

STOMATAL RESISTANCE

When C_3 and C_4 plants are compared under the same growing conditions, the C_4 plants often have a higher stomatal resistance and a lower transpiration rate than the C_3 plants. The experiment of Rachidi et al. (1993a) has already been mentioned in Chapters 8 and 10. They grew sunflower (*Helianthus annuus* L., C_3) and grain sorghum [*S. bicolor* (L.) Moench, C_4] on a silt loam soil near Manhattan, Kansas, under rain-fed conditions for 2 years. Sorghum had an average stomatal resistance that was 1.7 times higher than that of sunflower. This resulted in the transpiration rate of sunflower being 14% higher than that of sorghum. Others have found similar results, when C_3 and C_4 plants have been compared (Rawson and Begg 1977; Allen et al. 1985, p. 16).

Taken together (higher productivity of C_4 plants and higher stomatal resistance of C_4 plants), we can see why the C_4 plants have a higher WUE than C_3 plants. Both the numerator and denominator are changed in Equation 11.1 by the physiological characteristics of C_4 plants, and they result in a high WUE. The information in this section also indicates that the WUE of C_3 plants might be increased by increasing the CO_2 concentration. That is, C_3 plants might have water use efficiencies similar to C_4 plants, if the CO_2 concentration in the atmosphere is increased. Let us now look at the data for WUE, when C_3 and C_4 plants have been exposed to elevated levels of CO_2.

WATER USE EFFICIENCY UNDER ELEVATED CO_2

C_3 CROP (WHEAT)

Chaudhuri et al. (1990) grew winter wheat (*T. aestivum* L. cv. Newton), a C_3 crop, in the field under ambient (340 μmol/mol) and elevated levels (485, 660, 825 μmol/mol) of CO_2 during three growing seasons (1984–1987), in 16 underground boxes, as described in Chapters 1 and 6. The soil was a silt loam. Water in half of the boxes was maintained at $0.38\,m^3/m^3$ (high water level) and in the other half between 0.14 and $0.25\,m^3/m^3$ (low water level). Boxes were weighed to determine the amount of water used by transpiration. Grain yield of the high-water-level wheat grown under ambient CO_2 was about the same as the grain yield of low-water-level wheat grown at the highest level of CO_2 (825 μmol/mol) (3 year means: 725 and 707 g/m^2, respectively). Water transpired, indicated by water used (determined by weighing the underground boxes), increased as the level of CO_2 increased (Figure 11.1), although the amount varied by year. High-water-level wheat grown with 825 μmol/mol CO_2 transpired more water than low-water-level wheat grown under ambient levels of CO_2 (3 year means: 453 and 370 L/m^3, respectively). This was probably because leaf growth increased

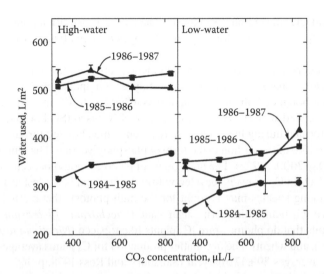

FIGURE 11.1 Winter wheat water use from planting to harvest supplied with high and low water levels as affected by CO_2 concentration during a 3 year study. Vertical bars = ±standard deviation. Only half of each bar is drawn, for clarity. (Reprinted from Chaudhuri, U.N., Kirkham, M.B., and Kanemasu, E.T., *Agron. J.*, 82, 637, 1990, Fig. 2. With permission. Copyright 1990, American Society of Agronomy.)

under the elevated CO_2 concentration, resulting in more water transpired. However, Chaudhuri et al. (1990) did not measure leaf area. Other studies have shown little change in crop ET in response to rising CO_2, because the effect of increase in individual leaf resistance is approximately offset by the increase in crop leaf area (crop surfaces available for transpiration) (Allen et al. 1985, p. 25; see also Chapter 10). Rising CO_2 is likely to result in little reduction of crop ET unless breeders provide crops that produce less leaf area under high CO_2 (Allen et al. 1985, p. 25).

The amount of water required to produce a gram of grain, as related to CO_2 concentration, was calculated from water used and grain yield for each CO_2 level by year (Chaudhuri et al. 1990). As noted, WUE is the reciprocal of the WR. The WR for grain production decreased as CO_2 concentration increased (Figure 11.2). During the 3 years of the study, water required to produce a gram of grain under high CO_2 (825 µmol/mol) and low-water-level conditions was less (547 mL/g) than that required to produce a gram of grain under high-water-level conditions at ambient CO_2 (642 mL/g). Under high water levels, the WR was reduced by 29% when the CO_2 level was raised from ambient (3-year-average WR = 642 mL/g) to 825 µmol/mol (WR = 458 mL/g). Under low water levels, the WR was reduced by 31% when the CO_2 level was raised from ambient (WR = 797 mL/g) to 825 µmol/mol (WR = 547 mL/g). The results showed that the WUE of wheat was increased, or the WR was decreased, under elevated CO_2 at both water levels.

The fact that water used by the wheat was not decreased under elevated CO_2, but WR was decreased, was an important result. Stomatal resistance increased under the elevated CO_2 (Chaudhuri et al. 1986b, 1987; the data are given in Tables 7.2 and 7.3 along with the water potential data). The elevated CO_2 increased the grain yield both under the high water and low water conditions (Chaudhuri et al. 1990). So the denominator in Equation 11.2 increased (decreasing WR), and stomatal closure decreased the numerator in Equation 11.2 (decreasing the WR). Together, the overall WR was decreased. This suggests that for a C_3 plant like wheat, WR is reduced under elevated CO_2 both because of an increase in biomass production and a decrease in water transpired.

C_4 Crops (Sorghum and Big Bluestem)

In 1984, Chaudhuri et al. (1986a) grew grain sorghum [*S. bicolor* (L.) Moench, hybrid Asgro Correl], a C_4 crop, under well-watered conditions in the field under ambient (330 µmol/mol) and

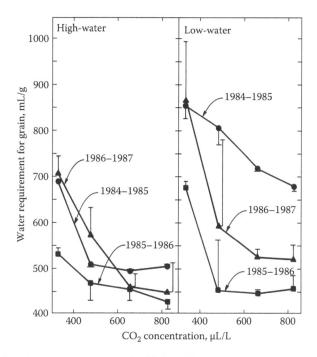

FIGURE 11.2 WR for winter wheat grown under high and low water levels as affected by CO_2 concentration during a 3 year study. Vertical bars = ±standard deviation. Only half of each bar is drawn, for clarity. (Reprinted from Chaudhuri, U.N., Kirkham, M.B., and Kanemasu, E.T., *Agron. J.*, 82, 637, 1990, Fig. 3. With permission. Copyright 1990, American Society of Agronomy.)

elevated levels (485, 660, 795 μmol/mol) of CO_2 in the 16 underground boxes, as described above and in Chapters 1 and 6. The WUE increased with increasing levels of CO_2 (Table 11.4). When the CO_2 concentration was increased by 2.4 times (330–795 μmol/mol), the WUE increased, or the amount of water needed to produce a gram of grain decreased. For example, it took only 776 g of water to produce a gram of grain at 795 μmol/mol CO_2, while, at 330 μmol/mol CO_2, it

TABLE 11.4
Water Used to Produce a Gram of Different Yield Components of Sorghum (g of H_2O/g) (i.e., the WR) as Affected by Ambient (330 μmol/mol) and Enriched Atmospheric CO_2

CO₂ (μmol/mol)	Leaf	Stem	Chaff	Vegetative Shoot[a]	Grain	Root	Total Phytomass[b]
330	1490	1111	2968	525	1090	3893	325
485	1475	774	2805	430	961	3802	276
660	1273	750	2142	388	826	2907	242
795	1303	776	2097	395	776	2682	238

Source: Reprinted from *Agric. Forest Meteorol.*, 37, Chaudhuri, U.N. et al., Effect of carbon dioxide on sorghum yield, root growth, and water use, 109–122, Table VI, Copyright 1986a, with permission from Elsevier.

Note: Experiment was done in 1984.

[a] Vegetative shoot includes leaf, stem, and chaff.

[b] Total phytomass includes vegetative shoot, grain, and root.

took 1090 g of water to produce a gram of grain. Therefore, the results suggested that with higher levels of CO_2 in the atmosphere, more sorghum grain can be produced with the same amount of water.

Using the data from Chaudhuri et al. (1986a) and Chaudhuri et al. (1990), we can calculate what the WRs of wheat and sorghum are a quarter century later, based on the CO_2 concentration in atmosphere in 2008, which is 385 μmol/mol (Dlugokencky 2009) and using a linear relationship for the data presented in Figure 11.2 (for wheat) and Table 11.4 (for sorghum). Wheat and sorghum are now using 41 and 43 mL, respectively, less water to produce a gram of grain than they were in the 1984–1987 period. If we assume that the WR of wheat grain under well-watered conditions is 642 mL/g (average of the values for ambient conditions, 1984–1987; Figure 11.2), then 41/642 = 0.06, or the WR to produce a gram of wheat grain has been reduced by 6% due to the 55 μmol/mol increase in CO_2 since 1984. If we assume that the WR of sorghum grain under well-watered conditions is 1090 mL/g (Table 11.4), then 43/1090 = 0.04, or the WR to produce a gram of sorghum grain has been reduced by 4% due to the 55 μmol/mol increase in CO_2. These may seem like small percentages. But spread over the large acreage of wheat and sorghum grown in Kansas, for example, the more efficient use of water now (2010) compared to the mid-1980s should have a large impact. In 2009, the acres of wheat and sorghum harvested in Kansas were 8,800,000 acres (3,520,000 ha) and 2,700,000 acres (1,080,000 ha), respectively (Kansas Department of Agriculture 2009).

Results for WR, similar to those for sorghum and wheat, were observed for big bluestem (*Andropogon gerardii* Vitman), a C_4 rangeland plant in the mid-continental grasslands of North America (Kirkham et al. 1991). It grew in a tallgrass prairie on a silty clay loam kept at a high water level (field capacity) or a low water level (half field capacity). Sixteen cylindrical plastic chambers were placed on the prairie to maintain two levels of CO_2 (337 μmol/mol and double that amount or 658 μmol/mol) over a full growing season. Photosynthesis and transpiration were determined with a portable leaf photosynthetic system. WR was calculated by dividing leaf transpiration rate (μmol/m²/s) by leaf photosynthetic rate (μmol/m²/s). This WR is based on leaf measurements, instead of measurements of water use and growth taken during an entire season. It is sometimes called the "instantaneous WR." Its inverse is called the "photosynthetic WUE" or the ratio of photosynthetic CO_2 exchange rate to transpiration rate (Allen et al. 1985, p. 16). Doubled CO_2 reduced transpiration rate of big bluestem by 25% and 35% under the high and low water treatments, respectively. CO_2 concentration did not affect the rate of photosynthesis under either water treatment. Under the high water treatment, the WRs averaged 162 and 250 μmol H_2O/μmol CO_2 during the season for the high and the low CO_2 concentrations, respectively, and, under the low water treatment, the values were 165 and 248 μmol H_2O/μmol CO_2, respectively. Thus, elevated CO_2 reduced the leaf WR by about 35% for both watering regimes. The results suggested that as the CO_2 concentration in the atmosphere increases, big bluestem on the tallgrass prairie in the Great Plains will require less water for the same productivity.

Under both the high and low water levels, stomatal resistance of the big bluestem was increased by elevated CO_2, which reduced the transpiration rate (Kirkham et al. 1991). So in Equation 11.2, when we consider the data for big bluestem, we know the numerator was changed because of the reduction in transpiration under elevated CO_2. However, under both the high and low water levels, elevated CO_2 did not increase photosynthetic rate for big bluestem. Therefore, the decrease in WR for this C_4 plant came from changing only the numerator in Equation 11.2.

C_3 AND C_4 PLANTS GROWN TOGETHER

In the work reported in the previous two sections (Chaudhuri et al. 1990, 1986a; Kirkham et al. 1991), the WUE of C_3 and C_4 plants grown together was not determined. He et al. (1992) in Kansas did measure photosynthetic rate and transpiration rate of big bluestem (C_4) and Kentucky bluegrass (*Poa pratensis* L.) (C_3) growing together in a tallgrass prairie in Kansas under two concentrations of CO_2 (337 µmol/mol or ambient and 658 µmol/mol or twice ambient). As noted in Chapter 10 (see the section "C_3 and C_4 Plants Compared"), these plants often do not grow at the same time of year, because the hot summer temperatures in Kansas favor the growth of the warm-season grass, big bluestem, and the cooler spring and fall temperatures favor the growth of the cool-season grass, Kentucky bluegrass. However, during the fall months, both grasses were growing, when He et al. (1992) took measurements on them. They calculated WR by dividing transpiration rate (µmol/m²/s) by photosynthetic rate (µmol/m²/s) (the instantaneous WR). For big bluestem, the WRs averaged 161 and 228 µmol H_2O/µmol CO_2 between September 20 and October 23, 1989, for the high and low CO_2 concentrations, respectively. For Kentucky bluegrass, these values were 344 and 888 µmol H_2O/µmol CO_2 for the high and low CO_2 concentrations, respectively. Thus, higher CO_2 reduced the WRs by 41.6% and 158% for big bluestem and Kentucky bluegrass, respectively. The results suggest that as the CO_2 concentration in the atmosphere increases, big bluestem and Kentucky bluegrass on the tallgrass prairie in Kansas will have increased water use efficiencies and that the WUE of the C_3 grass will be increased more than the WUE of the C_4 grass. In this case, the WR of the C_3 plant was reduced more than that of the C_4 plant, because elevated CO_2 increased the photosynthetic rate of only the C_3 plant, not the C_4 plant. Doubled CO_2 reduced the transpiration rate of both grasses (see Figure 10.2), but big bluestem was affected more than Kentucky bluegrass.

MAIZE, COTTON, SOYBEANS, RYEGRASS, AND WHITE CLOVER

Allen et al. (1985, pp. 16–25) review WUE of crops grown under elevated CO_2. Almost all data show an increase in WUE when plants are exposed to elevated levels of CO_2. They compared crops grown under an ambient CO_2 concentration of 330 µmol/mol CO_2 with those grown under twice the ambient level, 660 µmol/mol. The crops included maize (corn) (C_4), cotton (*Gossypium hirsutum* L.) (C_3), and soybeans [*G. max* (L.) Merr.] (C_3). Only about a quarter of the increase in WUE of maize under elevated CO_2 was due to increased photosynthetic rate and about three-quarters was due to reduction in transpiration rate. In contrast, about 60% or more of the increase in WUE for the C_3 plants (cotton, soybean) was due to an increase in photosynthetic rate. In all crops (C_3 and C_4), the WUE was about doubled when the CO_2 concentration was doubled. Nijs et al. (1989) also found for perennial ryegrass (*Lolium perenne* L.) (C_3) and white clover (*Trifolium repens* L.) (C_3) that WUE was roughly doubled when the CO_2 concentration was almost doubled (from 358 to 626 µmol/mol CO_2). The calculations of Allen et al. (1985) support the notion that C_4 plants respond to increasing CO_2 mainly by increasing stomatal resistance and reducing the transpiration rate, while C_3 plants respond mainly by increasing their photosynthetic rate.

Leaf area has a large influence on WUE (Allen et al. 1985, p. 21). Using data from Jones et al. (1985) for soybeans, Allen et al. (1985) created Figure 11.3, which shows the effect of leaf area index on WUE. Leaf area index is defined as the area of leaves above unit area of ground taking only one side of each leaf into account (Kirkham 2005, p. 457). Jones et al. (1985) grew soybeans under two concentrations of CO_2 (330 and 800 µmol/mol) and two leaf area indexes (6.0 and 3.3). The highest WUE occurred under high CO_2 and low leaf area index, and the lowest WUE occurred under low CO_2 and high leaf area index. Higher leaf area indexes associated with elevated CO_2 tend to increase the canopy transpiration and reduce the increase in WUE that would be expected from canopies with similar leaf area indexes (Allen et al. 1985, p. 21).

The biggest effect of elevated CO_2 on WUE appears to be mediated through its effect on enhancing photosynthesis and growth (Allen et al. 1985, p. 23), at least for C_3 plants. A secondary effect is reducing ET per unit ground area of the crop because of increased leaf area. The relative contributions of crop transpiration and direct evaporation at the soil surface to WUE need further investigation (Allen et al. 1985, p. 23). Allen et al. (1985, p. 23) say that increase in atmospheric CO_2 may be the only direct way of increasing crop WUE within a given climatic regime.

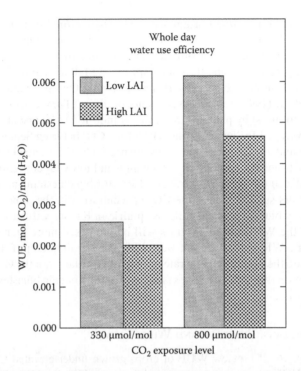

FIGURE 11.3 Effect of canopy leaf area on whole-day WUE of 'Bragg' soybeans grown at a daytime dry bulb air temperature of 31°C and a dewpoint air temperature of 21°C. The low leaf area index was 3.3 and the high leaf area index was 6.0. (Reprinted from Allen, Jr., L.H., Jones, P., and Jones, J.W., Rising atmospheric CO_2 and evapotranspiration, in *Advances in Evapotranspiration*, American Society of Agricultural Engineers, St. Joseph, MI, 1985, pp. 13–27, Fig. 1. With permission. Copyright 1985 American Society of Agricultural and Biological Engineers.)

TREES

Centritto et al. (1999b) grew cherry (*Prunus avium* L., a C_3 plant) from seed for two growing seasons in ambient (circa 350 µmol/mol) or elevated (ambient + circa 350 µmol/mol) CO_2 in six open-top chambers in a glasshouse at the University of Edinburgh in the United Kingdom. A further set of trees was maintained outside the chambers but inside the glasshouse to provide a control for the open-top chambers. The seedlings were in potting compost (a gravel, sand, peat, and loam-soil mixture). In the first growing season, gradual drought was imposed on rapidly growing cherry seedlings by withholding water for a 6 week drying cycle. In the second growing season, a rapid onset of drought was imposed at the height of the growing season on the seedlings, which had already experienced drought in the first growing season. Elevated CO_2 increased total dry-mass production in both watering regimes. Water loss did not differ in either well-watered or drought-stressed seedlings between elevated and ambient CO_2 (Figure 11.4a). Consequently, whole-plant WUE (the ratio of total dry mass produced to total water consumption) was increased under elevated CO_2 (Figure 11.4b).

Liang et al. (1995) determined the effect of elevated CO_2 on gas exchange and WUE of three temperate deciduous tree species [*Fagus crenata* Bl. (beech), *Gingko biloba* L. (ginkgo), and *Alnus firma* Sieb et Zucc. (alder)] under drought. Seedlings were grown within transparent open-top cabinets and maintained for 4 months at CO_2 concentrations of either 350 (ambient) or 700 µmol/mol (elevated). The plants were exposed to five watering regimes: leaf water potential, ψ_w, higher than −0.3 MPa (well watered), −0.5 and −0.8 MPa (moderate drought), −1.0 MPa, and lower than −1.2 MPa (extremely drought-stressed). Plants grew in pots filled with a mixture of vermiculite, volcanic granules, and native alluvial soil from Niigata, Japan. Increase in CO_2 concentration induced a 60%

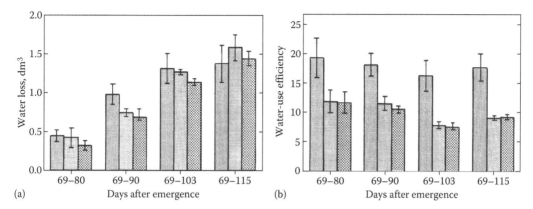

FIGURE 11.4 Total water loss (a) and plant WUE (g/kg) (b) of droughted cherry seedlings grown in ambient CO_2 (bar in center of each group), elevated CO_2 (bar at left in each group), or outside (control, bar on right in each group), during the first-growing-season drought cycle. Data are the means of three plants per treatment ± one standard error. (Centritto, M., Lee, H.S.J., and Jarvis, P.G., Interactive effects of elevated [CO_2] and drought on cherry (*Prunus avium*) seedlings. I. Growth, whole-plant water use efficiency and water loss, *New Phytol.*, 1999b, 141, 129–140, Fig. 7. Copyright Wiley-VCH Verlag GmbH & Co. KGaA. Reprinted with permission.)

average increase in net photosynthetic rate under well-watered conditions. The effect of the elevated level of CO_2 became more pronounced when drought stress was established, with an 80% average increase in the net photosynthetic rate at ψ_w as low as −0.8 MPa. Leaf conductance to water vapor transfer and transpiration rate (*E*) were decreased. Consequently, WUE increased under drought (Figure 11.5).

Centritto et al. (2002) examined the interactive effects of elevated CO_2 concentration and water stress on growth and physiology of 1-year-old peach (*P. persica* L.) seedlings grown in pots in open-top chambers with ambient (350 µmol/mol) or elevated (700 µmol/mol) CO_2. The pots contained a mixture of peat, sand, loam soil, grit, and lime fertilized with a nutrient solution. Water was withheld from half of the plants in each treatment for a 4 week drying cycle, to simulate a sudden and severe water stress during the phase of rapid plant growth. Throughout the growing season, seedlings in elevated CO_2 had a higher assimilation rate than seedlings in ambient CO_2, and this caused an increase in total dry mass of about 33%. Stomatal conductance, total water uptake, leaf area, and leaf number were unaffected by elevated CO_2. Because seedlings in the two CO_2 treatments had similar transpiration rates despite large differences in total dry mass, WUE of well-watered and water-stressed seedlings grown in elevated CO_2 was an average of 51% and 63% higher, respectively, than WUE of seedlings grown in ambient CO_2 (Figure 11.6).

Ceulemans and Mousseau (1994) reviewed literature on WUE of woody plants under elevated CO_2. They reported that plants grown with enriched CO_2 have a greater WUE than plants grown at preindustrial or ambient CO_2 concentrations. They also pointed out that, in water-limited environments, drought stress in plants might be ameliorated under elevated atmospheric CO_2. They said that because plant water use can be reduced by increased WUE with elevated CO_2, plants growing with enriched CO_2 would be more able to compensate for water stress.

In another review, Eamus (1991) presented data for WUE of 19 different species, many of them trees. He stated that the overwhelming majority of data support the view that elevated CO_2 increases WUE. But most research has been done over the short term and in controlled environments. He concluded that more studies are needed, particularly of mature trees where entire branches of trees in a woodland or forest are subjected to elevated CO_2 using a "branch cuvette" technique. These studies need to be done over long periods to determine seasonal variations of WUE.

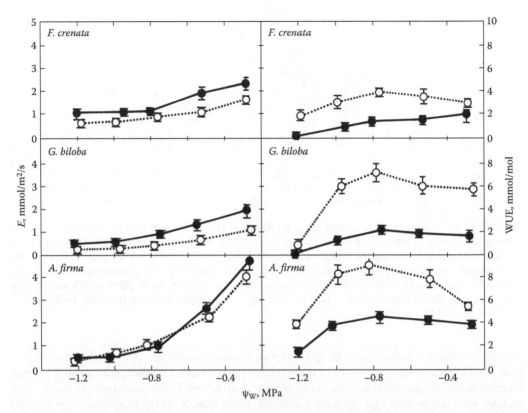

FIGURE 11.5 Response of transpiration rate (E, left) and WUE (WUE, right) to leaf water potential (ψ_w) for *F. crenata* Bl. (beech), *G. biloba* L. (ginkgo), and *A. firma* Sieb et Zucc. (alder) seedlings at 350 μmol/mol CO_2 (solid lines with closed symbols) and at 700 μmol/mol CO_2 (dotted lines with open symbols). Each point represents the mean of five replicate plants from each treatment. Each bar represents the standard error. (With kind permission from Springer Science+Business Media: *Photosynthetica*, Interactions of elevated CO_2 and drought stress in gas exchange and water-use efficiency in three temperate deciduous tree species, 31, 1995, 529–539, Liang, N., Maruyama, K., and Huang, Y., Fig. 3.)

CRASSULACEAN ACID METABOLISM SPECIES

Cui et al. (1993) determined CO_2 uptake, water vapor conductance, and WUE of *Opuntia ficus-indica* Mill. (Indian fig), a Crassulacean acid metabolism (CAM) species, at CO_2 concentrations of 370, 520, and 720 μmol/mol in open-top chambers during a 23 week period. Daily net CO_2 uptake for both basal and first-daughter cladodes was higher at higher CO_2 concentrations (Figure 11.7). Nine weeks after planting, daily net CO_2 at 520 and 720 μmol/mol was 76% and 98% higher, respectively, than at 370 μmol/mol. Eight weeks after daughter cladodes emerged, their daily net CO_2 uptake was 35% and 49% higher at 520 and 720 μmol/mol, respectively, than at 370 μmol/mol. CO_2 uptake for basal cladodes and first-daughter cladodes occurred chiefly at night (Figure 11.8a and b). Water vapor conductance was greatest at night, as is the case for CAM plants that open their stomata primarily at night (Salisbury and Ross 1978, p. 145). At 720 μmol/mol CO_2, water vapor conductance tended to be lower for basal cladodes (Figure 11.8c) and first-daughter cladodes (Figure 11.8d) during the early morning. Daily net CO_2 uptake that occurred during the daytime increased as the CO_2 concentration was increased for both basal and first-daughter cladodes (Figure 11.9a and b). Daily WUE was 88% higher under elevated CO_2 for basal cladodes (Figure 11.9c) and 57% higher for daughter cladodes (Figure 11.9d). During the first 9 weeks after planting, WUE of basal cladodes at 720 μmol/mol CO_2 was double that at

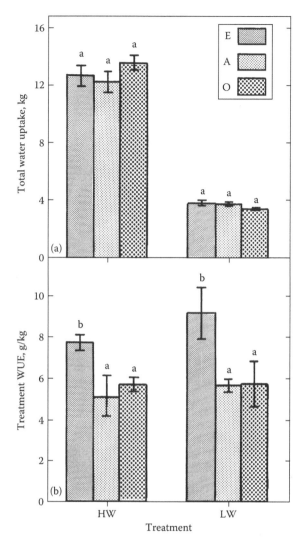

FIGURE 11.6 Mean (a) total water uptake (kg) and (b) whole-plant WUE (WUE; dry mass increment/total water use) of peach seedlings between days 370 and 399. *Abbreviations:* E, elevated CO_2; A, ambient CO_2; O, outside plot; HW, well watered; and LW, water stressed. Values are means of five plants per treatment ± 1 standard error of the mean. Different letters above columns indicate significant differences at P less than 0.05. (Reprinted from Centritto, M., Lucas, M.E., and Jarvis, P.G., *Tree Physiol.*, Fig. 5, 2002 by permission of the Oxford University Press.)

370 μmol/mol CO_2 (Figure 11.9c). For first-daughter cladodes, daily WUE increased with elevated CO_2. Twenty weeks after emergence, their WUE was 43% and 75% higher in 520 and 720 μmol/mol CO_2, respectively, compared with the 370 μmol/mol CO_2 (Figure 11.9d). The increased WUE was primarily because of increased daily net CO_2 uptake at higher CO_2 concentrations (Cui et al. 1993). The increased WUE was associated with higher biomass productivity of *O. ficus-indica*, which shows its potential in arid and semiarid regions, where it is extensively cultivated for fodder and fruit. The higher WUE together with increased CO_2 uptake under elevated CO_2 should enhance the productivity and economic usefulness of this species as atmospheric CO_2 concentrations increase in the future (Cui et al. 1993).

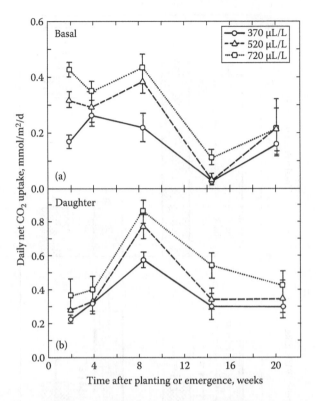

FIGURE 11.7 Daily net CO_2 uptake versus time after planting for basal cladodes (a) and versus time after emergence for first-daughter cladodes (b) grown under three CO_2 concentrations. Data are means ± standard error ($n = 5$ plants). (Reprinted from Cui, M., Miller, P.M., and Nobel, P.S., *Plant Physiol.*, 103, 519, 1993, Fig. 3. With permission. Copyright 1993 American Society of Plant Biologists.)

WATER USE EFFICIENCY OF FOSSIL AND HERBARIUM PLANTS

A large amount of time exists between the origin of land plants and the evolution of leaves with a flat blade (Kenrick 2001). For at least the first 40 million years of their existence, land plants were leafless or had only small spine-like appendages. The widespread appearance of megaphyll leaves, with their branched veins and planate form, did not occur until the close of the Devonian period about 360 million years ago (Beerling et al. 2001). (See Figure 9.10 for a figure of the geological time scale, which shows the Devonian period.) Beerling et al. (2001) proposed an explanation for this long delay. Drawing on anatomical and environmental data from the fossil record, they concluded that leaves with a broad lamina evolved in response to a massive drop in atmospheric CO_2 during the Late Devonian period. The evolution of the megaphylls coincides with a drop of around 90% in atmospheric CO_2 concentration (Beerling et al. 2001; Kenrick 2001). The fall was driven by plants. The evolution of land flora and, in particular, the advent of large plants, had a major impact on the carbon cycle by drawing CO_2 out of the atmosphere and locking it up in trunks and leaves (Kenrick 2001). Also important was the evolution of roots. The physical and chemical effects of root systems on rocks and soils increase rates of weathering, which is thought to have been responsible for removing large quantities of CO_2 from the Devonian atmosphere (Kenrick 2001).

Early Devonian land plants possessed few stomata. They occurred on leafless plants that possessed short, cylindrical, aerial stems (called "axes") or were on only minute leaves (microphylls) (Beerling et al. 2001). A planate leaf with the same stomatal characteristics would have suffered lethal overheating, because of greater interception of solar energy and low transpiration. When planate leaves first appeared in the Late Devonian and subsequently diversified in the Carboniferous

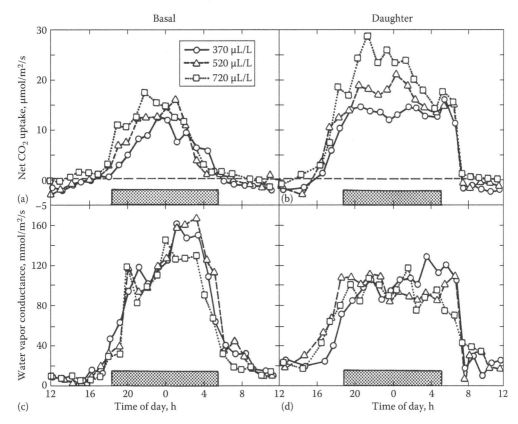

FIGURE 11.8 Daily time course of the instantaneous rate of net CO_2 uptake (a and b) and water vapor conductance (c and d) for basal and first-daughter cladodes 9 weeks after planting (June 4, 1992) and on first-daughter cladodes 8 weeks after emergence (June 18, 1992) and grown under three CO_2 concentrations. Data are means for five plants. Standard error values averaged 1.1 μmol/m²/s for CO_2 uptake and 0.19 mmol/m²/s for water vapor conductance. Stippled bars indicate night. (Reprinted from Cui, M. et al., *Plant Physiol.*, 103, 519, 1993, Fig. 1. With permission. Copyright 1993, American Society of Plant Biologists.)

period, they possessed substantially higher stomatal densitites. (The Carboniferous period is what the combination of the Pennsylvanian and Mississippian periods is called.) The evolution of planate leaves could only have occurred after an increase in stomatal density, allowing higher transpiration rates that were sufficient to maintain cool and viable leaf temperatures (Beerling et al. 2001).

Beerling et al. (2001) computed water use efficiencies (carbon gain per unit of water lost) using modeled gas exchange characteristics and compared the modeled values with independent estimates obtained using the stable carbon isotope composition ($\delta^{13}C$) of fossilized Palaeozoic terrestrial organic matter and a well-validated model relating plant WUE to $\delta^{13}C$ (Farquhar et al. 1989). (See Chapter 6 and the section "Carbon Isotope Ratios of Plants with Different Photosynthetic Pathways," which explains how $\delta^{13}C$ is determined and how it relates to WUE. We remember that there is an inverse relation between WUE and discrimination in C_3 plants. The lower the ratio or discrimination, the higher is the WUE.) Calculated this way, both approaches showed similar values and a pattern of declining plant WUE between the Late Silurian period (410–415 million years ago) and the Carboniferous (300–292 million years ago) (Figure 11.10). (Dates given by various authors for the times of these periods vary somewhat.)

The reason that the WUE dropped was probably due to the increased transpirational cooling that occurred with higher stomatal densities. The earliest preserved fossil megaphyll leaf cuticles date to the Carboniferous and have higher stomatal densities (e.g., 1000 mm⁻²) than early land plant

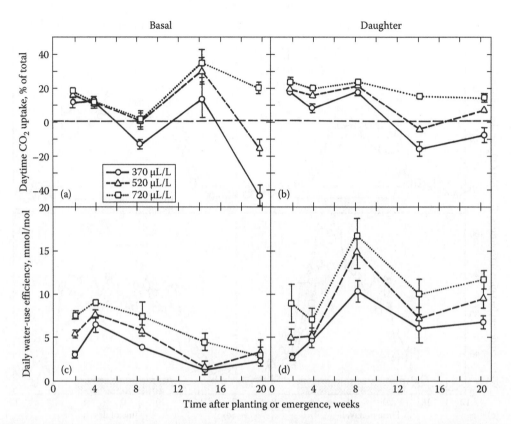

FIGURE 11.9 Daytime net CO_2 uptake as a percentage of total daily net CO_2 uptake (a and b) and daily WUE (c and d) versus time for basal cladodes (a and c) and first-daughter cladodes (b and d) and grown under three CO_2 concentrations. Data are the means ± SE ($n = 5$ plants). (Reprinted from Cui, M. et al., *Plant Physiol.*, 103, 519, 1993, Fig. 4. With permission. Copyright 1993, American Society of Plant Biologists.)

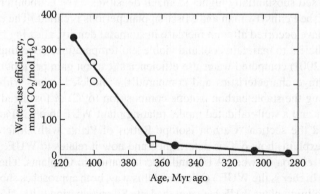

FIGURE 11.10 Simulated and isotopically derived changes in leaf and axis WUE during the Late Palaeozoic era. Age is given in million years. Black circles, estimate from soil organic $\delta^{13}C$; open circles, simulated Early Devonian axis; and open squares, simulated Early Carboniferous planate leaf. [An "axis" is a short, cylindrical aerial stem on early leafless plants.] (Reprinted by permission from Macmillan Publishers Ltd., *Nature*, Beerling, D.J., Osborne, C.P., and Chaloner, W.G., Evolution of leaf-form in land plants linked to atmospheric CO_2 decline in the Late Palaeozoic era, 410, 352–354, Fig. 2, Copyright 2001.)

axes (e.g., $10\,mm^{-2}$), a difference of two orders of magnitude (Beerling et al. 2001). Simulations for Late Devonian/Carboniferous leaves indicate transpiration rates sufficient to cool leaf temperatures well below the lethal range, even in the warm low latitudes (Figure 11.11). The reduced stomatal resistance of these planate leaves benefited photosynthetic productivity under the unusually low atmospheric CO_2/O_2 ratio by increasing CO_2 diffusion into leaf mesophyll, and thereby minimizing photorespiratory CO_2 evolution. This cooling was necessary for the evolution of laminate leaves.

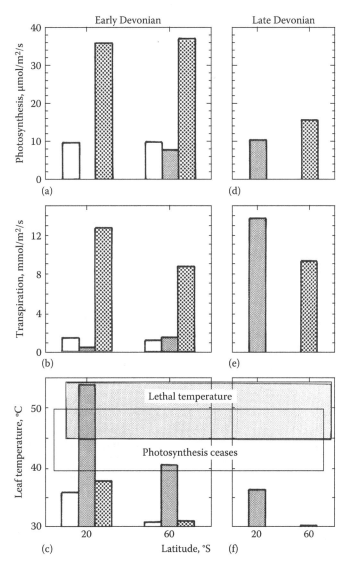

FIGURE 11.11 Simulated biophysical properties of leaves and axes in the Late Paleozoic environment. Midday photosynthetic rates (a), transpiration rates (b), and temperatures (c) for an axis with a low (white bars) stomatal density and planate leaves with high (stippled bars) or low (gray bars) stomatal density operating in a high-p_{CO_2} Early Devonian atmospheric at 60°S and 20°S. The absence of the low stomatal density planate leaf in (a) is because its temperature is lethal (see c). (d–f) Adjacent boxes show the same ecophysiological traits for a planate leaf with a high stomatal density in the lower p_{CO_2} atmosphere of the Late Devonian. Horizontal boxes in (c) and (f) indicate the temperature range at which CO_2 uptake by photosynthesis ceases and the lethal range for extant warm temperature and tropical species. (Reprinted by permission from Macmillan Publishers Ltd., *Nature*, Beerling, D.J., Osborne, C.P., and Chaloner, W.G., Evolution of leaf-form in land plants linked to atmospheric CO_2 decline in the Late Palaeozoic era, 410, 352–354, Fig. 1, Copyright 2001.)

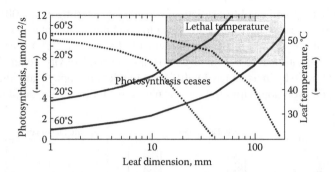

FIGURE 11.12 Simulated biophysical properties of Early Devonian planate leaves in relation to width. Dotted line, photosynthesis; solid line, leaf temperature. Shown for two palaeolatitudes. (Reprinted by permission from Macmillan Publishers Ltd., *Nature*, Beerling, D.J., Osborne, C.P., and Chaloner, W.G., Evolution of leaf-form in land plants linked to atmospheric CO_2 decline in the Late Palaeozoic era, 410, 352–354, Fig. 3, Copyright 2001.)

The likelihood of high-temperature injury and collapse of photosynthetic productivity in Early Devonian plants would have been critically dependent upon the ultimate lobe width of photosynthetic structures ("axes"), because this strongly influences boundary layer resistance and convective heat dissipation, that is, cooling capacity (Beerling et al. 2001). Beerling et al. (2001) calculated that widths of 50–100 mm at high latitudes, and much less than 20 mm in equatorial regions, are maximal before temperatures increase (Figure 11.12, solid lines) to a point where photosynthesis ceases. The earliest "proto-leaves" should, therefore, have been rather narrow, especially for plants in warm tropical regions, and this observation is backed up by the fossil record (Beerling et al. 2001).

Peñuelas and Azcón-Bieto (1992) measured $\delta^{13}C$ (they called it $\Delta^{13}C$) in 14 herbarium species of trees, shrubs, and herbs that had been collected during the previous 240 years from wet and dry habitats on the mountains and in the plains in Catalonia, Spain. The specimens were examined for changes in $\delta^{13}C$ to infer changes in WUE in response to rising CO_2 concentration. The $\delta^{13}C$ values were 19.91 (standard error of 0.32) for 1750–1760; 19.86 (SE = 0.21) for 1850–1890; and 19.95 (SE = 0.29) for 1925–1950 and decreased significantly to 18.87 (SE = 0.31) for 1982–1988. More irregular changes were found in $\delta^{13}C$ values of two C_4 species, but they also suggested a decrease in discrimination in recent decades. While Peñuelas and Azcón-Bieto (1992) presented no data on WUE, there is an inverse relation between discrimination and WUE of C_3 plants (as noted in the third paragraph of this section). Peñuelas and Azcón-Bieto (1992) said that their results suggest that either carbon assimilation rates have increased or stomatal conductance has decreased; therefore, there has been an increase in WUE over the last few decades. They conclude that the global rise in CO_2 already is influencing vegetation.

SUMMARY

Most data show that elevated atmospheric CO_2 increases WUE of plants, which is the photosynthetic production of plants divided by the water used to produce that amount of dry matter. Conversely, elevated CO_2 decreases the WR, which is the reciprocal of WUE. These results have been found for plants with all three types of photosynthetic systems (C_3, C_4, and CAM) and for a diversity of plants (annual crops and perennial trees).

Using the data from our experiments during the 1984–1987 period (the CO_2 concentration in the atmosphere in 1984 was 330 µmol/mol), we can calculate what the WRs of wheat and sorghum are a quarter century later, based on the CO_2 concentration in the atmosphere in 2008 (385 µmol/mol in 2008). Wheat and sorghum are now using 41 and 43 mL, respectively, less water to produce a gram of grain than they were in the 1984–1987 period. The WRs to produce a gram of wheat grain and sorghum grain have been reduced by 6% and 4%, respectively, due to the 55 μmol/mol increase in CO_2 since 1984. These may seem like small percentages. But spread over the large acreage of wheat and sorghum grown in Kansas, for example, the more efficient use of water now (2010) compared to the mid-1980s should have a large impact. The increasing level of atmospheric CO_2 concentration offers an opportunity to increase WUE of plants that other methods, such as increasing the fertility of soil, have failed to accomplish.

12 Elevated Atmospheric Carbon Dioxide: C_3 and C_4 Plants

INTRODUCTION

In previous chapters we have compared, in passing, plants with the C_3 and C_4 photosynthetic systems. In Chapter 6 (chapter on roots), we learned about carbon isotope ratios of plants with different photosynthetic pathways. In this chapter, we specifically compare the two photosynthetic systems, the advantage of C_4 photosynthesis, which results in greater dry matter production, and the evolution of C_4 photosynthesis. Then we compare different kinds of C_3 and C_4 plants (crop plants, grassland plants, marsh plants, and tropical grasses) grown under elevated CO_2.

C_3 VERSUS C_4 PHOTOSYNTHESIS

The pathway of carbon in C_3 plants can be found in biochemical or plant physiology textbooks (e.g., Conn and Stumpf 1963, pp. 296–297; Salisbury and Ross 1978, p. 139; Malkin and Niyogi 2000, pp. 568–628). It is sometimes called the Calvin cycle after Melvin Calvin, who won the Nobel Prize in Chemistry in 1961 for the study of chemical steps that take place during photosynthesis (Ståhle 1971). (For a biography of Calvin, see Appendix.) Calvin and his colleagues were able to determine the pathway because of two developments. First, in 1946, carbon-14 was made available in appreciable amounts from the Atomic Energy Commission. Second, the technique of paper chromatography was in full development and provided a means for separating the large number of cell constituents that occur in a plant. With these tools, Calvin's group was able to identify the early stable intermediates in the path of carbon from CO_2 to glucose (Conn and Stumpf 1963, p. 294).

We remember in C_3 photosynthesis that 3-phosphoglyceric acid (3-PGA) is the first product formed in the fixation of CO_2, but it does not form directly from three CO_2 molecules. Instead, 3-PGA forms in two concerted steps from the reaction of CO_2 with a five-carbon sugar, ribulose-1,5-bisphosphate (RuBP). The carboyxlation of the C_5 sugar produces a C_6 intermediate that is immediately cleaved into two molecules of 3-PGA. The enzyme that catalyzes this reaction is ribulose bisphosphate carboyxlase/oxygenase or Rubisco (Malkin and Niyogi 2000, pp. 610–611).

Figure 12.1 depicts a simplified scheme of the key biochemical elements in C_4 photosynthesis common to all C_4 species (Hatch 1992, p. 337). As shown in Figure 12.1, in all groups of C_4 plants atmospheric CO_2 is fixed in the mesophyll cells by reacting with phospho*enol*pyruvate (PEP) to give oxaloacetate (OAA), which then is converted to larger pools of malate or asparatate. Depending on the species, one or the other of these C_4 acids is then transferred to the adjacent bundle sheath cells, where it undergoes decarboxylation to give CO_2, which is refixed by the photosynthetic carbon reduction (PCR) cycle exclusively located in these bundle sheath cells. After decarboxylation, the residual three-carbon compound returns to the mesophyll cells where it serves as the precursor of pyruvate, which is converted to the primary CO_2 acceptor PEP (Hatch 1992, p. 338).

ADVANTAGE OF C_4 PHOTOSYNTHESIS

The C_4 photosynthetic system is an advantage to plants (Hatch 1992, p. 338). In the Calvin cycle in C_3 photosynthesis, there is, in addition to the usual carboxylation reaction catalyzed by the enzyme Rubisco (Hatch 1992, p. 335) giving 3-PGA, an oxygenase reaction also catalyzed by the

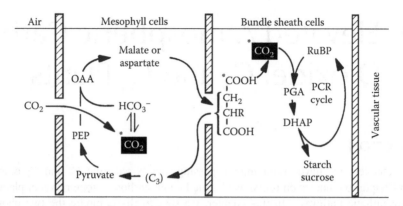

FIGURE 12.1 Typical C_4 leaf anatomy and a simplified scheme depicting the key biochemical elements of C_4 photosynthesis common to all C_4 species. The compound at the left in the "bundle sheath cells" is either malic acid or aspartic acid, where the R stands for OH in malic acid and NH_2 in aspartic acid. *Abbreviations*: OAA, oxaloacetate; PEP, phospho*enol*pyruvate; RuBP, ribulose-1,5-bisphosphate; PGA, phosphoglycerate; DHAP, dihydroxyacetone-phosphate; PCR cycle, photosynthetic carbon reduction cycle. (Reprinted from Hatch, M.D., *Plant Cell Physiol.*, Fig. 4, 1992 by permission of Oxford University Press.)

enzyme Rubisco, in which O_2 reacts with RuBP to give 3-PGA plus phosphoglycolic acid. This phosphoglycolate is metabolized through the so-called phosphorespiratory cycle to appear finally at 3-PGA, but only after one-quarter of the carbon is lost as CO_2 (Hatch 1992, p. 338). (We talked about photorespiration in Chapter 11.) This oxygenase reaction and associated metabolism have an adverse effect on the efficiency of photosynthesis in C_3 plants. For C_3 plants in air, the CO_2 to O_2 ratio at the site of Rubisco during maximum photosynthesis in leaves is about 0.025 (CO_2/O_2 in cell = 0.025). Because CO_2 and O_2 are strictly competitive in these reactions, the extent of oxygenase activity depends largely on the ratio of CO_2 to O_2, and with this ratio of about 0.025 the ratio of carboxylase activity to oxygenase activity is about 2.5–1. This gives a rate of CO_2 loss by photorespiration equivalent to 20% of the gross rate of photosynthetic CO_2 assimilation. Under these conditions, the quantum yield for C_3 plants, which is a measure of the efficiency of light utilization, is about 0.05 moles CO_2 per mol quanta of light absorbed. It requires about a 10-fold increase in this ratio of CO_2 to O_2 to eliminate essentially the oxygenase reaction and reduce the rate of photorespiratory CO_2 loss to less than 2% of the gross photosynthesis. Notably, under these conditions, the quantum yield is markedly increased to about 0.08 mol CO_2 (mol quanta)$^{-1}$. Of course, this is precisely the situation that is believed to prevail in the bundle sheath cells of C_4 plants, with CO_2 being sufficiently high to prevent oxygenase activity (Hatch 1992, p. 338).

Keeping in mind that the function of C_4 photosynthesis at the cell level is to eliminate the oxygenase reaction and associated photorespiration, it is interesting to compare the photosynthetic performance of C_4 and C_3 leaves (Hatch 1992, p. 339). As a broad generalization, C_4 leaves show higher rates of photosynthesis than C_3 leaves when each is measured at their respective temperature optima, and this remains so through the light range up to full sunlight (Figure 12.2). With few exceptions, C_3 plants fail to match even the slowest of the C_4 plants. However, C_3 plants perform similarly to C_4 plants if photosynthesis is measured under conditions where photorespiration is eliminated, i.e., under conditions where O_2 is reduced or CO_2 is increased relative to the levels of O_2 and CO_2 that the plants are normally exposed to (Hatch 1992, p. 339).

The basis for the differences between the photosynthetic performance of C_3 and C_4 leaves is further clarified by looking at photosynthetic efficiency in the light-limited region (Hatch 1992, pp. 339–340). As shown in Figure 12.3, there is a moderate range of quantum yields (a measure of photosynthetic efficiency) for different C_4 plants, but values are unaffected by varying temperature over a wide range. By contrast, the quantum yield for C_3 plants varies with temperature in such a way that the quantum yield for C_3 plants is higher than that for C_4 plants at lower temperature,

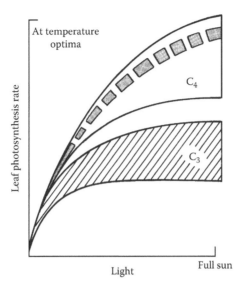

FIGURE 12.2 Range of leaf photosynthesis rates for C_3 and C_4 plants at respective temperature optimum for individual species. Dotted line shows response for C_3 plants measured in the absence of photorespiration. (Reprinted from Hatch, M.D., *Plant Cell Physiol.*, Fig. 5, 1992 by permission of Oxford University Press.)

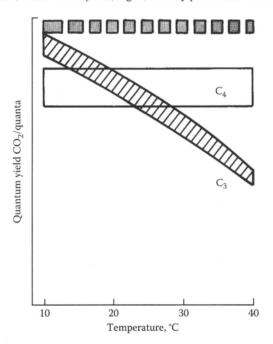

FIGURE 12.3 Range of quantum yields for C_3 and C_4 species measured at limiting light. Dotted line is for C_3 plants measured under conditions where photorespiration is negligible. Quantum yield is a measure of photosynthetic efficiency. (Reprinted from Hatch, M.D., *Plant Cell Physiol.*, Fig. 6, 1992 by permission of Oxford University Press.)

but less than for C_4 plants at temperatures above 25°C–30°C. Again, when the quantum yield for C_3 leaves is measured under conditions where photorespiration is eliminated, it is unaffected by temperature and it is higher than for C_4 leaves throughout the temperature range. This temperature effect on the efficiency of C_3 photosynthesis is largely explained by the fact that the oxygenase reaction catalyzed by Rubisco is disproportionately increased with increasing temperature relative to

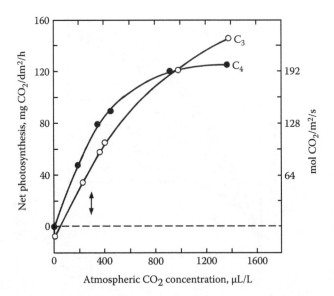

FIGURE 12.4 CO_2 dependency of leaf net CO_2 photosynthesis in a typical C_4 plant in comparison to a typical C_3 plant. The C_4 CO_2 response curve also is independent of O_2 concentration, whereas the C_3 CO_2 response curve is not independent of O_2. (From Black, C.C., Jr., Effects of CO_2 concentration on photosynthesis and respiration of C_4 and CAM plants, in Enoch, H.Z., and Kimball, B.A., Eds., *Carbon Dioxide Enrichment of Greenhouse Crops. Volume II. Physiology, Yield, and Economics*, CRC Press, Inc., Boca Raton, FL, 1986, pp. 29–40, Fig. 3. With permission.)

the carboyxlation reaction. Because the oxygenase reaction is the basis of photorespiration (as noted above), photorespiration increases disproportionately relative to photosynthesis as temperature increases, thus causing the progressive decline in the efficiency of C_3 photosynthesis (Hatch 1992, p. 340).

Elevating CO_2 from a level of 350 µmol/mol reduces the rate of photorespiration. For example, a doubling of 350 µmol/mol CO_2 [which is anticipated to occur within the next 100 years (Ehleringer et al. 1991)], will result in approximately a 50% reduction in the rate of photorespiration. Increasing an atmospheric CO_2 level of 350 µmol/mol fivefold would nearly eliminate photorespiratory activity in C_3 plants (Ehleringer et al. 1991).

Figure 12.4 shows the effect that enriching the atmosphere with CO_2 has on the photosynthetic rates of C_3 and C_4 plants (Black 1986, pp. 32–33). Enrichment above 350–400 µmol/mol of CO_2 has a diminished effect on net CO_2 fixation for C_4 plants. An inflection in the C_4 photosynthesis response curve to CO_2 occurs slightly above the arrow in Figure 12.4, which was the atmospheric CO_2 concentration when Black wrote his article in 1986 (350 µmol/mol). This inflection occurs because the C_4 cycle is a mechanism for enriching the CO_2 supply in the bundle sheath cells, where the C_3 cycle is carried out (Black 1986, p. 33). Because within the C_4 tissue ribulose-1,5-bisphosphate carboxylase is effectively saturated by current-day air CO_2 levels, enrichment of the atmosphere with CO_2 has a diminishing effect on C_4 photosynthesis (Black 1986, p. 33). We see in Figure 12.4 that C_3 photosynthesis shows no inflection point and increases as the atmospheric CO_2 concentration increases.

DRY MATTER PRODUCTION OF C_3 AND C_4 PLANTS

These differences in the photosynthetic capacity at the leaf level do give C_4 plants a potential advantage in terms of growth and dry matter production. The maximum recorded rates of dry matter production of C_3 and C_4 crops and pasture plants in the field range from 13 to 43 g/m²/day (average: 26 g/m²/day) and 37 to 57 g/m²/day (average: 49 g/m²/day), respectively (Hatch 1992, p. 340). The range of values barely overlaps, and dry matter production for C$_4$ species averages nearly twice that for C$_3$ species (Hatch 1992, p. 340). It should be noted that C$_4$ plants are also more efficient in water use (water loss per unit of carbon gain) compared with C$_3$ plants (Hatch 1992, p. 340). (We discussed this topic in Chapter 11 on water use efficiency.) We shall study the effects of elevated levels of CO$_2$ on growth in the final chapter.

Literature published in the mid-1970s, which was only about 10 years after the C$_4$ photosynthetic pathway had been discovered, suggested that the productivity of C$_3$ and C$_4$ species was almost indistinguishable. However, close inspection of the data on which these estimates were made showed a number of errors in determining productivity (Monteith 1978). When all unreliable figures were discarded, the maximum growth rate for stands of C$_3$ crops fell in the range of 34–39 g/m^2/day compared with 50–54 g/m^2/day for C$_4$ crops. The final dry weight of the C$_3$ and C$_4$ crops was strongly correlated with the length of the growing season, presumably because this period is a good measure of the amount of radiation intercepted by the foliage (Monteith 1978) (Figure 12.5). The seasonal mean growth rates (defined as standing dry weight at harvest divided by the length of the growing season) for the C$_3$ group in Figure 12.5 (plants cited in top left of figure) are 13.0 ± 1.6 g/m^2/day and for the C$_4$ group (top middle of figure) it is 22.0 ± 3.6 g/m^2/day. The mean photosynthetic efficiency for the C$_4$ group is 2.0% of total solar radiation compared with 1.4% for the C$_3$ group. The difference of 40% in the efficiency of solar energy conversion is significant in terms of food production (Monteith 1978).

We digress here for a discussion of solar energy conversion. The bulk of energy conversion on earth is carried out by oxygen-evolving plants (higher plants and algae). The main functional pigments, chlorophyll a and the accessory pigments, absorb all wavelengths shorter than 700 nm—or about half of the solar spectrum (Kok 1976, p. 846). Under optimal conditions, photosynthesis can convert up to 20% of absorbed radiant energy into chemical energy (Kok 1967, p. 30). However, under natural conditions, ≤1% of the solar energy reaching the earth's surface is routed through the plant kingdom and supports life, the remainder being wasted as heat (Kok 1976, p. 846). Unfortunately, farmers do not achieve this 20% conversion for yield. The best achieved crop yields may approach one-third of the maximum possible yield (c. 6%), and more normally these figures

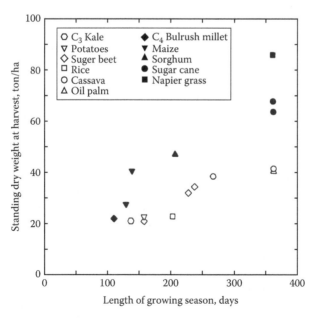

FIGURE 12.5 Standing dry weight of crops at harvest in relation to length of growing season. C$_3$ plants are the open symbols, and C$_4$ plants are the closed symbols. (Reprinted from Monteith, J.L., *Exp. Agric.*, 14, 1, 1978, Fig. 1. With permission. Copyright 1978, Cambridge University Press.)

are closer to only one-tenth of it (2%) (Kok 1967, p. 31). We see that the 2% conversion relates to C_4 crops (Monteith 1978), and C_3 crops have a conversion (1.4%) closer to plants growing under natural, unmanaged conditions (about 1%). Still the total energy conversion by photosynthesis (either C_3 or C_4) exceeds by many times the total industrial output of man (Kok 1976, p. 846).

The highest corn (*Zea mays* L.) yields ever achieved came from the fields of Francis Childs in Marshall County, Iowa (Martin 2008). He was the first farmer in a controlled contest to exceed 400 bushels per acre, achieving 405 bushels per acre in 2001 and 442 the next year (Martin 2008). Childs recorded yields of 577 bushels per acre in strips of less than 10 acre crops required for contests. We cannot convert the bushels per acre into dry matter yield, as cited in the previous paragraph, because we do not know the dry matter production (we only know the reproductive yield) or the length of Childs' growing seasons. [Assuming that a bushel of corn weighs 56 lb, which is commonly done, we do know that 357 bushels per acre = 22.8 metric ton/ha (Luellen 1998), and this conversion can be used to put the values, cited above, in metric tons/ha.] Average corn yields have climbed steeply to 153 bushels per acre in 2007 from 26.5 in 1932 (Martin 2008). It is interesting to note that these high yields occurred with a C_4 plant (corn). Many factors control yield, and CO_2 concentration in the atmosphere is a minor one compared to water and fertility. However, if environmental conditions are optimal for corn production, corn yields may be leveling off as the CO_2 concentration in the atmosphere increases (Figure 12.4).

EVOLUTION OF C_4 PLANTS

The following question arises: Why and when did C_4 plants evolve (Hatch 1992, p. 340)? Inspection of an evolutionary tree of angiosperms shows that there are 18 families (Ehleringer et al. 1991) in eleven orders (Hatch 1992, p. 340) containing C_4 species. They are widely distributed in the various evolutionary arms and are among the most recently evolved. To put this number (18 families) in perspective, let us review plant systematics and tally the number of plant families (Bailey 1974, pp. 55–56). There are four great divisions in the vegetable community (i.e., plants): Thallophyhta (the thallophytes); Bryophyta (mosses and liverworts); Pteridophyta (ferns and fern allies); and Spermatophyta (the spermatophytes or seed plants). The Spermatophyta are subdivided into two subdivisions: gymosperms and angiosperms. The angiosperm subdivision is divided into two classes: Monocotyleonae and Dicotyledonae. The Monocotyledonae are not divided into subclasses, but the Dicotyledonae are divided into two subclasses: the Polypetalae (petals or sepals distinct or separate when present) and the Gamopetalae (petals or sepals united). Each subclass is in turned divided into relatively large groups called orders. [Fernald (1950) lists 46 orders of plants in the central and northeastern United States and Canada.] Orders in turn are comprised of families. There are nearly 300 families of Spermatophyta of which about 50 belong in the Monocotyledonae (Bailey 1974, p. 56). The 18 families that have the C_4 type of photosynthesis are still few compared to the total number of plant families (300). If we consider species instead of families, C_4 species comprise only 3% of vascular plant species, but they account for some 25% of terrestrial photosynthesis (Edwards et al. 2010).

The C_4 pathway is a relatively new arrival on the evolutionary scene, and it must have evolved separately on many occasions (Hatch 1992, p. 340). Grasses with the C_4 type of photosynthesis rose to ecological dominance 3–8 million years ago (Edwards et al. 2010). It seems likely that the evolutionary pressure to develop C_4 photosynthesis as an option to C_3 photosynthesis probably resulted from a dramatic decline in the atmospheric CO_2 concentration as recently as 30–50 million years ago. To put this event in perspective, it should be noted that photosynthesis probably evolved about 3000 million years ago (Hatch 1992, p. 340; Edwards et al. 2010), and that the now ubiquitous angiosperms appeared about 120 million years ago in the Cretaceous Period (Friend and Guralnik 1959), during the latter part of the Mesozoic Era (Foster and Gifford 1959, p. 444). The current evidence is that atmospheric CO_2 levels were about 10 times current levels as recently as 120 million years ago (in the Cretaceous), but this declined to something close to present-day levels of 300 ppm

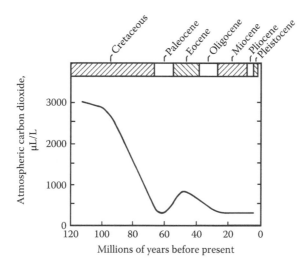

FIGURE 12.6 Modeled change in atmospheric CO_2 concentration during the past 100 million years. [In the figure, the Cretaceous is a Period. The Paleocene, Eocene, and Oligocene are Epochs within the Paleogene Period, and the Miocene, Pliocene, and Pleistocene are Epochs within the Neogene Period (Spencer 1962, p. 12–17).] (Reprinted from *Trends Ecol. Evol.*, 6, Ehleringer, J.R., Sage, R.F., Flanagan, L.B., and Pearcy, R.W., Climate change and the evolution of C_4 photosynthesis, 95–99, Copyright 1991, Fig. 5, with permission from Elsevier.)

CO_2 [when Hatch (1992) wrote his article] by 30–60 million years ago (Hatch 1992, pp. 340–341) (Figure 12.6). Because flowering plants lived in an atmosphere of CO_2 as high as 3000 ppm, they can adapt to concentrations much higher than now present.

We review here briefly the geological time scale as it relates to C_4 plants, as shown in Figure 12.6. For a figure of the entire geological time scale, see Figure 9.10. The time units of historical geology are era, period, epoch, and age. An era is the largest grouping of periods of geologic time. Each era covers many millions of years and each is characterized by the existence of certain major groups or forms of life (Spencer 1962, p. 13). Periods are subdivisions of a geological era, in which rock strata and fossils form a definite sequence (Friend and Guralnik 1959). The rock units formed during a period are called systems (Spencer 1962, pp. 13–14). The periods are subdivided into epochs, and the rock units formed during an epoch are called *series*. Epochs are in turn subdivided into ages, when it is felt that the chronology is sufficiently detailed to justify a time unit this small (Spencer 1962, p. 14). The Cretaceous (Figure 12.6, top, first word at left) is a period that occurred between 130 and 60 million years ago. The Paleogene Period, which lasted 42 million years, contains the Paleocene, Eocene, and Oligocene Epochs (Figure 12.6, top, second, third, and fourth word from left, respectively). The Neogene is the most recent period. It started 28 million years ago. It contains the Miocene, Pliocene, Pleistocene (Figure 12.6, top, fifth, sixth, and seventh word from left, respectively), and Recent Epochs. The Pleistocene Epoch was approximately 10,000–1,000,000 years ago (Friend and Guralnik 1959, p. 606; Spencer 1962, p. 12). The Recent Epoch is the last 10,000 years (Spencer 1962, pp. 12–17).

Figure 12.6 shows that atmospheric CO_2 levels declined abruptly at the end of the Cretaceous Period and during the Paleocene Epoch. Atmospheric CO_2 levels have remained relatively low over the past 50 million years (Ehleringer et al. 1991), being near current atmospheric levels in the Paleocene Epoch and again in the Miocene Epoch, and about double current levels during the Eocene and Oligocene Epochs (Figure 12.6). Atmospheric O_2 levels are thought to have remained relatively constant since the Cretaceous (Ehleringer et al. 1991), even though, of course, in current times the CO_2 concentration in the atmosphere is increasing (see Figures 1.1 through 1.3). C_4 photosynthesis may have first appeared in the Paleocene, but may have remained suppressed during the periods of elevated CO_2 levels that are thought to have occurred in the Eocene and Oligocene Epochs. However, by the Miocene Epoch, when CO_2 levels were reduced, atmospheric conditions would have again favored C_4 photosynthesis. There is strong evidence that C_4 species flourished and dominated many grasslands by the late Miocene Epoch (Ehleringer et al. 1991). Historical records show that the world's major grasslands developed during the Miocene (Ehleringer et al. 1991), and many grasses have the C_4 photosynthetic pathway. Further reductions in CO_2 levels during the Pleistocene would have stimulated the spread of C_4 plants. C_4 photosynthesis has no identifiable advantage in an environment in which atmospheric CO_2 levels are significantly above the current ambient levels [350 µmol/mol, when Ehleringer et al. (1991) wrote their paper]. It is now questionable whether the proliferation of C_4 plants will continue in an anthropogenically altered atmosphere (Ehleringer et al. 1991). [We note that the increase in CO_2 since the beginning of the industrial revolution (200 ppm in 1787 to 385 in 2008) (Dlugokencky 2009) is trivial compared to the changes in the past, which have been one order of magnitude in difference (Figure 12.6, 3000–300 ppm).]

We point out in passing that not only may have photosynthesis changed over geological time, but also respiration. Gonzàlez-Meler et al. (2009) found that respiration rates of *Arabidopsis thaliana* plants grown at the concentration of CO_2 in the Pleistocene (200 µmol/mol) were lower than plants grown at current levels of CO_2 (370 µmol/mol). The reduction in respiration was due to a reduced activity of the alternative pathway of respiration. Plant respiration has both the regular pathway for the production of energy that involves cytochrome c (Siedow and Day 2000, pp. 687–694), but also an alternative pathway that consumes photosynthate without producing chemical energy as effectively as respiration through the normal cytochrome pathway (Siedow and Day 2000, pp. 696–698). [See the December 2009 issue of *P. Plantarum* for a Special Issue on alternative oxidase (Arnholdt-Schmitt 2009).] The results of Gonzàlez-Meler et al. (2009) suggest that alternative-pathway-mediated responses of respiration to changes in atmospheric CO_2 may have enhanced survival of plants at low CO_2 levels in past times.

The presence of the C_4 photosynthetic pathway leads to a markedly different response of net photosynthesis to changes in atmospheric CO_2 or O_2 concentration than that found in C_3 plants (Ehleringer et al. 1991). As noted before, at low CO_2 concentrations, C_4 plants typically have higher photosynthetic rates than C_3 plants. But C_4 photosynthesis becomes saturated at concentrations above 350 µmol/mol CO_2, whereas C_3 photosynthesis does not (Figure 12.7a). [The abscissa in Figure 12.7 shows intercellular CO_2 concentration, not atmospheric CO_2 concentration. While intercellular CO_2 is lower than the CO_2 concentration in the surrounding air, the intercellular CO_2 concentration increases as the external CO_2 concentration increases, as shown experimentally by He et al. (1992) and Imai and Okamoto-Sato (1991). The results of the latter are discussed below.] Also, because of Rubisco oxygenation and subsequent photorespiration, the quantum yield or light-use efficiency of C_3 plants is strongly dependent on CO_2 (Figure 12.7b). In C_4 plants, the light-use efficiency is essentially independent of atmospheric CO_2, but does not attain the maximum levels of C_3 plants because of the extra ATP (energy) costs of the CO_2-concentration mechanism. Under changing atmospheric CO_2 levels, though, low CO_2 levels should favor C_4 photosynthesis whereas high CO_2 levels should favor C_3 photosynthesis. Studies of growth and competition between C_3 and C_4 plants under elevated CO_2 conditions have confirmed this observation (Ehleringer et al. 1991). Nie et al. (1992b) studied the frequency of a C_3 and C_4 grasses in a tallgrass prairie during two growing seasons under elevated CO_2 (twice ambient or an average concentration of 687 µmol/mol; ambient averaged 346 µmol/mol). At the end of the experiment under well-watered conditions, there were more C_3 (*Poa pratensis* L. or Kentucky bluegrass) plants in the elevated CO_2 treatment than in the ambient CO_2 treatment. The frequency of the major C_4 plants on the prairie (*Andropogon gerardii* Vitman or big bluestem; *A. scoparius* Michx. or little bluestem; and *Sorghastrum nutans* Nash. or Indiangrass) was not affected by CO_2.

The geological data in times closer to the present (last 20,000 years) also indicate that C_4 plants have a physiological advantage over C_3 plants under low atmospheric CO_2 concentrations (Robinson 1994b). Figure 12.8a shows that C_4 plants have a higher photosynthetic rate compared to C_3 plants

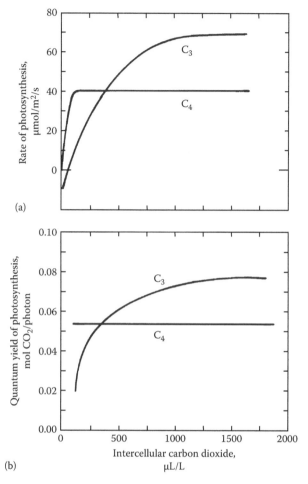

FIGURE 12.7 Effect of CO_2 concentration on the absolute rate of photosynthesis (a) and the quantum yield of photosynthesis or light-use efficiency (b) in C_3 and C_4 plants. (Reprinted from *Trends Ecol. Evol.*, 6, Ehleringer, J.R., Sage, R.F., Flanagan, L.B., and Pearcy, R.W., Climate change and the evolution of C_4 photosynthesis, 95–99, Copyright 1991, Fig. 3, with permission from Elsevier.)

at CO_2 values characteristic of the Pleistocene glaciations (190–240 µmol/mol CO_2). Experimental work shows that, at a CO_2 concentration on the order of 200 µmol/mol, C_3 plants have a lower photosynthetic rate than C_4 plants (Figures 12.4 and 12.7a). Hence, one would expect C_4 plants to expand during the ice ages. [The ice ages, also called the glacial epoch, were the time during the Pleistocene, when a large part of the Northern Hemisphere was covered with glaciers (Friend and Guralnik 1959).] As noted before, C_4 physiology is inefficient at low temperatures. Note in Figure 12.3 that, at about 10°C–15°C, C_3 plants have a higher quantum yield than C_4 plants. But CO_2 concentration may have been more important than temperature in the proliferation of C_4 plants. Historical $\delta^{13}C$ data may offer this evidence (Robinson 1994b). Figure 12.8b shows $\delta^{13}C$ for a 20,000 year southern Indian peat core, and these data are juxtaposed with historical CO_2 values as reconstructed from ice cores (the bottom part of Figure 12.8b; the two different symbols used in the bottom part of Figure 12.8b, diamonds and rectangles, are from two different data sets). As would be anticipated, if C_4 plants displaced isotopically lighter C_3 competitors under low CO_2 concentrations, $\delta^{13}C$ and CO_2 are inversely related. The data in Figure 12.8b show near-pure C_4 peat −12.8 per mill (−12.8‰) at the last glacial maximum (18,000 before present), when the CO_2 concentration was 180–190 µmol/mol (Figure 12.8a). The C_4 peat indicates that C_4 plants might have dominated at the last glacial maximum.

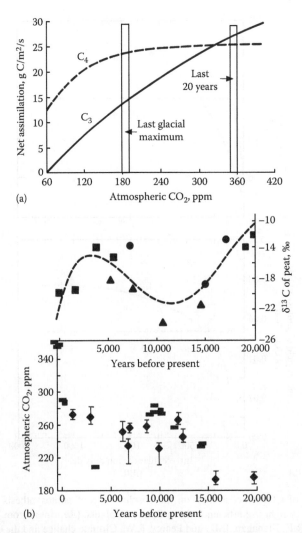

FIGURE 12.8 (a) Predicted net photosynthesis assuming 25°C air temperature, 2.5 kPa ambient partial pressure of water vapor, 10 mol/m²/s boundary layer conductance, C_3 and C_4 V_{max} for Rubisco, respectively, of 100 and 30 μmol/m²/s, and 1.64 mmol/m²/s photo flux density (about 80% of full sunlight). (b) Atmospheric CO_2 concentrations from two Antarctic ice cores as related to $\delta^{13}C$ measurements from a southern Indian peat core. (Reprinted by permission from Macmillan Publishers Ltd. *Nature*, Robinson, J.M. Atmospheric CO_2 and plants, 368, 105, 1994b, unnumbered figure, Copyright 1994. With permission.)

We now turn to some specific experiments in which the water relations of C_3 and C_4 plants have been compared under elevated CO_2 in the same experiment. We recognize that we have compared C_3 and C_4 plants in other chapters, and we shall not repeat those comparisons here.

C_3 AND C_4 PLANTS UNDER ELEVATED CO_2

CROP PLANTS

Imai and Okamoto-Sato (1991) studied the effects of elevated CO_2 in the atmosphere on gas exchanges of two C_3 crops [*Oryza sativa* L. cv. Nipponbare, rice; *Glycine max* (L.) Merr. cv. Bonminori, soybean) and two C_4 crops (*Echinochloa frumentacea* Link cv. Hidaakadbie, Japanese millet; *Eleusine coracana* Gaertn. cv. Sakase III, finger millet). Plants were grown in pots of soil

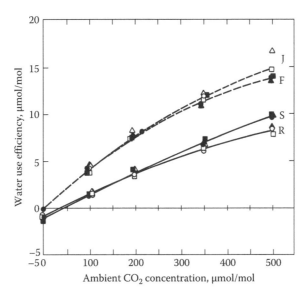

FIGURE 12.9   Effects of (ambient) CO_2 concentration and temperature on water use efficiency of crop plants. R, rice; S, soybean; J, Japanese millet; F, Finger millet. Circles (open and closed): 33°C; triangles (open and closed): 28°C; squares (open and closed): 23°C. (Reprinted from Imai, K. and Okamoto-Sato, M., *Jpn. J. Crop Sci.*, 60, 139, 1991, Fig. 2. With permission. Copyright 1991, Crop Science Society of Japan.)

(the soil was not identified) in growth cabinets with an ambient CO_2 air concentration of 500 μmol/ mol CO_2. Measurements of gas exchange were made with an open airflow gas exchange system using an infrared CO_2 analyzer. During measurements, the CO_2 concentration in the leaf chamber was changed from 0 to 500 μmol/mol CO_2 by adding 10% CO_2 from a cylinder to CO_2-free air. The leaf chamber had a water jacket in which water was circulated and this controlled the leaf temperature. Leaf temperatures were set at 23°C, 28°C, or 33°C. Rates of CO_2 exchange and transpiration of leaves at the three different temperatures were measured. The water use efficiency was expressed as the ratio of the CO_2 exchange rate to the transpiration rate.

As the CO_2 increased, water use efficiency increased (Figure 12.9). The water use efficiency of the two C_4 plants was always higher than that of the two C_3 plants. The intercellular CO_2 concentration increased with increasing ambient CO_2 concentration (Figure 12.10). The intercellular CO_2 concentrations were higher in the two C_3 plants than the two C_4 plants, and these data agree with those of He et al. (1992). At 500 μmol/mol ambient CO_2, the stomatal conductance of the two C_3 plants was higher than that of the two C_4 plants at all leaf temperatures (Figure 12.11). Imai and Okamoto-Sato (1991) concluded that the pronounced stomatal closure in C_4 plants at the high ambient CO_2 concentration (500 μmol/mol) contributed to their high water use efficiency and might account for the increased dry matter production of C_4 crops compared to C_3 crops that is documented in the literature and discussed above under "Dry Matter Production of C_3 and C_4 Plants."

GRASSLAND PLANTS

During two growing seasons (1989 and 1990), Polley et al. (1994) studied C_3 and C_4 plants competing with each other in a grassland in the southwest of the United States. The C_4 plant was the perennial grass *Schizachyrium scoparium* (Michx.) Nash [little bluestem—also named *A. scoparius* Michx., the name it usually goes by (Fernald 1950, p. 232; Pohl 1968, p. 220)], and the C_3 plant was the woody legume invader into the grassland, *Prosopis glandulosa* Torr. var. *glandulosa* (honey mesquite). Their objective was to determine effects of historical increases in atmospheric CO_2 concentration on competition between the two plants. The abundance of woody mesquite has increased

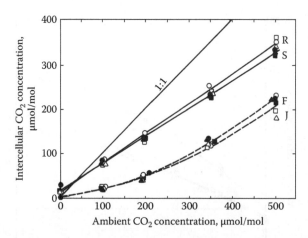

FIGURE 12.10 Relationships between ambient and intercellular CO_2 concentrations at three different temperatures. R, rice; S, soybean; J, Japanese millet; F, Finger millet. Circles (open and closed): 33°C; triangles (open and closed): 28°C; squares (open and closed): 23°C. (Reprinted from Imai, K. and Okamoto-Sato, M., *Jpn. J. Crop Sci.*, 60, 139, 1991, Fig. 3. With permission. Copyright 1991, Crop Science Society of Japan.)

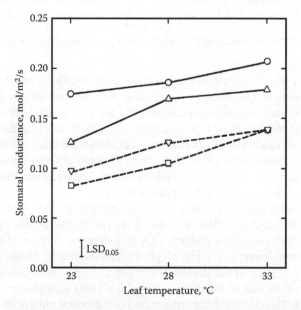

FIGURE 12.11 Effect of temperature on stomatal conductance to CO_2 transfer at 500 μmol/mol ambient CO_2. Circles: rice; triangles: soybean; squares: Japanese millet; upside-down triangles: finger millet. (Reprinted from Imai, K. and Okamoto-Sato, M., *Jpn. J. Crop Sci.*, 60, 139, 1991, Fig. 4. With permission. Copyright 1991, Crop Science Society of Japan.)

on C_4 grasslands in the southwestern United States during the past 150 years. Accompanying the increased density of mesquite is a 27% rise in atmospheric CO_2 from about 275 μmol/mol at the beginning of the nineteenth century to 350 μmol/mol, when Polley et al. (1994) published their work.

They grew the plants in 43 compartments that were each 38 m long. The soil under the compartments was a fine sandy loam. They imposed two watering regimes: well-watered (plants kept at field capacity) and dry (plants watered when soil was at 65% of field capacity). A stable CO_2 concentration gradient from about 340 to 200 μmol/mol was maintained during the daylight hours by automatically varying the rate of airflow through the compartments in response to changes in incident photosynthetic photon flux density and CO_2 depletion by the enclosed plants. The compartments were planted with various combinations of the two plants. Additional compartments were seeded to corn (Z. *mays* L. cv. Gaspe yellow flint). Leaf gas exchange measurements were made with an infrared gas analyzer. Water use efficiency, which they called the intrinsic water use efficiency, was determined by dividing the net CO_2 assimilation by stomatal conductance. Discrimination by *P. glandulosa* against $^{13}CO_2$ was determined from $\delta^{13}C$ values of leaves and air measured early in the growing season (April 1990). (Discrimination by the *S. scoparium* is not reported.) Discrimination by the Z. *mays* leaves (C$_4$) harvested in September 1990 was used to calculate the $\delta^{13}C$ of air to which the *P. glandulosa* plants harvested late in the growing season (October 1990) was exposed. The stable carbon isotope composition of Z. *mays* served as a proxy for the $^{13}C/^{12}C$ of chamber air. The proxy was necessary, because the chambers could not be opened. The opening would have mixed outside air with the air depleted of CO_2. Polley et al. (1994) point out that the isotopic composition of atmospheric CO_2 can be inferred from C$_4$ plants like Z. *mays*.

Water use efficiency for the C$_4$ *S. scoparium* was higher than that of the C$_3$ *P. glandulosa* (Figure 12.12). The water use efficiency of *S. scoparium* and *P. glandulosa* increased 89% and 65%, respectively, from 225 to 345 µmol/mol CO_2. A higher net CO_2 assimilation accounted for the CO_2-induced increase in the water use efficiency of *P. glandulosa*. Both an increase in the net CO_2 assimilation and a decline in stomatal conductance contributed to the higher water use efficiency in *S. scoparium*. The $\delta^{13}C$ of chamber air measured in April 1990 (Figure 12.13a) and of Z. *mays* leaves harvested in September 1990 (Figure 12.13b), a proxy for the $\delta^{13}C$ of chamber air, increased linearly (became more enriched in ^{13}C) as photosynthesis depleted CO_2 concentration from about 340 to 200 µmol/mol. The increase occurred in both well-watered and drought-stressed *P. glandulosa*. The values for $\delta^{13}C$ for the C$_3$ and C$_4$ plants shown in Figure 12.13 agree with the data that we saw in the chapter on roots. Plants with the C$_3$ pathway of carbon assimilation have a $^{13}C/^{12}C$ ratio about 20‰ less than that in the atmosphere, while plants with the C$_4$ pathway have a ratio that is lower than the atmosphere by about 10‰ (Farquhar et al. 1982). The near-pure C$_4$ peat shown in Figure 12.8 during the last glacial maximum has a value not too far from the −10‰ to −13‰ values

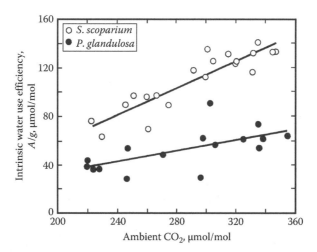

FIGURE 12.12 Intrinsic water use efficiency (leaf net assimilation, *A*, divided by stomatal conductance to water, *g*) of the C$_4$ *S. scoparium* and C$_3$ *P. glandulosa* as a function of the daytime CO_2 concentration at which plants were grown. Lines are linear regressions of single gas exchange measurements for *S. scoparium* ($r^2=0.81$, $P<0.0001$, $n=20$) and *P. glandulosa* plants ($r^2=0.42$, $P<0.005$, $n=17$) vs. CO_2 concentration. (Reprinted from Polley, H.W., Johnson, H.B., and Mayeux, H.S., *Ecology*, 75, 976, 1994, Fig. 2. With permission. Copyright 1994, Ecological Society of America.)

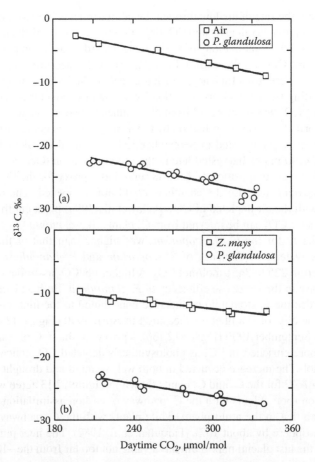

FIGURE 12.13 The stable C isotope composition ($\delta^{13}C$) of air and upper leaves from C_3 *P. glandulosa* and C_4 *Z. mays* plants along a daytime gradient of subambient CO_2 concentration. Lines are linear regressions of $\delta^{13}C$ on instantaneous (air) or mean daytime CO_2 concentration $[CO_2]$ (plant leaves) in (a) April 1990 ($r^2 = 0.99$, $P < 0.0001$, $n = 6$) and May 1990 (*P. glandulosa*, $r^2 = 0.85$, $P < 0.0001$, $n = 18$, well-watered plants) and (b) September 1990 (*Z. mays*, $r^2 = 0.89$, $P < 0.0001$, $n = 10$ plants) and October 1990 (*P. glandulosa*; $r^2 = 0.86$, $P < 0.0001$, $n = 12$, droughted plants). (Reprinted from Polley, H.W., Johnson, H.B., and Mayeux, H.S., *Ecology*, 75, 976, 1994, Fig. 3. With permission. Copyright 1994, Ecological Society of America.)

shown for *Z. mays* in Figure 12.13b. All the data confirm that the $\delta^{13}C$ values for C_4 plants are less negative than those for C_3 plants and that, as the CO_2 concentration increases in the atmosphere, the $\delta^{13}C$ values of C_3 and C_4 plants gets more negative.

Polley et al. (1994) concluded that the higher water use efficiency of honey mesquite may have benefited its establishment in arid areas by rendering the shrub better able to survive periods of water deprivation. They suggested that it could respond rapidly to rising CO_2 concentrations and spread throughout a grassland. Because the photosynthetic rate of C_3 plants increases as the CO_2 concentration in the air increases, as noted above (Figure 12.4), the rising CO_2 concentration over the past 150 years may have preferentially enhanced growth of C_3 shrubs like mesquite relative to the C_4 species that dominated grasslands in the southwestern United Stares at the beginning of the nineteenth century.

MARSH PLANTS

Curtis et al. (1990) compared the growth of roots of two different communities of perennial marsh plants growing with elevated CO_2 concentrations under field conditions: a C_3 sedge (*Scirpus olneyi*

Gray; *Scirpus* is also known as bulrush) and a C_4 grass [*Spartina patens* (Ait.) Muhl.; common names: salt meadow grass, salt marsh grass, or high-water grass] grown under elevated CO_2. The plants grew in a brackish marsh of the Rhode River, a subestuary of the Chesapeake Bay, which is an inlet from the Atlantic Ocean along the eastern coast of the United States. Within each community, circular plots with open-top chambers were established in April 1987 with three different treatments and the treatments continued until December 1988. Chambers were exposed to elevated CO_2 (681 μmol/mol in the 1988 season) or the ambient level (343 μmol/mol in 1988). Plots with no chambers were called the control treatment. The substrate underlying these marsh communities was mainly organic peat, consisting of a zone of dense roots and rhizomes (both living and dead) to a depth of about 0.5 m. Below this zone, the peat became progressively more decomposed and fluid in composition. After 1 year of treatment (April 1988), peat cores were extracted from each plot and repacked with commercial peat moss. Roots and rhizomes were then allowed to grow back into this peat substrate. Roots and rhizomes were extracted from regrowth cores between November 30 and December 3, 1988.

There were marked differences in root growth between the two communities. In the *Scirpus* community, root growth was evenly distributed throughout the 15 cm profile (Figure 12.14). Exposure to elevated CO_2 had a pronounced effect on root growth in this C_3 species, resulting in an 83% increase in total root biomass. In the *Spartina* community, root growth was unevenly distributed, with 70%–80% of the biomass occurring in the top 5 cm and less than 4% occurring between 10 and 15 cm. There was no response by this C_4 community to elevated CO_2 (Figure 12.15). Curtis et al. (1990) point out that, as a group, C_4 species respond less to elevated CO_2 than do C_3 species, and

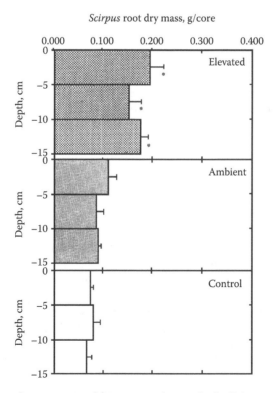

FIGURE 12.14 Dry mass of roots recovered from regrowth cores in the *Scirpus* community under elevated, ambient, and control treatments. Control treatments had no chamber over them. Narrow bars indicate ± 1 standard error; *$P < 0.05$ (significant difference between elevated and ambient). (Reprinted from Curtis, P.S., Balduman, L.M., Drake, B.G., and Whighan, D.F., *Ecology*, 71, 2001, 1990, Fig. 1. With permission. Copyright 1990, Ecological Society of America.)

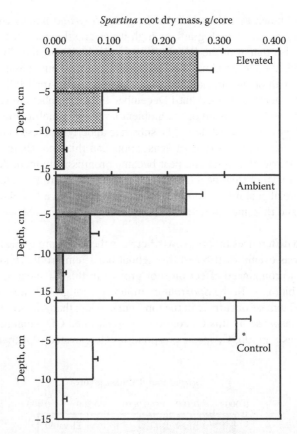

FIGURE 12.15 Dry mass of roots recovered from regrowth cores in the *Spartina* community under elevated, ambient, and control treatments. Control treatments had no chamber over them. Narrow bars indicate ±1 standard error; $*P < 0.05$ (significant difference between ambient and control). (Reprinted from Curtis, P.S., Balduman, L.M., Drake, B.G., and Whighan, D.F., *Ecology*, 71, 2001, 1990, Fig. 2. With permission. Copyright 1990, Ecological Society of America.)

their results are consistent with this observation. Curtis et al. (1990) did not report shoot growth for the two species.

TROPICAL GRASSES

Ghannoum et al. (1997) in Australia studied both shoot and root growth of *Panicum* under elevated CO_2. They chose this genus, because it contains species with the C_3 and C_4 photosynthetic pathways, many of which are used in the Australian pastoral industry. A number of the C_3 *Panicum* species are adapted to tropical climates, thus offering an opportunity for the comparison of the effects of CO_2 enrichment on species with different photosynthetic pathways but belonging to the same genus, while grown under similar conditions. The two tropical grasses that they studied were *Panicum laxum* (C_3) and *P. antidotale* (C_4). They gave no common names for the plants. The common name of the genus *Panicum* is panic-grass (Fernald 1950), and it is a great genus with more than 400 species widely spread over the globe (Bailey 1974). The plants grew in pots with a soil collected from the bush at Mt. Tomah, New South Wales. (They do not give the name of the soil.) After sowing, the pots were transferred into two naturally lit greenhouses at 350 or 700 μmol/mol CO_2. Half of each of the two glasshouses was covered with shade cloths cutting out 60% of incident sunlight. Ample water and nutrients were supplied to the plants, which were harvested at the end of the vegetative period and before maturity.

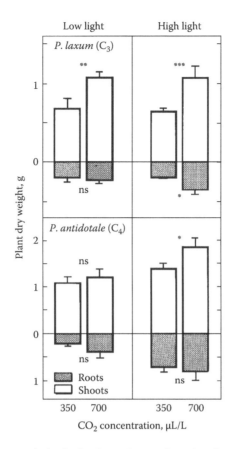

FIGURE 12.16 Dry matter accumulation in the shoots (open columns) and roots (solid columns) of *P. laxum* (C₃) and *P. antidotale* (C₄) grown at 350 and 700 μmol/mol CO_2 concentration and receiving 40% (low light) or 100% (high light) of direct sunlight. Each column represents the mean of four replications and each error bar represents one standard error. Significance levels for the CO_2 treatment are shown: ***$P < 0.001$; **$P < 0.01$; *$P < 0.05$; ns, not significant. (Reprinted from Ghannoum, O., von Caemmerer, S., Barlow, E.W.R., and Conroy, J.P., *Aust. J. Plant Physiol.*, 24, 227, 1997, Fig. 1. With permission. Copyright 1997, CSIRO Publishing.)

Total plant dry weight was enhanced by growth at elevated CO_2 concentration for both light treatments (shade and full sunlight) in *P. laxum* (C₃) (Figure 12.16, top), while enhancement occurred only in the high light treatment in *P. antidotale* (C₄) (Figure 12.16, bottom). Root growth was not increased by elevated CO_2, except for the C₃ plant under high light. They concluded that lack of response of C₄ plants to elevated CO_2 may be due to light limitation.

SUMMARY

Flowering plants evolved about 120 million years ago during the Cretaceous, when the atmospheric CO_2 concentration was 3000 μmol/mol. The concentration of CO_2 in the air declined after the Cretaceous, and C₄ plants apparently evolved when atmospheric CO_2 concentrations were lower (e.g., 180 μmol/mol CO_2) than they are now (385 μmol/mol, as of 2008). As the CO_2 concentration increases from 0 to 400 μmol/mol, the photosynthetic rate of C₄ plants increases. Above about 400 μmol/mol CO_2, the photosynthetic rate of C₄ plants often tends to level off. Because ambient CO_2 concentration (385 μmol/mol in 2008) apparently may be near the optimal for C₄ plants (about 400 μmol/mol), augmenting the CO_2 concentration may not always result in increased growth for C₄ plants, as it does for C₃ plants. However, the photosynthetic rate of C₃ plants increases and shows no sign of leveling off as the CO_2 concentration increases. At current levels of CO_2, the dry matter production of C_4 plants is higher than that of C_3 plants. When C_3 and C_4 plants are compared in the same experiment, the water use efficiency of the C_4 plants is greater than that of the C_3 plants. The $\delta^{13}C$ of C_3 plants is more negative (about $-20\permil$) than that of C_4 plants (about $-11\permil$), and the $\delta^{13}C$ value gets more negative as the CO_2 concentration increases for both C_3 and C_4 plants.

APPENDIX

BIOGRAPHY OF MELVIN CALVIN

Melvin Calvin, U.S. chemist and winner of the 1961 Nobel Prize in Chemistry for research into photosynthesis, was born April 8, 1911, in St. Paul, Minnesota. He has been the only plant physiologist to be named Nobel laureate. He received a BS degree from the Michigan College of Mining and Technology, Houghton, in 1931 and a PhD from the University of Minnesota in 1935. He then spent 2 years (1935–1937) at the University of Manchester, England, as a fellow of the Rockefeller foundation in the laboratory of Michael Polyani (Goodman 1971), the scientific and cultural giant (Benson 1997). At Manchester, he was introduced to coordination chemistry, and it became a central theme of his scientific life (Calvin 1996).

Calvin himself describes his time with Polyani (Calvin 1996) as follows: "After I finished my graduate studies at the University of Minnesota, I was fortunate enough to become a postdoctoral student of Professor Michael Polanyi at the University of Manchester. This was a most fortunate encounter for my whole scientific career, because in Manchester, working for Polyani, I became aware of the freedom of thought that allowed me to undertake work in any area of science that seemed appropriate to the questions I was faced with. I was not limited to physical chemistry, organic chemistry, biochemistry, or biology, but encompassed them all to some degree."

By good fortune he was invited in 1937 by Professor Gilbert N. Lewis to join the Department of Chemistry of the University of California, Berkeley, as instructor, to collaborate on Lewis's study of the color of molecules (Benson 1997). [Gilbert Newton Lewis (1875–1946) was a U.S. chemist whose realization of the significance of the electron pair in molecular structure laid the foundation for a revised theory of valence (Latimer 1971).] Lewis also gave him the freedom of scientific exploration as had Polyani (Calvin 1996). At Berkeley, Calvin became a colleague of Ernest Lawrence (Moses 1997). [Ernest Orlando Lawrence (1901–1958) was a U.S. physicist and winner of the 1939 Nobel Prize in Physics for his invention of the cyclotron.] Lawrence founded the Radiation Laboratory at Berkeley in 1931 (Moses 1997). It was in this laboratory that Sam Ruben and Martin Kamen discovered ^{14}C in 1940 (Moses 1997). Research on the path of carbon in photosynthesis started before World War II, but the war put a stop to the research (Moses 1997). By the time the war was over, Ruben had died in a laboratory accident and Kamen had left Berkeley for political reasons (Moses 1997). But the availability of ^{14}C had grown enormously as the result of its production in nuclear reactors. Late in 1945, Lawrence, who then in effect controlled access to ^{14}C, suggested to Calvin that he make use of it, both to synthesize labeled compounds for medical research and to resume the photosynthesis exploration. In the postwar days of "big" science in the style of the Manhattan Project (in which the University of California had been involved), Lawrence made available to Calvin the funding and laboratory space that he needed (Moses 1997). [The Manhattan Project was $2,000,000,000 expended upon industrial plants and laboratories to produce the first atomic bomb, which was tested on July 16, 1945 (Freidel 1971). It got its name from the Manhattan District, a division of the U.S. Army Corps of Engineers, established in 1942, which produced the atomic bomb (Friend and Guralnik 1959).]

The definitive experiments that determined the path of carbon in photosynthesis were done after the war in the period 1946–1956 (Calvin 1996). In these studies, Calvin showed the chemical pathways that are followed in green plants during the conversion of stored light energy, CO_2, and water into the thousands of compounds essential for their growth. The unraveling of the intricate details of photosynthesis is significant, because all life on earth is ultimately dependent upon the conversion of sunlight to useable chemical energy (Goodman 1971). His work takes on even more importance today, as methods to mimic photosynthesis synthetically are under active investigation to produce fuels.

Calvin himself describes the "turning point" that allowed him to find the path of carbon in photosynthesis (Calvin 1996). An elusive aspect of the research was to determine the cyclic character of the path of carbon. The turning point was, indeed, just a "moment". He was sitting in his car, probably parked in a red zone, while his wife (Genevieve Jemtegaard, daughter of Norwegian emigrant parents; they had two daughters and a son) was on an errand. He says, "I had for some months some basic information from the laboratory that was incompatible with everything that, up until then, I knew about the cyclic nature of the process. While sitting at the wheel of the car, the recognition of the missing compound occurred just like that—quite suddenly. Suddenly, also, in a matter of seconds, the complete cyclic character of the path of carbon became apparent to me. But the original recognition of PGA and how it got there, and how the CO$_2$ acceptor might be regenerated, all occurred in the matter of 30 s. So, there is such a thing as inspiration, I suppose, but one has to be ready for it. I don't know what made me ready at that moment, except I didn't have anything else to do but sit and wait. And perhaps that in itself has some moral" (Calvin 1996). He goes on further to say, "My personal philosophy perhaps is expressed best in the following statement: 'There is no such thing as pure science. By this I mean that physics impinges on astronomy, on the one hand, and chemistry and biology, on the other. The synthesis of a really new concept requires some sort of union in one [sic] mind of the pertinent aspects of several disciplines...It's no trick to get the right answer when you have all the data. The real creative trick is to get the right answer when you have only half of the data in hand, and half of it is wrong, and you don't know which half. When you get the right answer under those circumstances, you are doing something creative' " (Calvin 1996).

After he and his associates determined the path of carbon, he returned to his earlier interests in chelates and other problems in physical organic chemistry (Moses 1997). He branched out into the biochemistry of learning, for a short while using the flatworm *Planaria*. After they had been taught to solve simple maze problems, it was claimed that the animals could be chopped in half, whereupon the head portion grew a new tail, the tail end a new head with both reconstructed animals remembering their erstwhile "parent's" training (Moses 1997). Feeding trained *Planaria* to their naïve cannibalistic relatives was said to confer new skills. There was talk of feeding minced professors to the students (Moses 1997). He also was deeply involved with origin-of-life questions even before probes were sent to the moon and Mars. Early in the 1970s, he established a group to work on cancer. In the 1980s, he considered the use of certain plant oils as substitute diesel fuels should the world experience a severe petroleum shortage (Moses 1997).

Calvin was known for impressive presentations of his research and ideas. At one meeting of the American Association for the Advancement of Science, he began his symposium lecture on the mechanism of photosynthesis in his usual hesitating manner (Benson 1997). But he ended with a crescendo of rapid-fire supportive information, and the audience was deeply impressed with his solution of a central problem of plant physiology. Even Professor C.B. van Niel was so touched that he jumped from his seat in the front row and, with tears in his eyes, congratulated Calvin for the brilliant solution of the age-old problem (Benson 1997). [Cornelis Bernardus van Niel (1897–1985) was a Dutch–American microbiologist who, in 1930, showed that two small groups of bacteria, the purple and green bacteria, can also use light energy (not just higher plants, eukaryotic algae, and blue-green algae) for the synthesis of cell materials (Stanier et al. 1963, p. 273).]

Melvin Calvin was a fearless scientist, totally unafraid to venture into new fields like carcinogenesis, origin of life, immunochemistry, petroleum production from plants, farming, moon rock analysis, and development of novel synthetic biomembrane models for plant photosystems (Benson 1997). His love of plants shone daily by the ever-present small flower or fir twig in his lapel placed there by his wife (Benson 1997). He joined the American Society of Plant Physiologists (ASPP) in 1950 and served the society as its president in 1963–1964, the only Nobel Laureate to serve as the society's president (Benson 1997). He received ASPP's Stephen Hales Prize in 1956 "for charting the path of carbon in photosynthesis" (Benson 1997). In 1971 he was elected president of the American Chemical Society. In addition to the Nobel Prize, he also held a National Medal of Science. He was a member of the National Academy of Sciences and a foreign member of the Royal Society (Benson 1997).

For an interview that Frank B. Salisbury had with Melvin Calvin on January 4, 1977, see the textbook that Salisbury coauthored with Cleon Ross (Salisbury and Ross 1978, pp. 152–154). Calvin tells how he got into science, his involvement with laboratory work, the price one has to pay to go into science, and the rewards of science.

A press release from the University of California said that Calvin died at age 85 on January 8, 1997, in Berkeley, California, after years of failing health.

13 Elevated Atmospheric Carbon Dioxide: Plant Anatomy

INTRODUCTION

The objective of this chapter is to report the effects of elevated levels of CO_2 on plant anatomy. In particular, we focus on leaves and xylem tissue (including wood), because they are intimately associated with water loss and water transport. The effect of elevated CO_2 on roots and stomata is discussed in Chapter 6 (roots), Chapter 8 (stomatal conductance), and Chapter 9 (stomatal density).

LEAVES

Radoglou and Jarvis (1990b) investigated four poplar clones under ambient (350 μmol/mol) and double the ambient (700 μmol/mol) atmospheric CO_2 concentrations. The clones were as follows: Columbia River (CoR) (*Populus trichocarpa* Torr. & Gray V-24); Beaupre (Bea) (*P. trichocarpa* Torr. & Gray V-235 × *P. deltoides* Bartr. ex Marsh S. 1-173); Robusta (Rob) [*P. euramericana* (Dode) Guinier]; and Raspalje (Ras) (*P. trichocarpa* Torr. & Gray V-235 × *P. deltoides* Bartr. ex Marsh S. 1-173). Cuttings were planted in pots containing a mixture of loam, peat, and perlite and were grown in open-top chambers inside a glasshouse for 92 days.

Plants of all the clones grown in high CO_2 were taller and had more branches (Figure 13.1a). Mortensen (1994a) also found that elevated CO_2 (twice the ambient level) increased the number of lateral shoots in *Betula pendula* Roth. (European white birch), as well as stem diameter. The number of poplar leaves per plant (Figure 13.1b) and the total leaf area per plant (Figure 13.1c) were higher in elevated CO_2. In general, high CO_2 concentration had no effect on individual leaf size (Figure 13.1d), and the increase in leaf area in high CO_2 concentration was brought about by the increase in leaf number rather than leaf size. Total biomass at the end of the growth period (92 days) was larger for plants grown in high CO_2, and the mean increase of the four clones in high CO_2 was 45%, which is similar to increases for other C_3 tree species (Radoglou and Jarvis 1990b).

There was no difference in the number of cells per unit leaf area (Figure 13.2a), dry weight (Figure 13.2b), or fresh weight (Figure 13.2c) between the CO_2 treatments. But there was a difference in the number of cells per unit dry weight among clones. This was mainly because Rob, which had more cells per unit dry weight, had very small cells.

All four clones exhibited an increase in total leaf thickness at the higher CO_2 concentration (Figure 13.3). The increase averaged 12% for the four clones. Both upper and lower epidermis tended to be thicker in leaves that developed in high CO_2. Palisade and spongy parenchyma were thicker with elevated CO_2 (Figure 13.3), especially the spongy parenchyma. These adaptations of leaves under elevated CO_2 are xeromorphic traits (Fahn and Cutler 1992, p. 50; Kirkham 2005, p. 363), and this suggests that leaves under elevated CO_2 are more drought resistant.

Differences in leaf thickness were reflected in differences in the ratio of leaf weight (W) to leaf area (A) or the ratio W/A (Radoglou and Jarvis 1990b). We take a moment to comment about this ratio. It often is incorrectly called the "specific leaf weight" (Jarvis 1985). Specific leaf weight always equals 1.0 (unitless). In the international system of units, "specific" has a particular definition. It is a physical quantity and should be restricted to the meaning "divided by the mass." Thus, "specific leaf weight" is a nonsensical quantity. The weight of a leaf divided by its mass usually will be unity.

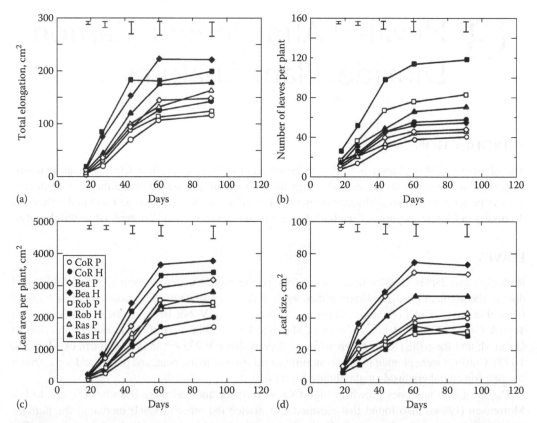

FIGURE 13.1 Effects of present ($P = 350\,\mu mol/mol$) and high ($H = 700\,\mu mol/mol$) CO_2 concentrations on stem elongation per plant (total length of stem and branches) (a), on number of leaves per plant (b), on total leaf area per plant (c), and on leaf size (d) of four poplar clones: CoR, Bea, Rob, and Ras. Vertical bars indicate ± 1 standard error of the mean (degrees of freedom = 17). Each point is the mean of three plants. (Reprinted from Radoglou, K.M. and Jarvis, P.G., *Ann. Bot.*, Fig. 1, 1990b by permission of Oxford University Press.)

Leaves of all the clones grown in the high CO_2 concentration had a larger dry weight per unit leaf area than those grown in the ambient CO_2 concentration (Figure 13.4a). Because there were no differences in the number of cells per unit leaf area in the high CO_2 treatment (Figure 13.2a), but the mesophyll cells and the intercellular spaces (as observed in leaf cross sections) were larger, it appeared that cell enlargement was more sensitive to CO_2 than cell division. Mesophyll cell size determines the internal leaf surface area available for the absorption of CO_2 (Radoglou and Jarvis 1990b). Thus, the thickness of the mesophyll may be related directly to photosynthesis.

We now comment on cell enlargement and cell division in a leaf. Leaf development has two distinct phases. In the first, cell number increases logarithmically from the time that the leaf is in the bud until it unfolds (Kirkham et al. 1972). Unfolding marks the beginning of the second phase in which increase in cell number with time gradually declines to zero. During the latter part of leaf development, cell expansion predominates. Kirkham et al. (1972) found that, in radish (*Raphanus sativus* L.) cotyledons, the critical turgor pressure for increased cell division (5–6 bar) was greater than that for increased cell enlargement (3 bar). The DNA extracted from the radish cotyledons gave a measure of cell number. The results suggested that increase in cell enlargement occurs at lower turgor pressures than those required for increased cell division. Increased division (increase in DNA content), with resulting cell expansion and growth, appears to take place only when turgor pressure is high. In the work by Radoglou and Jarvis (1990b), cell numbers were not affected by CO_2 treatment, and this indicated that DNA replication was not affected by the elevated CO_2, but cell enlargement was.

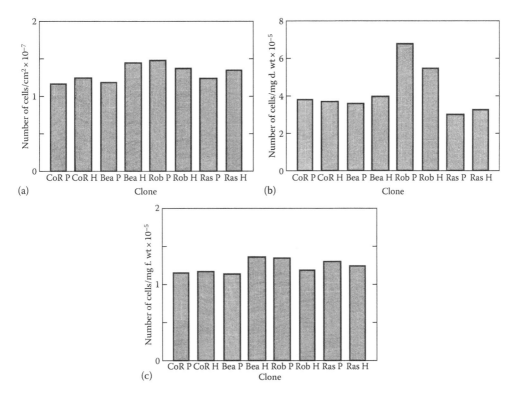

FIGURE 13.2 Response of number of cells per unit leaf area (a), per fresh weight (b), and per dry weight (c) to present ($P = 350\,\mu mol/mol$) and high ($H = 700\,\mu mol/mol$) CO_2 concentration of four poplar clones: CoR, Bea, Rob, and Ras. The $\pm$ standard error values of the mean (degrees of freedom = 17) were 0.20×10^{-7}, 0.19×10^{-5}, and 0.76×10^{-5}, respectively. (Reprinted from Radoglou, K.M. and Jarvis, P.G., *Ann. Bot.*, Fig. 3, 1990b by permission of Oxford University Press.)

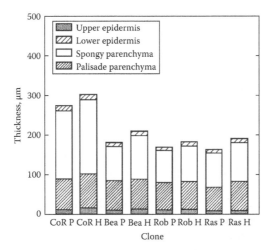

FIGURE 13.3 Effects of present ($P = 350\,\mu mol/mol$) and high ($H = 700\,\mu mol/mol$) CO_2 concentration on thickness of the total leaf, upper epidermis, palisade, spongy parenchyma, and lower epidermis of four poplar clones: CoR, Bea, Rob, and Ras. The $\pm$ standard error values of the mean (degrees of freedom = 17) were 3.5, 0.15, 0.37, 1.13, and 0.10, respectively. Each point is the mean of three plants. Four sections were sampled from each of two leaves per plant. (Reprinted from Radoglou, K.M. and Jarvis, P.G., *Ann. Bot.*, Fig. 4, 1990b by permission of Oxford University Press.)

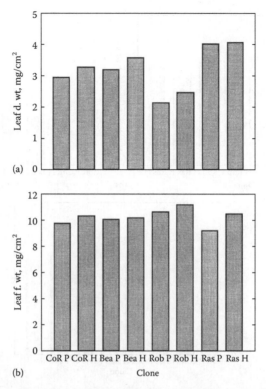

(a)

(b)

FIGURE 13.4 The response of leaf dry weight (a) and fresh weight (b) to present ($P = 350\,\mu$mol/mol) and high ($H = 700\,\mu$mol/mol) CO_2 concentration of four poplar clones: CoR, Bea, Rob, and Ras. The $\pm$ standard error values of the mean (degrees of freedom = 17) were 0.256 and 0.104, respectively. (Reprinted from Radoglou, K.M. and Jarvis, P.G., *Ann. Bot.*, Fig. 6, 1990b by permission of Oxford University Press.)

Hovenden and Schimanski (2000) also found that the leaf weight per unit leaf area (W/A) was increased at a higher level of CO_2 compared to a lower level. They grew clones of four different genotypes of southern beech [*Nothofagus cunninghamii* (Hook.) Oerst.] at ambient (370 µmol/mol CO_2) and depleted (170 µmol/mol) levels of CO_2. The genus *Nothofagus* has a well-preserved fossil record stretching back 50 million years. The oldest fossil leaves have greater stomatal density than leaves of extant plants. This species has survived different atmospheric CO_2 concentrations. The lowest, about 160 µmol/mol CO_2, occurred 15,000–20,000 years ago. Hovenden and Schimanski (2000) wanted to study the way the plants responded to changes in atmospheric CO_2—from the past low to the current 370 µmol/mol.

The genotypes were labeled according to the number of the site where they collected the cuttings (sites 3 and 7) and the number of the tree at that site. So the labels were 3/2, 3/5, 3/6, and 7/6. The cuttings, grown in controlled environmental cabinets, were put individually in pots containing a mixture of equal volumes of sand and pine bark, to which a slow release fertilizer was added. In addition to measurements of leaf area and leaf weight, they also measured the number of stomata and epidermal cells of recently expanded leaves. *N. cunninghamii* is hypostomatous, so only abaxial surfaces were examined. Stomatal index was calculated as [(stomatal density)/(stomatal density + epidermal cell density)] × 100.

Leaf area and leaf weight were not consistently reduced for all genotypes at the lower CO_2 concentration (Figure 13.5, top and middle). However, specific leaf weight, which they defined as leaf area (A) divided by leaf weight (W) [the reciprocal of W/A, which Radoglou and Jarvis (1990b) determined], was increased at the lower CO_2 concentration (Figure 13.5, bottom) for all genotypes. So the data of Radoglou and Jarvis (1990b) and Hovenden and Schimanski (2000)

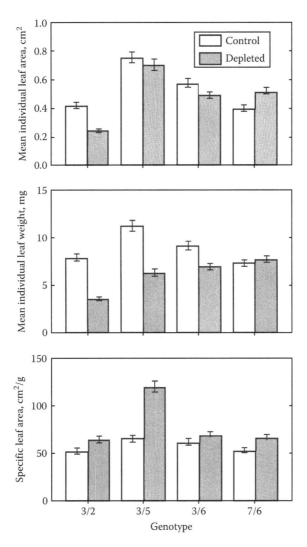

FIGURE 13.5 Gross leaf morphological characters of four genotypes of *N. cunninghamii* grown in control (370 μmol CO_2/mol, open columns) and depleted (170 μmol CO_2/mol, shaded columns) conditions. Error bars indicate the standard error. (Reprinted from Hovenden, M.J. and Schimanski, L.J., *Aust. J. Plant Physiol.*, 27, 281, 2000, Fig. 2. With permission. Copyright 2000, CSIRO Publishing.)

agree. The ratio *W/A* is increased under elevated CO_2, and its reciprocal, *A/W*, is decreased under elevated CO_2. Stomatal index, number of epidermal cells, and number of stomata differed among the southern-beech genotypes (Figure 13.6, top, middle, and bottom, respectively), but the effect of CO_2 was not consistent across genotypes. Hovenden and Schimanski (2000) concluded that the most universal response to low CO_2 concentration in terms of leaf characters was the increase in specific leaf area (or decrease in *W/A*). In the past, growth at a low CO_2 concentration probably had an effect on photosynthetic rate, and this was translated into thinner leaves (Figure 13.5, bottom).

Thomas and Bazzaz (1996) determined the effects of elevated CO_2 on leaf shape of dandelion (*Taraxacum officinale* Weber). They chose dandelion because it has "heteroblastic" leaf development. Esau (1977, p. 341) defines *heteroblasty*. If leaf ontogeny varies during different stages of development, so that one can distinguish between juvenile and adult leaf forms, the phenomenon is called *heteroblasty* (as contrasted with *homoblasty* when leaves are all alike). In dandelion, the first

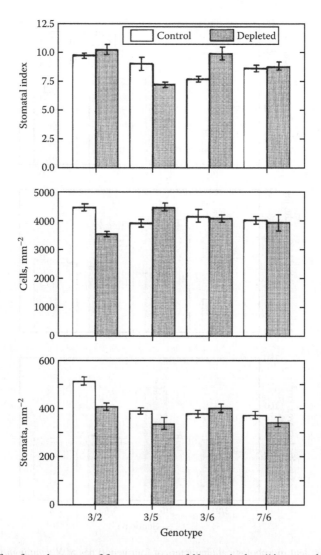

FIGURE 13.6 Leaf surface characters of four genotypes of *N. cunninghamii* in control (370 μmol CO_2/mol, open columns) and depleted (170 μmol CO_2/mol) conditions. Error bars indicate the standard error. (Reprinted from Hovenden, M.J. and Schimanski, L.J., *Aust. J. Plant Physiol.*, 27, 281, 2000, Fig. 3. With permission. Copyright 2000, CSIRO Publishing.)

few sets of leaves possess entire margins, while (under high light) subsequently produced leaves show a characteristic dentition pattern from which the plant's common name is derived (dandelion: from the French "dent de lion" or "lion's tooth") (Friend and Guralnik 1959). They grew the dandelions in pots with pure vermiculite fertilized weekly and placed them in greenhouses at two different CO_2 levels: 350 (ambient) and 700 μmol/mol. Leaf area was measured with a leaf area meter and leaf shape was analyzed on the basis of digitized images. They calculated a "dissection index" (DI), standardized so that a circular leaf would have a value of 1.0 or $DI = (P/2A)(A/\pi)^{1/2}$, where P is leaf perimeter and A is leaf area.

As found by Radoglou and Jarvis (1990b) and Hovenden and Schimanski (2000), specific leaf weight (the W/A of Radoglou and Jarvis) was higher at elevated CO_2. Leaf area increased under elevated CO_2 (Figure 13.7, top). Thomas and Bazzaz (1996) pointed out that in previous studies plants grown at elevated CO_2 have been found to produce a greater number of leaves per unit time than plants grown at ambient CO_2. [Note that Chaudhuri et al. (1989) found more wheat leaves at

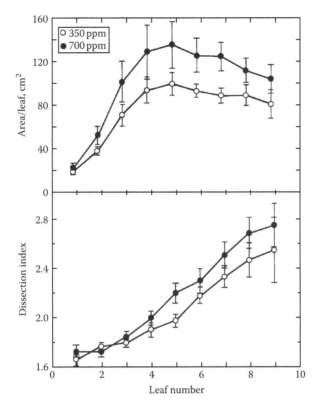

FIGURE 13.7 Individual leaf area and DI in relation to leaf number for individually grown *T. officinale* at ambient and elevated CO_2 levels. DI is calculated as the ratio of leaf perimeter to the square root of leaf area, standardized so that $DI = 1$ for a circular leaf. All values are plotted ± standard error. (Reprinted from Thomas, S.C. and Bazzaz, F.A., *Am. J. Bot.*, 83, 106, 1996, Fig. 4. With permission. Copyright 1996, Botanical Society of America.)

elevated CO_2 than at ambient CO_2. See Table 14.2 on Phenology.] So Thomas and Bazzaz (1996) investigated the possibility that observed changes in leaf shape (DI) were due to differences in developmental order. For leaf numbers 3–9, there was a consistent trend toward higher DI values at elevated CO_2 (Figure 13.7, bottom). Their results indicate that growth at elevated CO_2 accelerated the development of "mature" or "high light" type leaves in dandelion. They hypothesized that changes in the relative concentrations of leaf carbohydrates versus mineral nutrients affected patterns of cell proliferation and expansion that determine leaf shape. Although they did not assay leaf carbohydrates, previous studies have found increases in leaf carbohydrates under elevated CO_2 in herbaceous plants. Thomas and Bazzaz (1996) noted that the toothier dandelion leaves under elevated CO_2 will have effects on canopy roughness, boundary layer conductance, and canopy-level optical properties.

COMPARISON OF C₃ AND C₄ LEAVES

Dippery et al. (1995) compared leaf area of C_3 and C_4 plants grown in growth chambers with CO_2 partial pressures of 15 Pa (below the Pleistocene minimum), 27 Pa (preindustrial), 35 Pa (current), and 70 Pa (predicted for future).

We pause to consider the Pleistocene Epoch, which Dippery et al. (1995) discussed. It is before our current Recent Epoch. See Figure 9.10 for a diagrammatic representation of the geological time scale. We also mention the Pleistocene in Chapter 12 (see Figure 12.6). The Cenozoic, shown in the geological time scale in Figure 9.10, is divided into the Quaternary Period and the Tertiary

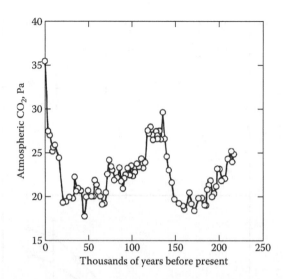

FIGURE 13.8 Atmospheric partial pressures of CO_2 during the last 220,000 years based on ice cores from the Vostok station in Antarctica. (With kind permission from Springer Science+Business Media: *Oecologia*, Effects of low and elevated CO_2 on C_3 and C_4 annuals. II. Photosynthesis and leaf biochemistry, 101, 1995, 21–28, Tissue, D.T., Griffin, K.L., Thomas, R.B., and Strain, B.R., Fig. 1.)

Period. The Quaternary Period is divided into the Pleistocene Epoch and the Recent Epoch (the last 10,000 years). The Pleistocene Epoch is considered to be the period of the "ice ages," when formation of large ice sheets on the continents began. (We mention the ice ages in the section "Evolution of C_4 Plants" in Chapter 12.) It started 1–2 million years ago (Spencer 1962, pp. 12, 386). Ice core data indicate that CO_2 partial pressure was as low as 18 Pa during the Last Glacial Maximum of the Pleistocene Epoch (Figure 13.8) (Tissue et al. 1995; Dippery et al. 1995).

Dippery et al. (1995) grew two annual weeds, *Abutilon theophrasti* Medic. (velvet leaf) (C_3) and *Amaranthus retroflexus* L. (pigweed) (C_4), from seed in pots containing gravel, Turface, and topsoil (not named). Plants were harvested 7, 14, 28, and 35 days after planting. Leaf area was measured with a leaf area meter. Specific leaf mass (SLM) was calculated as the ratio of leaf biomass to leaf area. [Dippery et al. (1995) use the term "mass" instead of "weight," but in this chapter we are using mass and weight interchangeably.] Relative growth rate (RGR) was calculated using the following equation: RGR=NAR×LAR, where NAR is the net assimilation rate (change in biomass per unit leaf area per unit time) and LAR is the leaf area ratio. Leaf area ratio was calculated as the ratio of leaf area to total plant biomass.

After 35 days of growth, CO_2 had no effect on the RGR in the C_4 plant (Figure 13.9, bottom), but the C_3 plant grown with 70 Pa CO_2 had a greater RGR after 7 days (Figure 13.9, top). In 15 Pa, the C_3 plants had lower RGRs at the beginning of the experiment. They also had a lower leaf area and SLM than the C_3 plants grown with higher concentrations of CO_2 (Figure 13.10, upper left and lower left, respectively), and they aborted reproduction at this low CO_2 concentration. Some consider 15 Pa CO_2 as the minimum CO_2 level necessary for photosynthesis in C_3 plants (Dippery et al. 1995). Leaf area of the C_4 plants under the four CO_2 concentrations was similar (Figure 13.10, upper right), but the SLM of the C_4 plants grown with 15 Pa CO_2 was lower than the SLMs of plants grown at the three higher CO_2 concentrations (Figure 13.10, lower right). Thinner leaves at low CO_2 may serve to increase the rate of CO_2 diffusion to the chloroplasts (Tissue et al. 1995). The results of Dippery et al. (1995) suggest that C_4 species may have been more competitive against C_3 species during periods of low atmospheric CO_2 such as the Pleistocene.

In a follow-up paper to that of Dippery et al. (1995), Tissue et al. (1995) also compared *A. theophrasti* and *A. retroflexus* grown under the same CO_2 concentrations (15, 27, 35, and 70 Pa CO_2). Tissue et al. (1995) measured photosynthetic rate with an open-type gas exchange system.

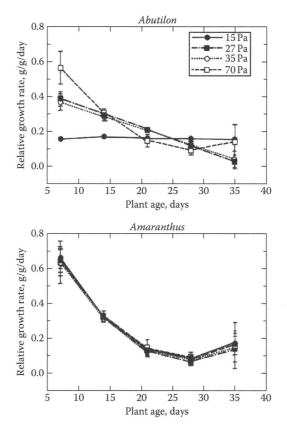

FIGURE 13.9 RGR as a function of plant age and CO_2 partial pressure of *A. theophrasti* (C_3) and *A. retroflexus* (C_4). The 95% confidence intervals (error bars) are also shown. (With kind permission from Springer Science+Business Media: *Oecologia*, Effects of low and elevated CO_2 on C_3 and C_4 annuals. I. Growth and biomass allocation, 101, 1995, 13–20, Dippery, J.K., Tissue, D.T., Thomas, R.B., and Strain, B.R., Fig. 4.)

Using this system, net assimilation versus the internal CO_2 partial pressure (A–C_i) curves was measured on fully developed leaves of plants 22–26 days after emergence. The system also recorded the CO_2 concentration of the atmosphere (C_a). The photosynthetic rate of the C_4 plant was similar under the four CO_2 concentrations, but the photosynthetic rate of the C_3 plant increased as the CO_2 concentration increased (Figure 13.11). The C_i/C_a ratio was higher in the C_3 plant grown at 15 Pa CO_2 compared with the other CO_2 concentrations (Figure 13.12a). In the C_4 plant, the C_i/C_a ratios were variable but, in general, were not different among CO_2 treatments (Figure 13.12b). In the C_3 plant, increasing the C_i concentration increased CO_2 for photosynthesis and increased photosynthetic rate (Figure 13.11). In the C_4 plant, which efficiently recycles CO_2 within the leaf, increasing the C_i had little effect on photosynthesis. Kirkham et al. (1991) also found that the C_i/C_a ratio did not differ in a C_4 plant (*Andropogon gerardii* Vitman, big bluestem) grown under ambient CO_2 (337 µmol/mol) and elevated CO_2 (658 µmol/mol) and exposed to two watering regimes: field capacity (Figure 13.13) or half field capacity (Figure 13.14).

LEAF ULTRASTRUCTURE

The first detectable effects of elevated CO_2 are most likely to be found during early plant development when leaf expansion and the chloroplasts are vulnerable to a changing external environment. Therefore, Robertson and Leech (1995) examined the changes in cell and chloroplast development of juvenile wheat (*Triticum aestivum* L. cv. Hereward) leaf tissue under a high and low

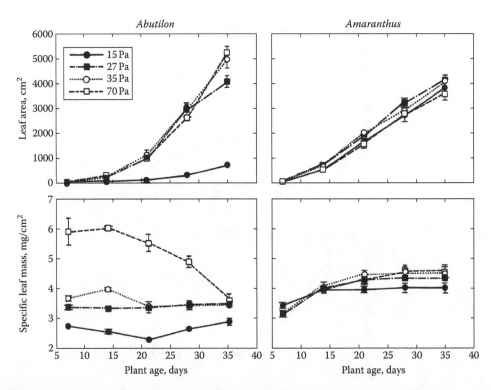

FIGURE 13.10 Leaf area and SLM as a function of plant age and CO_2 partial pressure for *A. theophrasti* (C_3) and *A. retroflexus* (C_4). Symbols represent means, and error bars represent ± standard error. (With kind permission from Springer Science+Business Media: *Oecologia*, Effects of low and elevated CO_2 on C_3 and C_4 annuals. I. Growth and biomass allocation, 101, 1995, 13–20, Dippery, J.K., Tissue, D.T., Thomas, R.B., and Strain, B.R., Fig. 5.)

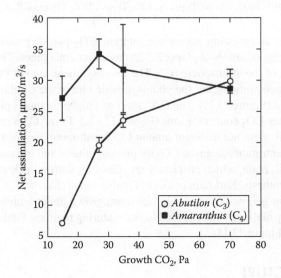

FIGURE 13.11 Net assimilation rate measured under ambient growth conditions for *A. theophrasti* (C_3) (open circles) and *A. retroflexus* (C_4) (closed squares) after 22–26 days at 15, 27, 35, or 70 Pa CO_2. Values are means ± standard errors ($n=3$). (With kind permission from Springer Science+Business Media: *Oecologia*, Effects of low and elevated CO_2 on C_3 and C_4 annuals. II. Photosynthesis and leaf biochemistry, 101, 1995, 21–28, Tissue, D.T., Griffin, K.L., Thomas, R.B., and Strain, B.R., Fig. 2.)

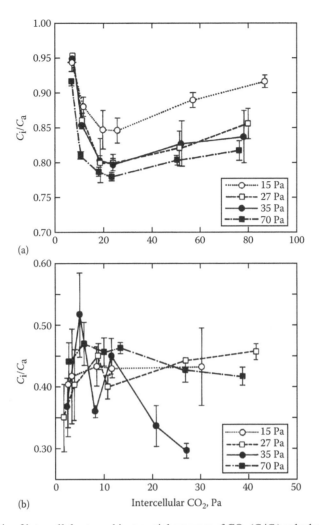

FIGURE 13.12 Ratio of intercellular to ambient partial pressure of CO_2 (C_i/C_a) calculated from A–C_i curves for (a) *A. theophrasti* (C_3) and (b) *A. retroflexus* (C_4) after 22–26 days at 15, 27, 35, or 70 Pa CO_2. Values are means ± standard errors ($n=3$). Note different scales for *Abutilon* and *Amaranthus*. (With kind permission from Springer Science+Business Media: *Oecologia*, Effects of low and elevated CO_2 on C_3 and C_4 annuals. II. Photosynthesis and leaf biochemistry, 101, 1995, 21–28, Tissue, D.T., Griffin, K.L., Thomas, R.B., and Strain, B.R., Fig. 5.)

concentration of CO_2. Plants grew in pots with fertilized sand. The pots were placed in controlled environmental cabinets with either ambient (350 μmol/mol) or elevated (650 μmol/mol) CO_2. The first leaf after 7 days of growth was harvested. Slices of leaf tissue (1 mm in thickness) were cut at a distance of 3–3.5 cm from the base of the leaf. They were fixed in glutaraldehyde followed by treatment with EDTA to ensure cell separation when tissue samples were teased apart on glass slides. Mesophyll cell plan area (μm²) and chloroplast face plan area (μm²) were measured from these cell populations. [The word "plan" means a diagram showing the arrangement in horizontal section of a structure (Friend and Guralnik 1959).] Tissue samples also were stained for starch with iodine, and variation in staining intensities between elevated CO_2 and ambient tissue was estimated using gray scale measurements on image analysis software.

Both mesophyll cell plan areas and individual chloroplast face plan areas were greater in elevated CO_2 (Figure 13.15b and d) compared to ambient CO_2 (Figure 13.15a and c). Thus, there was

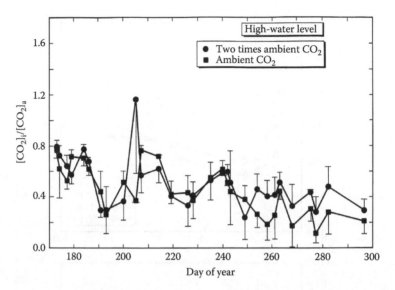

FIGURE 13.13 Ratio of intercellular to ambient CO_2 concentrations of big-bluestem leaves with two atmospheric CO_2 levels and grown under a high water level (field capacity). Vertical bars = ±standard deviation. Only half of each bar is drawn for clarity. (From Kirkham, M.B. et al., *Crop Sci.*, 31, 1589, 1991. Unpublished data.)

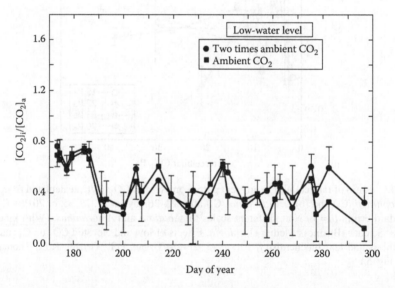

FIGURE 13.14 Ratio of intercellular to ambient CO_2 concentrations of big-bluestem leaves with two atmospheric CO_2 levels and grown under a low water level (half field capacity). Vertical bars = ±standard deviation. Only half of each bar is drawn for clarity. (From Kirkham, M.B. et al., *Crop Sci.*, 31, 1589, 1991. Unpublished data.)

acceleration in mesophyll cell and chloroplast expansion in young wheat leaves grown in elevated CO_2. Robertson and Leech (1995) estimated that the young wheat leaf cells from elevated CO_2-grown tissue at 7 days of age were about 12 h ahead of cells from comparable ambient CO_2-grown tissue. They suggested that this enhanced development in the wheat leaf in elevated CO_2 is facilitated by increased respiratory activity. They found that in elevated CO_2, the numbers of mitochondria per unit cell area in the youngest wheat leaf cells were more than doubled compared with ambient CO_2-grown tissue. Although increases in mesophyll cell area and chloroplast area were evident in elevated CO_2-grown tissue, chloroplast number per mesophyll cell was not changed.

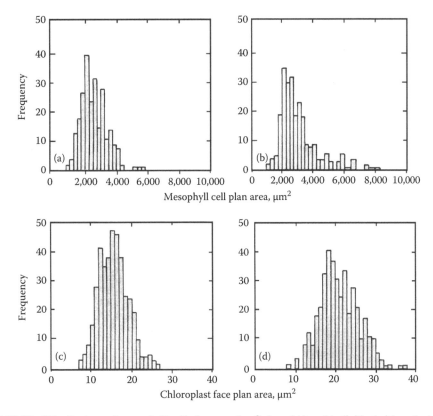

FIGURE 13.15 Distributions of mesophyll cell plan area (μm^2) (a and b) and individual chloroplast face plan area (μm^2) (c and d) measured in cell suspensions from the first leaf of 7-day-old wheat plants grown in ambient (350 $\mu mol/mol$) (a and c) or elevated (650 $\mu mol/mol$) CO_2 (b and d). For both elevated CO_2- and ambient CO_2-grown tissue, 250 individual cells were measured for mesophyll cell plan area (50 cells were measured from within each of five leaf samples), 400 individual chloroplasts for chloroplast face plan area (eight chloroplasts in 10 cells were measured from tissue within each of five leaf samples). (Reprinted from Robertson, E.J. and Leech, R.M., *Plant Physiol.*, 107, 63, 1995, Fig. 1. With permission. Copyright 1995, American Society of Plant Biologists.)

Robertson and Leech (1995) found a reduction in starch accumulation in the young wheat leaves grown in the elevated CO_2. They suggested that this may be related to the source-sink demands. Rapidly growing seedlings under elevated CO_2 are typically source limited. Therefore, additional carbohydrate resulting from enhanced CO_2 assimilation is immediately utilized for growth. In older wheat leaves (4 weeks old), they found much larger starch deposits in tissue grown in elevated CO_2. Thus, age of the plant under elevated CO_2 is an important factor that influences carbohydrate status.

Sanità de Toppi et al. (2003) studied the anatomy of stems and leaves of three legumes grown in free-air CO_2 enrichment (FACE) rings under elevated CO_2 (560 $\mu mol/mol$). Control plants grew under ambient CO_2 (concentration not stated). The plants were *Trifolium repens* L. (white clover), *Vicia hybrida* L. (common name unknown), and *V. sativa* L. (spring vetch). Leaf and stem samples were cut and observed under a microscope. They were stained to identify lignin. They measured vessel lumen diameter. In stems of *T. repens*, vascular bundles in CO_2-enriched plants had reduced lignin deposition of secondary walls. In stems of *V. sativa* and *V. hybrida*, no differences existed in lignification. However, in both of these species, spring xylem in stems had larger vessels in ambient CO_2 compared to elevated CO_2. Under ambient CO_2, *V. hybrida* and *V. sativa* had vessel diameters in stems of 63.5 ± 3.1 and $50.4 \pm 2.4\,\mu m$, respectively. Under elevated CO_2, *V. hybrida* and *V. sativa* had vessel diameters in stems of 45.6 ± 3.8 and $42.5 \pm 3.7\,\mu m$, respectively. In leaves, no differences in lignification and vessel diameter existed between ambient and elevated CO_2 plants. The data of Sanità de Toppi et al. (2003) show that effects of elevated CO_2 on the histology of the xylem tissue are species and organ specific.

WOOD ANATOMY AND DENSITY UNDER VARIABLE CO_2 CONCENTRATIONS

It is important to study wood anatomy and density because they determine hydraulic conductivity and mechanical strength (Zanne et al. 2010). Hättenschwiler et al. (1996) studied wood density as affected by CO_2 concentration. Four-year-old Norway spruce (*Picea abies* Karst.) trees (0.7 m tall) were planted in containers in early April 1993. The soil was a natural podsol from an old-growth montane spruce-fir forest near Lucerne, Switzerland. The profile was reestablished in the pots using horizons of the same depth as found in the field (15 cm B horizon, 5 cm A horizon, 3 cm raw humus, and 1.5 cm litter layer). Plants were exposed in growth chambers to 280 (i.e., preindustrial), 420, or 560 cm³/m³ (or 280, 420, or 560 μmol/mol) atmospheric CO_2. Stem discs of each individual tree were taken at harvest in mid-November 1995 and analyzed using x-ray densitometry. They measured a continuous wood density profile (Figure 13.16) along two radii for each of 162 samples.

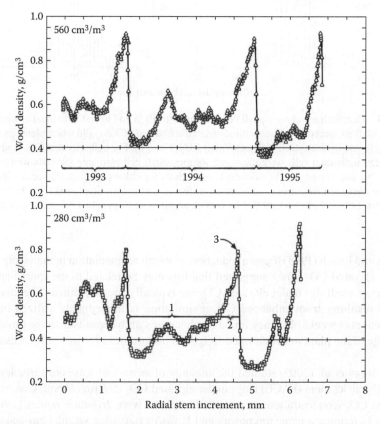

FIGURE 13.16 Examples of density profiles of the 3 year tree ring sequence formed under treatment conditions. Two profiles of the same spruce clone grown at 280 cm³ CO_2/m³ (lower panel) and at 560 cm³ CO_2/m³ (upper panel) at the same intermediate N deposition rate of 30 kg/ha/year. Tree properties are indicated as follows: total RW (1 + 2), early-wood width (1), late-wood width (2), and maximum late-wood density (3). (Reprinted from Hättenschwiler, S., Schweingruber, F.H., and Körner, Ch.: Tree ring responses to elevated CO_2 and increased N deposition in *Picea abies*, *Plant Cell Environ.*, 1996, 19, 1369–1378, Fig. 1. Copyright Wiley-VCH Verlag GmbH & Co. KGaA. With permission.)

These measurements provided two values (two radii) for six parameters for each tree ring: total ring width (RW), early-wood width, late-wood width, mean early-wood density, mean late-wood density, and maximum late-wood density.

Mean early-wood density was increased by CO_2 concentration (Figure 13.16). In 1995, it was 8% greater at the two higher CO_2 concentrations than at $280 \, cm^3 \, CO_2 \, m^{-3}$. Mean late-wood density was not affected by CO_2 concentration. The higher early-wood densities measured at elevated CO_2 concentration may have been caused by more rapid cell wall growth of early-wood tracheids or by the production of more but smaller early-wood cells. Whatever the cause, the higher mean early-wood density combined with the increased percentage of late-wood in trees grown at the two higher CO_2 concentrations led to a 12% increase in wood density compared to that measured in trees grown at $280 \, cm^3 \, CO_2 \, m^{-3}$. Higher stem wood density translates into higher specific gravity of wood. As specific gravity increases, wood strength properties usually increase because internal stresses are distributed among more molecular material. This is likely to increase the resistance of trees to external forces such as wind or snow loading and to improve the quality of timber (Hättenschwiler et al. 1996).

WOOD ANATOMY UNDER VARIABLE PRECIPITATION AND AMBIENT CO_2

Fonti and García-González (2004) studied the wood of European chestnut (*Castanea sativa* Mill.; also called Eurasian or Spanish chestnut) to see its relation to climatic variables. They did their work under ambient CO_2, but the results are instructive to see how wood changes with the environment, especially precipitation. Anatomical features, especially those related to vessels, bear ecologically relevant information. Yet few studies have been done with angiosperms to determine the relationship between tree growth and environmental conditions. Most work has been done with conifers. Only angiosperms have vessels and, in ring-porous wood, the vessels formed early in the spring have a wider diameter than those formed later in the season (Kirkham 2005, pp. 218–219). European chestnut is a ring-porous tree species, previously unstudied for dendroecological analyses. They obtained stem discs from chestnut trees that were 45–52 years old in the southern part of the Swiss Alps. Anatomical features of 51 trees were determined under a binocular microscope using an automatic image analysis system. Climatic data between 1956 and 1995 were collected from nearby weather stations at Lugano and Locarno, both in Switzerland. They performed a response function analysis. This is based on a multiple regression using climatic records as predictors and growth indices as predictants. The analysis gives a "response function" (no unit). Before computing the response functions, climate–growth relationships were explored initially by computing Pearson's correlation coefficients for the same data set. The response functions did not substantially differ from the correlation functions. So the response functions can be considered to show the correlation between climatic variables (temperature and rainfall) and wood anatomy.

The response functions for RW and mean vessel lumen area (MVA) are shown in Figure 13.17 (all 51 samples combined). RW (as well as late wood) responded positively to both the May temperature (dotted bars, Figure 13.17, left-hand side) and the July rainfall (gray bars, Figure 13.17, left-hand side) of the current year (the most recent ring), and they also responded slightly to the temperature in the current March. For MVA, the climate explained 76% of the total variance. MVA was negatively related to temperature in the current March (dotted bars, Figure 13.17, right-hand side), and to some extent, in the current February (i.e., just before the beginning of the growing season). This means that a high temperature in March results in small early wood vessels and vice versa. Figure 13.17 shows that the most significant climatic variables are July precipitation for RW and March temperature for MVA. Fonti and García-González (2004) concluded that mean vessel area of the ring-porous chestnut contains ecophysiological information that can be used for research in dendrochronology. Verheyden et al. (2005) also found a close association between vessel features and precipitation in a tropical tree without distinct growth rings, mangrove (*Rhizophora mucronata* Lam.), which provided evidence for a climatic driving force for the intra-annual variability of the vessel features.

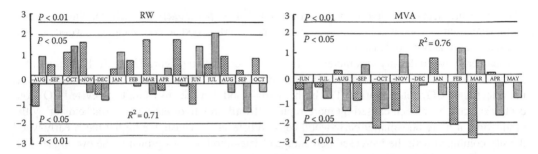

FIGURE 13.17 Charts show response functions of European chestnut (*C. sativa*) RWs and mean vessel area: dotted bars, temperature; gray bars, precipitation. Climatic data from 1956 to 1995 come from Lugano's meteorological station: RW, ring width; MVA, mean vessel area. The "negative" sign in front of the month indicates that the data are from the previous year (ring from the previous year). If no "negative" sign, the data are for the ring of the current year. The response functions, obtained from a program that statistically analyzes the relationship between climate and tree ring variation, are similar to correlation coefficients, but with values greater than one. (Reprinted from Fonti, P. and García-González, I.: Suitability of chestnut earlywood vessel chronologies for ecological studies, *New Phytol.*, 2004, 163, 77–86, Fig. 4. Copyright Wiley-VCH Verlag GmbH & Co. KGaA. With permission.)

INTERNAL LEAF CHARACTERISTICS OF C$_3$ AND C$_4$ PLANTS UNDER AMBIENT CO$_2$

Because of the differential response of C$_3$ and C$_4$ species to elevated CO$_2$, it is important to compare their anatomies, which we do in this section but with no reference to elevated CO$_2$. As noted in Chapter 6, when we talked of carbon isotope discrimination by C$_3$ and C$_4$ plants, there are different photosynthetic types with C$_4$ photosynthesis. We have not yet defined these different photosynthetic types, which we now do. We focus on the grasses, because the different photosynthetic types in C$_4$ plants are represented in the grass family. Grasses are monocotyledons ("monocots"), and the last common ancestor of dicotyledons ("dicots") and monocots existed about 150–300 million years ago (Eckardt 2007). The evolution of grasses using C$_4$ photosynthesis and their sudden rise to ecological dominance occurred 3–8 million years ago (Edwards et al. 2010). (We talk of the evolution of C$_4$ photosynthesis in Chapter 12.) Starting at least 105,000 years ago, humans relied on grass seeds (probably for food and making grinding and pounding implements), including those of sorghum grasses (*Sorghum* spp.) (Mercader 2009).

The domestication of wheat around 10,000 years ago marked a dramatic turn in the development and evolution of human civilization, as it enabled the transition from a hunter-gatherer and nomadic-pastoral society to a more sedentary, agrarian one (Eckardt 2010). Two of the most important traits in the evolution of bread wheat and other cultivated grasses were an increase in grain size and the development of non-shattering seeds. The increase in grain size was important for germination and growth of seedlings in cultivated fields. Non-shattering seeds are a hallmark of domestication, and they prevent natural seed dispersal and allow humans to harvest seeds at optimal times (Eckardt 2010). Gegas et al. (2010) showed that the grain shape and size are independent traits in both modern varieties and primitive wheat species, and they are under the control of distinct genetic components. Wheat domestication has transformed a long and thin primitive grain into a wider and shorter modern grain (Gegas et al. 2010).

We now say a few words about the terminology used in describing the grass family. The family used to be called the Gramineae family. It is now called the Poaceae family. Family names end in -aceae. In the old classification systems, the grass family and a few others were spelled differently (Leguminosae and Cruciferae are two other examples). When revised classification systems were proposed, they included changes in these few family names so that all would have the same spelling. A common genus was chosen for the new name, for example, *Poa* was used to form Poaceae, Leguminosae was replaced by Fabaceae (*Faba* being the common genus), and Cruciferae was replaced by Brassicaceae (*Brassica* being the common genus). The Cronquist (1981) system and similar revisions are now widely accepted, and new spellings of these family names are used by most taxonomists. They appear in books published since the late 1970s, like the *Atlas of the Flora of the Great Plains* (Barkley 1977) (Eileen K. Schofield, Senior Editor, Personal communication, Agricultural Communications, Kansas State University, February 13, 1992).

Of all the world's flowering plants, the grasses are the most important to man and agriculture (Eckardt 2008b). Grasses with approximately 10,000 species have evolved over the last 60 million years, and they provide over half of the world's caloric intake (Kellogg and Buell 2009). Although relatively small compared with other flowering plant families, the grass family surpasses all others in economic importance (maize, wheat, and rice alone account for a major portion of food calories consumed worldwide) and in ecological importance as well (grasses cover more than 20% of the earth's land surface and often dominate temperate and tropical habitats) (Eckardt 2008b). Grasses are the source of the principal foods not only for man but also for his domestic animals (Pohl 1968). Without the grasses, agriculture would be virtually impossible. Grain, sugar, syrup, spice, paper, perfume, pasture, oil, timber, and a thousand other items of daily use are products of various grasses. They hold the hills, plains, and mountains against the destructive erosive forces of wind and water. In the end, they form the sod that covers the dead (Pohl 1968).

Grasses, like corn (or maize, *Zea mays* L.) and sugarcane (*Saccharum officinarum* L.), are used as biofuels. One C_4 grass, *Miscanthus* × *giganteus* (it has no common name, except people do refer to it as "Miscanthus" for a common name; it is a tall perennial grass), is under much study in the Midwest of the United States as a bioenergy crop, because it can maintain photosynthetic activity at temperatures 6°C below the minimum for maize, which allows it to have a longer growing season in cool climates (Wang et al. 2008). In side-by-side large-scale trials of these two C_4 crops in the U.S. Corn Belt, Miscanthus was 59% more productive than grain corn (Dohleman and Long 2009). The average leaf area of Miscanthus was double that of maize, with the result that canopy photosynthesis was 44% higher in Miscanthus, corresponding closely to the biomass differences. The results indicated that the full potential of C_4 photosynthetic productivity is not achieved by modern temperate maize cultivars (Dohleman and Long 2009).

The grasses are all members of the great group of flowering plants that we call the monocotyledons. We know one of the big divisions in the grasses is their type of photosynthetic system: C_3 or C_4. Corn, sugarcane, millet (*Pennisetum glacum* R. Br.—sometimes called *P. americanum* Auth.), and sorghum [*Sorghum bicolor* (L.) Moench] have the C_4 photosynthetic system, while wheat (*T. aestivum* L.), oats (*Avena sativa* L.), rice (*Oryza sativa* L.), and bamboo (*Bambusa* sp.) have the C_3 photosynthetic system (Salisbury and Ross 1978, p. 140). The grass family is organized taxonomically into subfamilies, which are further divided into tribes.

Now, let us again review photosynthesis. In C_3 photosynthesis, we remember that CO_2 and water (H_2O) join with ribulose 1,5-bisphosphate to form 3-phosphoglycerate (3-PGA) (a 3-carbon compound). The enzyme involved is Rubisco (ribulose-1,5-bisphosphate carboxylase/oxygenase). In C_4 photosynthesis, HCO_3^- and phospho*enol*pyruvate (PEP) join to form a C_4 acid, oxaloacetic acid or oxacetate (a 4-carbon compound) and $H_2PO_4^-$ (Salisbury and Ross 1978, pp. 139–140; Malkin and Niyogi 2000, pp. 615, 621, 623). The enzyme involved is PEP carboxylase. In C_4 plants, the pathway involves an initial HCO_3^- fixation in the outer mesophyll cells, transfer of an organic acid from mesophyll to bundle sheath cells, and release of CO_2 for subsequent refixation in the Calvin cycle (Malkin and Niyogi 2000, p. 621). All C_4 fixation pathways begin in the mesophyll, where PEP carboxylase converts PEP to oxaloacetate. The 1-carbon substrate for this enzyme is not CO_2 but HCO_3^-. The aqueous equilibrium of these two species strongly favors HCO_3^- ion over gaseous CO_2, allowing for a more efficient initial fixation step. Moreover, PEP carboxylase cannot fix oxygen, which has a three-dimensional structure similar to that of CO_2 but not HCO_3^- (Malkin and Niyogi

2000, p. 621). The subsequent metabolism of oxaloacetate differs among species; but in all cases, a C_4 acid is transported to the bundle sheath cells, where it is decarboxylated. The CO_2 released is refixed by Rubisco. The resulting C_3 acid is then transported back to the mesophyll cells for regeneration of PEP (Malkin and Niyogi 2000, pp. 621–622).

Three variations of C_4 photosynthesis are known, differing in the C_4 acids transported between mesophyll and bundle sheath cells as well as the mechanism of decarboxylation. The reactions are as follows (Malkin and Niyogi 2000, p. 622):

$$\text{Malate} + NADP^+ \rightarrow \text{pyruvate} + CO_2 + NADPH \text{ (called } NADP^+ - \text{malic enzyme reaction)}$$

$$\text{Malate} + NAD^+ \rightarrow \text{pyruvate} + CO_2 + NADH \text{ (called } NAD^+ - \text{malic enzyme reaction)}$$

$$\text{Oxaloacetate} + ATP \rightarrow PEP + ADP + CO_2 \text{ (called PEP carboxykinase reaction)}$$

In the above reactions, the abbreviations are as follows [definitions come from Cammack (2006) and Conn and Stumpf (1963)]:

$NADP^+$ = oxidized nicotinamide-adenine dinucleotide phosphate. The positive charge refers to the nicotinamide ring; the overall charge is negative, owing to the two phosphate groups, which are negatively charged at neutral pH. [*Note:* Cammack (2006) has a mistake in this definition, because "phosphate" does not appear.]

NAD^+ = oxidized nicotinamide-adenine dinucleotide. The positive charge refers to the nicotinamide ring; the overall charge is negative, owing to the two phosphate groups, which are negatively charged at neutral pH.

$NADPH$ = nicotinamide-adenine dinucleotide phosphate (reduced). The corresponding oxidized state is $NADP^+$.

$NADH$ = nicotinamide-adenine dinucleotide (reduced). The corresponding oxidized state is NAD^+.

ATP = adenosine 5′-triphosphate.

ADP = adenosine 5′-diphosphate.

PEP = phospho*enol*pyruvate.

These reactions take place in the bundle sheath cells. The three variations in the C_4 photosynthesis are listed in Table 13.1 (Malkin and Niyogi 2000, p. 622). In all three cases, CO_2 is released in the bundle sheath cell and is refixed by the Rubisco localized in the chloroplasts of these cells.

TABLE 13.1
Variations in C_4 Photosynthesis

C_4 Acid Transported to Bundle Sheath Cells	C_3 Acid Transported to Bundle Sheath Cells	Decarboxylase	Plant Examples
Malate	Pyruvate	$NADP^+$-malic acid	Maize, sugarcane
Aspartate	Alanine	NAD^+-malic enzyme	Millet
Aspartate	Alanine, PEP, or pyruvate	PEP carboxykinase	*Panicum maximum*

Source: Reprinted from Malkin, R. and Niyogi, K., Photosynthesis, in Buchanan, B.B., Gruissem, W., and Jones, R.L., Eds., *Biochemistry and Molecular Biology of Plants*, American Society of Plant Physiologists, Rockville, MD, 2000, pp. 568–628, Table 12.10. With permission from the American Society of Plant Biologists, Rockville, MD.

In the chapter on roots, we studied $\delta^{13}C$ values and saw that they fell into two discrete ranges that are correlated with either the C_3 or the C_4 photosynthetic pathway. Discrimination against ^{13}C occurs during diffusion of CO_2 into the leaf and during the enzymatic conversion of dissolved CO_2 in the cytoplasm to carbohydrates (Hattersley 1982). In C_3 plants, the discrimination associated with RuP2 (ribulose-1,5-bisphosphate) carboxylation is recognized as the major fractionation process. [Hattersley (1982) refers to Rubisco as RuP2.] If one assumes that the fractionation by RuP2 carboxylase is constant, whatever the $\delta^{13}C$ value of the CO_2 pool available to this enzyme, then the difference between the $\delta^{13}C$ value of the atmosphere and that of the plant tissue is largely dependent on the extent to which the isotopic composition of the CO_2 pool available to RuP2 carboxylase can equilibrate with the isotopic composition of atmospheric CO_2 (Hattersley 1982). In C_3 plants during steady-state photosynthesis, this extent will vary according to photosynthetic rate, photorespiratory and dark respiratory rates, and stomatal conductance.

Hattersley (1982) uses terminology that has been widely accepted and that we should know. He pointed out that in C_4 plants, the $CO_2-HCO_3^-$ pool at the site of RuP2 carboxylation is not in the mesophyll (PCA tissue) (primary carbon assimilation) but in PCR (photosynthetic carbon reduction) of "Kranz" tissue. [As noted in Chapter 6, the unique leaf structure in C_4 leaves is called Kranz anatomy. Kranz is the German word for "garland" or "wreath," and the word refers to the characteristic bundle sheaths ("garlands") that surround the vascular tissue of C_4 plants and where C_3 photosynthesis takes place.] The PCR and PCA terminology was introduced by Hattersley et al. (1977) to refer to the two chlorenchymatous tissues involved in C_4 photosynthesis in a functional sense. Anatomically, the terms are usually, but not always, equivalent to bundle sheath and mesophyll tissues, respectively (Hattersley 1982). The isotopic composition of this pool will depend not only upon the same factors listed for C_3 plants (see the previous paragraph) but also upon (i) fractionation associated with the $CO_2-HCO_3^-$ equilibrium, (ii) fractionation associated with PEP carboxylation, and (iii) conductance of PCR (Kranz) tissue to CO_2 and HCO_3^- and, therefore, rates of $CO_2-HCO_3^-$ leakage from PCR to PCA (mesophyll) tissue (Hattersley 1982).

One of the principal functions of C_4 plant leaf anatomy is to provide a "CO_2-tight" PCR compartment that facilitates the concentration of CO_2 at the site of RuP2 carboxylase. If PCR tissue were totally "CO_2-tight," its $CO_2-HCO_3^-$ pool would be unable to equilibrate with any pool external to it, no fractionation by RuP2 carboxylase would occur, and C_4 plant $\delta^{13}C$ values would be less negative than C_3 plant values. This is because the net discrimination against ^{13}C associated with the $CO_2-HCO_3^-$ equilibration and with PEP carboxylation is much less than that associated with RuP2 carboxylation.

Hattersley (1982) labeled the three variations in C_4 photosynthesis, shown above in the table from Malkin and Niyogi (2000), as NADP-ME (ME stands for malic enzyme), NAD-ME, and PCK (PEP carboxykinase), respectively. [For an example of the use of these terms in a study of the three subtypes, with no elevated CO_2, see Carmo-Silva et al. (2009).] Hattersley (1982) gives the $\delta^{13}C$ values for 31 grass species representing the three different C_4 photosynthesis variations. The average $\delta^{13}C$ for the NADP-ME type was −11.4‰; for the NAD-ME type, it was −12.7‰; and for the PCK type, it was −12.0‰.

Now, we turn to a paper by Dengler et al. (1994), who compared differences in leaf anatomy of 125 C_3 and C_4 grasses under ambient CO_2 concentrations. The subfamilies that they studied were Panicoideae, Chloridoideae, Arundinoideae, Pooideae, and Bambusoideae. They studied the anatomies of plants with the three variations of C_4 photosynthesis: NADP-ME, NAD-ME, and PCK. Each biochemical type is typically associated with a distinctive suite of structural characteristics. Dengler et al. (1994) said that biochemical assays of some grass species reveal exceptions to the three anatomical–biochemical variations of C_4 plants. For example, species of *Eragrostis*, *Panicum*, and *Enneapogon* are PCK-like structurally, but NAD-ME biochemically. So they added a fourth group to the C_4 plants: PCK-like NAD-ME. The data in the figures that

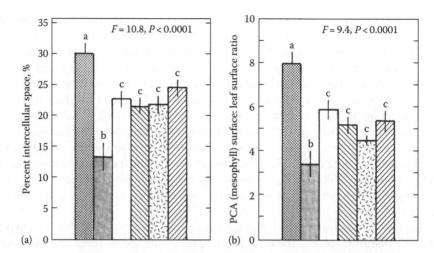

FIGURE 13.18 Comparison of means (±standard error) of C_3 Panicoideae, Arundinoideae, and Pooideae combined [C_3 (dotted bar), C_3 Bambusoideae (gray bar), and C_4 NADP-ME (open bar), NAD-ME (right-slanting striations), PCK-like NAD-ME (stippled bar), and PCK (left-slanting striations)] species for intercellular space volume as a proportion of PCA (or C_3 mesophyll) tissue (a) and ratio of internal surface area (PCA or C_3 mesophyll) surface exposed to intercellular space to external surface area (leaf surface) (b). Means with different letters are significantly different at $P < 0.004$. Significance of the difference among means of all groups is reported by the F-statistic in the analysis of variance. (Reprinted from Dengler, N.G., Dengler, R.E., Donnelly, P.M., and Hattersley, P.W., *Ann. Bot.*, Fig. 19, 1994 by permission of Oxford University Press.)

they presented for anatomical characteristics compare six groups: typical C_3 plants; the C_3 subfamily Bambusoideae, which had characteristics different than "classical" C_3 types, NADP-ME (C_4), NAD-ME (C_4), PCK-like NAD-ME (C_4), and PCK (C_4). Quantitative anatomical characteristics, including cross-sectional areas (volumes) of all tissues and perimeters (surface areas), of leaves of the 125 grass species were measured in transverse sections and photographed under a microscope.

The proportion of mesophyll tissue occupied by intercellular space was lower for C_4 species than for non-bambusoid C_3 species, but it did not differ among C_4 types (Figure 13.18a). C_3 bambusoid species had less intercellular space than did non-bambusoid C_3 species.

Mean values for PCA (mesophyll) tissue surface exposed to intercellular space expressed per unit leaf surface were lower for C_4 species as a whole than for non-bambusoid C_3 species, and means did not differ among C_4 types (Figure 13.18b). This ratio was higher in the non-bambusoid C_3 species group than in bambusoid species. When PCA (mesophyll) surface area was expressed per unit PCA tissue volume, values for the NADP-ME and NAD-ME types were lower than for other C_4 types, and those for bambusoid species were lower than for other C_3 species (Figure 13.19a).

Absolute values for the PCR (bundle sheath) total external surface did not differ between C_3 species as a whole and C_4 species as a whole, although values were lower for the NADP-ME type than for other C_4 types and for bambusoid species than for other C_3 species (Figure 13.19b). The proportion of total PCR (bundle sheath) surface area that was exposed to intercellular space was lower for C_4 than for non-bambusoid C_3 species, and lower in the NAD-ME than in the PCK type (Figure 13.19c). The ratio of PCR (bundle sheath) tissue surface adjacent to intercellular space:tissue volume ratio (Figure 13.19d) was strikingly lower than the comparable value for PCA tissue (Figure 13.19a) and was lower for C_4 than for non-bambusoid C_3 species, and, among C_4 types, was lowest in the NAD-ME type (Figure 13.19d).

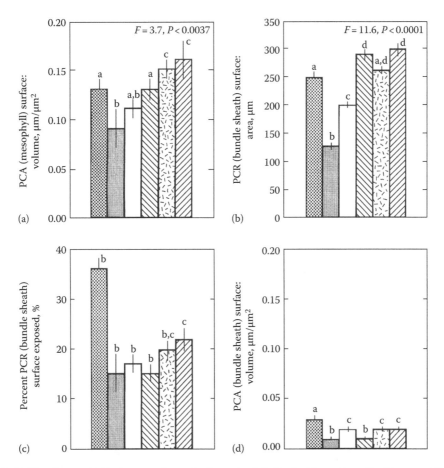

FIGURE 13.19 Comparison of means (±standard error) of C_3 Panicoideae, Arundinoideae, and Pooideae combined [C_3 (dotted bar), C_3 Bambusoideae (gray bar), and C_4 NADP-ME (open bar), NAD-ME (right-slanting striations), PCK-like NAD-ME (stippled bar), and PCK (left-slanting striations)] species for surface area exposed to intercellular space of PCA (or C_3 mesophyll): PCA tissue volume (a), surface area of the outer tangential wall of PCR tissue (b), proportion of outer tangential wall of PCR exposed:intercellular space (c), and ratio of surface area exposed to intercellular space of PCR (or C_3 bundle sheath) to tissue volume (d). Means with different letters are significantly different at $P < 0.004$. Significance of the difference among means of all groups is reported by the F-statistic in the analysis of variance. (Reprinted from Dengler, N.G., Dengler, R.E., Donnelly, P.M., and Hattersley, P.W., *Ann. Bot.*, Fig. 20, 1994 by permission of Oxford University Press.)

The mean cross-sectional area of vascular tissue per vein was lower in C_4 species than in non-bambusoid C_3 species (Figure 13.20a). The C_3 bambusoid species group had less vascular tissue per vein than the C_3 non-bambusoid species group. The ratio of chlorenchymatous tissue [PCA (mesophyll) plus PCR (bundle sheath)]:vascular tissue did not differ between C_3 and C_4 types, and it did not differ among C_4 types (Figure 13.20b).

In sum, Dengler et al. (1994) found that both mesophyll and bundle sheath tissues of C_4 species have less surface area exposed to intercellular space than do C_3 species. In C_4 species, the ratio of PCR (bundle sheath) tissue surface adjacent to intercellular space:tissue volume ratio is lower than the comparable value for PCA tissue. Anatomical features of the C_4 plants reduce apoplastic leakage of CO_2 from bundle sheath to intercellular space. No new diagnostic characters emerged from their study of leaf anatomy that can be used to distinguish one biochemical type from another.

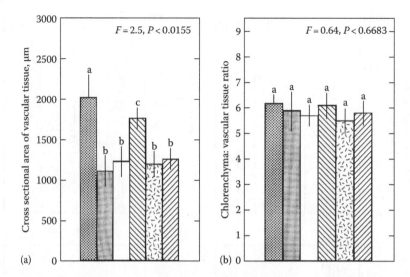

FIGURE 13.20 Comparison of means (±standard error) of C_3 Panicoideae, Arundinoideae, and Pooideae combined [C_3 (dotted bar), C_3 Bambusoideae (gray bar), and C_4 NADP-ME (open bar), NAD-ME (right-slanting striations), PCK-like NAD-ME (stippled bar), and PCK (left-slanting striations)] species for cross-sectional area of vascular tissue per vein (a) and ratio of total chlorenchyma [PCA (or C_3 mesophyll) plus PCR (or C_3 bundle sheath)] to vascular tissue (b). Means with different letters are significantly different at $P < 0.004$. Significance of the difference among means of all groups is reported by the F-statistic in the analysis of variance. (Reprinted from Dengler, N.G., Dengler, R.E., Donnelly, P.M., and Hattersley, P.W., *Ann. Bot.*, Fig. 21, 1994 by permission of Oxford University Press.)

SUMMARY

Data are consistent in showing that the ratio of leaf weight to leaf area (incorrectly called "specific leaf weight") increases as CO_2 concentration increases. That is, leaves are thicker under elevated CO_2. This has been observed for both C_3 and C_4 leaves. Thinner leaves at low CO_2 may serve to increase the rate of CO_2 diffusion to the chloroplasts. Thick leaves tend to be more xeromorphic than thin leaves, and this suggests that leaves under elevated CO_2 are more drought resistant than leaves grown under the ambient level of CO_2. Rising atmospheric CO_2 concentration results in denser wood. There is indication that elevated CO_2 reduces the diameter of vessels and tracheids in xylem tissue. When leaf anatomy of C_3 and C_4 plants is compared under ambient CO_2, both mesophyll and bundle sheath tissues of C_4 plants have less surface area exposed to intercellular space than C_3 plants.

14 Elevated Atmospheric Carbon Dioxide: Phenology

INTRODUCTION

The objective of this chapter is to report the effects of elevated levels of CO_2 on the phenology of plant growth. While phenology is not directly related to plant–water relations, we mention it because plant–water relations may indirectly be affected if phenology is changed by elevated CO_2. For example, if elevated CO_2 delays reproductive development, as we found for wheat (described in this chapter), then a plant is going to stay in the vegetative state longer. This means that there is more time for water to be lost from the leaves.

PHENOLOGY

The dictionary defines *phenology* (a shortened form of *phenomenology*) as, "The science of the relations between climate and periodic biological phenomena, as the migrations and breeding of birds, the fruiting of plants, etc." (*Webster's Collegiate Dictionary* 1939). The word comes from the Greek *phainein*, which means "to show." When discussing phenology in this chapter, we shall focus on crop plants, in particular wheat (*Triticum aestivum* L.), because that is the crop we have studied most closely concerning elevated CO_2 (Chaudhuri et al. 1986, 1987, 1989). For crops, phenological events are dates of sowing, dates of germination or emergence, dates of tillering, dates of shooting, and dates of heading. Even though the word phenology was coined in 1850 (Hopp and Vittum 1978), it is not a new science (Waggoner 1977). Phenological calendars were used several thousand years ago in China and ancient Rome to guide farming operations. Old-time New England farmers planted corn (*Zea mays* L.) when elm (*Ulmus* sp.) leaves were the size of a squirrel's ear. By that time, the soil was warm enough for corn seeds to germinate. Books about the phenology of crops have also been written (Aitken 1974; Lieth 1974).

We first should consider why every phase of every higher plant is related to the annual march of climatic events (Heslop-Harrison 1969). Primitive systems, in the historical and evolutionary sense, were based on the alternation of haploid and diploid generations. They lived under uniform conditions and there was no periodicity. Control was only endogenous, not environmental. This pattern still is seen in many plants, for example, the flora of the great tropical rain forests. If there are rhythms, they are innate. Different times of the year cannot be detected because they are similar.

As one moves away from the tropics, climate becomes variable and involves a transition between unfavorable (too wet, too dry, too hot, too cold) and favorable periods. The amplitude of physiological tolerance must be great enough to match the climatic changes. The plant must anticipate that there will be an unfavorable period and compensate in advance, so it can survive. Therefore, it must develop a timing mechanism. Since internal clocks are not good enough, the plant must seize upon accurate environmental clocks. The most reliable clocks are photoperiodic ones. The variation of day length with year is a precise mathematic curve related to latitude. Temperature varies cyclically, but is a less dependable clock because it is affected by weather. Photoperiod, with temperature as secondary, is the major controller of plant development. Annuals, which have only 1 year in which to complete their life cycles, are especially sensitive to photoperiod. If any phase is not precisely timed, the plant succumbs (Heslop-Harrison 1969). Therefore, annuals, like wheat, refer constantly to the photoperiodic clock.

Let us now consider wheat phenology from a physiological point of view.

WHEAT PHYSIOLOGY AND PHENOLOGY

STAGES OF GROWTH

Wheat is a determinate crop with distinct phenological phases as depicted by the different growth-stage scales, namely Feekes (Large 1954), Robertson (1968), Haun (1973), and Zadoks–Chang–Konzak (1974). The Robertson (1968) scale is rarely used by agronomists, but the other three are often used. Six definite phenostages may be morphologically distinguished during the growing season (Nuttonson 1955): emergence, tillering, booting, heading, flowering, and ripening. Depending on the objective of a given study (e.g., harvest data prediction, plant pathological investigations, computer modeling), these phenophases may be subdivided further (Large 1954; West and Wein 1971), or combined into fewer phases (Nuttonson 1955). The Feekes scale goes from 1 to 11; the Haun scale has growth units divided into decimal fractions between 0.0 and 14.0; and the Zadoks–Chang–Konzak scale goes from 1 to 100.

LIGHT AND TEMPERATURE

The completion of the life cycle of wheat (grain formation) depends upon both light (photoperiod) and temperature (McKinney and Sando 1935; Nuttonson 1948, 1955; Riddell and Gries 1958a,b; Riddell et al. 1958; Weibel 1958; Ahrens and Loomis 1963; Krekule 1964; Syme 1973; Halse and Weir 1974). Photoperiodically, wheat is classified as a long-day plant. That is, it requires long days (short nights) before it will flower. "Obligate" winter wheats need a cold period (vernalization) before they will flower. [See Chouard (1960) for a general review of vernalization.] Spring wheats do not need to be vernalized. The development of "intermediate" wheats is hastened by a cold period, but the plants eventually will produce seed even if not given a cold treatment. Spring wheats complete their life cycle quickly when given long days and temperatures of 20°C or above until harvest. Winter wheats grow most rapidly when given short days and low temperatures during the early stages of growth and long days and high temperatures during later stages (McKinney and Sando 1935).

WATER

Photoperiodic control ensures that the date of flowering is relatively independent of water supply or date of germination. This is especially appropriate for plants in semiarid parts of the world because calendar date rather than germination date is the better predictor of the end of the wet season (Fischer and Turner 1978). In this context, it is interesting to note that Nuttonson (1955) found that, for winter wheats, which are often grown in semiarid regions, date of maturity could be predicted just as well using March 1 for a starting date as using the date of emergence. The increase in adaptation to drier areas (shorter wet seasons) achieved by modern plant breeders in crops such as wheat, barley (*Hordeum vulgare* L.), sorghum [*Sorghum bicolor* (L.) Moench], and cotton (*Gossypium hirsutum* L.) has been closely linked to earlier flowering as a result of the lessening of sensitivity to photoperiod (Fischer and Turner 1978).

However, water stress does have an effect on the phenological development of wheat. Angus and Moncur (1977) watered wheat until floral initiation and then stressed the plants to −5, or −10, or −15, etc. bar and then rewatered. They found that with severe water stress (−25 to −40 bar) there was about a day's delay in the anthesis for each day's delay in watering. With moderate stress (down to −15 bar), dates of anthesis were ahead of well-watered controls. They concluded that water stress was an important factor affecting the phasic development of wheat, and, although it was less important an influence than day length or temperature, it should be included in developmental models if they are to predict the timing of wheat within an accuracy of a few days.

Nuttonson (1955) reported work by Russians that showed that moist conditions lengthened the growing period. They grew wheat in containers with soil moisture levels of 20%, 40%, and 60%

of the soil moisture-holding capacity. An increase of moisture from 20% to 40% lengthened the sown-to-headed period by 2–3 days and the headed-to-ripe period by 6–8 days. Transition from the 40%–60% moisture level caused further lengthening of the growing period by 1–2 days. Feyerherm (1977) hypothesized that excess soil moisture slows maturity because it causes a larger lag between soil and ambient air temperatures than occurs under drier moisture conditions. This results in slower plant development.

Nuttonson (1955) also observed that day-degree summations, the most commonly used unit to correlate phenological development and temperature, were consistently lower for wheats in regions with less than 18 in. (46 cm) precipitation and higher for wheats in regions with precipitation over 18 in. This held true for summations obtained from a considerable number of areas of different latitudes and altitudes (different temperatures and light regimes). He concluded that, in addition to latitude and altitude, moisture conditions were a factor in determining the day-degree requirements of a given variety of wheat.

CARBON DIOXIDE

Carbon dioxide affects phasic development, too. We shall consider CO_2 in detail, in particular our results, after this general introduction concerning phenology of wheat. Gifford (1977), in Australia, grew a spring variety of wheat at three different concentrations of CO_2 (ambient, or 280, 480, and 130 μmol/mol) from germination to maturity in naturally lit growth chambers under winter and summer light conditions. The date on which 50% of the main shoot ears reached anthesis was about 2 weeks earlier for the CO_2-depleted treatment than for the control under winter conditions, and this was related to two to three fewer leaves emerging on the main shoots in that treatment. But for the summer experiment, nine leaves were produced per main shoot for all treatments, and there was no difference observed among the dates of anthesis. His results suggest that CO_2 concentration should be monitored to predict accurately anthesis date of wheat.

MINERAL NUTRITION

Chemical fertility does not appear to have an effect on the length of the growing period of wheat (Nuttonson 1955). However, some Russian workers showed that phosphorus fertilizers accelerated ripening by 1–2 days (Nuttonson 1955). Russians also found that high osmotic pressure of the soil solution shortened the growing period (Nuttonson 1955). When the osmotic pressure of the soil solution was raised by 1–2 atm (about 1–2 bar) with salts of NaCl and Na_2SO_4, the ripening of wheat was shortened by 1–3 days. These results suggest that effects of fertilizers on plant development are related to water stress (see section "Water"), rather than to specific effects of individual elements.

LEAF NUMBER

Cereals must produce a minimum number of leaves before they will flower (Whyte 1948). Observations made by Purvis (1934) on the spike of rye (*Secale cereale* L.) primordia indicated that a minimal number of leaves must be formed before the differentiation of flower initials occurred. Under short-day conditions, this was about 22, irrespective of variety and temperature, while under long-day conditions, the minimal number was less for plants germinated at low rather than at high temperatures, namely, 12 for the former and 22–25 for the latter.

For wheat, Williams and Williams (1968) found that short-day stem apices at the beginning of floral induction were long and had many more foliar members than long-day apices. There were 13 foliage leaves in short days and seven or eight leaves in long days. Riddell and Gries (1958b) showed that low temperatures (vernalization) decreased the number of nodes differentiated prior to the appearance of the spike. Vernalization of late spring wheat varieties made them flower at the same time as early varieties. The late-maturing spring wheat varieties, in being more responsive to vernalization than the early varieties, more closely paralleled the response of winter wheat to a low-temperature treatment. As long ago as 1923, Vavilov, cited by Aitken (1974), noted a relationship between the time of flowering and vegetative nodes on the shoot. He said, "This character is in connection with early maturity; the more there are nodes, the more the race is late-ripening."

VARIETIES

Photoperiod (or day length) response is an important determinant of the adaptability of a particular variety to a given area (Martinić 1973), as has been shown by the haphazard nature of the early introduction of primarily northern European wheats into Australia (Goutzamanis and Connor 1977). Although wheat is classified as a long-day plant, there exists a wide varietal difference in the photoperiodic response (Nuttonson 1955), which is under genetic control (Pugsley 1963, 1965, 1966). [Normal functioning of the gametes depends upon the photoperiod (Sircar 1948).] European varieties have been regarded as more sensitive to photoperiod than the established Australian varieties (Syme 1968).

In general, it has been observed that varieties from northern regions are more sensitive to day length than varieties from more southern regions. For example, Wall and Cartwright (1974) found that north temperate varieties in long days had a mean spikelet number of 15 and they headed in 40–50 days, while in short days they had a mean of 24 spikelets per ear and failed to head in 100 days. They observed that the effects of vernalization on the tropical varieties were related to maturity class and ranged from negligible in early varieties to decreases of 20 days to heading and seven spikelets in late varieties in long days. Results with Indian early and late varieties have been similar (Sircar 1948).

The development of photoperiodic insensitive (or day-neutral) varieties allows adaptation to a great range of environments. For this reason, this has been one of the objectives in the wheat breeding program at the Centro Internacional de Mejoramient de Maíz y Trigo (CIMMYT 1975), near Mexico City.

DEVERNALIZATION

An agronomic problem in obligate winter wheats is devernalization. In many environments, a temporary period of high temperature may partially or completely annul any degree of vernalization already obtained. Because vernalization requires oxygen (Chouard 1960), waterlogging in a field in a wet winter produces anaerobic conditions that favor devernalization. Seeds in a heavy clay soil are more liable to devernalization through anaerobic conditions than those in light, well-aerated soils. [Not only does compaction encourage devernalization, it also reduces the emergence rate and the subsequent development of the wheat plants (Nuttonson 1955).] The effectiveness of treatments leading to devernalization (high temperature, anaerobic conditions) depends on the interval between the end of the vernalization chilling and the beginning of devernalization. If plants are devernalized, revernalization may be achieved by another chilling.

YIELD

It is commonly observed that early varieties of wheat yield less than late varieties (Rawson 1970, 1971; Arnon 1975). Shortening of the developmental stages under the influence of high temperatures is usually accompanied by a lowering in the grain yield. In Wall and Cartwright's study (1974), effects of temperature varied with variety and with vernalization treatment. Early varieties and vernalized later varieties headed earlier and had fewer spikelets in the warmer regimes, while unvernalized later varieties tended to have more spikelets and headed later. In Oklahoma and Kansas, wheat breeders are in a dilemma. If they breed for early varieties, summer droughts can be avoided, but yielding ability will be poor. If they breed for late varieties, yielding ability is high, but yield is often reduced, not only by droughts, but also by high temperatures and high winds that shrivel the grain.

METHODS TO DETERMINE CROP DEVELOPMENT

Of all the factors that affect phenology, temperature, light, and photoperiod have been most studied. They are all positively correlated with the phenological development of a crop. The techniques that have been used by agricultural phenologists to determine crop development are the following:

Thermal constants (or heat sums) (Reamur 1740, as cited by Waggoner 1974a, 1977)
Remainder index (Nuttonson 1955; Wielgolaski 1974)
Photo-thermal units (Nuttonson 1955)
Solar units (Fitzpatrick 1973; Baron et al. 1975)
Hours of bright sunshine (Smith 1967)
Solar-thermal units (Caprio 1974)
Open-water evaporation sums (Thornthwaite 1948; Fitzpatrick and Nix 1969)
Physiological indexes (Livingston 1916a,b)
Exponential indexes (Price 1911; Livingston and Livingston 1913)

Of the quantitative techniques used to predict phenological development, the remainder index is the most widely used (Waggoner 1974a). The basis for this technique is that a critical number of day-degree units must be accumulated before the crop can pass from one phenostage to the next. Nuttonson (1955), in a study of the phenology of 'Marquis' spring wheat and other crops grown over a wide range of latitudes throughout North America, found that a multiple of the average length of day and the summation of day-degrees, referred to as photo-thermal units, provided a less variable unit of measurement of the interval between phenological events than the use of day-degree summation alone or the use of the number of days. Some phenological stages of wheat have been shown to be sensitive to photoperiod, for example, tillering (Grantham 1917; Chinoy and Nanda 1951; Darwinkel et al. 1977) and stem extension (Chinoy and Nanda 1951). For these stages, in particular, the use of photo-thermal units (day degrees × day length) give an improved phenometric index compared to day-degree summations alone (Goutzamanis and Connor 1977). The interaction between day length and temperature probably explains the success of the ΣE_o (open pan evaporation) technique (Nix and Fitzpatrick 1969; Fitzpatrick 1973), because both temperature and day length are determinants of pan evaporation (Goutzamanis and Connor 1977). However, not all workers have found ΣE_o to give superior results (Goutzamanis and Connor 1977). In their model of wheat growth, Goutzamanis and Connor (1977) used heat sums above a base temperature of 4°C for all phases except stem extension, in which case photo-thermal units were employed.

Q_{10}, DEGREE-DAY CONCEPT AND HEAT UNITS

Before we look at our results with elevated CO_2 and wheat phenology, let us first describe how heat (or temperature) affects reactions (such as the rate of development of a plant).

The rate of reactions in an organism increases with the temperature up to a maximum (Giese 1962, p. 199). With further increase in temperature, the rate of activity declines. Therefore, for meaningful comparisons of the effects of temperature on the rates of various biological processes, the rates of reaction must be compared for the same temperature interval. Usually, one determines the ratio of the rate of activity at a given temperature to the rate at a temperature 10°C lower. This ratio, called the *temperature coefficient*, is designated Q_{10}.

Four different kinds of temperature coefficients have been found for cell processes, each corresponding to a distinctive mechanism of action.

1. Those that are about 2
2. Those that are considerably higher than 2
3. Those that are less than 2
4. Those that are about 1

In comparable thermobiological reactions as well as in thermochemical reactions, the rate increases about twofold (i.e., 100%) for each 10°C rise in temperature, a relation that is called the *van't Hoff rule*. The fundamental cause underlying this principle is perhaps the same for both types of reactions. One possibility suggests itself, namely a comparable rise in kinetic energy of the system. However, calculation shows that a rise of 10°C in temperature increases the average kinetic energy of the molecules only to a very limited extent—equal to $10/T$, where T is the absolute temperature. At room temperature, the absolute temperature is about 300°; therefore, $10/T$ would be 10/300 or 3.3%. This trifling increase in the average kinetic energy of the molecules by itself is quite inadequate to explain the twofold rise in the rate of reaction that follows a 10°C rise in temperature (Giese 1962, p. 199).

However, in 1889, Arrhenius proposed a plausible explanation for a twofold (or greater) rise in the rate of a thermochemical reaction with an increase in temperature of 10°C (Giese 1962, p. 199). (For a biography of Svante Arrhenius, see Appendix.) He pointed out that not all the molecules in a population of a given kind of molecules, at a given temperature, have the same amount of kinetic energy. Following random collisions, some of the molecules accumulate more energy than others. Some become energy-rich, some energy-poor, and most are in the middle class between. Arrhenius suggested that it is the energy-rich molecules that are most likely to react on collision, and that the rate of reaction might, therefore, depend upon the concentration of such molecules. A molecule must have a certain amount of kinetic energy to undergo a thermochemical reaction. This critical amount of energy has been called the *energy of activation*. Arrhenius found means for determining experimentally this energy of activation. He then used the Maxwell–Boltzmann distribution law to determine the number of molecules possessing the energy of activation before and after a 10°C rise in temperature. The calculation shows that several times as many activated molecules are present at the higher temperature as at a temperature 10°C lower. It seems most likely, therefore, that the increase in the rate of thermochemical reactions with a rise in temperature is attributable to the disproportional increase in the fraction of molecules with the required energy of activation for a particular reaction (Giese 1962, pp. 199–200).

Thermochemical and thermobiological reactions follow the van't Hoff rule with a Q_{10} of about 2. For example, the Q_{10} for the thermochemical reaction of the activity of malt amylase is 2.2 (Giese 1962, p. 200). The temperature coefficient (Q_{10}) for denaturation of proteins is of a different order of magnitude from that of thermochemical reactions. The coefficient is very high. For example, protein coagulation of egg albumin has a Q_{10} of 635 (Giese 1962, p. 200). It must be realized that proteins are highly organized molecules and that denaturation involves the change from this highly organized structure to a much more random one. For this reason, even a small rise in temperature in the range that causes denaturation has a marked effect out of all proportion to the increase in the kinetic energy of the system (Giese 1962, p. 202). The Q_{10}'s for the rates of diffusion and conductivity (movement of electrolytes in an electric field) are less than 2.0. For example, the Q_{10} is 1.27 for the diffusion of electrolytes, 1.37 for the diffusion of sugar, and 1.41 for the diffusion of dextrin (Giese 1962, p. 202). The Q_{10} for conductivity is also between 1 and 2. For example, the Q_{10} of the conductivity of KCl in water is 1.22–1.24 (Giese 1962, p. 200). The Q_{10} for photochemical and photobiological reactions is only slightly greater than 1.0, indicating that the rate of a photochemical reaction is little influenced by heat. The Q_{10} for the bleaching of visual purple in the eye by yellow light is 1.0, and the Q_{10} for photosynthesis at low light intensity is 1.06 (Giese 1962, p. 200). The Q_{10} of photosynthesis at high light intensity is a thermobiological reaction and is affected by temperature, and its Q_{10} varies between 1.6 and 4.3 (Giese 1962, p. 200).

The van't Hoff rule is widely used to compare the activity of different thermobiological reactions. Levitt (1980) gives values for many different types of reactions as affected, for example, by chilling (Levitt 1980, pp. 43–46) and freezing (Levitt 1980, pp. 250–253). When it is inconvenient to perform experiments at temperature intervals 10 degrees apart, the data are gathered for smaller or larger temperature intervals, and the temperature coefficient may be determined by use of the van't Hoff equation (Giese 1962, p. 204):

$$Q_{10} = \left(\frac{k_2}{k_1} \right)^{[10/(t_2 - t_1)]} \tag{14.1}$$

where
k_2 is the rate of the reaction at temperature t_2
k_1 is the rate of the reaction at temperature t_1

This equation is more convenient to handle in the form

$$\log Q_{10} = \left[\frac{10}{(t_2 - t_1)} \right] \left[\log \left(\frac{k_2}{k_1} \right) \right] \tag{14.2}$$

Example

What is the temperature coefficient, given the following data?

    Rate of reaction at 30°C = 5 days
    Rate of reaction at 20°C = 10 days

A constant amount of 10 g is produced in the reaction. The amount could be any number, but a value has to be inserted into the equation in order to solve it. In the following example, "X" could be inserted instead of "10 g," and one would still get the right answer.

Answer

We know $t_1 = 30°C$, $t_2 = 20°C$, $k_1 = 5$ days, $k_2 = 10$ days.

$$\log Q_{10} = \left[\frac{10}{(t_2 - t_1)} \right] \left[\log \left(\frac{k_2}{k_1} \right) \right]$$

$$= \frac{10}{(20° - 30°)} \log \left\{ \left[\left(\frac{10\,g}{10\,days} \right) \middle/ \left(\frac{10\,g}{5\,days} \right) \right] \right\}$$

$$= -1 \log \left(\frac{1}{2} \right) = -1\,(-0.301)$$

$$= 0.301; \quad Q_{10} = 2.0$$

The Q_{10} had to be 2.0 by the way we set up the problem (10°C difference with the time taking twice as long at the lower temperature).

Another Example

A colleague and his students at Kansas State University looked at the pollen development of two varieties of wheat ("Anhui 11" and "Yangmai 1") under two temperatures, 12.5°C (t_1) and 8.5°C (t_2). Variety Anhui 11 had pollen mature in 25.8 days at t_1 and 38.4 days at t_2. Variety Yangmai 1 had pollen mature in 23.8 days at t_1 and 39.8 days at t_2. For the "constant amount," we shall use the words "pollen maturation."

The Q_{10} for Anhui 11 was

$$\log Q_{10} = \frac{10}{(8.5° - 12.5°)} \log\left\{\left[\left(\frac{\text{pollen maturation}}{38.4\,\text{days}}\right) \middle/ \left(\frac{\text{pollen maturation}}{25.8\,\text{days})}\right)\right]\right\}$$

$$= \left(\frac{10}{-4}\right) \log\left(\frac{25.8}{38.4}\right)$$

$$= (-2.5)\,(-0.173)$$

$$= 0.4324; \quad Q_{10} = 2.7$$

For Yangmai 1

$$\log Q_{10} = \frac{10}{(8.5° - 12.5°)} \left\{\left[\log\left(\frac{\text{pollen maturation}}{39.8\,\text{days}}\right) \middle/ \left(\frac{\text{pollen maturation}}{23.8\,\text{days})}\right)\right]\right\}$$

$$= \left(\frac{10}{-4}\right)\left[\log\left(\frac{23.8}{39.8}\right)\right]$$

$$= -2.5\,(-0.223)$$

$$= 0.558; \quad Q_{10} = 3.6$$

The Q_{10} for Yangmai 1 was greater than that for Anhui 11.

The degree-day concept dates back over 200 years (Chang 1968, p. 77). It holds that the growth of a plant is dependent upon the total amount of heat to which it was subjected during its life time. A degree-day is a measurement of the departure of the mean daily temperature above the minimum threshold temperature for a plant. If the minimum temperature is 50°F, and the mean daily temperature is 60°F, then we would have 10 degree-days using °F. The minimum threshold temperature varies with plants [e.g., 40°F or 4.4°C for peas (*Pisum sativum* L.); 50°F or 10.0°C for corn (*Z. mays* L.), and 55°F or 12.8°C for citrus fruits)] (Chang 1968, p. 77). The degree-day concept assumes that the relationship between growth and temperature is linear instead of logarithmic as predicted by the van't Hoff law. In spite of its lack of theoretical soundness, the simplicity of the degree-day method has been widely used to guide agricultural operations and the planning of land usage (Chang 1968, p. 77).

The degree-day concept, also called the heat-unit concept, must be considered as a simplification of the net results of many temperature-dependent physiological and biochemical processes through an entire subperiod or season of development (Lowry 1969, p. 193). It is as if the net results of photosynthesis, respiration, transpiration, and many other even more complex processes and reactions, could be regarded as the result of some temperature-sensitive biochemical "master reaction," whose rate rises as temperature rises (Lowry 1969, p. 193). The dangers inherent in simplifying things to this extent are obvious, but for purposes of approximating seasonal temperature effects on plant growth and development, this simplification has been useful and has been employed in a class of procedures known collectively as the "heat unit (HU)" concept. Equation 14.3 shows the HU concept (Lowry 1969, p. 194):

$$HU = \sum_{N} \left\{ \left[\frac{(T_M + T_m)}{2} \right] - T_t \right\}, \quad \text{where} \quad \left[\frac{(T_M + T_m)}{2} \right] > T_t \tag{14.3}$$

where HU is the number of heat units accumulated during N days. This equation is employed with daily temperature maximums, T_M, minimums, T_m, and threshold temperature, T_t. The results are expressed as "growing degree-days" and the units are either °F-day or °C-day.

GROWTH STAGES OF WHEAT AS AFFECTED BY ELEVATED CO_2

As noted in Chapters 1, 6, and 11, we used closed-top chambers to control the CO_2 concentrations at 340 (ambient), 485, 660, and 825 µmol/mol during a 3 year period (1984–1987) in which we grew winter wheat (Chaudhuri et al. 1986–1989). The chambers were placed over underground boxes that could be pulled out of the ground and weighed to determine water lost. Water in half of the boxes, which contained silt loam soil, was maintained at a high water level (field capacity: 0.38 m³/m³) and water in the other half was maintained at a low water level (half field capacity). Water stress treatments began in the spring, and the wheat was kept watered until then.

Carbon dioxide was applied about 2 weeks after planting, which occurred on October 19, 1984, October 7, 1985, and September 26, 1986. In the first season (1984–1985), the chambers were left on the plants during the winter months. However, for an unknown reason, the plants grew poorly in the spring (Chaudhuri et al. 1986, p. 13). Perhaps, the damage may have been the result of ice formation (Chaudhuri et al. 1986, p. 33) because the plants got too cold due of the lack of a protective snow cover or the damage occurred due to the high conductance of the metal rhizotron structure in the 1984–1985 season (Chaudhuri et al. 1987, p. 16). In that first year, wheat had to be transplanted into the chambers in March 1985 due to the poor stand. In the second (1985–1986) (Chaudhuri et al. 1987) and third (1986–1987) (Chaudhuri et al. 1989) growing seasons, the chambers were removed at the beginning of the winter dormancy period (mid-November), and the chambers were placed back on the wheat in early March, when CO_2 application began in the last 2 years. It was assumed that discontinuing CO_2 enrichment during the winter months did not affect the performance of wheat, because little growth occurs during the dormant period (Chaudhuri et al. 1987, p. 16). Straw was placed over the wheat during the winters of 1985–1986 and 1986–1987. Harvests were on June 27, 1985, June 9, 1986, and June 8, 1987.

Chaudhuri et al. (1989, p. 11) (Table 14.1) show thermal (heat) units required during the 3 year experiment with wheat under elevated CO_2 to reach the different stages of growth. Due to physical limitations of the experiment (difficulty in removing the chamber often and limited plant area), certain growth stages (double ridge and milk-development stages) could not be determined with great accuracy for all the treatments. Thermal units were calculated from daily maximum and minimum air temperatures recorded inside the chambers when plants were enclosed in the chambers. Otherwise, when the chambers were not over the rhizotrons during the winter months, air temperatures were collected from the surrounding area.

The emergence and appearances of the leaves and tillers for the plants grown inside the chambers were ahead by a few days compared to the plants in the open field (Chaudhuri et al. 1986, p. 33, 1987, p. 21). The other growth stages were also a few days ahead for the plants inside the chambers compared to the open field, irrespective of moisture and CO_2 treatments (Chaudhuri et al. 1987, p. 21). This may have been primarily due to higher air temperature (2°C–3°C) inside the chambers compared to the open field.

As noted, growth stages of the wheat plants grown under different CO_2 concentrations were not observed closely or frequently. However, visual observations indicated that the appearance of the head was delayed by a couple of days in the chambers treated with elevated levels of CO_2, particularly in the irrigated rhizotrons (Chaudhuri et al. 1986, p. 35). That is, the plants exposed to the ambient level of CO_2 were slightly advanced compared to the plants exposed to elevated levels of

TABLE 14.1

Approximate Days after Emergence, Dates, and Thermal Units from Emergence for the Appearance of Commonly Known Growth Stages of Winter Wheat (1984–1987)

Growth Stages	Days After Emergence			Dates (Month and Day)			Approx. Thermal Units (°C-Day)		
	84–85	85–86	86–87	84–85	85–86	86–87	84–85	85–86	86–87
Jointing	175	171	196	4.12	4.30	4.12	1205	1200	1197
Flag leaf	186	187	207	4.23	4.19	4.23	1430	1450	1396
Booting	192	193	213	4.29	4.25	4.29	1506	1550	1524
Anthesis	202	202	228	5.09	5.04	5.09	1707	1730	1738
Milk	215	211	233	5.22	5.13	5.19	1917	2000	2001
Ripe	249	229	245	6.20	6.05	6.20	2394	2350	2342
Harvest	...[a]	...	...	6.27	6.09	6.08	...	2445	...

Sources: Data from Chaudhuri, U.N. et al., Effect of elevated levels of CO_2 on winter wheat under two moisture regimes, *Response of Vegetation to Carbon Dioxide*, Research Report No. 040, U.S. Department of Energy, Carbon Dioxide Research Division, Office of Energy Research, Washington, DC, 1987, xiv+70pp; Chaudhuri, U.N. et al., Effect of elevated levels of CO_2 on winter wheat under two moisture regimes, *Response of Vegetation to Carbon Dioxide*, Research Report No. 050, U.S. Department of Energy, Carbon Dioxide Research Division, Office of Energy Research, Washington, DC, 1989, ix+49pp.

[a] Not reported.

CO_2 (Chaudhuri et al. 1987, p. 21). Leaves of the plants subjected to high CO_2 remained greener longer compared to leaves with ambient CO_2 (Chaudhuri et al. 1989, p. 12). Similar trends were observed for the plants under the moisture-stressed conditions. Plants under moisture-stressed conditions were slightly advanced compared to the well-watered plants.

Thermal units of about 173°C–260°C-day were accumulated before the CO_2 chambers were placed over the plants. Thermal units of about 170°C and 210°C-day were required for the appearance of the third leaf and the first tiller, respectively (Chaudhuri et al. 1989, p. 10). Irrespective of different planting dates in the 3 years, total thermal units accumulated from emergence to harvest were similar, about 2400°C-day. No apparent differences in terms of total days required from planting to harvest appeared among the different CO_2 concentrations.

The leaf growth rate expressed in numbers using the Haun scale (1973) was observed on a tagged leaf of the main stem. In the 1985–1986 season, the Haun scale was recorded after injection of CO_2 until the completion of flag leaf emergence (Table 14.2). (The scale in Table 14.2 stops before reproductive development, and only records vegetative growth.) The results showed that the main tiller under the highest CO_2 concentration (825 µmol/mol) produced two more leaves compared to the wheat under ambient CO_2 during the same period. Increased leaf numbers under enriched CO_2 environments may be partially associated with higher leaf temperatures resulting from increased stomatal resistance. Stomatal resistance of the wheat is given in Chapter 7 (Tables 7.2 and 7.3), along with the data for water potential.

In the 1985–1986 season, the Zadoks–Chang–Konzak scale was used between booting and late-dough stage. Table 14.3 indicates that under well-watered conditions different elevated levels of CO_2 (485, 660, and 825 µmol/mol) did not show any effects on the growth stages. However, the plants exposed to the ambient level of CO_2 were slightly advanced compared to the plants exposed to elevated levels of CO_2. Similar results were observed for the plants under the moisture-stressed conditions, except that by the end of the experiment (May 27, 1986; Table 14.3), as the concentration of CO_2 increased, development was increasingly retarded. As for the vegetative stages, plants in the reproductive stages under moisture-stressed conditions were advanced compared to the

TABLE 14.2
Leaf Growth of Winter Wheat as Affected by Enriched CO_2

CO_2 Conc. (μmol/mol)	10.28	11.1	11.5	11.11	11.18	4.9	4.15	4.22
			1985				1986	
340	4.6	5.4	5.7	6.3	8.1	9.9	10.9	12.0
485	4.5	5.7	6.2	8.1	9.5	11.7	12.8	13.5
660	4.3	5.7	6.3	8.7	10.0	11.9	12.9	13.7
825	4.7	5.8	6.7	9.2	10.7	12.1	13.2	13.9

Source: Data from Chaudhuri, U.N. et al., Effect of elevated levels of CO_2 on winter wheat under two moisture regimes, *Response of Vegetation to Carbon Dioxide*, Research Report No. 050, U.S. Department of Energy, Carbon Dioxide Research Division, Office of Energy Research, Washington, DC, 1989, ix + 49pp.

Note: The Haun scale has been used to document the stages for different dates (month followed by day) for the 1985–1986 growing season.

TABLE 14.3
Growth Stages of Winter Wheat as Affected by Enriched CO_2 for Two Moisture Levels

CO_2 Conc. (μmol/mol)	4.25	4.28	5.02	5.05	5.15	5.20	5.27
Well-watered							
340	44	47	56	59	72	75	86
485	43	44	54	57	71	76	81
660	43	45	53	55	71	76	82
825	42	44	52	54	68	75	80
Water-stressed							
340	44	54	57	61	78	78	95
485	45	53	57	60	76	77	92
660	46	51	58	59	74	77	88
825	48	50	57	58	74	76	85

Source: Data from Chaudhuri, U.N. et al., Effect of elevated levels of CO_2 on winter wheat under two moisture regimes, *Response of Vegetation to Carbon Dioxide*, Research Report No. 050, U.S. Department of Energy, Carbon Dioxide Research Division, Office of Energy Research, Washington, DC, 1989, ix + 49pp.

Note: The Zadoks–Chang–Konzak scale has been used to document the stages for different dates (month followed by day) for the 1985–1986 growing season. All dates are in 1986.

well-watered plants. The wheat plants under elevated CO_2 had more leaves than the plants under the ambient level (Table 14.2), and this indicated that plants were staying in the vegetative stages longer and producing more leaves before they entered reproductive development.

Data in the literature are contradictory concerning the effect of CO_2 on tillering (Chaudhuri et al. 1986, p. 4). During a 3 year period (1971–1973) in Mexico, Fischer and Aguilar (1976) added

CO_2 at four different periods of growth, which corresponded approximately to months 1, 2, 3, and 4 after planting. Floral initiation took place during period 1 and anthesis took place during period 3. Biggest yield increases came for CO_2 added in the second and third months and were associated with more heads per meter squared and grains per spikelet, hence more grains per meter squared. During the fourth month (grain filling), seed weight and grain yield were increased in 1972, but not in 1973. There was an absence of yield response to growth increases with CO_2 fertilization during tillering (period 1). This indicated that early growth and tillering were relatively unimportant. They felt that the increased head number with CO_2 fertilization was the result of increased tiller survival rather than increased tiller production.

Krenzer and Moss (1975) found no effect of CO_2 concentration on the vegetative or reproductive development rates of wheat between enriched and normal CO_2 levels. They applied CO_2 during one of three periods: (1) the vegetative period (germination to floral initiation), (2) the floral development period (floral initiation to anthesis), and (3) the grain development period (anthesis to maturity). Grain yield increased when CO_2 enrichment was added during the latter two periods. No effect of CO_2 enrichment on grain yield was observed when the additional CO_2 was given during the vegetative period. Carbon dioxide enrichment caused an increase in seed number per plant when applied during the second period, and in seed size when applied during the third period. Kendall et al. (1985) also found that CO_2 enrichment (1200 μmol/mol CO_2; ambient was 400 μmol/mol) of winter or spring wheat cultivars did not have any effect on growth before anthesis. These results suggest that response to CO_2 occurs when maximum net carbon dioxide exchange rates occur, which is at the later reproductive growth stage (Kanemasu and Hiebsch 1975).

Other studies do show an effect of high CO_2 on tillering. Havelka et al. (1984) applied extra CO_2 at three different periods: (1) jointing to anthesis, (2) jointing to maturity, and (3) anthesis to maturity. They reported that the 17% increase in grain yield that they observed with elevated CO_2 resulted from more seeds per meter squared, which was associated with an increase in heads per meter squared rather than more seeds per head, indicating an increase in the number of tillers. This was confirmed by increases in stem weight and total dry matter due to the high CO_2. The grain-yield increase could have been due also to increased tiller survival with enriched CO_2. Eissa and Baker (undated), Gifford (1977), Sionit et al. (1980, 1981a,b), and Combe (1981) also found increased tillering with augmented levels of CO_2. Combe (1981) reported that the number of tillers was increased (by 12%) by CO_2 treatment, but at harvest, plants grown with ambient levels of CO_2 had the same number of tillers as those grown with high CO_2. Apparently, the extra tillers that were produced due to CO_2 did not survive. In Combe's (1981) experiment, the increase in grain yield (30%–40%) was due to an increased number of seeds per head, and not to more heads or to increased seed size.

We (Chaudhuri et al. 1986, 1987, 1989) found that increasing levels of CO_2 delayed reproductive development (advancement to the next growth stage). Marc and Gifford (1984) found the opposite. They reported that CO_2 enrichment (ambient + 500 μmol/mol CO_2; ambient ranged between 350 and 390 μmol/mol) of wheat and sunflower (*Helianthus annuus* L.) advanced floral initiation by 1–3 days. But elevated CO_2 had no effect on the floral initiation of sorghum (a short-day plant). The sensitive period for stimulation of floral initiation commenced within 10 days before the formation of double ridges. The hastening of growth, as occurred in the wheat and sunflower that Marc and Gifford (1984) studied, may provide the potential for improving the number of crops that can be grown in a particular period of time (Wittwer 1983). Lanning et al. (2010) obtained yield data for 'Thatcher,' a hard red spring wheat (*T. aestivum* L.) grown at six sites in Montana between 1950 and 2007. They also got the weather data during these 57 years. They found that March temperature increased at all sites over this period, and there were earlier heading dates, which they associated with the warmer temperatures. They did not consider CO_2 in their study of archived data.

However, Bencze et al. (2004) confirmed our observations, i.e., that elevated CO_2 retards reproductive development. They grew two winter wheat varieties ('Mv Martina' and 'Mv Emma,'

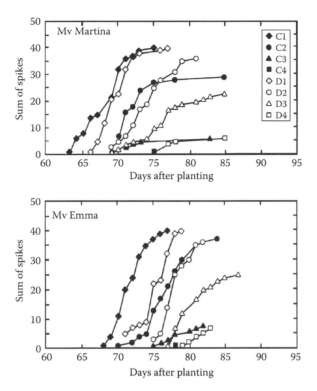

FIGURE 14.1 Cumulative number of successive spikes that emerged each day at ambient and doubled CO_2 level (C=control, D=doubled CO_2 level; 1–4=first to fourth spike). (Up to four spikes emerged per plant; the numbers on the y-axis show the total count for the 40 plants counted in each treatment.) (Reprinted from Bencze, S. et al., *Cereal Res. Commun.*, 32, 233, 2004, Fig. 1. With permission. Copyright 2004, Akadémiai Kiadó.)

commonly grown in Hungary) in growth chambers under ambient CO_2 (375 μmol/mol) and doubled atmospheric CO_2 (750 μmol/mol). Emma is a later ripening variety than Martina. The plants grew in pots with a mixture of garden soil and sand. There were four plants per pot. Forty plants from each variety and treatment were monitored during heading to collect data on the emergence of the first and further spikes. A low number of plants had three or more spikes.

Heading date was delayed by elevated CO_2 (Bencze et al. 2004). The main spike on Martina and Emma grown at elevated CO_2 appeared 1.5 and 3.5 days later, respectively, than it did in the control plants. Slafer and Rawson (1997) also reported a delay in the heading of wheat at high CO_2. Although there was a delay in the start of heading in Martina under elevated CO_2, the emergence of the second and third spikes began on the same day in both treatments (Figure 14.1: the solid symbols show the control plants, the open symbols show the plants with doubled CO_2). Plants of Emma started to produce subsequent spikes later at high CO_2 than at ambient CO_2. But in a few days, the plants grown at high CO_2 reached the control level as the process was faster (and the curve steeper) at high CO_2. Both varieties yielded more grain under the elevated CO_2 than the ambient CO_2. Bencze et al. (2004) concluded that the unfavorable delay of the heading date under high CO_2 might be counterbalanced by the faster production of additional spikes.

In the work of Bencze et al. (2004), the wheat was well watered. As noted above, Angus and Moncur (1977) found that water stress has an effect on the phenological development of wheat. In their work, with extreme water stress, anthesis was delayed but with moderate water stress, anthesis was advanced. However, they did not study elevated CO_2. The interaction of water stress and CO_2 on the phenological development of wheat needs further study. If we (Chaudhuri et al. 1986, 1987,

1989) had used a less severe water stress (the soil in our drought treatment was maintained at half field capacity), we might have found anthesis advanced under moderate stress and elevated CO_2.

PHENOLOGY UNDER ELEVATED CO_2

HORTICULTURAL CROPS

For many vegetable crops grown in greenhouses, elevated CO_2 up to about 1000 µmol/mol usually shortens the time taken until flowering begins (Murray 1997, p. 171). The difference in flowering time may range from a few days (e.g., begonia or *Begonia* sp.) to more than 15 days, as in *Saintpaulia ionantha* Wendl. (African violet). For some, the extent of advancement may depend on the CO_2 concentration. Thus, flowering is advanced by about 1 week in *Callistephus chinensis* Nees (China aster) plants growing at 600 µmol/mol CO_2, but by a shorter time at 900 µmol/mol. The response can also depend on population density, as in *Phlox drummondii* Hook (annual phlox). There might be no effect of elevated CO_2 on flowering time, as in *Salvia* (sage). Retardation of 1 or 2 days has been observed for *Chrysanthemum morifolium* Ramat. (florists' chrysanthemum) and *Nicotiana tabacum* L. (tobacco) (Murray 1997, p. 172).

In monoecious plants like *Cucumis sativus* L. (cucumber), fruit number depends on the number of female flowers. High CO_2 can alter the sex ratio to give more female flowers in cucumber (Baker and Enoch 1983). This occurs because high CO_2 encourages branching and there are more female flowers on the branches than on the main stem. There is an age window in the development of a fruit. During this period, the fruit is vulnerable to abortion. The level of source and sink imbalance at which fruits begin to be aborted is reached somewhat later by the high CO_2 plant than by the low CO_2 plant, and more fruit are enabled to pass through this developmental period without being aborted (Baker and Enoch 1983).

CROPS IN FACE EXPERIMENTS

Earliness has been one reason that growers have been encouraged to enrich their greenhouses with CO_2, because it will shorten the time to market and reduce heating fuel costs (Kimball et al. 2002). Even plants in the wild are being noted to flower earlier now than in the past (Miller-Rushing and Inouye 2009), when atmospheric CO_2 concentrations were lower. However, changes in phenology observed in FACE (free-air CO_2 enrichment) experiments have been small or nonexistent (Kimball et al. 2002). In the Arizona FACE facility, where the soil is a clay loam (Pinter et al. 1996), initially it was thought that FACE-grown spring wheat (*T. aestivum* L.) grown under elevated CO_2 was ready for harvest 1 week earlier than the controls (Pinter et al. 1996). However, it was found that the blowers in the FACE plots were causing the faster development compared to the ambient CO_2 plots that had no blowers. When blowers were added to the control plots, wheat development was accelerated by an average of only 0.4 days based on the difference in time between tillering and maturity (Kimball et al. 2002). Time between tillering and maturity of sorghum [*S. bicolor* (L.) Moench] was decreased by about 2 days in FACE plots with elevated CO_2 compared to control plots (Kimball et al. 2002). Potato (*Solanum tuberosum* L.) maturity was accelerated by only about a day, even though it appeared that senescence caused the leaf area to begin decreasing about 2 weeks earlier in the FACE plots (Kimball et al. 2002). In a FACE experiment in central Italy on a sandy-clay-loam soil, potato was grown for 3 years (1995, 1998, 1999) under ambient (not stated) and elevated levels of CO_2 (460, 560, and 660 µmol/mol in 1995 and 550 µmol/mol in 1998 and 1999). Crop phenology was not affected by elevated CO_2 (Bindi et al. 2006). At the Arizona FACE facility, grape (*Vitis* sp.) phenology was unaffected by growth at high CO_2. In several of these FACE experiments, secondary treatments such as moisture stress, low-soil nitrogen, and plant variety affected phenology more than did the high CO_2 (Kimball et al. 2002). Temperature is one of the most important factors driving phenological changes (Menzel and Sparks 2006).

In a FACE experiment in Japan, where rice (*Oryza sativa* L.) grew under 367 μmol/mol CO_2 (ambient) or 575 μmol/mol CO_2 (elevated) for three seasons, CO_2 enrichment slightly accelerated phenological development (Kobayashi et al. 2006). Heading was accelerated by about 2 days, and maturity was 2–3 days earlier. Although the accelerated development reduced the opportunity for the elevated CO_2 plants to capture light and to fix carbon, the difference was less than 2% of the whole growth duration (120–133 days), and hence the reduced growth duration did not have a large impact on the rice biomass and grain yield.

NATURAL ECOSYSTEMS

In natural as well as agricultural ecosystems, phenology is an important characteristic. At what time does most of the growth occur? At what time is the canopy closed? Phenology determines whether a plant has sufficient water for growth and yield and is important for the outcome of competition between different plant species (Dijkstra et al. 1996).

Annual plants may develop more quickly under elevated CO_2, reaching full leaf area, biomass, and flower and fruit production sooner than plants at ambient CO_2. Winter annuals from the California chaparral developed faster, completed their life cycles earlier, and died earlier as atmospheric CO_2 increased (Oechel and Strain 1985, p. 134). For the experimental weed *Arabidopsis thaliana* (L.) Heynh., flowering under elevated CO_2 depends on genotype and the amount of nutrients that are supplied to the plants. With low nutrient supply, flowering was shown to be accelerated by 9 days in one genotype (Murray 1997, p. 171), but not advanced in another genotype either with low or high nutrient supply (Murray 1997, p. 172). With 900 μmol/mol CO_2, the perennials *Panicum antidotale* and *P. decipiens* (in the grass family) are both advanced by about 1 week (Murray 1997, p. 171). However, flowering is not advanced in *P. tricanthum* growing with 900 μmol/mol CO_2. If the life cycle is shortened due to elevated CO_2, annuals under elevated CO_2 may not be more productive than those at normal CO_2, but maximum biomass is reached at an earlier time, possibly avoiding periods when drought stress becomes severe. In a competitive situation, early leaf and seed production could shift competitive relationships (Oechel and Strain 1985, p. 134).

There is, however, also evidence of either no effect or delayed senescence of some species under elevated CO_2. In a study of alpine vegetation, no difference in flowering behavior due to elevated CO_2 could be detected in *Carex curvula* (a sedge) (Körner et al. 1996). The arctic cotton grass, *Eriophorum vaginatum*, retained maximal photosynthetic capacity later into the fall when grown at high CO_2. Delayed leaf senescence potentially extends the growing season and increases production. But in arctic and temperate ecosystems, it may also subject plants to greater risk of tissue damage by frost before removal of organic and inorganic nutrients prior to leaf abscission (Oechel and Strain 1985, p. 134). Phenology, and its coincidence with weather conditions, is of crucial importance to the understanding of the response of terrestrial ecosystems to elevated CO_2 (Dijkstra et al. 1996).

In southern France, Roy et al. (1996) studied Mediterranean plants grown at 350 or 700 μmol/mol CO_2 in three different experiments. In one experiment, among the 17 species for which fruit set time was recorded, it occurred earlier (by 1–4 weeks) for three species and later (by 1–3 weeks) for four species. In another experiment, among the five species that were analyzed for the time of flower emergence, three had a delayed emergence by 1–3 weeks while two were only marginally affected. The third experiment also showed a species-specific effect of CO_2 on reproductive phenology. No trends due to the family or to biological type could be drawn from any of the three experiments. Much additional research is needed on the influence of CO_2 on phenological patterns in different ecosystems (Koch and Mooney 1996).

Because of shifts in phenology due to elevated CO_2, the time of herbivore feeding could be changed. This could affect productivity and competitive ability of plants. Plants that show slowed seasonal physiological senescence at elevated levels of CO_2 at the end of the season could experience major changes in herbivory (Oechel and Strain 1985, p. 140).

Forest Trees

The responses of the phenology of forest trees to elevated CO_2 are highly species-specific (Jach et al. 2001). For example, elevated CO_2 has been shown to affect bud phenology of two coniferous trees species, *Picea sitchensis* Carr. (Sitka spruce) and *Pinus sylvestris* L. (Scotch pine) in a contrasting way. In *P. sitchensis*, the phenological response to a doubling of CO_2 concentration was modified by nutrient availability (Lee and Jarvis 1996). When seedlings receiving ambient CO_2 and limited nutrient applications rates (i.e., a low-N or a medium-N nutrient treatment) were compared with their counterparts in elevated CO_2, those grown in elevated CO_2 flushed later in spring and senesced earlier in autumn. So, overall, the growing season was reduced. However, this effect of growth in elevated CO_2 was ameliorated by increased nutrient application rates. The nutritional impact might completely mask the effect of elevated atmospheric CO_2 concentrations on phenology. Because elevated CO_2 will delay bud burst and advance bud set of *P. sitchensis*, the likelihood of late spring or early autumn frost damage is reduced (Jach et al. 2001).

However, results were different for *P. sylvestris* (Jach et al. 2001). Spring bud development was hastened under elevated CO_2 in 3-year-old *P. sylvestris* seedlings in both the first and second years of treatment. Bud burst was advanced in the elevated CO_2 treatment during both years, by 6 days during the first year of treatment and by 9 days during the second year of treatment. Bud set in autumn of the first year of CO_2 treatment was hastened by elevated CO_2. Buds overwintered and started the second year of treatment when they were already in an early stage of growth (i.e., slightly swollen). After the dormancy period, the elevated CO_2 stimulated the buds to develop faster as compared with the ambient CO_2 treatment (Jach et al. 2001).

In sweet chestnut seedlings (*Castanea sativa* Mill.), bud burst was delayed by 2–3 weeks when plants grew in elevated (twice-ambient) CO_2. Leaves of sweet chestnut also senesced earlier in elevated CO_2, and leaf fall began 1 week earlier (Mousseau et al. 1996).

Shifts in growth phenology of shoots and in allocation patterns under elevated CO_2 may have important implications for wood characteristics (Jach et al. 2001). Elevated CO_2 may change wood characteristics by altering the time of transition from early wood to late wood. It has been hypothesized that in *Pinus* spp. the transition from early wood to the production of late wood occurs when the strong growth sinks of shoot development are reduced. Excess photosynthates are then allocated to the bole (the tree trunk), and secondary cell-wall thickening is promoted. In *P. taeda* L. (loblolly pine) an increase in the annual ring width is due primarily to a sensitive response of early wood to elevated CO_2 (Jach et al. 2001).

Shifts in carbon assimilation and growth may be translated into differences in xylem growth. Changes in xylem differentiation under elevated CO_2 have been reported. In radiata pine (also called Monterey pine; *P. radiata* D. Don), tracheid-wall thickening increased with augmented CO_2, and consequently wood density increased (Jach et al. 2001). However, not all studies show an effect of elevated CO_2 on wood density in radiata pine. Elevated CO_2 has been reported to alter wood quality by increasing xylem vessel size and number in *Quercus robur* L. (English oak) and by increasing wall thickness in *Prunus avium* L. (sweet cheery). An increased lignin to nitrogen ratio and a higher wood density in *P. abies* Karst. (Norway spruce) under elevated CO_2 conditions have been observed (Jach et al. 2001).

C_3 versus C_4 Plants

Bazzaz et al. (1985a, p. 162) state that the life cycles of most C_3 species are speeded up at higher CO_2 concentrations, and the life cycles of C_4 species are apparently slowed down. However, not all data confirm this observation. Hesketh and Hellmers (1973) studied changes in floral phenology of C_3 and C_4 plants under elevated CO_2. Floral initiation was greatly delayed in four sorghum [*S. bicolor* (L.) Moench] cultivars and slightly delayed in three other crops (*Z. mays* L. or corn, C_4; *H. annuus* L. or sunflower, C_3; and *G. hirsutum* L. or cotton, C_3), when grown in atmospheres containing approximately 1000 μmol/mol CO_2 as compared with plants grown in ambient air. The high CO_2 concentration delayed the flowering of cotton by 1.4–3.2 nodes. At the end of 40 days, no heads were evident on the sorghum plants. Plants then were moved into the ambient CO_2 greenhouse, and heads appeared shortly thereafter in plants grown in ambient CO_2. The plants grown in CO_2-enriched air developed 9.7 more leaves before heading, suggesting that floral initiation occurred after plants were placed in the greenhouse with ambient CO_2. In contrast to the data of Hesketh and Hellmers (1973), Reddy et al. (2000) reported that the developmental rates of cotton were not affected by the doubling of atmospheric CO_2 concentration.

Bazzaz et al. (1985a, p. 162) give a table showing the developmental rate of plants as affected by elevated CO_2. It shows that development of *Amaranthus* (amaranth, C_4), *Datura* (jimsonweed, C_3), and *Phlox* (phlox, C_3) is accelerated by elevated CO_2, while that of *Sorghum* (C_4) and *Abutilon* (Indian mallow, C_3) is delayed and unchanged, respectively.

EFFECT OF PHOTOPERIOD

Reekie et al. (1994) did a study to determine if the effect of elevated CO_2 on flowering phenology is a function of the photoperiodic response of the species involved. Four long-day plants, *Achillea millefolium* L. (milfoil), *C. chinensi* Nees. (China aster), *Campanula isophylla* Moretti (common name unknown), and *Trachelium caeruleum* L. (throatwort is the name of the genus), and 4 short-day plants, *Dendranthema grandiflora* (authority and common name unknown), *Kalanchoe blossfeldiana* von Poellnitz (kalanchoe), *Pharbitis nil* Choisy (now called *Ipomoea nil* Roth) (Japanese morning-glory; Salisbury and Ross 1978, p. 336), and *Xanthium pensylvanicum* Wallr. (cocklebur is the name of the genus), were grown under inductive photoperiods (9 h for short day and 17 h for long day) at either 350 or 1000 μmol/mol CO_2. The time of visible flower bud formation, flower opening, and final plant biomass were assessed.

Elevated CO_2 advanced flower opening in all 4 long-day species and delayed flowering in all 4 short-day species (Reekie et al. 1994). In the long-day species, the effect of CO_2 was primarily on bud initiation. All four species formed buds earlier at high CO_2. Bud development, the difference in time between flower opening and bud initiation, was advanced in only one long-day species, *C. chinensis*. Mixed results were obtained for the short-day species. Elevated CO_2 had no effect on bud initiation but delayed bud development in *Dendranthema* and *Kalanchoe*. In *Xanthium*, bud initiation rather than bud development was delayed. Data on bud initiation and development were not obtained for *Pharbitis*. The negative effect of CO_2 upon phenology in the short-day species was not associated with negative effects on growth. Elevated CO_2 increased plant size in both long-day and short-day species. The results of Reekie et al. (1994) suggest that different aspects of the flowering process are affected by elevated CO_2 and that the specific effect may depend upon the photoperiodic response of the particular species.

QUESTIONS TO BE ANSWERED ABOUT ELEVATED CO_2 AND PHENOLOGY

Even though there have been several studies concerning the effects of elevated CO_2 on phenology, still more work needs to be done. I have compiled the following list of questions concerning what we need to know about elevated CO_2 and phenology. Most of the questions focus on wheat, the major crop in Kansas.

1. How is flowering date affected by variable CO_2 concentrations? Is flowering of obligate winter wheats hastened if they are given elevated CO_2 in winter?
2. How is vernalization affected by elevated CO_2? Are vernalized wheat plants more responsive to elevated levels of CO_2 than nonvernalized plants?
3. Relating flowering dates to degree days implies process, such as van't Hoff's law (Price 1911; Waggoner 1974a). Process should be studied. That is, how does elevated CO_2 affect

the van't Hoff relationship? In this respect, we need official meteorological stations that record not only temperature and precipitation but also CO_2 concentration. Many phenological data are needed so standard errors and other statistics can be calculated (West and Wein 1971; Waggoner 1974a,b).

4. What does the growing tip of wheat look like as it changes from a vegetative tip to a flowering tip under elevated CO_2? How does CO_2 concentration affect the timing (and appearance) of this change? Careful morphological studies of the plants should be carried out at different official weather locations.

5. How do stresses (wind, water, temperature, salinity) affect the morphology of the tip under elevated CO_2? Specimens and photographs should be collected that can be preserved for future reference.

6. Should "standard varieties" be designated, so they can be compared around the world under elevated CO_2? In 1955, Nuttonson used 'Kharkov' as a standard winter wheat variety and 'Thatcher' as a standard spring wheat variety. In the state of Oklahoma, in the 1970s, the wheat breeder designated 'Scout 66' as a standard variety for comparison purposes (Smith 1976). The same varieties need to be studied in different locations as well as different varieties in the same location.

7. Can we get early, high-yielding varieties of wheat with elevated CO_2? Can we manipulate elevated CO_2 to speed up growth without reducing yields?

8. What is the relation between soil type and rate of development of wheat under elevated CO_2?

9. Stress hastens maturity of wheat (Cohen 1971). Does elevated CO_2 hasten this maturity even further or does elevated CO_2 slow it down?

SUMMARY

Horticultural crops grown in greenhouses with elevated CO_2 often are advanced in development, and this shortens their growing season. However, changes in phenology of agronomic crops (spring wheat, sorghum, potato, grape, rice) grown in FACE have usually been small or nonexistent. Our work with winter wheat in enclosed chambers (Chaudhuri et al. 1986, 1987, 1989) showed that elevated CO_2 delayed development of wheat in the reproductive stages. That is, plants grown under ambient CO_2 (340 μmol/mol CO_2) advanced through reproductive growth stages faster than did plants grown under elevated CO_2. Wheat plants grown with 825 μmol/mol CO_2 had two more leaves than plants under the ambient level, and this indicated that plants were staying in the vegetative stages longer and producing more leaves before they entered reproductive development. Even though some experiments confirm our results, others do not. These other studies report no effect of CO_2 on wheat vegetative-growth stages, and in the reproductive stage, hastening of floral development has been documented under elevated CO_2. Results probably depend upon variety and environmental conditions (e.g., level of drought). In our studies, plants grown with elevated CO_2 under well-watered and dry conditions responded the same—i.e., reproductive development was retarded by elevated CO_2.

Annual plants grown in natural ecosystems may develop more quickly under elevated CO_2 than plants at ambient CO_2. There is also evidence of no effect or delayed senescence on annuals. The responses of the phenology of forest trees to elevated CO_2 are highly species specific. Spring bud development can be hastened, delayed, or not changed by elevated CO_2. Shifts in phenology may change wood characteristics by altering the time of transition from early wood to late wood and xylem growth. These changes result in higher wood density under elevated CO_2. The life cycles of both C_3 and C_4 plants can be affected by elevated CO_2, and no general conclusion can be made about the effect of elevated CO_2 on their phenology. Each can be hastened or delayed by elevated CO_2. Photoperiod interacts with CO_2. Data in one experiment show that elevated CO_2 advances flowering in long-day species and delays flowering in short-day species.

APPENDIX

BIOGRAPHY OF SVANTE AUGUST ARRHENIUS

Arrhenius, a Swedish chemist, was one of the founders of physical chemistry. He was born on February 19, 1859, near Uppsala. As a child at school, he distinguished himself in mathematics. At the age of 17, he entered the University of Uppsala, devoting himself to physics. Because he found that the instruction in physics there left something to be desired in those days, in 1881 he went to study under Erik Edlund at Stockholm (da Costa Andrade and Oesper 1971).

In 1884, he took his doctor's degree at Uppsala with a thesis on galvanic conduction in electrolytes. [*Galvanic* means producing an electric current, especially from a battery (Friend and Guralnik 1959)]. This thesis, whose development later earned Arrhenius the Nobel Prize, was awarded the lowest "note" that could be granted without a definite refusal. It is divided into two parts, the first dealing with the experimental determination of the electrical conductivity of extremely dilute solutions, and the second with the theory of electrolytic conductivity. This second part discusses his theory of electrolytic dissociation and activity coefficients. The revolutionary nature of the theory, which required, for instance, that free sodium ions should exist in a solution of sodium chloride, prevented its being speedily accepted, and it may be said that it was neglected rather than actively opposed (da Costa Andrade and Oesper 1971).

Arrhenius himself reported an anecdote that reveals the attitude toward his views when they were new: "I came to my professor, Cleve, whom I admired very much, and I said, 'I have a new theory of electrical conductivity as a cause of chemical reactions.' He said, 'This is very interesting,' and then he said, 'Good-bye.' He explained to me later that he knew very well that there are so many different theories formed, and that they are almost all certain to be wrong, for after a short time they disappeared; and therefore, by using the statistical manner of forming his ideas, he concluded that my theory also would not exist long" (da Costa Andrade and Oesper 1971).

Arrhenius's thesis won him the support of Wilhelm Ostwald (1853–1932; German chemist; received Nobel Prize in Chemistry, 1909), in whose laboratory he worked in 1886, and in 1887 he likewise entered into a close friendship with Jacobus H. van't Hoff. [See Kirkham (2005, pp. 310–311) for a biography of van't Hoff.] The theories of van't Hoff on osmotic pressure and of Ostwald on the affinity of acids agreed well with the views of Arrhenius. These three friends worked together unselfishly for the new doctrines of physical chemistry, which ultimately won general acceptance. Arrhenius had a genius for friendship and, in the 40 years between 1887 and his death, he met practically all the great men of science and won their affection and regard (da Costa Andrade and Oesper 1971).

In 1891, Arrhenius declined a professorship at Giessen, Germany, and was appointed lecturer at the University in Stockholm. In 1895, he became a professor of physics there. From 1887 to 1902, he also filled the office of rector of the university. Many students from abroad as well as from Sweden worked under his direction. Arrhenius was a man of wide interest, and about this time he began to turn his attention to problems of the chemistry of living matter. In 1904, he delivered at the University of California a course of lectures designed to illustrate the applications of the methods of physical chemistry in explaining the reactions of toxins and antitoxins. [This was the year before Ludwig Boltzmann delivered lectures at the University of California. See Kirkham (2005, p. 421) for Boltzmann's lectures in 1905 at the University of California.] Arrhenius's lectures were published in 1907 under the title *Immunochemistry* (da Costa Andrade and Oesper 1971).

Arrhenius was also much occupied with problems of cosmogony. [*Cosmogony* is defined as the origin or generation of the universe (Friend and Guralnik 1959).] In his *Worlds in the Making* (1908), an English translation of *Das Werden der Welten* (1907), he combated the generally accepted doctrine that the universe is tending to what Rudolf Clausius [see Kirkham (2005, p. 311) for a biography of Clausius] termed *Wärmetod* (death of heat) through exhaustion of all sources of heat and motion. Arrhenius suggested that, by virtue of a mechanism that maintains available energy, the universe is self-renovating. He further put forward the idea that life is universally diffused and is constantly emitted from all habitable worlds in the form of spores that traverse space for years or ages. The majority of them are ultimately destroyed by the heat of some blazing star. But a few spores find a resting place on bodies that have reached the habitable stage. He was one of the first to stress the important part that the pressure of light must play in cosmic physics. He pointed out that the repulsion of the tails of comets from the sun could be explained by this pressure. Astronomical problems, especially the question of the habitability of Mars, are discussed in his *Destinies of the Stars* (1918) (da Costa Andrade and Oesper 1971).

Arrhenius received universal acknowledgement as one of the great men of this time when he was awarded the Nobel Prize in Chemistry in 1903 "in recognition of the special services rendered by him to the development of chemistry by his electrolytic theory of association." In 1905, after refusing a full professorship and private laboratory in Berlin, he was appointed director of the Nobel Institute for Physical Chemistry at Experimentalfältet, just outside Stockholm, a post which he held until a few months before his death on October 2, 1927. He was buried in Uppsala. In 1910, he was made a foreign member of the Royal society. He received numerous honorary doctorates as well as the Davy medal of the Royal society and the Faraday medal of the Chemical society. He enjoyed visiting colleagues abroad and receiving his contemporaries and students at his home. A genial humor characterized his talks and his private conversation, and he was welcome at scientific discussions in any land (da Costa Andrade and Oesper 1971).

15 Elevated Atmospheric Carbon Dioxide: Growth and Yield

INTRODUCTION

In this last chapter, we look at the effects of elevated CO_2 on growth and yield under well-watered conditions. We define "growth" as "an increase in size, or weight" or "the amount of this [increase]" (Friend and Guralnik 1959). We define "yield" as the "aggregate of products resulting from the growth or cultivation of a crop and usually expressed in quantity per area" (Barnes and Beard 1992b). Yield commonly refers to the product consumed or marketed. In Chapter 1, we considered the effects of elevated levels of CO_2 on growth and yield of plants under drought, and in Chapter 6 the focus was root growth. In this chapter, our interest will be on aboveground growth and yield. We devote a chapter to growth and yield, because they integrate all factors that affect a plant during its life cycle. In any plant–water relations experiment, a measure of growth (e.g., height, biomass) should be included (Kirkham 2005, p. 6).

Before we begin, we cite a classic paper (Kimball 1983) concerning elevated CO_2 and plant yield. Bruce A. Kimball is a soil physicist with the U.S. Department of Agriculture in Maricopa, Arizona, and a pioneer in high-CO_2 plant research (Stafford 2007). (See Appendix for a biography of Bruce Kimball.) In his paper, he analyzed more than 430 observations of the yield of 37 species grown with CO_2 enrichment, which were published in more than 70 reports of experiments carried out in a 64 year period beginning in 1918. For "yield," he considered both marketable yield (e.g., grain) and vegetative yield (i.e., dry weight or height). The exact concentrations of CO_2 in the different studies that he investigated varied, and he reports them, as well as the type of yield under consideration, in extensive appendixes. When all types of yield were averaged together, his analysis showed that, with a doubling of atmospheric CO_2 concentration, yield probably will increase by 33%, with a 99.9% confidence interval from 24% to 43%. Increases in grain yield were similar to increases in vegetative yield. Review papers published since Kimball's (1983) seminal paper confirm his observations that growth is usually increased under elevated CO_2 (Bazzaz et al. 1985b; Strain and Cure 1985a; Drake and Leadley 1991; Kimball et al. 1993; Idso and Idso 1994; Lawlor and Mitchell 2000).

In the experiments reported in this chapter, plants have been well watered, unless otherwise stated. As in previous chapters, we consider different types of plants, including agronomic plants, horticultural plants, pasture plants, marsh plants, weeds, cacti, and trees. Now let us look at individual crops, beginning with wheat, one of the two most important food crops in the world, the other one being rice. Wheat is used to make bread, the staff of life and regarded as the basic food (Friend and Guralnik 1959). More wheat is produced than rice, but more people eat rice as their chief food (Heyne and Sherman 1971).

WHEAT

Amthor (2001) analyzed 50 studies (156 experiments) concerning the effects of CO_2 concentration [CO_2] on wheat (*Triticum aestivum* L.) yield. Studies were divided into five categories based on the method of controlling [CO_2]: laboratory chamber, greenhouse, closed-top field chamber, open-top field chamber, and a free-air (chamber-less) field CO_2 enrichment (free-air CO_2 enrichment or FACE) system. Three studies, all conducted in greenhouses, included subambient [CO_2] (less than 330 ppm, the unit that Amthor used). A [CO_2]-yield curve was estimated by pooling data from

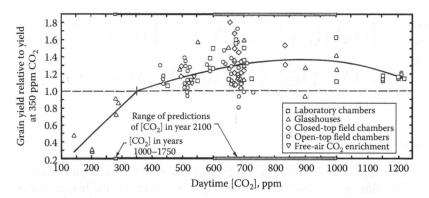

FIGURE 15.1 Effect of [CO_2] on wheat yield relative to yield at 350 ppm CO_2 for various methods of controlling [CO_2]. Results shown are for wheat supplied with ample water and nutrients and with most favorable temperature and [O_3] in each study. When the "ambient-[CO_2]" treatment differed from 350 ppm in an experiment, a linear relationship was assumed between yield and [CO_2] and the experimental [CO_2]s were adjusted to a 350 ppm reference value. For example, in one experiment, yield was 11.9% greater with a 180 ppm [CO_2] increase from 370 to 550 ppm and it was assumed here that the same 11.9% yield gain would have occurred with a 180 ppm increase in [CO_2] from 350 to 530 ppm, so a relative yield of 1.119 is plotted at 530 ppm CO_2 for this experiment. Deviations of ambient-[CO_2] treatments from 350 ppm were small so the assumption of a linear effect of [CO_2] on yield had limited impact on this analysis. The reference point of unit relative yield at 350 ppm CO_2 is marked with a vertical line ("|"). The solid line left of "|" is the least-squares straight-line fit to subambient-[CO_2] experiments. The solid line right of "|" is the least-squares cubic fit to superambient-[CO_2] experiments up to 1250 ppm CO_2. Two experiments were off scale; they had relative yields of 2.11 at 900 ppm CO_2 and 2.05 at 735 ppm CO_2, respectively. They were included in defining the cubic equation. There was no mechanistic basis for the discontinuity in the slopes of the lines summarizing subambient and superambient results as they met at 350 ppm CO_2. (Reprinted from *Field Crops Res.*, 73, Amthor, J.S., Effects of atmospheric CO_2 concentration on wheat yield: Review of results from experiments using various approaches to control CO_2 concentration, 1–34, Fig. 1, Copyright 2001, with permission from Elsevier.)

multiple experiments (Figure 15.1). To do this, he calculated the ratio of yield in subambient- or superambient-CO_2 concentration (in the range of 140–1250 ppm) to yield in near-ambient (i.e., about 350 ppm) CO_2 concentration. The data came from 113 experiments. Those 113 ratios of yield were derived only from experiments using ample water and nutrients, and ambient (or most beneficial) air temperature, in order to focus on effects of CO_2 concentration rather than interactions between CO_2 concentration and other factors. He fit a least-squares straight-line equation, forced through the point of unit relative yield at 350 ppm CO_2, to results from subambient-[CO_2] experiments (straight line in Figure 15.1). It indicated a 0.34% decline in yield per ppm [CO_2] reduction from 350 ppm CO_2. By extrapolation, zero yield would occur at 50–60 ppm CO_2, which happens to approximate the [CO_2]-compensation point for C_3 photosynthesis with normal atmospheric oxygen concentration.

All five methods of controlling [CO_2] provided yield data in the range of 350–1250 ppm [CO_2]. He fit a least-squares cubic equation, forced through the point of unit relative yield at 350 ppm CO_2, to the superambient-CO_2 yield data pooled for all the methods (curved line in Figure 15.1). The equation indicated that yield was maximal at about 890 ppm CO_2, with a 37% increase in yield relative to 350 ppm CO_2. The cubic equation indicated that yield increased about 31% when [CO_2] increased from 350 to 700 ppm, a value similar to 33% determined by Kimball (1983) for a doubling of the CO_2 concentration. Taken together, Amthor's data indicated that effects of [CO_2] on wheat yield were strongly positive at subambient-CO_2 concentrations, positive and weakening at superambient-CO_2 concentrations up to about 900 ppm, and negative at CO_2 concentrations greater than 1000–2000 ppm.

Neales and Nicholls (1978) pointed out that response to CO_2 concentration depends upon age of the wheat plants. They grew wheat (*T. aestivum* L. cv. Pinnacle) in pots with a soil mix (sand, peat, loam) that were placed in a growth cabinet at the following CO_2 levels: 200, 300, 400, 600, and 800 volumes per million (vpm). They measured net assimilation rate (E_A) and harvested the plants on days 10, 14, 17, 21, and 24 after germination to determine leaf area (A) and dry weight (W) of the leaves, stem, and roots. They calculated relative growth rate (R_W, mg/mg/day) and relative leaf growth rate (R_L, cm²/cm²/day). The effects of CO_2 concentration on the various growth parameters interacted with plant age. For instance, in the 10-day-old plants, relative growth rate and net assimilation rate were increased by 35% and 55%, respectively, by an increase in CO_2 levels from 200 to 800 vpm. But these two growth parameters were reduced by 44% and 16%, respectively, in 24-day-old plants over the same interval of CO_2 concentration (Figure 15.2a and b). However, E_A increased with CO_2 concentration for 10-day-old plants and did not change with 24-day-old plants (Figure 15.2c). They said that the absence of a sustained, long-term effect of CO_2 enrichment on growth rate in their experiment was "paradoxical." Most experiments show a sustained increase in growth during the growth cycle. They suggested that one possible explanation was that, in their wheat plants, assimilatory capacity exceeded demand for assimilate. Robertson and Leech (1995), whose work was cited in the chapter on anatomy (Chapter 13), also found that age affected their

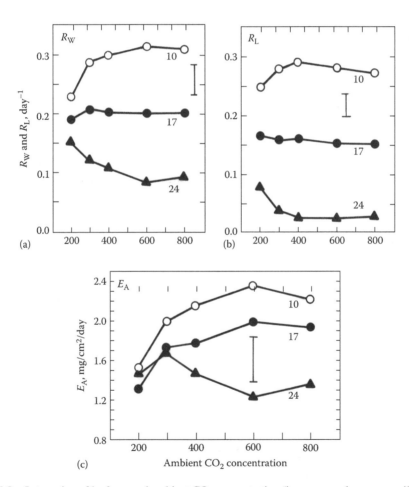

FIGURE 15.2 Interaction of leaf age and ambient CO_2 concentration (in vpm or volumes per million) in the determination of (a) relative growth rate (R_W), (b) relative leaf area growth rate (R_L), and (c) net assimilation rate (E_A). Error bars indicate the magnitude of least significant differences between means ($P=0.05$). The numbering of the three curves of each part of the figure indicates the plant age, in days. (Reprinted from Neales, T.F. and Nicholls, A.O., *Aust. J. Plant Physiol.*, 5, 45, 1978, Fig. 5. With permission. Copyright 1978, CSIRO Publishing.)

results, because the carbohydrate status of plants changed with age. Neales and Nicholls (1978) concluded that, due to the short-time effects they observed, long-term experiments, such as the one by Gifford (1977), are more useful, when one is attempting to determine the growth of wheat in response to global increases in CO_2 levels. As reported in Chapter 14, Gifford (1977) grew wheat for up to 140 days after planting under three concentrations of CO_2: subambient (130 vpm CO_2), ambient (280 vpm CO_2), and superambient (480 vpm CO_2), and dry matter and grain yield increased with increasing CO_2 concentration.

In another short-term experiment (24 days) with wheat (*T. aestivum* L. cv. Ritmo), Šantrůček et al. (1994) found that dry weights were slightly more under elevated CO_2 (700 μmol/mol) than ambient (350 μmol/mol). The CO_2 assimilation rate was higher under elevated CO_2 compared to ambient only until the 18th day after planting, and it decreased between the 18th and 23rd day. They hypothesized that the lower assimilation rate after the 18th day under elevated CO_2 was due to a limited regeneration of ribulose bisphosphate in the Calvin cycle.

Response of wheat to CO_2 concentration also depends upon plant competition. Du Cloux et al. (1987) grew winter wheat (*T. aestvium* L. cv. Capitole) in growth chambers under two CO_2 concentrations and two plant densities. One chamber contained the normal CO_2 concentration or 330 mm^3/dm^3 (300 μmol/mol) and the other had a doubled concentration or 660 mm^3/dm^3 (660 μmol/mol). Three-day-old seedlings were planted in pots at two plant densities: 5 plants per pot or 1 plant per pot for the high- and low-density treatments, respectively. In the high-density experiment, plants were harvested 23, 30, and 37 days after germination, and in the low-density experiment, plants were harvested 23, 38, and 72 days after germination. Growth was followed in two ways: by harvesting plant material on these days, and by using gas exchange measurements of photosynthesis and respiration. They put the gas exchange data into an equation to calculate what they called the "continuous growth curve." They corrected the data for soil respiration, which they estimated. As shown in Figure 15.3, results for the two methods did not agree. Gas exchange measurements resulted in an overestimation of dry weight. They attributed this difference, in part, to wrong estimations of respiration. The ratio of the dry weight at 660 μmol/mol CO_2 to that

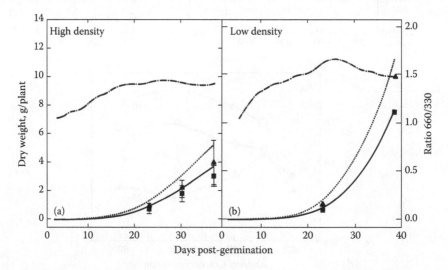

FIGURE 15.3 Continuous growth curves obtained from the integration of the net daily CO_2 uptake of whole canopies: solid line, 330 mm^3/dm^3 CO_2 and dotted line, 660 m^3/dm^3 CO_2. Dry weights from harvests (samples of 15 and 5 plants in the high- and the low-density experiment, respectively): solid square, 330 mm^3/dm^3 CO_2 and solid triangle, 660 mm^3/dm^3 CO_2. Ratio of high CO_2 to control growth curves: line at top of (a) and (b) with dots and dashes: (a) 200 plants/m^2; (b) 40 plants/m^2. Vertical bars show standard deviations. (Reprinted from Du Cloux, H.C., André, M., Daguenet, A., and Massimino, J., *J. Exp. Bot.*, Fig. 1, 1987 by permission of Oxford University Press.)

at 330 μmol/mol CO_2 was greater than one (right-hand side ordinate; dotted and dashed line at the top in Figure 15.3a and b), indicating that the elevated CO_2 increased dry weight. The ratio reached a maximum of 1.5 for the high-density plants and a maximum of 1.7 for the low-density plants. This implied that an isolated plant might respond to elevated atmospheric CO_2 more than plants planted closely together. For an isolated plant, leaves can expand and photosynthesize without competition from neighboring leaves. Under field conditions, to obtain the full effect of elevated CO_2 on plant growth, the data suggested that seeds should be planted in a wide spacing rather than in narrow spacing.

RICE

Baker et al. (1990b) grew rice (*Oryza sativa* L. cv. IR30) in paddy culture in outdoor, naturally sunlit controlled-environment, plant growth chambers in Gainesville, Florida. They exposed the plants throughout the season to subambient (160 and 250 μmol/mol), ambient (330 μmol/mol), or superambient (500, 660, 900 μmol/mol) CO_2 concentrations. The plants grew in a fine sand soil. They sampled the plants at 19, 30, 44, 58, 71, 86, and 110 days after planting. Shoot biomass increased with increasing CO_2 concentration from 160 to 500 μmol/mol CO_2 (Figure 15.4). By the end of the growing season, shoot biomass in the superambient treatments was roughly double that in either of the subambient treatments. The response of tillering to CO_2 was similar to that of biomass accumulation. There was an increase in tillering from 160 to 500 μmol/mol CO_2 (Figure 15.5). In all CO_2 treatments, rice plants attained the greatest number of fertile tillers by 30 days

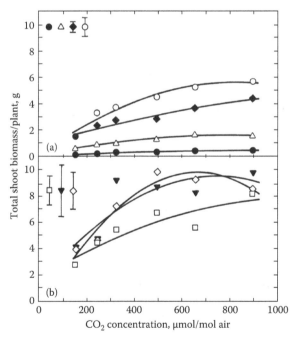

FIGURE 15.4 Mean shoot biomass of rice plants grown to maturity in paddy culture in controlled-environment chambers at subambient (160 and 250), ambient (330), and superambient (500, 660, and 900 μmol/mol air) CO_2 concentrations, at (a) 19 (closed circles), 30 (open upright triangles), 44 (closed diamonds), and 58 (open circles) days after planting and (b) 71 (open squares), 86 (closed upside down triangles), and 100 (open diamonds) after planting. Standard error (vertical bars) for each sampling date is for each of the means and was obtained by pooling each of the within-chamber variances. (Reprinted from Baker, J.T. et al., *J. Agric. Sci.*, 115, 313, 1990b, Fig. 1. With permission. Copyright 1990, Cambridge University Press.)

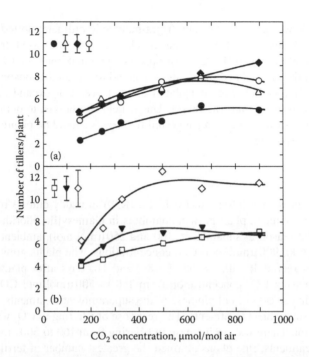

FIGURE 15.5 Mean number of tillers/plant of rice grown to maturity in paddy culture in controlled-environment chambers at subambient (160 and 250), ambient (330), and superambient (500, 660, and 900 μmol/mol air) CO_2 concentrations, at (a) 19 (closed circles), 30 (open triangles), 44 (closed diamonds), and 58 (open circles) days after planting and (b) 71 (open squares), 86 (upside down solid triangles), and 110 (open diamonds) days after planting. Standard error (vertical bars) for each sampling date is for each of the means and was obtained by pooling each of the within-chamber variances. (Reprinted from Baker, J.T. et al., *J. Agric. Sci.*, 115, 313, 1990b, Fig. 3. With permission. Copyright 1990, Cambridge University Press.)

after planting. The rise in number of tillers at the end of the growing season (Figure 15.5) was due to the production of small, ineffective tillers during and following grain filling. Doubling the CO_2 concentration from 330 to 660 μmol/mol resulted in a 32% increase in grain yield (Baker et al. 1990b). A subsequent paper by Baker et al. (1992) on rice confirmed the earlier results. Dry weight of the aboveground biomass increased with increasing CO_2 concentration from the subambient concentration of 160 μmol/mol CO_2 to the superambient concentration of 500 μmol/mol with only small differences among the superambient treatments (Figure 15.6).

Wild and cultivated rice respond differently to elevated CO_2. The wild lines of rice are weeds, and they compete with cultivated rice in the field. Ziska et al. (2010) evaluated the growth and reproduction of cultivated rice (*O. sativa* L.; Clearfield, CL161) and red or weedy rice (*O. sativa* L. biotype Stuttgart, StgS, a troublesome weed in rice fields in Arkansas) in monoculture and at two competitive densities (8 and 16 plants/m²) using CO_2 concentrations that corresponded to the 1940s, current levels, and one projected for the middle of the twenty-first century (300, 400, and 500 μmol/mol, respectively). Competition was determined by using plant relative yield for biomass and seed yield as a function of CO_2. In monoculture, increasing CO_2 above the 300 μmol/mol mid-twentieth century level resulted in an increase in both seed yield and biomass at harvest for the wild rice biotype, but only a small increase in biomass for the cultivated line. At maturity, plant relative yield increased for the wild rice, but decreased for the cultivated line as a function of competitive density. The basis for the CO_2-induced changes in the competitive ability of the wild rice was unclear to the authors. They pointed out that competitive outcomes are determined early in growth and, although leaf area was not monitored, the leaf area of the weed may have increased faster than that of the crop. With the ability to intercept more sun, the weed could grow faster than the crop. The authors

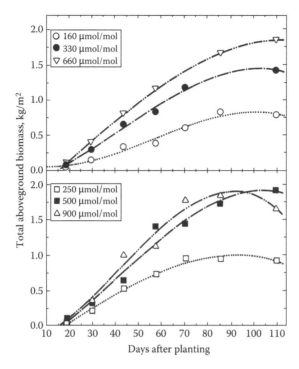

FIGURE 15.6 Seasonal trends in total aboveground biomass for rice plants grown to maturity in subambient (160 and 250), ambient (330), and superambient (500, 660, and 900 μmol/mol) CO_2 concentration during the day. (Baker, J.T., Laugel, F., Boote, K.J., and Allen Jr., L.H.: Effects of daytime carbon dioxide concentration on dark respiration in rice, *Plant Cell Environ.*, 1992, 15, 231–239, Fig. 3. Copyright Wiley-VCH Verlag GmbH & Co. KGaA. Reproduced with permission.)

concluded that red rice, a pernicious weed in rice cultivation, demonstrated a stronger response to rising CO_2 than its commercial counterpart, with greater competitive ability, and subsequent negative effects on cultivated rice yields.

BARLEY

Weigel et al. (1994) grew two cultivars of spring barley (*Hordeum vulgare* L. cv. Alexis and cv. Arena), as well as two cultivars of spring wheat (*T. aestivum* L. cv. Star and cv. Turbo), during a season under ambient CO_2 (384 ppm) and elevated CO_2 (471, 551, 624, and 718 ppm) in open-top chambers (OTCs) to determine the effect of elevated CO_2 on biomass. Plants grew in pots with peat and soil. They sampled the plants four times during the season: stem elongation stage (35–36 days after planting); at the end of booting (57–58 days after planting); at anthesis (70–72 days after planting); and at maturity (99–101 days after planting). Shoot biomass of barley increased with increasing CO_2 concentration, and CO_2-related growth stimulation diminished only at the highest treatment level (Figure 15.7, right). Shoot biomass of wheat also increased with increasing CO_2 concentration (Figure 15.7, left). (Data for 624 ppm CO_2 are missing from the figures, due to malfunctioning of the system at this concentration). At harvest, maximal increase in dry matter amounted to about 70%–110% and 30%–40% for barley and wheat, respectively. Tillering increased between 384 and 551 ppm CO_2 for both species (Figure 15.8). Yield (Figure 15.9) increased as the CO_2 concentration increased. Maximal CO_2-related increase in yield ranged between 4.1 g per pot for the wheat cultivar Star and 11.0 g per pot for the barley cultivar Arena. On a relative basis, the yield increase was 52% and 89% for the two barley cultivars (Alexis and Arena, respectively) and

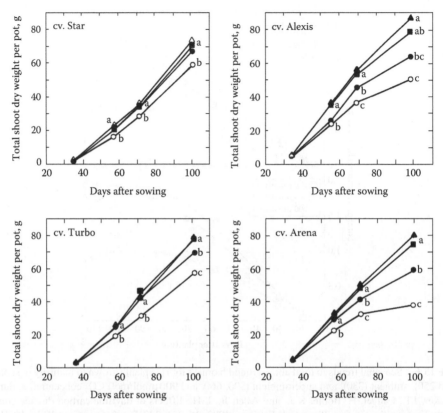

FIGURE 15.7 Mean shoot biomass of wheat ('Star' and 'Turbo') and barley plants ('Alexis' and 'Arena') grown in open-top chambers until maturity at different CO_2 concentrations (open circles, 384 ppm; closed circles, 471 ppm; closed squares, 551 ppm; closed triangles, 718 ppm). Means of one sampling date showing different letters differ significantly ($P < 0.05$). (Reprinted from *Agric. Ecosyst. Environ.*, 48, Weigel, H.J., Manderscheid, R., Jäger, H.-J., and Mejer, G.J., Effects of season-long CO_2 enrichment on cereals: I. Growth performance and yield, 231–240, Fig. 2, Copyright 1994, with permission from Elsevier.)

19% and 27% for the two wheat cultivars (Star and Turbo, respectively). Their study showed that approximately doubling the CO_2 concentration (384–718 ppm) increased grain yield for both barley and wheat cultivars.

Data of Ingvardsen and Veierskov (1994) with barley (*H. vulgare* L. cv. Digger) confirmed those of Weigel et al. (1994). They grew barley for 22 days at either ambient CO_2 concentration in a greenhouse (ambient concentration not given) or in enclosed chambers with a CO_2 enrichment of up to 675 µmol/mol CO_2 (675 ppm) above the ambient concentration (probably about 1035 ppm, if ambient were 360 ppm). Plants grew in pots with a commercial potting mix. Shoot dry weight was greater at the elevated CO_2 concentration than at the ambient concentration (Figure 15.10).

OATS

Because little attention has been given to plant growth at CO_2 concentrations that prevailed in the past, Polley et al. (1992) studied the growth of monocultures and mixtures of oats (*Avena sativa* L.) and wild mustard [*Brassica kaber* (DC.) 'Wheeler'; it is also called charlock (Bailey 1974, p. 438)], a common weed of cultivated lands, at subambient-CO_2 concentrations. Both are C_3 plants. They grew the plants in a 38-m-long controlled-environment chamber in a greenhouse along a gradient of CO_2 concentrations from about 330 µmol/mol to a minimum of 150 µmol/mol CO_2. Photosynthesis depleted the CO_2 concentration in the 38 m chamber from the concentration of the greenhouse

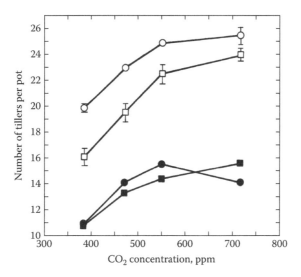

FIGURE 15.8 Mean number of tillers per pot of wheat plants ('Star,' closed squares and 'Turbo,' closed circles) and barley plants ('Alexis,' open squares and 'Arena,' open circles) at anthesis after growth in open-top chambers at different CO_2 concentrations. Bars represent standard error when it exceeds the size of the symbol ($n=7–8$). (Reprinted from *Agric. Ecosyst. Environ.*, 48, Weigel, H.J., Manderscheid, R., Jäger, H.-J., and Mejer, G.J., Effects of season-long CO_2 enrichment on cereals: I. Growth performance and yield, 231–240, Fig. 3, Copyright 1994, with permission from Elsevier.)

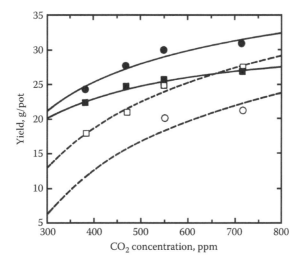

FIGURE 15.9 Dose–response relationship of season-long CO_2 fumigation in open-top chambers on mean grain yield per pot of wheat plants ('Star,' closed squares and 'Turbo,' closed circles) and barley plants ('Alexis,' open squares and 'Arena,' open circles). (Reprinted from *Agric. Ecosyst. Environ.*, 48, Weigel, H.J., Manderscheid, R., Jäger, H.-J., and Mejer, G.J., Effects of season-long CO_2 enrichment on cereals: I. Growth performance and yield, 231–240, Fig. 5, Copyright 1994, with permission from Elsevier.)

air (about 350 µmol/mol) to a minimum that depended on leaf area of the enclosed plants. The CO_2-concentration gradient was maintained during daylight by varying the rate of air flow through the chamber. The soil was a sandy loam. Aboveground biomass per plant of oats and wild mustard increased 198% and 106%, respectively, between CO_2 concentration of 150 and 330 µmol/mol (Figure 15.11). The CO_2-induced increase in aboveground biomass of plants of each species did not

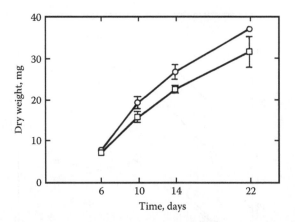

FIGURE 15.10 Top dry weight of barley plants at days 6, 10, 14, and 22 after planting. The plants were grown with ambient air (open squares) or air supplemented with 670 μmol/mol CO_2 (open circles). Data are means of three experiments. Vertical bars denote standard deviations. (Ingvardsen, C. and Veierskov, B., *J. Exp. Bot.*, Fig. 2, 1994 by permission of Oxford University Press.)

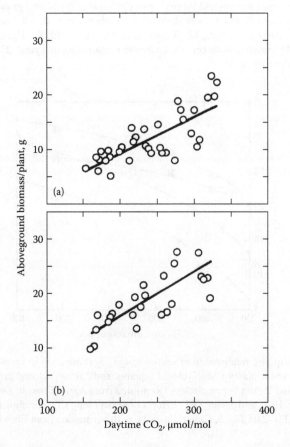

FIGURE 15.11 Aboveground biomass of oat (*A. sativa*) (a) and wild mustard (*B. kaber*) (b) plants as a function of daytime CO_2 concentration for the final 4 weeks of growth along a gradient of subambient-CO_2 concentration. Solid lines are linear regression fits to data for wild mustard from monocultures and 3:1 and 2:2 wild mustard–oats mixtures ($r^2=0.56$, $P<0.001$, $n=26$) and for oats across monocultures and all wild mustard–oats mixtures ($r^2=0.65$, $P<0.0001$, $n=35$). (Reprinted from Polley, H.W. et al., *Int. J. Plant Sci.*, 153, 453, 1992, Fig. 4. With permission. Copyright 1992, University of Chicago Press.)

vary among monocultures and mixtures. These large responses of oats and wild mustard to increasing subambient-CO_2 concentrations suggested that past increases in atmospheric CO_2 concentration may have greatly increased the productivity of C_3 plants.

SOYBEAN

Soybean [*Glycine max* (L.) Merr.] is the most important dicotyledonous crop, and in terms of total global grain production, it is the fourth most important grain crop after maize, rice, and wheat (Zhu et al. 2010). [Even though the fruit of soybean and all other legumes is a pod and not a true grain, which is the fruit of cereals (Robbins et al. 1964, pp. 252–262), it still is called a "grain" by producers of soybean.] Studies show that growth of soybean is increased under elevated CO_2 (Allen et al. 1987; Miglietta et al. 1993b; Xu et al. 1994; Prior and Rogers 1995; Ziska and Bunce 1995; Zhu et al. 2010). When data are combined from different experiments (Allen et al. 1987), they predict a 32% increase in soybean seed yield with a doubling of CO_2 from 315 (CO_2 concentration in the air in 1958) to 630 ppm (Figure 15.12). Due to global CO_2 increases, yields may have increased by 13% from about 1800 AD (when CO_2 concentrations were probably in a range of 260–285 ppm) to 1986 (when CO_2 concentration was 345 ppm).

In a FACE experiment at Urbana, Illinois, in 2002, soybean grew under 380 μmol/mol CO_2 (ambient) or 550 μmol/mol CO_2 (elevated). Yield for the 380 and 550 μmol/mol CO_2 treatments was 4.60 and 5.29 t/ha, respectively (Zhu et al. 2010). Under elevated CO_2, yield increased by 15%, while biomass increased by 18%, and photosynthesis increased by 23%. The greater percent increase in photosynthesis compared to yield reflects an increase in respiration associated with the greater biomass, and it may also indicate a lack of adequate sink capacity to utilize fully the increased supply of photosynthate. Nevertheless, elevated CO_2 resulted in a large increase in yield (Zhu et al. 2010). We shall return to more FACE experiments in the section entitled "Free-Air CO_2 Exchange Studies" in this chapter.

Elevated CO_2 not only increases aboveground growth, but it also increases mass and number of soybean nodules. When soybean seeds inoculated with *Bradyrhizobium japonicum* were grown in a sandy loam soil in growth chambers at near ambient CO_2 (400 μmol/mol) or elevated CO_2

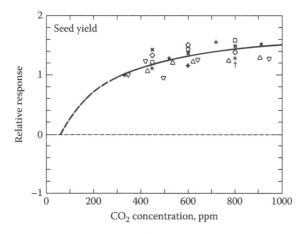

FIGURE 15.12 Rectangular hyperbola response of normalized soybean seed yield data to carbon dioxide concentration for 'Ransom,' 'Bragg,' and 'Forrest' cultivars of soybean. Data obtained from outdoor controlled-environment chamber experiments at three locations: Gainesville, Florida, 1981 (squares), 1982 (crosses); Mississippi State, Mississippi, 1981 (diamonds), 1982 (pluses); Clemson, South Carolina, 1982 (asterisk); and from open-top chamber experiments at Raleigh, North Carolina, 1980 (stars); 1981 (triangles); 1982 (inverted triangles). All data were normalized to 330–340 ppm = 1.0. (Reproduced from Allen, L.H. et al., *Global Biogeochem. Cycles* 1, 1, 1987, Fig. 2. With permission. Copyright 1987, American Geophysical Union.)

(800 μmol/mol), elevated CO_2 increased mass and number of soybean nodules by 63% and 50%, respectively, while shoot dry weight was increased by 30% (Prévost et al. 2010). Oikawa et al. (2010) saw that elevated CO_2 increased seed production in nodulated soybean. The increase in seed production was explained by an increase in the rate of dry mass production during the reproductive period.

COTTON

Reddy et al. (1992) grew cotton (*Gossypium hirsutum* L.) plants from the seedling stage to maturity in naturally lighted plant growth chambers that maintained CO_2 concentrations between ambient (350 μL/L CO_2) and twice the ambient level (700 μL/L CO_2) to determine the effect of elevated CO_2 on fruiting branches and fruiting sites. Cotton plants produce three branch types from the main stem. These are vegetative branches and primary and secondary fruiting branches. The vegetative branches, usually produced below node six, behave much like the main stem in that they may also produce fruiting branches. Primary fruiting branches are the first branches that produce squares at each node, while secondary fruiting branches are arbitrarily designated any subsequently formed fruiting branches at any main stem or vegetative branch node (Reddy et al. 1992).

Plants grown in ambient CO_2 produced fewer vegetative branches than plants grown with 700 μL/L CO_2. The number of primary fruiting branches was not influenced by CO_2 concentration (Figure 15.13). However, the number of secondary branches increased on plants grown in high CO_2. The time required to produce fruiting-branch nodes decreased slightly as the atmospheric CO_2 concentration increased, resulting in more fruiting sites per branch at elevated CO_2 (Figure 15.14). The combined effect of more secondary branches and more rapid node production caused about 40% more fruiting sites to be formed on the plants grown with 700 μL/L CO_2 than the plants grown with 350 μL/L CO_2. Most of the additional CO_2-induced fruiting sites were produced on the lower branches (Figure 15.14).

HORTICULTURAL CROPS

Nederhoff and Buitelaar (1992) studied the effect of two levels of CO_2 on fruit production of eggplant (*Solanum melongena* L. cv. Cosmos) grown in compartments in a greenhouse under two CO_2 levels: 413 and 663 μmol/mol. Plants grew in rockwool. Seeds were planted on October 20, 1990,

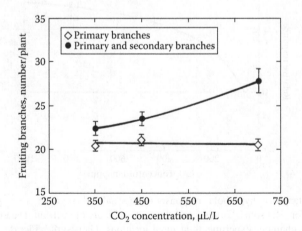

FIGURE 15.13 Effect of atmospheric CO_2 on number of primary and secondary fruiting branches produced by Pima cotton when grown at optimum temperature (c. 26.4°C). Standard errors of the means are shown when they are larger than the symbol size. (Reprinted from Reddy, K.R. et al., *World Resour. Rev.*, 4, 348, 1992, Fig. 5. With permission. Copyright 1992, World Resource Review.)

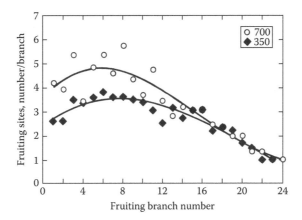

FIGURE 15.14 Influence of CO_2 enrichment on number of fruiting sites produced by each fruiting branch grown at optimum temperature (26.4°C). The first fruiting branch was produced at node six on the main stem. (Reprinted from Reddy, K.R. et al., *World Resour. Rev.*, 4, 348, 1992, Fig. 6. With permission. Copyright 1992, World Resource Review.)

and CO_2 treatments began on December 10, 1990. Fruits were harvested every week from February 11 until July 18, 1991 (23 times). The first harvest was designated as Week 7 of the experiment, and the last harvest was designated as Week 29. The production rate was always higher at high CO_2 than at low CO_2 (Figure 15.15). At the final harvest (July 18), cumulative production was 24% higher at elevated CO_2 than at low CO_2. The maturation period and fruit quality (firmness) did not differ between the two CO_2 treatments. The recommended CO_2 level in greenhouse cultivation

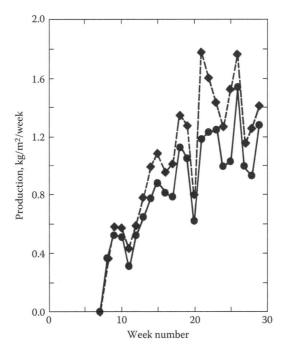

FIGURE 15.15 Rate of fruit production (kg/m²/week) of eggplant at low CO_2 (continuous line) and high CO_2 (dashed line). Week 7 was February 11, 1991, and Week 29 (last harvest) was July 18, 1991. (Reproduced from Nederhoff, E.M. and Buitelaar, K., *J. Hort. Sci.* 67, 805, 1992, Fig. 3. With permission. Copyright 1992, The Invicta Press.)

is about 700–900 μmol/mol (Nederhoff and Buitelaar 1992). [This value is a little lower than the value of 1200 μmol/mol suggested by Northen and Northen (1973, p. 40), as mentioned in the section "Carbon Dioxide Enrichment in Greenhouses" in Chapter 10.] Nederhoff and Buitelaar (1992) concluded that eggplants should grow at these high CO_2 concentrations for maximum yield.

Radoglou and Jarvis (1993) grew broad bean (*Vicia faba* L.) in open-top growth chambers inside a greenhouse that they maintained at ambient (350 μmol/mol) and elevated (700 μmol/mol) CO_2 concentrations. The effects of CO_2 enrichment on growth after germination were determined for 45 days. There was no effect of elevated CO_2 on dry weight of the seedlings during this early phase. The results contrasted for those of green (snap) bean (*Phaseolus vulgaris* L.) grown in the same facility, which had a greater total dry mass 17 days after CO_2 enrichment compared to those grown at ambient CO_2. The authors attributed the lack of response of broad bean seedlings to elevated CO_2 early in their growth period to the presence and longevity of the large cotyledons that provided an available substrate for growth.

In a 2 year study (1989–1990), Hartz et al. (1991) grew cucumber (*Cucumis sativus* L. cv. Dasher II), summer squash (*Cucurbita pepo* L. cv. Gold Bar), and tomato (*Lycopersicon esculentum* Mill. cv. Bingo) in polyethylene tunnels maintained at ambient CO_2 or enriched to between 700 and 1000 μmol/ mol CO_2. (Tomato was studied only in 1989.) Enrichment began immediately after crop establishment and continued for about 4 weeks. The enrichment increased plant dry weight in both years of the study. At harvest in 1989, enriched cucumber, squash, and tomato plants yielded 30%, 20%, and 32% more fruit, respectively. In 1990, cucumber and squash yields were increased by 20% and 16%, respectively, by elevated CO_2. In another study with tomato, in which three isogenic genotypes varying in their resistance to root-feeding nematodes were investigated, elevated CO_2 (750 μmol/mol; ambient was 370 μmol/mol) increased height and biomass of all three (Sun et al. 2010).

Root crops are an important component of the world's food supply, especially in developing countries where food supplies are frequently marginal (Biswas et al. 1996). Increases in atmospheric CO_2 usually lead to increases in plant growth and yield, but little is known about the response of root crops to CO_2 enrichment under field conditions. Consequently, Biswas et al. (1996) studied the effect of elevated CO_2 on the growth of sweet potato [*Ipomoea batatas* (L.) Lamarck cv. Georgia Jet] for 2 years (1984 and 1985). Even though sweet potato is a horticultural crop in the United States, it is a staple in tropical and subtropical Asia and Africa. There were five treatments consisting of an open field plot with no chamber and four plots with OTCs containing CO_2 levels of ambient (354 μmol/mol) and 75, 150, and 300 μmol/mol CO_2 above ambient. The soil was a sandy loam. Harvest began 92 and 89 days after planting in 1984 and 1985, respectively. Yield of storage roots increased with increasing CO_2 in both years (Figure 15.16a). Within OTCs, CO_2 enrichment from about 360–660 μmol/mol led to increases in total storage-root fresh weight per plant of 46% and 75% in 1984 and 1985, respectively. The number of storage roots increased with CO_2 enrichment by 47% and 37% in 1984 and 1985, respectively (Figure 15.16b). The mean fresh weight of individual storage roots increased by 29% in 1985 but not in 1984 (Figure 15.16c). Thus, the enhancement of storage root yield with CO_2 enrichment was due to an increase in the number of storage roots in 1984 and to increases in both the number and size of storage roots in 1985. Root crops generally show greater increases in yield with a doubling of atmospheric CO_2 than do other types of crops, because the storage roots can act as sinks for the increased carbohydrate produced (Biswas et al. 1996). The yield increases for sweet potato of 46% (1984) and 75% (1985) at approximately double the ambient CO_2 concentration exceeded the average increase in yield of 33% reported by Kimball (1983) for a wide variety of agricultural crops.

Idso and Kimball (1989) grew carrot (*Daucus carota* L. var. *sativa* cv. Red Cored Chantenay) and radish (*Raphanus sativus* L. cv. Cherry Belle) from seed in OTCs at 340 or 640 μmol/mol CO_2. They harvested the plants four different times (weeks 4, 5, 6, and 7 after planting) to determine whole-plant dry weights. Although they did not impose different temperature treatments, they obtained natural variation in air temperatures throughout the growth period, which they monitored.

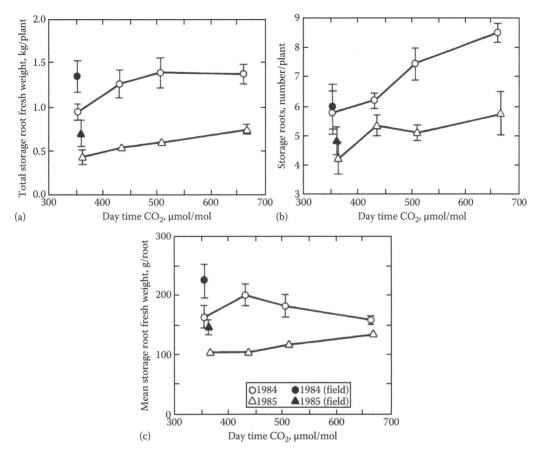

FIGURE 15.16 Total storage-root fresh weight (a), number of storage roots (b), and mean storage-root fresh weight (c) for sweet potato exposed to CO_2-enrichment in open-top chambers. Culls (diameter less than 2.5 cm) were included in the total storage-root fresh weight, but omitted from the number of storage roots and the mean storage-root fresh weight. Field plants were grown outside of open-top chambers. Values are means (±standard error) of three replicate samples of 10 plants each. Error bars that appear to be missing are obscured by the symbols. (Reprinted from Biswas, P.K. et al., *Crop Sci.*, 36, 1234, 1996, Fig. 1. With permission. Copyright 1996, Crop Science Society of America.)

Through this natural variation, they were able to plot growth at temperatures ranging between 17°C and 32°C for the two CO_2 treatments. At 25°C, the productivity enhancement factor for the 300 μmol/mol increase in CO_2 was 1.5 and 2.0 for radish and carrot, respectively. They plotted a "growth modification factor" (1.0 = no enhancement in growth by CO_2) versus temperature, and by extrapolating their data to 1.0, they found no growth enhancement of CO_2 at 12.5°C for carrot and 11°C for radish. Curtis et al. (1994) grew another kind of radish, wild radish [*R. raphanistrum* L., a widely distributed weed, according to Bailey (1974, p. 439)], under ambient and twice-ambient CO_2. The elevated CO_2 led to increases in flower and seed production (by 22% and 13%, respectively). They did not report whole-plant dry weights.

Mortensen (1994b) in Ås, Norway (southeast of Trondheim in western Norway), grew eight different vegetables in pots that he placed in large open-top field chambers, which received either ambient CO_2 (355 μmol/mol) or elevated CO_2 (800–900 μmol/mol). Three of them were root crops: onion (*Allium cepa* L.); celery root [*Apium graveolens* L. var. *rapaceum* (Mill.) Gaud.]; and carrot (*D. carota* L. var. *sativa*). The other five were leafy vegetables: leek (*A. ampeloprasum* L. var. *porrum* Reg.), celery [*A. graveolens* var. *dulce* (Mill.) Pers.], Chinese cabbage (*B. pekinensis* Rupr.), lettuce (*Lactuca sativa* L.), and parsley (*Petroselinum crispum* Nym.). He grew only one cultivar of celery, celery root, and parsley. For the other crops, he grew two or more cultivars. Elevated CO_2 increased the fresh-weight yield by 23% in onion (two cultivars) and by 9% in carrot (three cultivars). The dry-weight yield was increased by 18% in lettuce (three cultivars), 19% in carrot, and 17% in parsley (one cultivar). The yields of leek (two cultivars), Chinese cabbage (three cultivars), celery (one cultivar), and celery root (one cultivar) were not affected by the CO_2 concentration. He stated two reasons for the relatively high proportion of the species that did not respond to CO_2 enrichment. First, the plants grew under natural environmental conditions. The temperature in West Norway is cool (13°C–15°C mean temperature), and at low temperatures, there is low metabolic activity. As Idso and Kimball (1989) pointed out, CO_2 ceases to have a growth stimulating effect at about 11°C–12°C. Second, he attributed the lack of response to CO_2 to the low air pollution in Norway. At high CO_2, stomata close and this excludes growth-reducing air pollutants. The level of air pollutants such a sulfur dioxide and nitrogen dioxide is low in West Norway compared with many other regions in Europe. We have not considered air pollution in this book, but it is an important factor when considering the effects of CO_2 on plant growth. Elevated CO_2 ameliorates the detrimental effects of ozone (Ainsworth et al. 2008; Tu et al. 2009; Kontunen-Soppela et al. 2010). Emission of plant-generated volatile organic compounds, like isoprene, is inhibited by high CO_2 (Rasulov et al. 2009).

Downtown and Grant (1994) grew two variegated ornamental species, oleander (*Nerium oleander* L.) and willow myrtle [*Agonis flexuosa* (Willd.) Sweet], in controlled-environmental cabinets under either an ambient level of CO_2 (partial pressure of 400 µbar CO_2) or with twice-ambient CO_2 (partial pressure of 800 µbar). The amount of variegation in the leaf affected results. Yellow portions of leaves showed little or no photosynthetic activity. A major effect of variegation was to slow overall plant growth compared with completely green plants. The relative growth response to high CO_2 was greater for the variegated plants compared to the completely green plants. After 5 months of growth under the two CO_2 treatments, total plant dry weight per pot of the variegated plants under elevated and ambient CO_2 was 20.4 and 12.5 g, respectively (63% increase in growth due to CO_2); for the completely green plants, these values were 120.8 and 89.7 g, respectively (35% increase in growth due to CO_2). Their data showed that the low levels of photosynthetic activity in the yellow portions of the variegated leaves were stimulated by high CO_2.

PASTURE AND GRASSLAND PLANTS

So far, we have been discussing the effects of elevated levels of CO_2 on plants used by humans for food or ornamental purposes. We now turn to pastures and natural grasslands, which are important agriculturally because they provide food for grazing animals. Newton et al. (1995) examined the growth of pasture plants under three different levels of CO_2: 350, 525, or 700 µmol/mol CO_2. Large turves (1 m×0.5 m×0.3 m) were removed intact from an established pasture near Palmerston North, New Zealand, where the soil was a sand. The pasture was at least 30 years old, and the principal species were *Lolium perenne* L. (C_3; perennial ryegrass; Poaceae or grass family), *Trifolium repens* L. (C_3; white clover; Fabaceae or pulse family), *Poa pratensis* L. (C_3; Kentucky bluegrass; Poaceae), *Cynosurus cristatus* L. (C_3; crested dog's-tail; Poaceae), *Agrostis capillaris* L. (C_3; the common name for the genus is bentgrass; Poaceae), *Anthoxanthum odoratum* L. (C_3; common name of the genus is sweet vernal grass; Poaceae), *Paspalum dilatatum* Poir. (C_4; dallis grass; Poaceae), and *Plantago lanceolata* L. [C_3; ribgrass; Plantaginacea or plantain family, a common and troublesome weed of grasslands (Fernald 1950, p. 1316)]. Three controlled-environment rooms provided the different CO_2 concentrations. The turves were grown through simulated seasonal changes in temperature and photoperiod taken from long-term records. Each calendar month was reduced to 3 weeks in a growth room. A yield cut was made after each 3 week period by harvesting to 2.5 cm above-ground level. After 261 days of exposure to CO_2 (i.e., after one simulated year), the cumulative total yield at 525 µmol/mol was 2% greater than ambient and the yield at 700 µmol/mol CO_2 was 10% greater (Figure 15.17). Although the change in total dry matter was small, the botanical

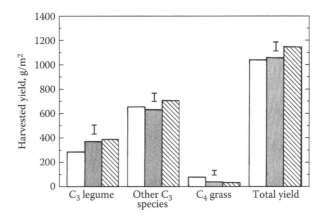

FIGURE 15.17 Total and component cumulative yields of pasture turves grown at 350 (open squares), 525 (closed squares), or 700 (right-slanting lines) μmol/mol CO_2 for 261 days. Error bar is the least significant difference ($P<0.05$). (Newton, P.C.D., Clark, H., Bell, C.C., Glasgow, E.M., Tate, K.R., Ross, D.J., and Yeates, G.W., Plant growth and soil processes in temperate grassland communities at elevated CO_2, *J. Biogeogr.*, 1995, 22, 235–240, Fig. 2. Copyright Wiley-VCH Verlag GmbH & Co. KGaA. With permission.)

composition was changed by CO_2 concentration. The proportion of the legume (*T. repens*, white clover) in the sward was greater at elevated CO_2 while the proportion of the C_4 species (*P. dilatatum*) declined (Figure 15.17). The data of Newton et al. (1995) suggest that species composition will change in pastures as the CO_2 concentration in the atmosphere increases. In a later paper, the same group in New Zealand (Clark et al. 1997) found that herbage harvested from turves taken from a *L. perenne*- and *T. repens*-dominated pasture was increased by 7% and 14% when the plants grew at 525 and 700 μmol/mol CO_2, respectively, compared to plants grown at 350 μmol/mol CO_2.

In a 3 year study, Bunce (1993) studied competition between alfalfa (*Medicago sativa* L.) and weeds (mainly C_4 grasses) grown at ambient CO_2 (350 μmol/mol) and twice-ambient CO_2 (700 μmol/mol). Averaged over the 3 years, alfalfa yield was 1.38 times higher at the doubled CO_2 concentration (yield was 395 g/m^2) compared to the ambient (286 g/m^2). Weed growth was also increased by elevated CO_2, but to a lesser extent. Weed yield was 1.20 times higher with doubled CO_2 (362 g/m^2) compared to ambient (301 g/m^2).

Baxter et al. (1994) studied the response of three species in upland grasslands in the United Kingdom to either ambient (340 μmol/mol) or elevated (680 μmol/mol) CO_2 concentrations in four outdoor OTCs. The plants were two perennial grasses (*A. capillaris* L. and *Festuca vivipara* L.) from the montane grasslands of Snowdonia (Snowdonia is the mass of Snowdon mountains found in Wales) and an arctic-alpine species, *P. alpina* L. The arctic-alpines are found at high altitude, including Snowdonia. Such species are of particular interest in relation to climate change studies, because it is these species, at the absolute limit of their ecological range, which may be the first to show the possible effects of climate change. Plants grew in pots with commercial compost. After 79 days, total dry weight in elevated CO_2 of *A. capillaris* was twice that at ambient CO_2 (Figure 15.18a). Both above- and belowground dry weight was greater at elevated CO_2 (Figure 15.18b and c). By 105 days, the shoot dry weight of *P. alpine*, when most plants were in flower, was 30% greater in plants grown at elevated CO_2 (Figure 15.18e), but root weight as not altered (Figure 15.18f). *F. vivipara* was grown for the longest time (189 days). The plants grew slowly and did not flower. Because few tillers were established, there were only two harvests for this species. There was no difference in total plant dry weight between CO_2 treatments after 65 days (Figure 15.18g). By 189 days, the whole-plant and shoot dry weights of plants grown at elevated CO_2 were about 50% of those of controls (Figure 15.18g and h). Root growth was unchanged in plants grown in elevated CO_2 (Figure 15.18i). Unlike *A. capillaris* or *P. alpina*, shoots of *F. vivipara* showed marked discoloration and premature

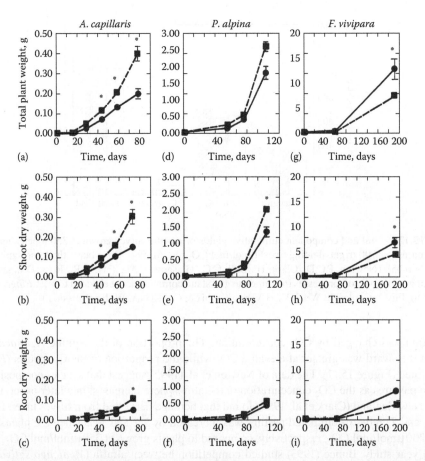

FIGURE 15.18   Total (a), shoot (b), and root (c) dry weight of *A. capillaris*, *P. alpina*, and *F. vivipara* grown at 340 (closed circles) or 680 (closed squares) μmol/mol CO_2. Data represent the mean±one standard error of seven (*Agrostis*) or six (*Poa* and *Festuca*) replicates per CO_2 treatment. * indicates significant difference at $P \leq 0.05$. The letters a–i in the figure are used to describe it in the text. (Baxter, R., Ashenden, T.W., Sparks, T.H., and Farrar, J.F., *J. Exp. Bot.*, Fig. 1, 1994 by permission of the Oxford University Press.)

senescence of leaves. The reason that *F. vivipara* grew less under elevated CO_2 is not known. The fact that the arctic-alpine species, *P. alpina*, showed increased growth under elevated CO_2 suggests that this species will not die out as the CO_2 concentration in the atmosphere increases.

Owensby et al. (1993) determined above- and belowground biomass production of a tallgrass prairie in Kansas exposed to ambient CO_2 (330–340 μmol/mol) and twice-ambient CO_2 concentrations in OTCs during three seasons (1989, 1990, and 1991). Dominant species were *Andropogon gerardii* (big bluestem; C_4), *A. scoparius* (little bluestem; C_4), *Sorghastrum nutans* (Indiangrass; C_4), and *P. pratensis* (Kentucky bluegrass; C_3). Aboveground biomass was estimated by periodic sampling throughout the growing seasons. In 1991, the chambers were taken down before the last harvest of *P. pratensis*, so biomass data for it are missing in 1991. Root production was estimated using root ingrowth bags, which remained in place throughout the growing season. The soil at the site was a silty clay loam. In addition to the chambered plots, there were un-chambered plots. Chambers received water that was equivalent to natural rainfall levels. Total precipitation (12 months) in 1989, 1990, and 1991 was 619, 785, and 669 mm, respectively. These amounts were −214, −50, and −166 mm below normal, respectively, for each year. Thus, each year had below-normal levels of rainfall, but 1989 was an extremely dry year. Midday xylem pressure potentials were measured weekly between May 30 and September 25, 1991, with a pressure chamber.

Elevated CO_2 increased the biomass of the chambered C_4 plant, *A. gerardii*, in 1989 and 1991, but not 1990 when the soil moisture was only slightly below the long-term average (Figure 15.19). Biomass of the C_3 plant was increased by elevated CO_2 in 1990, but not 1989. The lack of response of *P. pratensis* in 1989 was probably due to lack of water in that dry year. The biomass production for both *A. gerardii* and *P. pratensis* in 1989 was less than it was in 1990. *P. pratensis* is a cool-season grass and only grows in the cooler months of the spring and fall on the tallgrass prairie in Kansas (He et al. 1992). Rains are more prevalent in the spring and fall in Kansas. *P. pratensis* is not adapted to dry summer months in Kansas, as *A. gerardii* is, and *P. pratensis*'s low biomass in 1989 reflects this (Figure 15.19). Root growth was increased under the elevated-CO_2 treatment (Figure 15.20). Midday xylem pressure potentials indicated less moisture stress for the *A. gerardii* plants under elevated CO_2 compared to plants grown at ambient CO_2 (Figure 15.21). These results are consistent with data presented in Chapter 7. Both C_3 and C_4 plants on the tallgrass prairie increased growth under elevated CO_2, but the response was dependent upon soil moisture. The top biomass of the C_4 plant (*A. gerardii*) was not increased by elevated CO_2 in a year when soil moisture was near normal (Owensby et al. 1993).

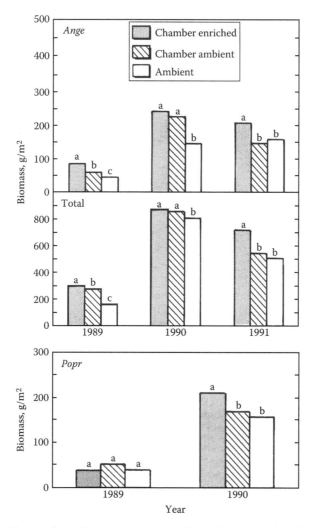

FIGURE 15.19 Peak biomass for indicated years and species and species groups for native tallgrass prairie exposed to twice-ambient and ambient CO_2 concentrations. Means with a common letter do not differ ($P < 0.10$). *Ange* = *A. gerardii*, *Popr* = *P. pratensis*. (Reprinted from Owensby, C.E. et al., *Ecol. Appl.*, 3, 644, 1993, Fig. 2. With permission. Copyright 1993, Ecological Society of America.)

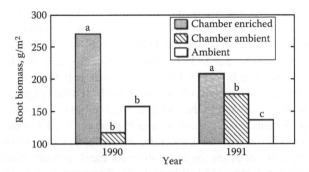

FIGURE 15.20 Root biomass in ingrowth bags to a 15 cm depth in a tallgrass prairie exposed to twice-ambient and ambient CO_2 concentrations. Data are means of four bags per plot in 1990 and eights bags per plot in 1991, replicated three times. Means with a common letter within year do not differ ($P < 0.10$). (Reprinted from Owensby, C.E. et al., *Ecol. Appl.*, 3, 644, 1993, Fig. 4. With permission. Copyright 1993, Ecological Society of America.)

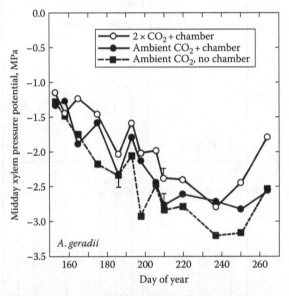

FIGURE 15.21 Midday xylem pressure potentials for *A. gerardii* exposed to ambient and twice-ambient CO_2 concentrations. Data for weekly estimates from May 30 to September 25, 1991. Only the largest standard error for each treatment is plotted. (Reprinted from Owensby, C.E. et al., *Ecol. Appl.*, 3, 644, 1993, Fig. 6. With permission. Copyright 1993, Ecological Society of America.)

Marks and Strain (1989) also studied the effect of elevated CO_2 on the growth of C_3 and C_4 plants grown together. They were aster (*Aster pilosus* Willd.; C_3) and broomsedge (*A. verginicus* L.; C_4). Aster is an herbaceous perennial that dominates old fields in the Piedmont plateau of North Carolina, United States, after 2–3 years of abandonment. It is replaced as the dominant species in successive years by broomsedge, a perennial bunchgrass. These species compete for limited water, resulting in diminished survival of aster when the species are grown together. Plants grew in pots with a standard substrate mixture. They were placed in controlled-environment chambers with two CO_2 concentrations (350 and 650 µmol/mol). After emergence, plants were thinned so that pots contained either two asters, two broomsedges, or one aster and one broomsedge. There were two watering regimes. Control pots were watered each day. Pots in the water-stressed treatment were allowed to dry until leaf xylem pressure potential at midday was approximately −1.5 MPa [wilting

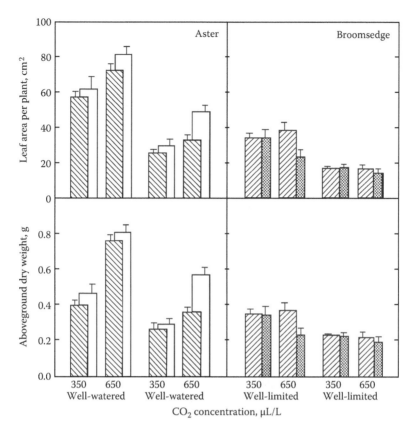

FIGURE 15.22   Total leaf area and aboveground dry weight for *A. pilosus* (C$_3$, aster) and *A. virginicus* (C$_4$, broomsedge) at the end of the water treatment (40 days after emergence). Key: striations slanted to right, aster grown with aster; open squares, aster grown with broomsedge; striations slanted to left, broomsedge grown with broomsedge; stipples, broomsedge grown with aster. Error bars are ± 1 standard error of the mean ($n = 10$). (Reprinted from Marks, S. and Strain, B.R., Effects of drought and CO$_2$ enrichment on competition between two old-field perennials, *New Phytol.*, 1989, 111, 181–186, Fig. 1. Copyright Wiley-VCH Verlag GmbH & Co. KGaA. With permission.)

point; see Kirkham (2005, p. 106)], when the plants were re-watered. Leaf area was measured with a leaf area meter. As expected, the well-watered plants grew more than the drought-stressed plants (Figure 15.22). Aster grown with itself (Figure 15.22, bars with striations slanted to the right) and aster grown with broomsedge (Figure 15.22, open bars) had greater leaf area and aboveground growth under elevated CO$_2$ than under ambient CO$_2$. However, the growth of broomsedge was not affected by CO$_2$ concentration. Marks and Strain (1989) conclude that, with future increases in atmospheric CO$_2$ concentration, aster (C$_3$) may delay broomsedge (C$_4$) dominance in old-field communities.

MARSH PLANTS

Arp and Drake (1991) studied the sedge, *Scirpus olneyi* Gray (*Scirpus* is bulrush), which was growing in the brackish marsh on the Chesapeake Bay, Maryland, United States. They exposed it to two CO$_2$ concentrations (ambient and ambient + 340 μmol/mol CO$_2$; ambient concentration was not stated) for four growing seasons. Plants with elevated CO$_2$ showed a 13.2% higher shoot density, and a 4.4% increase in aboveground biomass, compared to plants grown with ambient CO$_2$. Percentage senescence was reduced in high CO$_2$ (4.6%) when compared with ambient CO$_2$ (8.6%).

HERBS AND WEEDS

Gay and Hauck (1994) studied *L. temulentum* L. (a C_3 plant), also called ryegrass or darnel (Bailey 1974). Darnel is a poisonous weed resembling rye, often found in grain fields. This is not to be confused with *L. perenne* L. (perennial or English ryegrass) or *L. multiflorum* Lamarck (Italian ryegrass). [The word "darnel" probably is derived from the Old French *darnu*, which means "stupe-fied," so called from its supposed stupefying or intoxicating qualities (Friend and Guralnik 1959). In *L. temulentum*, the outer glume produces a narcotic poison (Bailey 1974).] Plants grew in controlled environments at two CO_2 concentrations (ambient CO_2 or 345 μmol/mol and elevated CO_2 or 575 μmol/mol) and at two light intensities (low photosynthetic photon flux density or 135 μmol/m^2/s and high photosynthetic photon flux density or 500 μmol/m^2/s). Plants were harvested every 2 or 3 days until about 40 days after sowing to determine leaf lengths and leaf growth rates. Leaf growth curves were fitted using a logistic equation. Leaf length (Figure 15.23a) and leaf growth rate (Figure 15.23b) were increased under elevated CO_2, but only under the low-light conditions. (For leaf length, where data points are shown in Figure 15.23a, compare the solid triangles for high CO_2 with the solid circles for low CO_2.) In high light, the effects of increasing CO_2 were small. They concluded that *L. temulentum* is limited in its ability to respond to increased CO_2 in high light, but it can respond in low light.

The genus *Rumex* [Polygonaceae (buckwheat or knotweed family); common name, dock or sorrel] consists of weedy herbs. Pearson and Brooks (1995) studied *Rumex obtusifolius* L. (broad-leaved dock; C_3), a perennial herb that grows to be 40–150 cm tall. In many countries, it is the most troublesome weed in intensively managed permanent grasslands. The possession of a large root system explains much of the persistence and aggressive nature of *R. obtusifolius*. They grew the plants in pots containing compost and grit placed in day-lit environmental chambers with air containing 350 or 600 μmol/mol CO_2. During the experiment, there were five flushes of new leaves, which they labeled A, B, C, D, and E. They determined dry weights of roots and shoots, and weighed senescent leaves. Data from gas exchange measurements were used to calculate the instantaneous water use efficiency (photosynthetic rate divided by transpiration rate). Plants grown under 600 μmol/mol CO_2 had higher aboveground (Figure 15.24a) and root (Figure 15.24b) dry weights than plants grown in 350 μmol/mol CO_2. For plants grown in 350 μmol/mol CO_2, water use efficiency was relatively constant over the first five measurement days and then declined (Figure 15.25). By contrast, plants

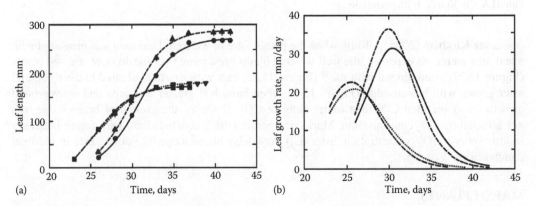

(a)   (b)   Time, days

FIGURE 15.23 (a) Time course of leaf length of four leaves of *L. temulentum* L., a type of ryegrass, grown in ambient CO_2 (345 μL/L) and low light (135 μmol/m^2/s) (solid circle and solid line); increased CO_2 (575 μL/L) and low light (solid triangle and short-dashed line); ambient CO_2 and high light (500 μmol/m^2/s), (solid square and long-dashed line); and high CO_2 and high light (upside down solid triangle and dotted line). Lines are fitted logistic curves and points are means. Standard errors are shown by vertical lines where larger than the symbol. (b) Leaf growth rates derived from the fitted lines in (a). (Gay, A.P. and Hauck, B., *J. Exp. Bot.*, Fig. 4, 1994 by permission of Oxford University Press.)

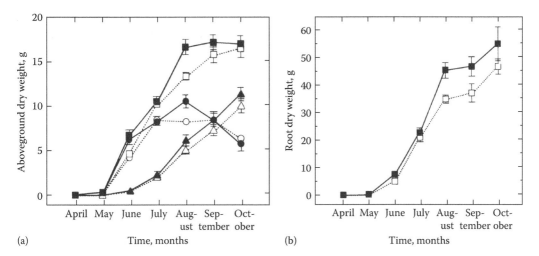

FIGURE 15.24 The average dry weight of component parts of *R. obtusifolius* grown in 350 (open circle, open square, open triangle) and 600 (closed circle, closed square, closed triangle) μmol/mol CO_2. (a) Green tissue, open circle and closed circle; senescent tissue, open triangle and closed triangle; total aboveground tissue (green + senescent), closed square and open square. (b) Roots, open and closed squares. The solid lines connect data points for the 600 μmol/mol CO_2 treatment and the dotted lines connect data points for the 350 μmol/mol CO_2 treatment. Symbols are shown ± one standard error of the mean ($n = 12$). (Reproduced from Pearson, M. and Brooks, G.L., *J. Exp. Bot.*, Fig. 7, 1995 by permission of Oxford University Press.)

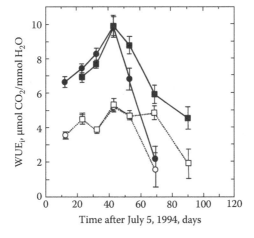

FIGURE 15.25 Changes in instantaneous water use efficiency for plants of *R. obtusifolius* grown in 350 (open circle, open square) and 600 (closed circle, closed square) μmol/mol CO_2. Differently shaped symbols represent leaves from different flushes. A-flush, open circle and closed circle; D-flush, open square and closed square. Note that for the first four measurement days, the open squares and the open circles fall on top of each other. Symbols are shown ± one standard error of the mean ($n = 16$). (Reproduced from Pearson, M. and Brooks, G.L., *J. Exp. Bot.*, Fig. 5, 1995 by permission of Oxford University Press.)

grown in 600 μmol/mol CO_2 showed an increase in water use efficiency before a rapid decrease. The increase in water use efficiency under elevated CO_2 is confirmed by many other studies (see Chapter 11). Pearson and Brooks (1995) concluded that the greater root proliferation and increased water use efficiency of *R. obtusifolius* under elevated CO_2 should give this weed a competitive advantage as the CO_2 concentration in the atmosphere increases.

Taylor et al. (1994) compared the growth of two herbs (both C_3 plants): *P. media* L. (in the plantain family or Plantaginaceae; common name hoary plantain) and *Anthyllis vulneraria* L. (in the legume family or Fabaceae; common name kidney vetch or woundwort, because of its ancient use in healing wounds). *A. vulneraria*, which grows to 20–30 cm tall, is grown for forage or planted as an ornamental, but it also occurs on waste ground (Fernald 1950, p. 896; Bailey 1974, p. 570). *P. media* grows in fields and lawns. Unlike *A. vulneraria*, *P. media* has leaves that are closely rosulate (in the form of a rosette) and usually flat on the ground (Fernald 1950, p. 1316). Linear variable differential transducers (LVDTs), suitable for nondestructive analysis of leaf extension, were used to study leaf extension rate of the two plants during a diurnal cycle with a 16 h photoperiod. For *P. media*, both daytime and nighttime rates of extension were increased in elevated CO_2, while for *A. vulneraria*, only nighttime leaf extension was increased, an effect due largely to a transient peak in growth occurring immediately at the start of the dark period (Figure 15.26). [The data for Figure 15.26 have been extracted from the paper by Ferris and Taylor (1994), which is cited also in Chapter 7.] Under elevated CO_2, *A. vulneraria* showed little response to CO_2 and this was reflected in cell wall properties that were unaltered by elevated CO_2. The increased leaf cell expansion of *P. media* under elevated CO_2 appeared to be the result of increased cell wall extensibility and wall loosening (Ferris and Taylor 1994; Taylor et al. 1994). (At the end of Chapter 7, we discuss cell wall extensibility as affected by elevated CO_2.)

Ward and Strain (1997) studied growth of the weed *Arabidopsis thaliana* L. (Brassicaceae or mustard family; common name is mouse-ear-cress; a C_3 plant) under different CO_2 concentrations. The weed is 5–45 cm high and has basal rosettes of petiolate leaves. It grows in dry fields, roadsides, and waste places and produces numerous seeds (Fernald 1950, pp. 710–711). Their objective was to determine if wild genotypes of *A. thaliana* from a range of elevations exhibited variation in growth and reproduction to three CO_2 concentrations: the low CO_2 of the Pleistocene (20 Pa), the current global CO_2 level (35 Pa), and the elevated CO_2 predicted for the future (70 Pa). Plants from different elevations were chosen because plants at high elevation are exposed to lower CO_2 partial pressure than plants at sea level. (See Chapter 9 for stomatal density in response to elevation and CO_2 concentration.) They obtained genotypes of *A. thaliana* from eight different elevations: 0 m (Seattle, Washington; genotype abbreviated SW); 50 m (Maidstone, Kent, England; MK); 50 m (Sidmouth, Devon, England; SD); 200 m (Columbia, Missouri, United States; CM); 225 m (Chagford, Devon, England; CD); 1200 m (Cape Verde Islands, Atlantic Ocean; CV); 1580 m (Kashmir governed by India; KI); and 3400 m (Pamiro-Alay, Tadjikistan; PT). The plants grew in pots with a mixture of gravel, turface, and vermiculite, which were placed in controlled growth chambers maintained at the three CO_2 partial pressures. Plants were harvested after 70 days of growth, and total dry weights

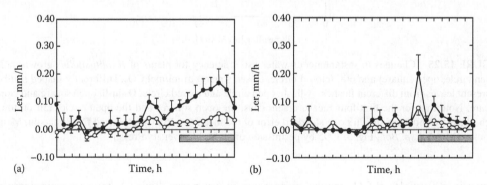

FIGURE 15.26   Leaf extension in elevated CO_2. Diurnal patterns of leaf extension for (a) *P. media* and (b) *A. vulneraria* exposed to either ambient, 350 μmol/mol CO_2 (open circles) or elevated, 590 μmol/mol CO_2 (closed circles). Black bar indicates dark period; n = at least 8. Standard errors are also shown. Data from Ferris and Taylor (1994). (Taylor, G., Ranasinghe, S., Bosac, C., Gardner, S.D.L., and Ferris, R., *J. Exp. Bot.*, Fig. 2, 1994 by permission of Oxford University Press.)

(sum of root, stem, leaf, and reproductive structures) were determined. The average number of seeds per plant (a silique is the pod of plants in the mustard family) was determined. Seed number per plant was calculated as the number of siliques per plant multiplied by the average number of seeds per silique. Genotypes showed similar responses to the CO_2 treatments (Figure 15.27). Total mass and reproductive mass increased as the CO_2 partial pressure increased. All plants had fewer siliques per plant at 20 Pa CO_2 compared with 35 Pa CO_2 (Figure 15.28a). In general, seed number per silique did not vary due to CO_2 concentration (Figure 15.28b), but seed number per plant increased as the CO_2 concentration increased, except for the SW genotype (Figure 15.28c). Even though the greatest increase in seed number was between 20 Pa and 35 Pa, the increase in seed number per plant with the elevated level of CO_2 (70 Pa) for five of the six genotypes suggests that *A. thaliana* is going to produce more seeds as the CO_2 concentration rises, and it will become more of a problem as a weed.

Bazzaz and Garbutt (1988) studied four annual weeds, two C_3 and two C_4, under elevated CO_2 concentrations. The plants were *Ambrosia artemisiifolia* L. (common ragweed; C_3; Asteraceae or composite family), *Abutilon theophrasti* Medic. (velvetleaf; C_3; Malvaceae or mallow family),

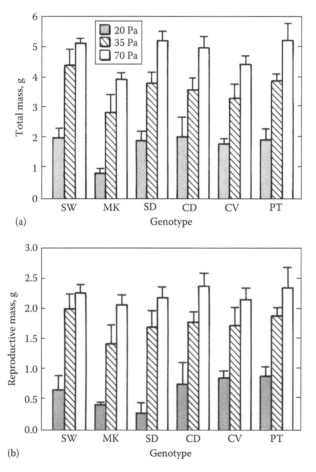

(a)

(b)

FIGURE 15.27 Effects of 20, 35, and 70 Pa CO_2 on (a) total mass and (b) reproductive mass for genotypes of *A. thaliana* from different elevations. Note that the order of genotypes from left to right corresponds to increasing elevation. Values are means ± standard errors ($n = 4$ to 8). (Reprinted from Ward, J.K. and Strain, B.R.: Effects of low and elevated CO_2 partial pressure on growth and reproduction of *Arabidopsis thaliana* from different elevations, *Plant Cell Environ.*, 1997, 20, 254–260, 1997, Fig. 1. Copyright Wiley-VCH Verlag GmbH & Co. KGaA. With permission.)

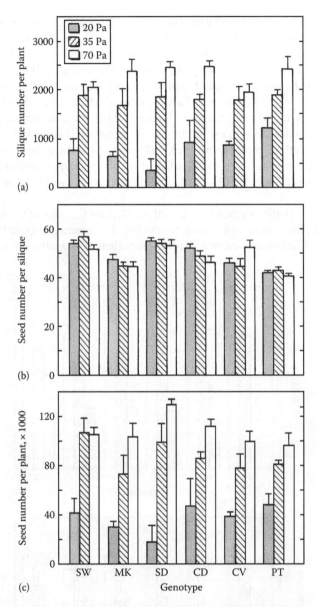

FIGURE 15.28 Effects of 20, 35, and 70 Pa CO_2 on (a) silique number per plant, (b) seed number per silique, and (c) seed number per plant (total seed number) for genotypes of *A. thaliana* from different elevations. Values are means ± standard error ($n = 4$ to 8 for silique number per plant and seed number per plant; $n = 30$ for seed number per silique). (Reprinted from Ward, J.K. and Strain, B.R.: Effects of low and elevated CO_2 partial pressure on growth and reproduction of *Arabidopsis thaliana* from different elevations, *Plant Cell Environ.*, 1997, 20, 254–260, 1997, Fig. 3. Copyright Wiley-VCH Verlag GmbH & Co. KGaA. With permission.)

Amaranthus retroflexus L. (pigweed; C_4; Amaranthaceae or amaranth family), and *Setaria faberii* Herrm. (common name of the genus is bristly foxtail; C_4; Poaceae family or grass family). Plants were germinated and then transplanted into pots with a mixture of soil, sand, and perlite that were placed in 12 growth chambers with CO_2 levels controlled at 350, 500, and 700 μL/L. Plants were harvested 72 days after transplanting, when aboveground biomass and seed biomass were determined. The biomass of *A. artemisiifolia* increased with elevated CO_2 levels (Figure 15.29a). Seed

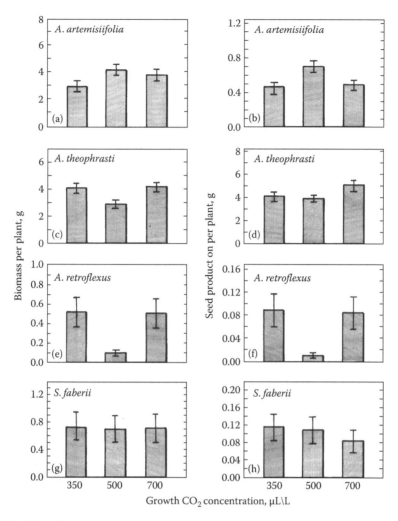

FIGURE 15.29 Mean biomass production and mean total seed mass of four annuals at three CO_2 levels. *Ambrosia* and *Abutilon* are C_3, and *Amaranthus* and *Setaria* are C_4. Bars indicate ± standard error. The letters a–h in the figure are used to describe it in the text. (Reprinted from Bazzaz, F.A. and Garbutt, K., *Ecology*, 69, 937, 1988, Fig. 2. With permission. Copyright 1988, Ecological Society of America.)

production in this species increased with CO_2 at 500 μL/L, but not at 700 μL/L (Figure 15.29b). The biomass of *A. theophrasti* declined slightly at a CO_2 level of 500 μL/L compared to the 350 μL/L level, and biomass at the low and high level of CO_2 were similar (Figure 15.29c). There was, however, an increase in its seed production at 700 μL/L CO_2 (Figure 15.29d). The biomass (Figure 15.29e) and seed production (Figure 15.29f) of *A. retroflexus* (C_4) declined by a large amount at a CO_2 concentration of 500 μL/L. *S. faberii* (C_4) showed no response to CO_2 (Figure 15.29g and h). Bazzaz and Garbutt (1988) had no explanation for the large decrease in growth at 500 μL/L CO_2 for *A. retroflexus*.

To investigate further these four annuals, Garbutt et al. (1990) grew the plants under the same conditions as in the study by Bazzaz and Garbutt (1988), and had the same CO_2 concentrations (350, 500, and 700 μL/L). They included a fifth annual weed, *Chenopodium album* L. (also called pigweed like *A. retroflexus*, but it is also known as lamb's quarters; C_3; Chenopodiaceae or goose-foot family). Plants were harvested six times throughout the experiment, and relative growth rate

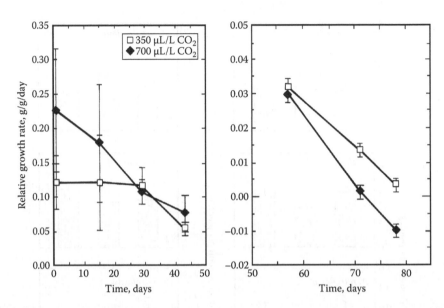

FIGURE 15.30 Change over time in relative growth rate of *Abutilon* at CO_2 levels of 350 and 700 μL/L. The graph has been split at day 50 for clarity (note change in scale). Vertical bars indicate 95% confidence intervals. (Reprinted from Garbutt, K., Williams, W.E., and Bazzaz, F.A., *Ecology*, 71, 1185, 1990, Fig. 1. With permission. Copyright 1990, Ecological Society of America.)

was calculated using a regression method. During the last three harvests, midday water potential of shoots was measured using a pressure chamber. The relative growth rate of *Amaranthus* was always higher at 700 μL/L CO_2 than at 350 μL/L. In both *Abutilon* and *Ambrosia*, the relative growth rate was greater at 700 μL/L CO_2 than at 350 μL/L CO_2 during the first half of the experimental period, but during the second half of the period, the reverse was true (Figure 15.30 for *Abutilon*). Reproductive biomass (seed, fruits, and flowers) increased with increasing CO_2 in *Amaranthus* (C_4), *Chenopodium* (C_3), and *Ambrosia* (C_3). *Abutilon* (C_3) showed a peak in its reproductive biomass at 500 μL/L CO_2, but reproductive biomass was the same at 350 and 700 μL/L CO_2. There was no change in the reproductive biomass of *Setaria* (C_4) under the three CO_2 concentrations, as Bazzaz and Garbutt (1988) found. Bazzaz and Garbutt (1988) state that the lack of a response of *Setaria* to CO_2 is expected from C_4 species, a common view (Tuba et al. 1993). However, even in the study by Garbutt et al. (1990), a C_4 plant, *Amaranthus*, grew more under elevated CO_2 compared to ambient CO_2. So both C_3 and C_4 plants do increase growth under elevated CO_2 depending upon the conditions of the experiment, and no generalization can be made saying that C_4 plants do not respond to elevated CO_2. Water potential increased (became less negative) with increasing CO_2 in *Amaranthus*, *Ambrosia*, and *Setaria* (Figure 15.31). In *Abutilon* and *Chenopodium*, water potential showed the same trend, but the differences were not significant. A higher water potential under elevated CO_2 is commonly observed (see Chapter 7).

Garbutt et al. (1990) do not give biomass data, as Bazzaz and Garbutt (1988) gave. So, we do not known if *A. retroflexus* in the study of Garbutt et al. (1990) had a decrease in growth at 500 μL/L, as reported by Bazzaz and Garbutt (1988). However, the leaf area, leaf biomass, and reproductive biomass of *A. retroflexus* was the same at 350 and 500 μL/L CO_2 in the study of Garbutt et al. (1990), which suggests that biomass was probably not reduced at 500 μL/L CO_2 as it was in the previous study. The unexplained reduction in biomass and seed production at 500 μL/L for *A. retroflexus* (Figure 15.29e and f), as reported by Bazzaz and Garbutt (1988), perhaps was a one-time event.

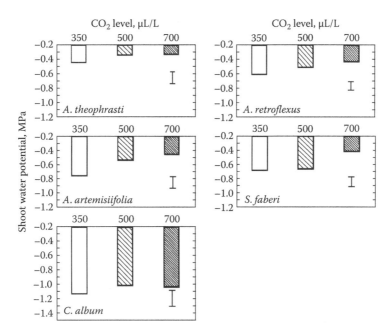

FIGURE 15.31 Shoot water potential at midday for five annual plants at three CO_2 levels (data lumped over all samples dates). Narrow vertical bars indicate ± 2 standard errors in a Student–Newman–Keuls multiple comparison. *Abutilon*, *Ambrosia*, and *Chenopodium* are C_3; *Amaranthus* and *Setaria* are C_4. (Reprinted from Garbutt, K., Williams, W.E., and Bazzaz, F.A., *Ecology*, 71, 1185, 1990, Fig. 2. With permission. Copyright 1990, Ecological Society of America.)

CAM PLANTS

Elevated CO_2 increases the growth of Crassulacean acid metabolism (CAM) plants. *Opuntia ficus-indica* (L.) Miller (prickly pear cactus) is a productive CAM plant cultivated in many countries (Nobel and Israel 1994). When Nobel and Israel (1994) grew it under doubled CO_2, the annual aboveground dry mass increased by up to 40%. In a follow-up study, North et al. (1995) monitored the development of its first-order daughter cladodes (flattened stem segments) under ambient (370 µL/L) and elevated (750 µL/L) atmospheric CO_2 concentrations. At four months, cladodes under the doubled CO_2 were 27% thicker, and the chlorenchyma was 31% thicker. [Chlorenchyma is parenchyma tissue containing chloroplasts—e.g., the leaf mesophyll (Esau 1977, p. 505).] At 16 months, the difference in cladode thickness diminished, but the chlorenchyma remained thicker under doubled CO_2, which may contribute to greater net CO_2 uptake for *O. ficus-indica* under elevated CO_2 concentrations.

Hylocereus undatus Britt. & Rose, commonly known as night-blooming cereus, is a CAM plant. It is the best known of all night-blooming cereuses and is widely cultivated in the tropics and subtropics, and in the tropics it makes a beautiful hedge plant (Bailey 1974). Other common names for it are pitahaya or dragon fruit, and its area of cultivation includes subtropical Israel (Weiss et al. 2009). In Israel, pitahayas are cultivated for fruit production under netting to prevent photoinhibition damage. Weiss et al. (2009) grew the plants in gravel fertilized with Hoagland's solution under greenhouse conditions with either elevated CO_2 (1000 µmol/mol) or ambient CO_2 (380 µmol/mol). When compared to the ambient treatment, CO_2 enrichment increased total daily net CO_2 uptake, nocturnal acid accumulation, shoot elongation, and total dry mass by 39%, 24%, 14%, and 6%, respectively. Weiss et al. (2009) conclude that, because pitahayas already grow under netting, it should be easy to stretch plastic film over the supporting structure during the night to enable enrichment with CO_2.

This should involve only a minimum extra cost, while, at the same time, it would make a substantial contribution to increasing fruit yield.

DECIDUOUS TREES

Idso and Kimball (1991a) grew sour orange trees (*Citrus aurantium* L.) in four open-top chambers supplied with ambient CO_2 or a CO_2-enriched atmosphere consisting of an extra 300 μmol/mol of air. After 2.5 years of growth under the two CO_2 treatments, soil cores were taken at different depths (20 cm increments) and at different distances from each tree. Fine roots were the most predominant type encountered. Large roots were so few that they could not obtain biomass data for them. The elevated CO_2 did not affect root growth from the 40–120 cm depth (Figure 15.32, top). However, the 0–40 cm depth did show an increase in growth due to elevated CO_2 (Figure 15.32, bottom). This depth contains most of the roots (70% of the roots of the ambient trees and 78% of the roots of the CO_2-enriched trees). They calculated that the fine-root biomass was enhanced by 175% in the CO_2-enriched trees, which was similar to the growth enhancement from elevated CO_2 for the total aboveground trunk plus branch volume (179%). Figure 15.33 shows the increase in trunk cross-sectional area of the trees grown under the two CO_2 treatments for 3 years (Idso and Kimball 1991b; Idso et al. 1991a). Idso and Kimball (1991a) suggested that one reason for the large increase in growth under elevated CO_2 is that sour orange trees do not close their stomata much, as the CO_2 content of the air is raised (Idso 1991). Measurements of photosynthesis of the sour orange trees (Idso and Kimball 1994a) showed enhancement of the photosynthetic rate of the CO_2-enriched trees with little reduction in stomatal conductance (Idso 1991; Idso et al. 1991b). The growth of the sour

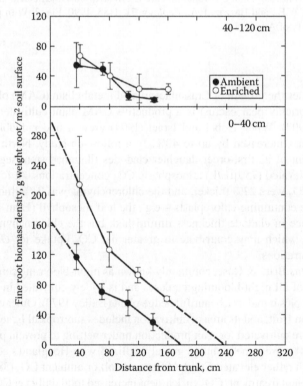

FIGURE 15.32 Fine-root biomass density versus distance from tree trunks of sour orange trees for the soil depth interval 40–120 cm (top) and the 0–40 cm layer (bottom). Vertical lines are probably standard deviations. (Reprinted from *Agric. Forest Meteorol.*, 55, Idso, S.B. and B.A. Kimball, Effects of two and a half years of atmospheric CO_2 enrichment on the root density distribution of three-year-old sour orange trees, 345–349, Fig. 1, Copyright 1991, with permission from Elsevier.)

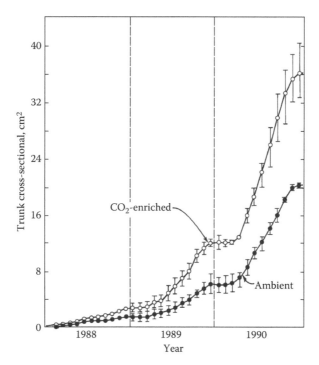

FIGURE 15.33 Mean trunk cross-sectional areas ± standard deviations for four sour orange trees growing in ambient air and four sour orange trees growing in air enriched with an additional 300 ppm of CO_2. (Reprinted from Idso, S.B. and Kimball, B.A. 1991b. Doubling CO_2 triples growth rate of sour orange trees. DOE Research Summary No. 13. Current Research Sponsored by the U.S. Department of Energy's Carbon Dioxide Research Program. Carbon Dioxide Information Analysis Center, Environmental Sciences Division, Oak Ridge National Laboratory, Oakridge, TN. 4pp, Fig. 1. With permission.)

orange trees was followed for 17 years under the two CO_2 treatments (Kimball et al. 2007). When summed over the duration of the experiment, there was an overall enhancement of 70% of total biomass production due to elevated CO_2. Thicker trunks and branches and more branches and roots were produced, but the root/shoot ratio was unaffected.

During 3 years, Bazzaz et al. (1993) studied the effect of elevated CO_2 on the growth of five co-occurring members of temperate forest communities: ash (*Fraxinus americana* L.); gray birch (*Betula populifolia* Marsh.); red maple (*Acer rubrum* L.); yellow birch (*B. alleghaniensis* Britt.); and striped maple (*A. pensylvanicum* L.). Plants grew in pots with a mixture of sand, perlite, and peat, which were placed in growth chambers controlled at either 350 or 700 μmol/mol CO_2. At each harvest (November of 1990, 1991, and 1992), biomass of leaves, stems, and roots was measured. After 3 years of growth, saplings in an elevated CO_2 environment were larger than those grown in ambient CO_2 (Figure 15.34). Most of the increase in growth due to elevated CO_2 occurred in the first year. The initial size advantage of elevated CO_2 plants at the end of year one may have eroded during years two and three because these larger plants grew relatively more slowly than the smaller, ambient CO_2 grown plants. The decrease in growth enhancement due to elevated CO_2 during years two and three suggests that forest productivity enhancement may not be sustained for long periods of time. Increase in carbon sequestering in terrestrial ecosystems may be less than anticipated. New Zealand has had forested areas for 100 million years, since the Mesozoic (Kelliher et al. 2000), and the ancient trees in them are not growing in size and are just maintaining their existence (David Whitehead, Landcare Research, Canterbury Agriculture and Science Center, Lincoln, New Zealand, personal communication, February 25, 2005).

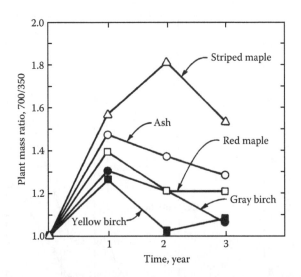

FIGURE 15.34 Time course of the ratio of total mass of plants grown in 700 μL/L CO_2 divided by that of plants grown in 350 μL/L CO_2 for five temperate tree species. (With kind permission from Springer Science+Business Media: *Oecologia*, CO_2-induced growth enhancements of co-occurring tree species decline at different rates, 96, 1993, 478–482, Bazzaz, F.A., Miao, S.L., and Wayne, P.M., Fig. 2.)

Miao (1995) studied the effect of elevated CO_2 on the growth of red oak (*Quercus rubra* L.). She grew seedlings in pots with soil, turface, and sand that she placed in one of two chambers in a greenhouse, set at either 350 (ambient) or 700 (elevated) μmol/mol CO_2. After 138 days of growth, she harvested the plants and determined dry weight of roots and shoots. Plants grown from larger seeds had greater growth than those grown from smaller seeds. In addition to differences in growth due to acorn size, biomass varied with CO_2 concentration. Total biomass of plants grown in elevated CO_2 was 47% greater than that of plants grown at ambient CO_2 (Figure 15.35). She suggested that red oak genotypes might be selected with big acorns, and plants from these acorns then will grow larger under elevated CO_2 than plants from small acorns.

Vivin et al. (1995) also found that oak grew more under elevated CO_2 compared to ambient CO_2. They placed germinated seedlings of *Q. robur* L. (English oak) in pots filled with peat and sand, which they placed in two transparent tunnels located in a greenhouse. One tunnel contained ambient CO_2 (350 μmol/mol CO_2) and the other one contained twice the ambient level (700 μmol/mol CO_2). At the end of the growing season, height of plants grown with elevated CO_2 was 28% greater than that of plants grown with ambient CO_2 (Figure 15.36).

While these and other studies with deciduous trees species, such as sweet chestnut (*Castanea sativa* Mill.) and beech (*Fagus sylvatica* L.), showed increased growth under elevated CO_2 (El Kohen et al. 1993), a FACE study with deciduous trees in Switzerland showed no stimulation in stem growth and leaf litter production after 4 year (2001–2004) exposure to elevated CO_2 (530 μmol/mol) (Körner et al. 2005). The forest contained *Quercus*, *Fagus*, *Acer*, *Carpinus*, and *Tilia*. However, the flux of carbon into the soil increased in response to elevated CO_2. Using a ^{13}C tracer, Körner et al. (2005) found that soil CO_2 concentrations were 44% higher in the elevated area as compared to the control area in 2002. Mycorrhizal fungi had 64% new carbon by the end of the first year. The tracer work showed a rapid flux of new carbon through the system. The increased soil CO_2 concentrations provided evidence for enhanced metabolic activity in soils under CO_2-enriched trees. A lack of response in stem growth or leaf litter production did not preclude a faster rate of belowground (root) production (Körner et al. 2005). Most studies of trees have been done under controlled-environmental conditions (in growth chambers or greenhouses) and on young trees. More long-term studies done under FACE conditions are needed to determine the effect of elevated CO_2 on

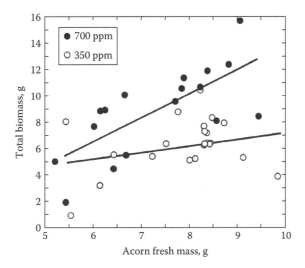

FIGURE 15.35   Scatter plots of final total biomass against initial across fresh mass for *Q. rubra* seedlings grown in 350 and 700 µL/L CO_2 treatments. (Reprinted from Miao, S.: Acorn mass and seedling growth in *Quercus rubra* in response to elevated CO_2, *J. Veg. Sci.*, 1995, 6, 697–700, Fig. 1. Copyright Wiley-VCH Verlag GmbH & Co. KGaA. With permission.)

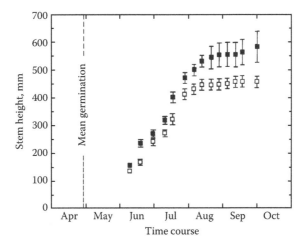

FIGURE 15.36   Time course of stem height in *Q. robur* seedlings grown under 350 or 700 µmol/mol CO_2 concentrations. Open squares, 350 µmol/mol CO_2; closed squares, 700 µmol/mol CO_2; $n=26$–40; vertical bars = ±standard error. Differences between treatments were significant ($P<0.05$) on all dates. (*Plant Physiol. Biochem.*, 33, Vivin, P., Gross, P., Aussenac, G., and Guehl, J.M., Whole-plant CO_2 exchange, carbon partitioning and growth in *Quercus robur* seedlings exposed to elevated CO_2, 201–211, Fig. 2, Copyright 1995, with permission from Elsevier.)

mature forests. We return to FACE studies later in this chapter. Also, because most of the work with elevated CO_2 has been done on temperate-zone trees, more research is needed on tropical plants (Hogan et al. 1991).

EVERGREEN TREES AND SHRUBS

Like the many studies with deciduous trees, studies with evergreen trees, such as pine (Garcia et al. 1994; Idso and Kimball 1994b), show that they grow more under elevated CO_2 compared to ambient levels. This is despite the fact that the stomatal conductance of conifers like pine appears to be little affected by elevated CO_2 (see Chapter 8). Centritto et al. (1999c,d) grew Sitka spruce [*Picea sitchensis* (Bong.) Carr.] in a mixture of sand, peat, and loam and placed the plants in OTCs with ambient (350 μmol/mol CO_2) or elevated CO_2 (ambient + 350 μmol/mol CO_2). Plants grown in elevated CO_2 were taller than plants grown at the ambient level of CO_2 (Figure 15.37) (Centritto et al. 1999c). The total dry mass and leaf dry mass of Sitka spruce saplings were higher in elevated CO_2 (about 40% for total dry mass and about 34% for leaf dry mass) than in ambient CO_2. The data were plotted logarithmically (Figure 15.38) (Centritto et al. 1999c), and the slopes in the figure give relative growth rates. The data indicated that during the first year, the relative growth rates of both total (Figure 15.38a) and leaf (Figure 15.38b) dry mass were larger in elevated CO_2 than in ambient

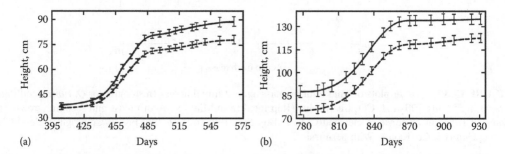

(a)　　Days　　　　　(b)　　　Days

FIGURE 15.37　Heights of Sitka spruce saplings grown in ambient (dashed line) or elevated (solid line) [CO_2], shown as days from the beginning of the experiment in the second growing season (a) (data are means of 100 plants per treatment ± one standard error of the mean), and the third growing season (b) (data are means of 40 plants per treatment ± one standard error of the mean). (Centritto, M., Lee, H.S.J., and Jarvis, P.G., *Tree Physiol.*, Fig. 1, 1999c of the Oxford University Press by permission of Oxford University Press.)

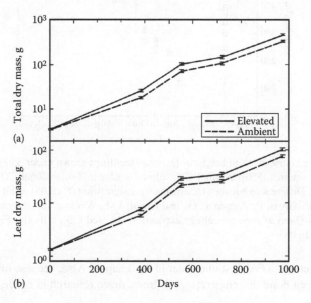

(a)

(b)　　　Days

FIGURE 15.38　Combined exponential total dry mass (a) and leaf dry mass (b) of Sitka spruce saplings versus days of exposure to ambient [CO_2] or elevated [CO_2]; the dry mass data are plotted on a logarithmic scale so that the slopes of the lines show the relative growth rates. Data are means of 20–40 plants per [CO_2] treatment ± standard error of the mean. (Reprinted from Centritto, M., Lee, H.S.J., and Jarvis, P.G., Increased growth in elevated [CO_2]: An early, short-term response? *Global Change Biol.*, 1999d, 5, 623–633, Fig. 1. Copyright Wiley-VCH Verlag GmbH & Co. KGaA. With permission.)

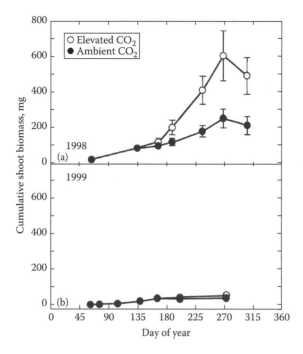

FIGURE 15.39 New shoot production of *L. tridentata* at ambient and elevated [CO_2] at the Nevada Desert FACE Facility (United States) in 1998 (a) and 1999 (b) for each CO_2 treatment. (Reprinted by permission from Macmillan Publishers Ltd. *Nature*, Smith, S.D., Huxman, T.E., Zitzer, S.F., Charlet, T.N., Housman, D.C., Coleman, J.S., Fenstermaker, L.K., Seemann, J.R., and Nowak, R.S., Elevated CO_2 increases productivity and invasive species success in an arid ecosystem, 408, 79–82, Fig. 1, Copyright 2000.)

CO_2, but the distance between the two curves remained constant (i.e., with the same slope at the same time), which indicated that the relative growth rates for total dry mass and leaf dry mass for the two treatments did not differ.

Smith et al. (2000) studied the growth of the evergreen shrub *Larrea tridenata* (creosote bush) under different CO_2 concentrations. Creosote bush is extremely drought tolerant and can absorb water from soil to −6.0 MPa (Kirkham 2005, p. 106). Plants grew in a FACE facility in an undisturbed Mojave Desert (arid) ecosystem in southern Nevada, United States. There were six plots: three maintained at 550 μmol/mol CO_2 since April, 1997 and three maintained at ambient CO_2. During the 1998 growing season (from March to September for perennials), the southwestern United States experienced an El Niño cycle during which the FACE facility received 309 mm of rain, which was 2.4-fold higher than the long-term average for the area. In contrast, the 1999 season was a dry La Niña cycle in which 107 mm of rain fell, and only 17 mm occurred before the onset of the growing season. The cumulative increase in new-shoot biomass for *L. tridentata* was higher (roughly double) at elevated CO_2 than at ambient CO_2 in 1998 (Figure 15.39a), but no difference in new shoot production occurred between CO_2 treatments in the dry year of 1999 (Figure 15.39b). Plant growth was restricted in the dry (normal) year. The results suggested that elevated CO_2 might increase the growth of this arid-region evergreen bush in wet years, but not in dry years. Because the desert is usually dry, the effect of elevated CO_2 may have a minimal effect on its growth.

TEMPERATURE

As noted in the Preface, this book does not consider the greenhouse effect (warming of the atmosphere by trace gases). It focuses on water, not other factors that affect plant growth under elevated CO_2, such as temperature. In 1983, Kimball (1983) pointed out that "theoretical models have predicted that the mean global air temperature will increase 3°C–4°C with a doubling of the CO_2 concentration … However, more recent 'earth experiments' indicate that the temperature rise will be less than 0.26°C for a doubling of CO_2 concentration." He concludes that, because this is a small increase, "the primary effect on agricultural production may only be that of the increased CO_2 concentration per se. The climate controversy is continuing and is beyond the scope of this paper." In this book, I have taken the same approach. I have considered only the effects of elevated CO_2 on soil and plant water relations, not responses related to global warming. It is important to recognize that elevated CO_2 has effects on plants without consideration of the greenhouse effect.

Every year, the American Meteorological Society documents the year's current weather and climate events from around the world in a publication called "State of the Climate" (Peterson and Baringer 2009b). It is published as a supplement to the *Bulletin of the American Meteorological Society*. At the time of the writing of this section of the book, the most recent edition was for the year 2008 (Peterson and Baringer 2009a). As documented in this edition, the global temperature had not risen in the past decade. Between January 1999 and December 2008, the HadCRUT3 dataset showed that the temperature change was +0.07°C ± 0.07°C/decade (Knight et al. 2009). HadCRUT3 stands for the "Hadley Centre/CRU gridded monthly temperatures dataset," and CRU stands for "Climate Research Unit." The Hadley Center is located in Exeter, Devon, United Kingdom. This dataset shows that the temperature of the earth has not increased in the past 10 years. Temperature records go back 100 years for weather stations throughout Kansas, and an analysis of them shows that there is no trend for temperature. The temperature in Kansas is not getting warmer or cooler (Mary C. Knapp, State Climatologist, Kansas State University, personal communication, May 12, 2010). Robinson et al. (2007) came to the conclusion that increased levels of atmospheric CO_2 in the twentieth and early twenty-first centuries have produced no deleterious effects upon the earth's weather and climate. For a review of global warming and cooling and their causes during the geological history of the earth, see Kutílek (2008) and Kutílek and Nielsen (2010).

The literature concerning the effects of temperature on plants grown under elevated CO_2 is extensive and beyond the scope of this book. For reviews, see the following: Baker and Allen (1993); Crawford and Wolfe (1999); Smith et al. (1999); Prasad et al. (2005); and Reddy et al. (2005). The focus of much research on climate change is the predicted warming in temperature (e.g., see Walthall 2009; Baethgen 2010; Ledig et al. 2010). Even subtle changes in temperature have effects on human health (Epstein et al. 1998; Epstein 2005).

As stated at the beginning of this chapter, the literature indicates that a doubling of the atmospheric CO_2 concentration generally increases plant growth by about 30% (Kimball 1983). Idso et al. (1987) demonstrated that this stimulatory effect of atmospheric CO_2 enrichment is strongly temperature dependent. Their results suggested that for a 3°C increase in mean surface air temperature, the growth enhancement factor for such a CO_2 increase rises from 1.30 to 1.56. Their results also indicated that atmospheric CO_2 enrichment tends to reduce plant growth at relatively cold air temperatures, i.e., below a daily mean air temperature of about 18.5°C.

It is important to study the interaction of elevated CO_2 with temperature on plant productivity (Long 1991). Results vary concerning the interaction. For grain legumes (kidney bean or *P. vulgaris* L. and peanut or *Arachis hypogaea* L.), elevated CO_2 increased yields due to increased photosynthesis and growth. However, at above optimum temperatures, the beneficial effects of elevated CO_2 were offset by negative effects of temperature on yield and yield components, leading to lower seed yield and poor seed quality (Prasad et al. 2002, 2003, 2005). For cotton (*Gossypium* sp.), elevated CO_2 increased productivity, because of improved CO_2 exchange rates, vegetative growth, and number of bolls formed. However, increases in temperature above the optimum for cotton decreased yields due to increased boll abscission and smaller boll size. Doubling CO_2 did not ameliorate the adverse effects of high temperature on the reproductive growth of cotton (Reddy et al. 2005).

In contrast to the results for grain legumes and cotton, Idso et al. (1989) found that when the ambient concentration of CO_2 (340 μmol/mol) was increased to 640 μmol/mol, debilitating effects of high temperatures (above 30°C) on the aquatic fern *Azolla* were reduced. They concluded that CO_2 enrichment may be capable of preventing the deaths of some plant species in situations where their demise is normally brought about by unduly high temperatures. More often than not, data show that the relative growth-enhancing effects of atmospheric CO_2 enrichment are the greatest when environmental stresses like high temperature are the most severe (Idso and Idso 1994).

Alonso et al. (2009) found that CO_2 enrichment (370 μmol/mol CO_2; ambient CO_2 concentration was 370 μmol/mol) decreased the ribulose-1,5-bisphosphate (RuBP) limited photosynthesis rates at lower temperatures (15°C–25°C) in wheat (*T. aestivum* L.) and increased ribulose-1,5-bisphosphate carboxylase/oxygenase (Rubisco)- and RuBP-limited rates at higher temperatures (30°C–35°C). They concluded that the effects of future increases in air CO_2 on Rubisco kinetics may improve the photosynthetic response of wheat to global warming. Another study done, also with wheat under elevated CO_2 (700 μmol/mol) and elevated temperature (ambient + 4°C) (Gutiérrez et al. 2009), showed that elevated CO_2 concentration increased crop dry matter at maturity by 12%–17%, while above-ambient temperatures did not affect dry matter yield. Growth under CO_2 enrichment increased dark- and light-adapted photochemical efficiencies. Gutiérrez et al. (2009) concluded that elevated levels of CO_2 can mitigate photochemical inhibition as a result of warm temperatures.

Few studies have considered plants grown under elevated CO_2 and varying water and temperature regimes. Andresen et al. (2010) found in a Danish FACE experiment that the effects of soil warming (1°C increase at the 2 cm depth) on microbial activity in a sandy soil with heathland plants were often counteracted or erased when combined with elevated CO_2 (510 μmol/mol) and drought (5% by volume in the dry plots; well-watered plots had 17% by volume water). They could not fully explain their results and concluded that the interactions of elevated CO_2 with other environmental variables (i.e., temperature and drought) are complex and need further study. They cannot be predicted from experiments that consider only one variable.

Qaderi et al. (2006) grew canola (*B. napus* L.) under two temperature regimes (22°C/18°C, day/night and 28°C/24°C, day/night) in controlled-environmental chambers at ambient (370 μmol/mol) and elevated (740 μmol/mol) CO_2 concentrations. One half of the plants were watered to field capacity and the other half were watered when they reached the wilting point. Drought-stressed plants grown under the higher temperature at ambient CO_2 had decreased stem height and diameter, leaf number and area, dry matter, and net CO_2 assimilation rate. Elevated CO_2 partially reversed the inhibitory effects of higher temperature and drought on leaf dry weight. The results showed that higher temperature and drought inhibited many plant processes, but elevated CO_2 partially mitigated some adverse effects.

Brakke and Allen (1995) investigated air temperature and available soil water on gas exchange of two different types of citrus (*Citrus*) seedlings grown under two CO_2 concentrations (330 and 840 μmol/mol). At the elevated CO_2 concentration, net canopy CO_2 uptake rates were double those that occurred at ambient CO_2 levels. When soil water was not readily available, the citrus seedlings were more sensitive to high air temperature, which resulted in a reduction of CO_2 uptake. The inhibitory effects of reduced available soil water on citrus CO_2 assimilation rates were lessened at the elevated atmospheric CO_2 level.

Stronach et al. (1994) grew groundnut (also called peanut; *A. hypogaea* L.) in soil (1.25 m depth) in greenhouses kept at two air temperatures (28°C and 32°C) and two atmospheric CO_2 concentrations (375 and 700 μmol/mol) and two soil moisture treatments (irrigated to field capacity or allowed to dry from 22 days after planting until the end of the experiment, 125 days after planting). Groundnut growing in the elevated CO_2 environment transpired less of the available soil water per unit biomass produced compared to groundnut grown with ambient CO_2. They found a temperature increase of 4°C had no effect on the water use of the crop. The implication of this observation is that the effect of increasing CO_2 on crop growth will be of net benefit even when allowing for a temperature rise of 4°C (28°C–32°C).

FREE-AIR CO$_2$ EXCHANGE STUDIES

Free-air CO$_2$ exchange (FACE) technology now allows studies of crops at elevated CO$_2$ under fully open-air field conditions (Long et al. 2006). (The term "FACE" also stands for "free-air CO$_2$ enrichment.") The FACE studies show that yield may be enhanced less than that determined in studies where plants have been grown in enclosures. From data in the literature, Long et al. (2006) calculated increases in yield due to elevated CO$_2$ (550 μmol/mol) for wheat and soybeans grown in enclosures or in FACE facilities. They assumed that the current CO$_2$ concentration in the atmosphere is now 380 μmol/mol. For wheat, they found a mean increase of 31% and 13% for enclosed and FACE studies, respectively. For soybeans, the increase was 32% and 14% for enclosed and FACE studies, respectively. Long et al. (2006) stated that the environmental constraints of chambers, even open-top chambers, prevent plants from being exposed to actual environmental conditions, as they are in FACE studies. Temperatures are higher in chambers, and the plastic enclosures inhibit transmission of sunlight.

Even though the increase due to elevated CO$_2$ may be less under FACE studies compared to enclosed studies, they still show that elevated CO$_2$ increases growth. A 13% or 14% increase in yield, spread over hundreds of acres on which each crop grows, would be large. In 2009 in Kansas, 8,800,000 acres (3,520,000 ha) of winter wheat were harvested with an average yield of 41 bushels per acre and a total production of 360,800,000 bushels (U.S. Department of Agriculture 2009a). If there were a 13% yield increase due to a 170 μmol/mol increase in CO$_2$ (380–550 μmol/mol), the yield would be 46.3 bushels per acre and the production would be 407,704,000 bushels. [A bushel of wheat weighs about 60 lb (Finney et al. 1987, p. 679) or 27.3 kg. Therefore, 1 bushel per acre = 60 lb/ acre = 68 kg/ha]. So, there would be an extra 46,904,000 bushels of wheat produced in a year due to the elevated CO$_2$. From one bushel of wheat, 48 loaves of bread can be made, if each loaf weighs 1 lb (0.45 kg) (James P. Shroyer, Department of Agronomy, Kansas State University, personal communication, September 28, 2009). This means an extra 2,251,392,000 loaves of bread would be available from Kansas wheat in a year, if the CO$_2$ concentration increased by 170 μmol/mol. In April, 2009, Kansas winter wheat was selling at $5.13 per bushel (U.S. Department of Agriculture 2009b). With the elevated CO$_2$, wheat farmers in Kansas would have gotten $240,617,520 for the extra wheat produced. We now turn from Kansas to the entire United States. In 2009 in the United States, 1,491,769,000 bushels (40,684,609,000 kg) of winter wheat were produced (U.S. Department of Agriculture 2009c). A 13% increase in yield due to elevated CO$_2$ (193,930,000 more bushels or 5,289,000,000 kg) would increase this number to 1,685,699,000 bushels (45,973,609,000 kg). Because about two-thirds of the U.S. wheat production enters into the export market (Briggle and Curtis 1987, p. 10), these extra bushels of wheat would help provide bread for hungry people in other nations.

HARVEST INDEX

Harvest index (HI) is defined as the "usable biomass as a fraction of the total biomass" (Rosenzweig and Hillel 1998, p. 159) or "the fraction of the total plant matter that is marketable" (Rosenzweig and Hillel 1998, p. 249). The "total biomass" or the "total plant matter" usually excludes the roots, so HI is calculated by getting weight of the marketable yield (e.g., grain in the case of a grain crop) and dividing that by the total aboveground biomass. For a grain crop, HI = grain yield/total shoot biomass (Hall and Ziska 2000). HI is a term used only in recent decades (since about 1965–1970). It is not defined in the "Glossary of Crop Science Terms" published in 1992 (Barnes and Beard 1992b). It is widely used by agronomists, because grain yield and aboveground growth can be distinguished and measured. It is a common factor calculated by crop-growth models, and the development of crop-growth models and the use of the term coincided. Crop-growth models could be developed when computers became widely available (mid-1960s). The term HI is rarely used by ecologists, because marketable yield is usually not of importance when natural ecosystems are being studied.

As noted in the introduction to this chapter, yield is increased about 33% on average for a doubling of CO$_2$ concentrations. HI is usually increased under elevated CO$_2$ (Strain and Cure 1985b,

p. 207), but by a much smaller percentage (Acock and Allen 1985, p. 71; Kimball 1985, p. 193). Cure (p. 105) reports changes in HI of different crops due to a doubling of CO_2 concentration (680 μmol/mol) compared with controls (300–350 μmol/mol CO_2). She gives means ± 95% confidence limits, as estimated by regression analysis. Where she gives no ± value, the data are based on only two data points and error degrees of freedom were 0. So she could not calculate confidence limits. The values for HI are as follows (because HI is a proportional variable already, these values were calculated as the simple difference between the value at elevated CO_2 and the value at the control level of CO_2):

Barley: +0.013
Corn: +0.043 ± 0.046
Potato: +0.019
Rice: +0.019 ± 0.006
Soybean: −0.05 ± 0.044
Wheat: +0.024 ± 0.023

The only negative value was for soybean. However, she (Cure 1985, p. 106) found one study done with soybean under water stress, when HI was increased under doubled CO_2. The values for soybean in that study were as follows (calculated the same way as described above):

Soybean, control (no water stress): +0.016
Soybean, water-stressed: +0.027

Pettersson et al. (1993) also showed that doubling CO_2 increased the HI of barley (from 0.44 at 350 μmol/mol CO_2 to 0.48 at 700 μmol/mol CO_2). A review of the literature for the HI of corn grown between 1900 and 1988, without reference to the level of CO_2 in the atmosphere, shows that its HI has varied from a low of 0.21 (in the period 1900–1930) to a high of 0.58–0.59 (in 1976) (Lorenz et al. 2010). The review indicates that the HI of corn does not appear to be related to time, because plants grown in 1930 had HIs as high as 0.52 and 0.54 (Lorenz et al. 2010).

Most people report that the HI of soybean is an exception to the general rule that HI of crops increases under elevated CO_2 (Kimball 1985, p. 193; Baker and Allen 1993, p. 254; Wittwer 1995, p. 90). A meta-analysis of soybean showed that plants grown under elevated CO_2 were unchanged in root-to-shoot ratios (see Chapter 6 for a discussion of root-to-shoot ratios), whether nodulated or not. But seed yield accounted for a smaller part of final dry weight of the soybeans, so HI was reduced by CO_2 (Hill et al. 2006). In a FACE (free-air CO_2 enrichment) experiment in Champaign, Illinois, with soybean grown during three seasons at elevated CO_2 (550 μmol/mol) and ambient CO_2 (372 μmol/mol), there was a consistent and significant, albeit small, decline in HI in all 3 years. HI was decreased by 3%, 2%, and 2% in 2001, 2002, and 2003, respectively (Ort et al. 2006).

However, rice also has been shown in one experiment to have a decreased HI under elevated CO_2. Rice was grown for 3 years (1998–2000) in a FACE facility in on Honshu Island, Japan. FACE plants were 200 μmol/mol above the ambient CO_2 concentration (average CO_2 concentrations for the ambient and elevated concentrations were 367 and 575 μmol/mol, respectively). The HI was reduced by about 2% by the elevated CO_2 treatment (Kobayashi et al. 2006).

A general explanation for the increases in grain yield associated with increases in HI is increased efficiency of carbohydrate partitioning (Hall and Ziska 2000; Lawlor and Mitchell 2000). Because soybean is the only species for which HI has been shown consistently to decrease at elevated CO_2, this raises questions about the efficiency of partitioning of carbohydrates in soybean leaves during reproductive growth (Cure 1985, p. 112). Due to the increasing CO_2 concentration in the air, it seems likely that plant varieties developed in the future will have higher HIs than those of the cultivars being tested today (Kimball 1985, p. 193).

Because HI does not increase by nearly as much as yield under doubled CO_2 (a few percent for HI versus a third or 33% for yield), there probably will be about a third more plant residue to manage with a doubling of the CO_2 level. The higher amounts of residue will help reduce erosion (Kimball 1985, p. 193). Plant residues are a natural resource for preventing erosion as well as for maintaining or increasing the organic matter content of the soil.

QUALITY

Few studies have been done to determine the effect of elevated CO_2 on food quality (Porter 2005). Early studies suggested that the quality of products of crops is little affected by elevated CO_2 (Lawlor and Mitchell 1991). But concerns have been raised that quality will be reduced, and, in particular, the protein in food will decrease (Stafford 2007). Some see no cause for alarm due to the slight decrease in nutrition. Many scientists feel that the global quantity of food in coming decades is more important than quality, especially in developing countries where the food supply must be maintained. Food production is the primary goal in these countries (Stafford 2007). The challenge to produce enough food for growing populations is always present. [See Kirkham (2005, pp. 1–6) for a discussion of the world's population growth.]

YIELD

In his extensive review of the literature, Kimball (1983) found that yield was increased on average by 33% by a doubling of CO_2 concentration in the air. Using this number, we can calculate how much yield has increased due to the increase in CO_2 in the 50 year period since Keeling started to monitor the CO_2 concentration at Mauna Loa, Hawaii, and in Antarctica at the South Pole. (See Chapter 1 for Keeling's studies.) As stated in the Preface, the CO_2 concentration in the atmosphere in 1958 was 316 ppm. In 2008, it was 385 ppm (Dlugokencky 2009). It has increased 69 ppm in 50 years. Using the 1958 value as our base, for a doubling of the CO_2 in 1958, we would have 316 ppm more in the air. Therefore, we are (69 ppm/316 ppm) × 100 = 21.8% of the way toward a doubling of the CO_2 in the atmosphere. Taking a third of this (yields are increased a third by a doubling of the CO_2 concentration), we get 21.8% × 0.33 = 7.2%. If other factors controlling plant growth are neglected, the calculation shows that yields were 7.2% more in 2008 than in 1958 due to the increase in CO_2 in the atmosphere.

SUMMARY

Most studies confirm the well-documented result that growth and yield are increased under elevated CO_2. Yield (reproductive or marketable yield as well as vegetative yield) increases about a third (33%) for a doubling of CO_2. The increase in growth under elevated CO_2 has been demonstrated for a wide variety of plants, including examples discussed in this chapter (agronomic plants, horticultural plants, grassland plants, marsh plants, weeds, cacti, deciduous trees, and evergreen trees). Plants with the C_4 type of photosynthesis can increase growth under elevated CO_2, in contrast to some papers that say that C_4 plants do not respond to elevated CO_2. However, C_4 plants sometimes do not increase growth under elevated CO_2, while C_3 plants almost always do. In one study reported in this chapter, when C_3 and C_4 plants were grown together, the C_3 plants grew more than the C_4 plants. That is, the C_3 plants were better competitors under elevated CO_2 than the C_4 plants. Reports of elevated CO_2 decreasing growth are rare and cannot be explained. Because HI does not increase by nearly as much as yield under doubled CO_2 (a few percent for HI versus about 33% for yield), there probably will be about a third more plant residue to manage with a doubling of the CO_2 level. More research is needed in FACE facilities compared to studies done in enclosed environments. Studies need to be done long term to show the full impact of elevated CO_2 during the life span of a plant. A calculation shows that, if yields are increased by a third with a doubling of the CO_2 in the atmosphere, then yields in 2008 were about 7% more than in 1958, when Charles Keeling first started to measure the CO_2 concentration in the atmosphere, due to increased CO_2.

APPENDIX

BIOGRAPHY OF BRUCE ARNOLD KIMBALL

Bruce Kimball, pioneering researcher in studying the effects of elevated CO_2 on crop growth, was born on September 27, 1941, in Aitkin, Minnesota, the son of Robert Clinton and Rica (Barneveld) Kimball (Marquis Who's Who 2008). He grew up on the family dairy farm about 30 mi from there. He got his BS degree in soils from the University of Minnesota in 1963, when he received the Caleb Dorr Award for being the highest ranking man in the College of Agriculture (Kissel 1988). He went to Iowa State University, Ames, Iowa, for his MS degree, which he got in 1965 in soil physics (under the direction of my father, Don Kirkham). There he met Laurel Sue Hanway, the daughter of Iowa State University's agronomist, John J. Hanway, who defined the growth stages of corn (*Zea mays* L.) (Hanway 1963) and showed that foliar feeding of soybeans at the reproductive stage dramatically increased their yield (Garcia and Hanway 1976). They married on August 20, 1966, and have three daughters, Britt, Megan, and Rica (Marquis Who's Who 2008). He continued his studies in soil physics at Cornell University, Ithaca, New York, where he received the PhD degree in 1970 under the direction of Edgar ("Ed") R. Lemon. For his dissertation research, he measured atmospheric pressure fluctuations to determine their effect on gas movement in soil. He concluded that under most normal cropping conditions with adequate moisture, molecular diffusion is the predominant mechanism of exchange between soil gases and the atmosphere (Kissel 1988).

In 1969, Dr. Kimball joined the staff of the U.S. Water Conservation Laboratory of the U.S. Department of Agriculture (USDA), Agricultural Research Service, in Phoenix, Arizona. There he extended his soil gas exchange studies to the movement of water vapor through mulches. Then, as a member of a team studying the evaporation process at the soil surface, he began measurements of soil heat flux, codeveloping new techniques for making the measurements. He made comparisons with theoretical predictions and found that the mutually compensating movements of latent heat on opposing gradients of temperature and water content led normal thermal conduction to be the main heat transfer mechanism under these conditions. Besides the soil heat flux techniques, Dr. Kimball developed other data collection or analysis techniques. These included two methods for smoothing data, fabrication of a conductivity-type CO_2 analyzer, and development of two new designs of psychrometers (Kissel 1988).

To assess the feasibility of water- and energy-conserving greenhouses, Dr. Kimball developed a versatile modular energy balance computer model of a greenhouse, in use by several research groups around the world. He also evaluated the growth response of tomatoes to CO_2 enrichment in closed, high-humidity greenhouses. These studies led to an assemblage and analysis of the voluminous literature on the effects of CO_2 on the yield of crops, with a resultant prediction that the expected doubling of the atmospheric CO_2 concentration during the next century will probably increase crop yields an average of 32% (Kimball 1983). The paper describing this work was identified as "key literature" by the U.S. Department of Energy. Together with H.Z. Enoch, Dr. Kimball edited a two-volume book on CO_2 enrichment of greenhouse crops (Enoch and Kimball 1986a,b). He also was chair of the editorial committee that prepared *Impact of Carbon Dioxide, Trace Gases, and Climate Change on Global Agriculture* (Kimball et al. 1990).

Since the mid-1980s, Dr. Kimball has led a team to study the effects of increasing atmospheric CO_2 concentration on the yield and water relations of crops (Kissel 1988). From 1990 to 2006, he was Research Leader of the Environmental and Plant Dynamics Research Group of the U.S. Water Conservation Laboratory in Phoenix (Marquis Who's Who 2008). In 2006, the USDA moved the headquarters to Maricopa, Arizona, where Dr. Kimball now has his office. In the late 2000s, he and his team carried out a "Hot Serial Cereal Experiment" in which infrared heaters were placed over wheat to see the effect of elevated temperature on the crop (personal communication, December 17, 2008).

Dr. Kimball served as associate editor for the *Soil Science Society of America Journal* from 1977 to 1982; for the *Transactions of the ASAE* (American Society of Agricultural Engineers) from 1984 to 1987 (Kissel 1988); and for *Global Change Biology* (Marquis Who's Who 2008). He has been named a Highly Cited Researcher in agriculture by the Institute for Scientific Information (Marquis Who's Who 2008). In 1987, he became Fellow of both the American Society of Agronomy and the Soil Science Society of America. He was on the Board of Directors of Soil Science Society of America between 1994 and 1997. His avocations are computers and biking (Marquis Who's Who 2008).

Epilogue

Water movement through the soil-plant-atmosphere system is a catenary process (van den Honert 1948). Each part of the chain (i.e., movement of water from the soil to the plant; movement from the plant to the atmosphere) is linked. Elevated CO_2 in the atmosphere changes this catenary process. Briefly, the process is as follows. The key factor is stomatal closure under elevated CO_2. Stomata are extremely sensitive to the concentration of CO_2 in the atmosphere and close when the concentration increases (Chapter 8). When the stomata close, transpiration is reduced (Chapter 10), resulting in more water in the soil (Chapter 5). With an increase in soil water content under elevated CO_2, the plants have more water for uptake, and this causes an increase in plant water potential (Chapter 7). Plant water potential and growth are directly related, and the higher the water potential, the more the growth. Thus, we see that in the equation for photosynthesis given in the Preface (CO_2 and H_2O joining to form sugar and oxygen), both CO_2 and H_2O are affected by elevated CO_2, and each factor favors increased growth. Because we have more CO_2 in the atmosphere, the reaction is pushed to the right, which results in more production of sugar and oxygen. With more water in the soil, the water potential of the plants is increased, and this favors growth too.

The major portal for entry of water into a plant is the root, and the major exit is the stoma. Consequently, in the book, we took a detailed look at both. Chapter 6 focused on roots, and Chapters 8 and 9 considered stomata. Root weight is increased with elevated concentrations of CO_2 in the atmosphere. In Chapter 6, we used the data that we obtained in Kansas between 1984 and 1987 for sorghum and winter wheat to calculate how much root weight has increased due to increasing CO_2 concentrations in the atmosphere. The concentration of CO_2 in the atmosphere has increased 55 μmol/mol between 1984 and 2008. With an increase of 55 μmol/mol CO_2 in the atmosphere, root weights of sorghum and winter wheat have increased 4 and 7 g/m^2, respectively.

Roots need at least 10% by volume air space in the soil to survive, which means that the minimum oxygen concentration in the soil needs to be about 2% because air contains 21% oxygen (Chapter 3). No such limiting (maximum) concentration has been established for CO_2. Because oxygen is more important for root growth than CO_2, most studies concerning the effects of soil water content and soil compaction on root growth have focused on oxygen in the soil rather than CO_2. The absence of extreme toxicity at high levels of CO_2 (24% in one study done with a sandy loam soil) indicated that CO_2 would rarely, if ever, limit root growth in soil. Excessive moisture and mechanical impedance apparently are more important in decreasing root growth than elevated levels of CO_2 in the soil. Short periods when oxygen is lacking in the soil (e.g., 3 min or less for cotton) can immediately reduce root growth. But growth returns to normal if oxygen is reapplied. For cotton, periods of anaerobiosis of longer than 30 min cause root death of cotton (Chapter 4).

Most data show that elevated CO_2 decreases stomatal conductance in plants, including annual crops, grassland species, weeds, and deciduous and evergreen trees with broad leaves (Chapter 8). When stomatal resistance (reciprocal of stomatal conductance) of C_3 and C_4 plants has been compared, elevated CO_2 increases stomatal resistance of both, and the resistance is higher in C_4 plants compared to C_3 plants. The stomatal conductance of conifers, like pines with needles for leaves, often appears to be little affected by elevated CO_2. This is perhaps because their stomata are sunken and are overtopped by other epidermal cells, characteristics that conserve water. The xeromorphic anatomy of the stomata of pines may allow CO_2 to enter for photosynthesis, but inhibit the exit of water.

Several authors have demonstrated, both experimentally and from fossil and herbarium material, an inverse relationship between stomatal density and increasing ambient CO_2 levels (Chapter 9). These investigations suggest that reductions in stomatal density are a physiological response by plants to increase water use efficiency, which may influence their growth and survival under moisture-limited conditions. However, other authors have not confirmed these observations and have suggested that variations in stomatal density may be more strongly influenced by factors other than CO_2, such as altitude and microclimatic conditions including light and moisture.

Elevated atmospheric CO_2 increases water use efficiency of plants (Chapter 11), which is the photosynthetic production of plants divided by the water used to produce that amount of dry matter. Conversely, elevated CO_2 decreases the water requirement, which is the reciprocal of water use efficiency. Using the data from our wheat and sorghum experiments during the 1984–1987 period (the CO_2 concentration in the atmosphere in 1984 was 330 μmol/mol), we can calculate what the water requirements of wheat and sorghum are a quarter century later, based on the CO_2 concentration in atmosphere in 2008 (385 μmol/mol in 2008). Wheat and sorghum are now using 41 and 43 mL, respectively, less water to produce a gram of grain than they were in the 1984–1987 period. The water requirements to produce a gram of wheat grain and sorghum grain have been reduced by 6% and 4%, respectively, due to the 55 μmol/mol increase in CO_2 since 1984. These may seem like small percentages. But spread over the large acreage of wheat and sorghum grown in Kansas (for example), the more efficient use of water now (2010) compared to the mid-1980s should have a large impact. The results suggest that concentrations of atmospheric CO_2, which continue to rise, may have major effects on the water use efficiency of crops. The increasing level of atmospheric CO_2 concentration offers an opportunity to increase water use efficiency of plants that other methods, such as increasing the fertility of soil, have failed to accomplish.

Growth of C_3 plants is favored over that of C_4 plants under current and increasing atmospheric CO_2 concentrations (Chapter 12). Flowering plants evolved about 120 million years ago during the Cretaceous, when the atmospheric CO_2 concentration was 3000 μmol/mol. The concentration of CO_2 in the air declined after the Cretaceous, and C_4 plants apparently evolved when atmospheric CO_2 concentrations were lower (e.g., 180 μmol/mol CO_2) than they are now (385 μmol/mol). As the CO_2 concentration increases from 0 to 400 μmol/mol, the photosynthetic rate of C_4 plants increases. Above about 400 μmol/mol CO_2, the photosynthetic rate of C_4 plants often tends to level off. Because ambient CO_2 concentration (385 μmol/mol) apparently may be near the optimal for C_4 plants (about 400 μmol/mol), augmenting the CO_2 concentration may not always result in increased growth for C_4 plants, as it does for C_3 plants. However, we cannot categorically say that plants with the C_4 type of photosynthesis will not increase growth under elevated CO_2, because experiments show that they do (Chapter 15). At current levels of CO_2, the dry matter production of C_4 plants is higher than that of C_3 plants (Chapter 12).

When drought occurs, elevated CO_2 often compensates for reduction of growth due to the drought stress (Chapter 1). The degree of compensation of drought-induced reduction in growth depends upon the level of CO_2, the species, the stage of plant growth, and the severity of the moisture stress. The data indicate that in a CO_2-increased world, C_3 plants will survive better under drought compared to a low-CO_2 world. At ambient CO_2 levels, C_4 plants have a higher photosynthetic rate than C_3 plants, both under well-watered and drought-stressed conditions. In general, the overall effect of elevated CO_2 is to make C_3 plants appear to respond the same as C_4 plants, as far as their photosynthetic rate is concerned. Data indicate that, as CO_2 concentration in the atmosphere increases, yield of C_4 plants will be maintained under drought and yield of C_3 plants will increase so that it will become more similar to that of C_4 plants under drought.

Elevated CO_2 has some beneficial effects on plant anatomy. Data are consistent in showing that the ratio of leaf weight to leaf area (incorrectly called "specific leaf weight") increases as CO_2 concentration increases (Chapter 13). That is, leaves are thicker under elevated CO_2. This has been observed for both C_3 and C_4 leaves. Thinner leaves at low CO_2 may serve to increase the rate of CO_2 diffusion to the chloroplasts. Thick leaves tend to be more xeromorphic than thin leaves, and this suggests that leaves under elevated CO_2 are more drought-resistant than leaves grown under the ambient level of CO_2. Rising atmospheric CO_2 concentration results in denser wood.

Because elevated CO_2 stimulates growth, yield is usually increased under elevated CO_2 (Chapter 15). In a classic paper, Kimball (1983) analyzed more than 430 observations of the yield of 37 species grown with CO_2 enrichment, which were published in more than 70 reports of experiments carried out in a 64 year period beginning in 1918. His analysis showed that, with a doubling of atmospheric CO_2 concentration, yield probably will increase by 33%. In this book, I have not cited papers dealing with models, except in passing. My focus has been on data published in refereed journal articles. However, Ken Caldeira at the Carnegie Institution for Science in Stanford, California (Meserve 2008–2009), has modeled plant growth under elevated CO_2. He and his colleagues (Govindasamy et al. 2002) have found that "doubling the amount of carbon dioxide while holding steady all other inputs—water, nutrients, and so forth—yields a 70% increase in plant growth, an obvious boon to agricultural productivity" (Levitt and Dubner 2009, p. 185). This is an even greater increase than that documented by Kimball (1983). The year-to-year increases in crop yields that have been observed during the last 50 years, since Charles Keeling first started to record the CO_2 concentration in the atmosphere in 1958, probably are related, in part, to the increased CO_2 concentration in the atmosphere. If other factors controlling plant growth are neglected, we can calculate that yields are 7% more in 2008 than in 1958 due to the increase in atmospheric CO_2. The elevated levels of CO_2 in the air probably are adding to our food security without our recognizing it.

We can list general conclusions about the effects of elevated CO_2 on soil and plant water relations, as follows:

1. Stomata close under elevated CO_2 (Chapter 8).
2. Transpiration is reduced under elevated CO_2 (Chapter 10).
3. Water in the soil is increased under elevated CO_2 (Chapter 5).
4. Plant water potential is increased under elevated CO_2 (Chapter 7).
5. Root weight and length are increased under elevated CO_2 (Chapter 6).
6. Water use efficiency is increased under elevated CO_2 (Chapter 11).
7. Under drought, elevated CO_2 often compensates for reductions in growth due to drought (Chapter 1).
8. Leaves are thicker under elevated CO_2 (Chapter 13).
9. Elevated CO_2 favors the growth of C_3 plants more than C_4 plants (Chapter 12).
10. Elevated CO_2 often accelerates the development of horticultural crops in greenhouses, but it can retard, advance, or have no effect on the phenology of plants in agricultural and natural ecosystems (Chapter 14).
11. Fossil and herbarium records indicates that stomatal density decreases with increasing CO_2 concentration, but not all experimental data confirm this observation (Chapter 9).
12. High concentrations of CO_2 in the soil do not inhibit root growth, as long as the roots have sufficient oxygen (Chapters 3 and 4).
13. Carbonated irrigation water added to soil to increase growth generally has no effect on growth of plants, because the CO_2 in the carbonated water dissipates quickly (Chapter 2).
14. Growth and yield of plants is increased under elevated CO_2 (Chapter 15).

References

Acock, B. and L.H. Allen, Jr. 1985. Crop responses to elevated carbon dioxide concentrations. In Strain, B.R. and Cure, J.D., Eds., *Direct Effects of Increasing Carbon Dioxide on Vegetation*. DOE/ER-0238. Washington, DC: Office of Energy Research, U.S. Department of Energy, pp. 53–97.

Agosti, D. 2003. Switzerland's role as a hotspot of type specimens. *Nature* 421: 889.

Ahrens, J.F. and W.E. Loomis. 1963. Floral induction and development in winter wheat. *Crop Sci.* 3: 463–466.

Ainsworth, E.A., A. Rogers, and A.D.B. Leakey. 2008. Targets for crop biotechnology in a future high-CO_2 and high-O_3 world. *Plant Physiol.* 147: 13–19.

Aitken, Y. 1974. *Flowering Time, Climate and Genotype*. Carlton, Victoria, Australia: Melbourne University Press, 193pp.

Akita, S. and D.N. Moss. 1972. Differential stomatal response between C_3 and C_4 species to atmospheric CO_2 concentration and light. *Crop Sci.* 12: 789–793.

Akita, S. and D.N. Moss. 1973. Photosynthetic response to CO_2 and light by maize and wheat leaves adjusted for constant stomatal aperture. *Crop Sci.* 13: 234–237.

Allen, L.H., Jr. 1973. *Crop Micrometeorology: A Wide-Row Light Penetration. B. Carbon Dioxide Enrichment and Diffusion*. Dissertation Abstracts International, Vol. 33(12). Order No. 73-14, 716, 392pp.

Allen, L.H., Jr. 1990. Plant responses to rising carbon dioxide and potential interactions with air pollutants. *J. Environ. Qual.* 19: 15–34.

Allen, R.E. 1990. *The Concise Oxford Dictionary of Current English*, 8th edn. Oxford: Clarendon Press, 1,454pp.

Allen, L.H., Jr., P. Jones, and J.W. Jones. 1985. Rising atmospheric CO_2 and evapotranspiration. In *Advances in Evapotranspiration*. St. Joseph, MI: American Society of Agricultural Engineers, pp. 13–27.

Allen, L.H., Jr., K.J. Boote, J.W. Jones, P.H. Jones, R.R. Valle, B. Acock, H.H. Rogers, and R.C. Dahlman. 1987. Response of vegetation to rising carbon dioxide: Photosynthesis, biomass, and seed yield of soybean. *Global Biogeochem. Cycles* 1: 1–14.

Allen, L.H., Jr., R.R. Valle, J.W. Mishoe, and J.W. Jones. 1994. Soybean leaf gas-exchange responses to carbon dioxide and water stress. *Agron. J.* 86: 625–636.

Allen, R.G., L.S. Pereira, D. Raes, and M. Smith. 1998. *Crop Evapotranspiration. Guidelines for Computing Crop Water Requirements*. FAO Irrigation and Drainage Paper 56. Rome: Food and Agriculture Organization of the United Nations, 300pp.

Alonso, A., P. Pérez, and R. Martínez-Carrasco. 2009. Growth in elevated CO_2 enhances temperature response of photosynthesis in wheat. *Physiol. Plant.* 135: 109–120.

Altevogt, A.S. and M.A. Celia. 2004. Numerical modeling of carbon dioxide in unsaturated soils due to deep subsurface leakage. *Water Resources Res.* 40, doi:10.1029/2003WR002848, 2004 (9pp).

American Society of Agricultural Engineers. 1985. *Advances in Evapotranspiration. Proceedings of the National Conference on Advances in Evapotranspiration*, December 16–17, 1985, Chicago, IL. St. Joseph, MI: American Society of Agricultural Engineers, 453pp.

American Society of Agronomy, Crop Science Society of America, and Soil Science Society of America. 1999. *Root Growth in Soil*. Two films developed by the Agricultural Research Service in collaboration with Auburn University Television in the late 1960's, made available by a special program of the Southern Regional Information Exchange Group-25 and the Agronomic Science Foundation. "Cotton Root Growth," 15 minutes and 52 seconds; "Time-Lapse Photography of Root Growth," 14 minutes and 49 seconds. Madison, WI: American Society of Agronomy, Crop Science Society of America, and Soil Science Society of America.

Amthor, J.S. 2001. Effects of atmospheric CO_2 concentration on wheat yield: Review of results from experiments using various approaches to control CO_2 concentration. *Field Crops Res.* 73: 1–34.

André, M. and H. Du Cloux. 1993. Interaction of CO_2 enrichment and water limitations on photosynthesis and water efficiency in wheat. *Plant Physiol. Biochem.* 31: 103–112.

Andresen, L.C., A. Michelsen, S. Jonasson, I.K. Schmidt, T.N. Mikkelsen, P. Ambus, and C. Beier. 2010. Plant nutrient mobilization in temperate heathland responds to elevated CO_2, temperature and drought. *Plant Soil* 328: 381–396.

Angus, J.F. and M.W. Moncur. 1977. Water stress and phenology in wheat. *Aust. J. Agric. Res.* 28: 177–181.

Apel, P. 1989. Influence of CO_2 on stomatal number. *Biol. Plant.* 31: 72–74.

Argo, W.R., J.A. Biernbaum, and W.C. Fonteno. 1996. Root medium carbon dioxide and oxygen partial pressures for container-grown chrysanthemums. *HortScience* 31: 385–388.

Arnholdt-Schmitt, B. 2009. Alternative oxidase (AOX) and stress tolerance—approaching a scientific hypothesis: Introduction. *Physiol. Plant.* 137: 314–315. [Special Issue on Alternative Oxidase; Vol. 137, Issue No. 4, December 2009, pp. 313–608.]

Arnon, I. 1975. Physiological principles of dryland crop production. In Gupta, U.S., Ed., *Physiological Aspects of Dryland Farming*. Barking, Essex, U.K.: Applied Science Publishers, Ltd., pp. 3–145.

Arp, W.J. 1991. Effects of source-sink relations on photosynthetic acclimation to elevated CO_2. *Plant Cell Environ.* 14: 869–875.

Arp, W.J. and B.G. Drake. 1991. Increased photosynthetic capacity of *Scirpus olneyi* after 4 years of exposure to elevated CO_2. *Plant Cell Environ.* 14: 1003–1006.

Assmann, S.M. 1993. Signal transduction in guard cells. *Annu. Rev. Cell Biol.* 9: 345–375.

Assmann, S.M. and X.-Q. Wang. 2001. From milliseconds to millions of years: Guard cells and environmental responses. *Curr. Opin. Plant Biol.* 4: 421–428.

Baethgen, W.E. 2010. Climate risk management for adaptation to climate variability and change. *Crop Sci.* 50: S-70–S-76.

Bailey, L.H. 1974. *Manual of Cultivated Plants*, Revised edn., 14th printing. New York: Macmillan Publishing Company, 1,116pp.

Baker, D.N. and H.Z. Enoch. 1983. Plant growth and development. In Lemon, E.R., Ed., *CO_2 and Plants. The Response of Plants to Rising Levels of Atmospheric Carbon Dioxide*. AAAS Selected Symposium 84. Boulder, CO: Westview Press, Inc. and Washington, DC: American Association for the Advancement of Science, pp. 107–130.

Baker, J.T., and L.H. Allen, Jr. 1993. Contrasting crop species responses to CO_2 and temperature: Rice, soybean and citrus. In Rozema, J., Lambers, H., van de Geijn, S.C., and Cambridge, M.L., Eds., *CO_2 and Biosphere*. Dordrecht, the Netherlands: Kluwer Academic Publishers, pp. 239–260. [The book is reprinted from *Vegetatio* 104/105, 1993.]

Baker, J.T., L.H. Allen, Jr., K.J. Boote, P. Jones, and J.W. Jones. 1990a. Rice photosynthesis and evapotranspiration in subambient, ambient, and superambient carbon dioxide concentrations. *Agron. J.* 82: 834–840.

Baker, J.T., L.H. Allen, Jr., and K.J. Boote. 1990b. Growth and yield responses of rice to carbon dioxide concentration. *J. Agric. Sci.* 115: 313–320.

Baker, J.T., F. Laugel, K.J. Boote, and L.H. Allen, Jr. 1992. Effects of daytime carbon dioxide concentration on dark respiration in rice. *Plant Cell Environ.* 15: 231–239.

Barkley, T.M., Ed. 1977. *Atlas of the Flora of the Great Plains*. Ames, IA: The Iowa State University Press, 600pp.

Barnes, R.F. and J.B. Beard, Eds. 1992. *Glossary of Crop Science Terms*. Madison, WI: Crop Science Society of America, 90pp.

Baron, V., C.F. Shaykewich, and R.I. Hamilton. 1975. Relation of corn maturity to climatic parameters. *Can. J. Soil Sci.* 55: 343–347.

Barrett, A.J. and R.M. Gifford. 1995. Acclimation of photosynthesis and growth by cotton to elevated CO_2: Interactions with severe phosphate deficiency and restricted rooting volume. *Aust. J. Plant Physiol.* 22: 955–963.

Basile, G., M. Arienzo, and A. Zena. 1993. Soil nutrient mobility in response to irrigation with carbon dioxide enriched water. *Commun. Soil Sci. Plant Anal.* 24: 1183–1195.

BassiriRad, H., V.P. Gutschick, and J. Lussenhop. 2001. Root system adjustments: Regulation of plant nutrient uptake and growth responses to elevated CO_2. *Oecologia* 126: 305–320.

Baxter, R., T.W. Ashenden, T.H. Sparks, and J.F. Farrar. 1994. Effects of elevated carbon dioxide on three montane grass species. I. Growth and dry matter partitioning. *J. Exp. Bot.* 45: 305–315.

Bazzaz, F.A. 1990. The response of natural ecosystems to the rising global CO_2 levels. *Annu. Rev. Ecol. Syst.* 21: 167–196.

Bazzaz, F.A. and K. Garbutt. 1988. The response of annuals in competitive neighborhoods: Effects of elevated CO_2. *Ecology* 69: 937–946.

Bazzaz, F.A., K. Garbutt, and W.E. Williams. 1985a. Effect of increased atmospheric carbon dioxide concentration on plant communities. In Strain, B.R. and Cure, J.D., Eds., *Direct Effects of Increasing Carbon Dioxide on Vegetation*. DOE/ER-0238. Washington, DC: United States Department of Energy, Office of Energy Research, Office of Basic Energy Sciences, Carbon Dioxide Research Division, pp. 155–170.

Bazzaz, F.A., K. Garbutt, and W.E. Williams. 1985b. *The Effect of Elevated Atmospheric CO_2 on Plant Communities*. DOE/EV/04329-5. Washington, DC: United States Department of Energy, Office of Energy Research, Office of Basic Energy Sciences, Carbon Dioxide Research Division, 23pp+16pp, appendix.

Bazzaz, F.A., S.L. Miao, and P.M. Wayne. 1993. CO_2-induced growth enhancements of co-occurring tree species decline at different rates. *Oecologia* 96: 478–482.

Beadle, C.L.O., P.G. Jarvis, and R.E. Nielson. 1979. Leaf conductance as related to xylem water potential and carbon dioxide concentrations in Sitka spruce. *Physiol. Plant.* 45: 158–166.

Beerling, D.J. and C.K. Kelly. 1997. Stomatal density responses of temperate woodland plants over the past seven decades of CO_2 increase: A comparison of Salisbury (1927) with contemporary data. *Am. J. Bot.* 84: 1572–1583.

Beerling, D.J., C.P. Osborne, and W.G. Chaloner. 2001. Evolution of leaf-form in land plants linked to atmospheric CO_2 decline in the Late Palaeozoic era. *Nature* 410: 352–354.

Behboudian, M.H. and R. Lai. 1995. Partitioning of photoassimilates in "Virosa" tomatoes under elevated CO_2 concentration. *J. Plant Physiol.* 147: 43–47.

Ben-Asher, J., G.E. Cardon, D. Peters, D.E. Rolston, J.W. Biggar, C.J. Phene, and J.E. Ephrath. 1994. Determining root activity distribution by measuring surface carbon dioxide fluxes. *Soil Sci. Soc. Am. J.* 58: 926–930.

Bencze, S., O. Veisz, and Z. Bedő. 2004. Effects of high atmospheric CO_2 on the morphological and heading characteristics of winter wheat. *Cereal Res. Commun.* 32: 233–240.

Benson, A. 1997. Melvin Calvin. *ASPP News* 24(2): 6. Rockville, MD: American Society of Plant Physiologists (now called the American Society of Plant Biologists).

Bergmann, D.C., W. Lukowitz, and C.R. Somerville. 2004. Stomatal development and pattern controlled by a MAPKK kinase. *Science* 304: 1494–1497.

Bernacchi, C.J., A.D.B. Leakey, L.E. Heady, P.B. Morgan, F.G. Dohleman, J.M. McGrath, K.M. Gillespie, V.E. Wittig, A. Rogers, S.P. Long, and D.R. Ort. 2006. Hourly and seasonal variation in photosynthesis and stomatal conductance for 3 years under fully open-air field conditions. *Plant Cell Environ.* 29: 2077–2090.

Bernstein, L. 1961. Osmotic adjustment of plants to saline media. I. Steady state. *Am. J. Bot.* 48: 909–918.

Bernstein, L. 1963. Osmotic adjustment of plants to saline media. II. Dynamic phase. *Am. J. Bot.* 50: 360–370.

Berntson, G.M. and F.A. Bazzaz. 1996a. Belowground positive and negative feedbacks on CO_2 growth enhancement. *Plant Soil* 187: 119–131.

Berntson, G.M. and F.A. Bazzaz. 1996b. The allometry of root production and loss in seedlings of *Acer rubrum* (Aceraceae) and *Betula papyrifera* (Betulaceae): Implications for root dynamics in elevated CO_2. *Am. J. Bot.* 83: 608–616.

Berntson, G.M. and F.A. Bazzaz. 1997. Elevated CO_2 and the magnitude and seasonal dynamics of root production and loss in *Betula papyrifera*. *Plant Soil* 190: 211–216.

Berntson, G.M. and F.I. Woodward. 1992. The root system architecture and development of *Senecio vulgaris* in elevated CO_2 and drought. *Func. Ecol.* 6: 324–333.

Berntson, G.M., K.D.M. McConnaughay, and F.A. Bazzaz. 1993. Elevated CO_2 alters deployment of roots in "small" growth containers. *Oecologia* 94: 558–564.

Berryman, C.A., D. Eamus, and G.A. Duff. 1994. Stomatal responses to a range of variables in two tropical tree species grown with CO_2 enrichment. *J. Exp. Bot.* 45: 539–546.

Bhattacharya, S., N.C. Bhattacharya, and B.R. Strain. 1985. Rooting of sweet potato stem cuttings under CO_2-enriched environment with IAA treatment. *HortScience* 20: 1109–1110.

Bhupinderpal-Singh, M.J. Hedley, and S. Saggar. 2005. In situ dynamics of recently allocated ^{14}C in pasture soil and soil solution collected with Rhizon Soil Moisture Samplers. *Aust. J. Soil Res.* 43: 659–666.

Bierhuizen, J.F. and R.O. Slatyer. 1964. Photosynthesis of cotton leaves under a range of environmental conditions in relation to internal and external diffusive resistances. *Aust. J. Biol. Sci.* 17: 348–359.

Billes, G., H. Rouhier, and P. Bottner. 1993. Modifications of the carbon and nitrogen allocations in the plant (*Triticum aestivum* L.) soil system in response to increased atmospheric CO_2 concentration. *Plant Soil* 157: 215–225.

Bindi, M., F. Miglietta, F. Vaccari, E. Magliulo, and A. Giuntoli. 2006. Growth and quality responses of potato to elevated $[CO_2]$. In Nösberger, J., Long, S.P., Norby, R.J., Stitt, M., Hendrey, G.R., and Blum, H., Eds., *Managed Ecosystems and CO_2. Case Studies, Processes, and Perspectives.* Berlin: Springer-Verlag, pp. 105–119.

Biswas, P.K., D.R. Hileman, P.P. Ghosh, N.C. Bhattacharya, and J.N. McCrimmon. 1996. Growth and yield responses of field-grown sweetpotato to elevated carbon dioxide. *Crop Sci.* 36: 1234–1239.

Björkman, O., M. Nobs, R. Pearcy, J. Boynton, and J. Berry. 1971. Characteristics of hybrids between C_3 and C_4 species of *Atriplex*. In Hatch, M.D., Osmond, C.B., and Slatyer, R.O., Eds., *Photosynthesis and Photorespiration.* New York: Wiley-Interscience, pp. 105–119.

Black, C.C., Jr. 1986. Effects of CO_2 concentration on photosynthesis and respiration of C_4 and CAM plants. In Enoch, H.Z. and Kimball, B.A., Eds., *Carbon Dioxide Enrichment of Greenhouse Crops. Volume II. Physiology, Yield, and Economics*. Boca Raton, FL: CRC Press, Inc., pp. 29–40.

Blum, A., R. Munns, J.B. Passioura, and N.C. Turner. 1996. Genetically engineered plants resistant to soil drying and salt stress: How to interpret osmotic relations? *Plant Physiol.* 110: 1051–1053.

Boetsch, J., J. Chin, M. Ling, and J. Croxdale. 1996. Elevated carbon dioxide affects the patterning of subsidiary cells in *Tradescantia* stomatal complexes. *J. Exp. Bot.* 47: 925–931 (plus cover).

Boussingault and Léwy [no initials of first names given in original publication]. 1853. Mémoire sur la composition de l'air confine dans la terre végétale. *Ann. Chim. Phys.* 37(3rd series): 7–50.

Bowen, I.S. 1926. The ratio of heat losses by conduction and by evaporation from any water surface. *Phys. Rev.* 27: 779–787.

Boyer, J.S. 1982. Plant productivity and environment. *Science* 218: 443–448+cover.

Boynton, D. and W. Reuther. 1938. Seasonal variation of oxygen and carbon dioxide in three different orchard soils during 1938 and its possible significance. *Proc. Am. Soc. Hort. Sci.* 36: 1–6.

Brakke, M. and L.H. Allen, Jr. 1995. Gas exchange of *Citrus* seedlings at different temperatures, vapor-pressure deficits, and soil water contents. *J. Am. Soc. Hort. Sci.* 120: 497–504.

Bray, E.A., J. Bailey-Serres, and E. Weretilnyk. 2000. Responses to abiotic stresses. In Buchanan, B.B., Gruissem, W., and Jones, R.L., Eds., *Biochemistry and Molecular Biology of Plants*. Rockville, MD: American Society of Plant Physiologists, pp. 1158–1203.

Bremer, D.J., J.M. Ham, and C.E. Owensby. 1996. Effect of elevated atmospheric carbon dioxide and open-top chambers on transpiration in a tallgrass prairie. *J. Environ. Qual.* 25: 691–701.

Briggle, L.W. and B.C. Curtis. 1987. Wheat worldwide. In Heyne, E.G., Ed., *Wheat and Wheat Improvement*, 2nd edn. Madison, WI: American Society of Agronomy, Crop Science Society of America, and Soil Science Society of America, pp. 1–32.

Briggs, L.J. and H.L. Shantz. 1913a. *The Water Requirement of Plants. I. Investigations in the Great Plains in 1910 and 1911*. U.S. Department of Agriculture Bureau of Plant Industry Bulletin No. 284. Washington, DC: U.S. Department of Agriculture, 49pp.

Briggs, L.J. and H.L. Shantz. 1913b. *The Water Requirement of Plants. II. A Review of the Literature*. U.S. Department of Agriculture Bureau of Plant Industry Bulletin No. 285. Washington, DC: U.S. Department of Agriculture, 96pp.

Briggs, L.J. and H.L. Shantz. 1917. The water requirement of plants as influenced by environment. *Proceedings of the Second Pan-American Science Congress* (Washington, DC), Vol. 3, pp. 95–107 (cited by Tanner and Sinclair, 1983).

Brown, P. 2005. Climate Change Pioneer Scientist. Charles Keeling. *The Guardian Weekly*, 173(3), July 8–14, 2005, p. 29.

Brown, H.T. and F. Escombe. 1900. Static diffusion of gases and liquids in relation to the assimilation of carbon and translocation in plants. *Phil. Trans. Roy. Soc. London* 193B: 223–291.

Brown, A.L.P., F.P. Day, and D.B. Stover. 2009. Fine root biomass estimates from minirhizotron imagery in a shrub ecosystem exposed to elevated CO_2. *Plant Soil* 317: 145–153.

Brun, L.J., E.T. Kanemasu, and W.L. Powers. 1972. Evapotranspiration from soybean and sorghum fields. *Agron. J.* 64: 145–148.

Buckingham, E. 1904. *Contributions to Our Knowledge of the Aeration of Soils*. U.S. Department of Agriculture Bureau of Soils Bulletin 25. Washington, DC: U.S. Government Printing Office, 52pp.

Bunce, J.A. 1992. Stomatal conductance, photosynthesis and respiration of temperate deciduous tree seedlings grown outdoors at an elevated concentration of carbon dioxide. *Plant Cell Environ.* 15: 541–549.

Bunce, J.A. 1993. Effects of doubled atmospheric carbon dioxide concentration on the responses of assimilation and conductance to humidity. *Plant Cell Environ.* 16: 189–197.

Bunce, J.A. 1993. Growth, survival, competition, and canopy carbon dioxide and water vapor exchange of first year alfalfa at an elevated CO_2 concentration. *Photosynthetica* 29: 557–565.

Bunce, J.A. 2004. Carbon dioxide effects on stomatal responses to the environment and water use by crops under field conditions. *Oecologia* 140: 1–10.

Bunce, J.A. 2007. Low carbon dioxide concentrations can reverse stomatal closure during water stress. *Physiol. Plant.* 130: 552–559.

Bunce, J.A. 2009. Use of the response of photosynthesis to oxygen to estimate mesophyll conductance to carbon dioxide in water-stressed soybean leaves. *Plant Cell Environ.* 32: 875–881.

Burkart, S., R. Manderscheid, and H.-J. Weigel. 2004. Interactive effects of elevated atmospheric CO_2 concentrations and plant available soil water content on canopy evapotranspiration and conductance of spring wheat. *Eur. J. Agron.* 21: 401–417.

Burnett, R.B., U.N. Chaudhuri, E.T. Kanemasu, and M.B. Kirkham. 1985. Sorghum at elevated levels of CO_2. *Response of Vegetation to Carbon Dioxide*. Research Report No. 024. Washington, DC: U.S. Department of Energy, Carbon Dioxide Research Division, Office of Energy Research. ix+47pp.

Buttery, B.R., C.S. Tan, R.I. Buzzell, J.D. Gaynor, and D.C. MacTavish. 1993. Stomatal numbers of soybean and response to water stress. *Plant Soil* 149: 283–288.

Cabrera, M.L., T.R. Kelley, O.C. Pancorbo, W.C. Merka, and S.A. Thompson. 1994. Ammonia volatilization and carbon dioxide emission from poultry litter: Effects of fractionation and storage time. *Commun. Soil Sci. Plant Anal.* 25: 2341–2353.

Calvin, M. 1996. Following the trail of light. *ASPP News* 23(2): 5–6. Rockville, MD: American Society of Plant Physiologists (now called the American Society of Plant Biologists).

Cammack, R. 2006. *Oxford Dictionary of Biochemistry and Molecular Biology*, Revised edn. Oxford, U.K.: Oxford University Press, 720pp.

Campbell, G.S. 1977. *An Introduction to Environmental Biophysics*. New York: Springer-Verlag, 159pp.

Campbell, G.S., W.D. Zollinger, and S.A. Taylor. 1966. Sample changer for thermocouple psychrometers: Construction and some applications. *Agron. J.* 58: 315–318.

Cannon, W.A. and E.E. Free. 1925. *Physiological Features of Roots, With Especial Reference to the Relation of Roots to Aeration of the Soil by William Austin Cannon. With a Chapter on Differences Between Nitrogen and Helium as Inert Gases in Anaerobic Experiments on Plants by Edward Elway Free*. Carnegie Institution of Washington Publication No. 368. Lancaster, PA: Intelligencer Printing Company, 168pp.

Caprio, J.M. 1974. The solar thermal unit concept in problems related to plant development and potential evapotranspiraton. In Lieth, H., Ed., *Phenology and Seasonality Modeling*. New York: Springer-Verlag, pp. 353–364.

Carmo-Silva, A.E., A. Francisco, S.J. Powers, A.J. Keys, L. Ascensão, M.A.J. Parry, and M.C. Arrabaça. 2009. Grasses of different C_4 subtypes reveal leaf traits related to drought tolerance in their natural habitats: Changes in structure, water potential, and amino acid content. *Am. J. Bot.* 96: 1222–1235.

Castonguay, Y., G. Thibault, P. Rochette, A. Bertrand, S. Rochefort, and J. Dionne. 2009. Physiological responses of annual bluegrass and creeping bentgrass to contrasted levels of O_2 and CO_2 at low temperatures. *Crop Sci.* 49: 671–689.

Cattell, J., Ed. 1955. *American Men of Science. A Biographical Directory*. Lancaster, PA: The Science Press, and New York, NY: R.R. Bowker Company, 2,180pp.

Causey, D., D.H. Janzen, A. Townsed Peterson, D. Vieglais, L. Krishtalka, J.H. Beach, and E.O. Wiley. 2004. Museum collections and taxonomy. *Science* 305: 1106–1107.

Centritto, M. 2005. Photosynthetic limitations and carbon partitioning in cherry in response to water deficit and elevated [CO_2]. *Agric. Ecosyst. Environ.* 106: 233–242.

Centritto, M., F. Magnani, H.S.J. Lee, and P.G. Jarvis. 1999a. Interactive effects of elevated [CO_2] and drought on cherry (*Prunus avium*) seedlings. II. Photosynthetic capacity and water relations. *New Phytol.* 141: 141–153.

Centritto, M., H.S.J. Lee, and P.G. Jarvis. 1999b. Interactive effects of elevated [CO_2] and drought on cherry (*Prunus avium*) seedlings. I. Growth, whole-plant water use efficiency and water loss. *New Phytol.* 141: 129–140.

Centritto, M., H.S.J. Lee, and P.G. Jarvis. 1999c. Long-term effects of elevated carbon dioxide concentration and provenance on four clones of Sitka spruce (*Picea sitchensis*). I. Plant growth, allocation and ontogeny. *Tree Physiol.* 19: 799–806.

Centritto, M., H.S.J. Lee, and P.G. Jarvis. 1999d. Increased growth in elevated [CO_2]: An early, short-term response? *Global Change Biol.* 5: 623–633.

Centritto, M., M.E. Lucas, and P.G. Jarvis. 2002. Gas exchange, biomass, and whole-plant water-use efficiency and water uptake of peach (*Prunus persica*) seedlings in response to elevated carbon dioxide concentration and water availability. *Tree Physiol.* 22: 699–706.

Centro Internacional de Mejoramient de Maíz y Trigo. 1975. *CIMMYT Report on Wheat Improvement 1975*. El Batan, Mexico: International Maize and Wheat Improvement Center, 207pp.

Ceulemans, R. and M. Mousseau. 1994. Effects of elevated atmospheric CO_2 on woody plants. Tansley Review No. 71. *New Phytol.* 127: 425–446.

Chalmers, N., D. Hawksworth, D.S. Ingram, G. Long, G.T. Prance, P.H. Raven, and L.E. Skog. 1990. In defence of taxonomy. *Nature* 347: 224.

Chang, J.-H. 1968. *Climate and Agriculture. An Ecological Survey*. Chicago, IL: Aldine Publishing Company, 304pp.

Chapin, F.S. III and R.W. Ruess. 2001. Carbon cycle. The roots of the matter. *Nature* 411: 749, 751, 752.

Chasan, R. 1995. New openings into stomata. *Plant Cell* 7: 1113–1115.

Chaudhuri, U.N., R.B. Burnett, M.B. Kirkham, and E.T. Kanemasu. 1986a. Effect of carbon dioxide on sorghum yield, root growth, and water use. *Agric. Forest Meteorol.* 37: 109–122.

Chaudhuri, U.N., R.B. Burnett, E.T. Kanemasu, and M.B. Kirkham. 1986b. Effect of elevated levels of CO_2 on winter wheat under two moisture regimes. *Response of Vegetation to Carbon Dioxide.* Research Report No. 029. Washington, DC: U.S. Department of Energy, Carbon Dioxide Research Division, Office of Energy Research. xii+77pp.

Chaudhuri, U.N., R.B. Burnett, E.T. Kanemasu, and M.B. Kirkham. 1987. Effect of elevated levels of CO_2 on winter wheat under two moisture regimes. *Response of Vegetation to Carbon Dioxide.* Research Report No. 040. Washington, DC: U.S. Department of Energy, Carbon Dioxide Research Division, Office of Energy Research. xiv+70pp.

Chaudhuri, U.N., E.T. Kanemasu, and M.B. Kirkham. 1989. Effect of elevated levels of CO_2 on winter wheat under two moisture regimes. *Response of Vegetation to Carbon Dioxide.* Research Report No. 050. Washington, DC: U.S. Department of Energy, Carbon Dioxide Research Division, Office of Energy Research, ix+49pp.

Chaudhuri, U.N., M.B. Kirkham, and E.T. Kanemasu. 1990a. Carbon dioxide and water level effects on yield and water use of winter wheat. *Agron. J.* 82: 637–641.

Chaudhuri, U.N., M.B. Kirkham, and E.T. Kanemasu. 1990b. Root growth of winter wheat under elevated carbon dioxide and drought. *Crop Sci.* 30: 853–857.

Chen, X.M., G.B. Begonia, D.M. Alm, and J.D. Hesketh. 1993. Responses of soybean leaf photosynthesis to CO_2 and drought. *Photosynthetica* 29: 447–454.

Chen, X.M., G.B. Begonia, and J.D. Hesketh. 1995. Soybean stomatal acclimation to long-term exposure to CO_2-enriched atmospheres. *Photosynthetica* 31: 51–57.

Chen, L.-Q., C.-S. Li, W.G. Chaloner, D.J. Beerling, Q.-G. Sun, M.E. Collinson, and P.L. Mitchell. 2001. Assessing the potential for the stomatal characters of extant and fossil *Ginkgo* leaves to signal atmospheric CO_2 change. *Am. J. Bot.* 88: 1309–1315.

Chinoy, J.J. and K.K. Nanda. 1951. Effect of vernalization and photoperiodic treatments on growth and development of crop plants. II. Varietal differences in stem elongation and tillering of wheat and their correlation with flowering under varying photoinductive and post-photoinductive treatments. *Physiol. Plant.* 4: 427–436.

Chouard, P. 1960. Vernalization and its relation to dormancy. *Annu. Rev. Plant Physiol.* 11: 191–238.

Clark, H., P.C.D. Newton, C.C. Bell, and E.M. Glasgow. 1997. Dry matter yield, leaf growth and population dynamics in *Lolium perenne*/*Trifolium repens*-dominated pasture turves exposed to two levels of elevated CO_2. *J. Appl. Ecol.* 34: 304–316.

Clifford, H.T., R.W. Rogers, and M.E. Dettmann. 1990. Where now for taxonomy? *Nature* 346: 602.

Cohen, D. 1971. Maximizing final yield when growth is limited by time or by limiting resources. *J. Theoret. Biol.* 33: 299–307.

Coleman, J.S. and F.A. Bazzaz. 1992. Effects of CO_2 and temperature on growth and resource use of co-occurring C_3 and C_4 annuals. *Ecology* 73: 1244–1259.

Combe, L. 1981. Effect of carbon dioxide and culture under artificial climate on the growth and yield of winter wheat. *Agronomie* 1: 177–186 (in French, English summary).

Conn, E.E. and P.F. Stumpf. 1963. *Outlines of Biochemistry.* New York: John Wiley & Sons, 391pp.

Cosgrove, D.J., E. Van Volkenburgh, and R.E. Cleland. 1984. Stress relaxation of cell walls and yield threshold for growth. Demonstration and measurement by micro-pressure probe and psychrometer techniques. *Planta* 1652: 46–54.

Cramer, M.D. and S.H. Lips. 1995. Enriched rhizosphere CO_2 concentrations can ameliorate the influence of salinity on hydroponically grown tomato plants. *Physiol. Plant.* 94: 425–432.

Crawford, R.M.M. and D.W. Wolfe. 1999. Temperature: Cellular to whole-plant and population responses. In Luo, Y. and Mooney, H.A., Eds., *Carbon Dioxide and Environmental Stress.* San Diego, CA: Academic Press, pp. 61–106.

Crocker, W. and L.I. Knight. 1908. Effect of illuminating gas and ethylene upon flowering carnations. *Bot. Gaz.* 46: 259–276.

Cronquist, A. 1981. *An Integrated System of Classification of Flowering Plants.* New York: Columbia University Press, 1,152pp.

Cui, M., P.M. Miller, and P.S. Nobel. 1993. CO_2 exchange and growth of the Crassulacean acid metabolism plant *Opuntia ficus-indica* under elevated CO_2 in open- top chambers. *Plant Physiol.* 103: 519–524.

Cummings, M.B. and C.H. Jones. 1918. *The Aerial Fertilization of Plants with Carbon Dioxide.* Bulletin No. 211. Burlington, VT: Vermont Agricultural Experiment Station (cited by Rogers et al., 1997).

Cure, J.D. 1985. Carbon dioxide doubling responses: A crop survey. In Strain, B.R. and Cure, J.D., Eds., *Direct Effects of Increasing Carbon Dioxide on Vegetation.* DOE/ER-0238. Washington, DC: United States Department of Energy, Office of Energy Research, Office of Basic Energy Sciences, Carbon Dioxide Research Division, pp. 99–116.

Cure, J.D. and B. Acock. 1986. Crop responses to carbon dioxide doubling: A literature survey. *Agric. Forest Meteorol.* 38: 127–145.

Curtis, P.S., L.M. Balduman, B.G. Drake, and D.F. Whigham. 1990. Elevated atmospheric CO_2 effects on belowground processes in C_3 and C_4 estuarine marsh communities. *Ecology* 71: 2001–2006.

Curtis, P.S., A.A. Snow, and A.S. Miller. 1994. Genotype-specific effects of elevated CO_2 on fecundity in wild radish (*Raphanus raphanistrum*). *Oecologia* 97: 100–105.

da Costa Andrade, E.N. and R.E. Oesper. 1971. Arrhenius, Svante August. *Encycl. Brit.* 2: 474–475.

Dai, Z., M.S.B. Ku, and G.E. Edwards. 1993. C_4 photosynthesis. The CO_2-concentration mechanism and photorespiration. *Plant Physiol.* 103: 83–90.

Dalton, R. 1999. U.S. universities find that demand for botanists exceeds supply. *Nature* 402: 109–110.

Dalton, R. 2007. Endangered collections. *Nature* 446: 605–606.

Daniels, F. and R.A. Alberty. 1966. *Physical Chemistry*, 3rd edn. New York: John Wiley & Sons, 767pp.

Daniels-Lake, B.J. and R.K. Prange. 2009. The interaction effect of carbon dioxide and ethylene in the storage atmosphere on potato fry color is dose-related. *HortScience* 44: 1641–1644.

Darwin, C. (assisted by F. Darwin). 1897. *The Power of Movement in Plants*. New York: D. Appleton and Company, 592pp.

Darwin, F. 1897. Observations on stomata by a new method. *Proc. Cambridge Phil. Soc.* 9: 303–308.

Darwin, F. 1898. Observations on stomata. *Phil. Trans. Roy. Soc. London* 190B: 532–621.

Darwin, F. and E.H. Acton. 1909. Stomatal transpiration observed with the horn-hygroscope. Section 118. In *Practical Physiology of Plants*, 3rd edn. London: Cambridge University Press, pp. 102–104.

Darwinkel, A., B.A. ten Hag, and J. Kuizenga. 1977. Effect of sowing date and seed rate on crop development and grain production of winter wheat. *Neth. J. Agric. Sci.* 25: 83–94.

Davies, W.J. and J. Zhang. 1991. Root signals and the regulation of growth and development of plants in drying soil. *Annu. Rev. Plant Physiol. Plant Mol. Biol.* 42: 55–76.

Davis, T.D. and J.R. Potter. 1983. High CO_2 applied to cuttings: Effect on rooting and subsequent growth in ornamental species. *HortScience* 18: 194–196.

Davis, T.D. and J.R. Potter. 1989. Relations between carbohydrate, water status and adventitious root formation in leafy pea cuttings rooted under various levels of atmospheric CO_2 and relative humidity. *Physiol. Plant.* 77: 185–190.

de Willigen, P., M. Heinen, and M.B. Kirkham. 2005. Transpiration and root water uptake. In Anderson, M.G., Ed., *Encyclopedia of Hydrological Sciences*. Chichester, U.K.: John Wiley & Sons, pp. 1055–1068.

de Wit, C.T. 1958. *Transpiration and Crop Yields*. Verslag Landbouwkundig Onderzoek No. 64.6. Wageningen, the Netherlands: Institute of Biological and Chemical Research on Field Crops and Herbage, 88pp.

Decagon Devices. 2006. *Leaf Porometer: Operator's Manual*, Version 1.0. Pullman, WA: Decagon Devices, 66pp.

Del Castillo, D., B. Acock, V.R. Reddy, and M.C. Acock. 1989. Elongation and branching of roots on soybean plants in a carbon dioxide-enriched aerial environment. *Agron. J.* 81: 692–695.

Dengler, N.G., R.E. Dengler, P.M. Donnelly, and P.W. Hattersley. 1994. Quantitative leaf anatomy of C_3 and C_4 grasses (Poaceae): Bundle sheath and mesophyll surface area relationships. *Ann. Bot.* 73: 241–255. (C_3 and C_4 are written in the title without subscripts.)

Derner, J.D., C.R. Tischler, H.W. Polley, and H.B. Johnson. 2004. Intergenerational above- and belowground responses of spring wheat (*Triticum aestivum* L.) to elevated CO_2. *Basic Appl. Ecol.* 5: 145–152.

Diamond, J. 2005. Obituary. Ernst Mayr (1904–2005). *Nature* 433: 700–701.

Diemer, M.W. 1994. Mid-season gas exchange of an alpine grassland under elevated CO_2. *Oecologia* 98: 429–435.

Dijkstra, P., S. Nonhebel, C. Grashoff, J. Goudriaan, and S.C. van de Geijn. 1996. Response of growth and CO_2 uptake of spring wheat and faba bean to CO_2 concentration under semifield conditions: Comparing results of field experiments and simulations. In Koch, G.W. and Mooney H.A., Eds., *Carbon Dioxide and Terrestrial Ecosystems*. San Diego, CA: Academic Press, pp. 251–264.

Dinsmore, K.J., U.M. Skiba, M.F. Billett, and R.M. Rees. 2009. Effect of water table on greenhouse gas emissions from peatland mesocosms. *Plant Soil* 318: 229–242.

Dippery, J.K., D.T. Tissue, R.B. Thomas, and B.R. Strain. 1995. Effects of low and elevated CO_2 on C_3 and C_4 annuals. I. Growth and biomass allocation. *Oecologia* 101: 13–20.

Dlugokencky, E.J. 2009. Carbon dioxide [in "State of the Climate in 2008"]. *Bull. Am. Meteorol. Soc.* 90(8): S34–S35.

Dohleman, F.G. and S.P. Long. 2009. More productive than maize in the Midwest: How does Miscanthus do it? *Plant Physiol.* 150: 2104–2115.

Doi, M. and K.-i. Shimazaki. 2008. The stomata of the fern *Adiantum capillus-veneris* do not respond to CO_2 in the dark and open by photosynthesis in guard cells. *Plant Physiol.* 147: 922–930.

Doley, D. 2004. Evolution of plant water relations research in Australia. *Func. Plant Biol.* 31: 405–414.

Domec, J.-C., S. Palmroth, E. Ward, C.A. Maier, M. Thérézien, and R. Oren. 2009. Acclimation of leaf hydraulic conductance and stomatal conductance of *Pinus taeda* (loblolly pine) to long-term growth in elevated CO_2 (free-air CO_2 enrichment) and N-fertilization. *Plant Cell Environ.* 32: 1500–1512.

Downtown, W.J.S. and W.J.R. Grant. 1994. Photosynthetic and growth responses of variegated ornamental species to elevated CO_2. *Aust. J. Plant Physiol.* 21: 273–279.

Drake, B.G. and P.W. Leadley. 1991. Canopy photosynthesis of crops and native plant communities exposed to long-term elevated CO_2. *Plant Cell Environ.* 14: 853–860.

Drake, B.G., M.A. Gonzàlez-Meler, and S.P. Long. 1997. More efficient plants: A consequence of rising atmospheric CO_2? *Annu. Rev. Plant Physiol. Plant Mol. Biol.* 48: 609–639.

Drennan, P.M. and P.S. Nobel. 1996. Temperature influences on root growth for *Encelia farinosa* (Asteraceae), *Pleuraphis rigida* (Poaceae), and *Agave deserti* (Agavaceae) under current and doubled CO_2 concentrations. *Am. J. Bot.* 83: 133–139.

Du Cloux, H.C., M. André, A. Daguenet, and J. Massimino. 1987. Wheat response to CO_2 enrichment: Growth and CO_2 exchanges at two plant densities. *J. Exp. Bot.* 38: 1421–1431.

Dueñas, C., M.C. Fernández, J. Carretero, E. Liger, and M. Pérez. 1995. Emissions of CO_2 from some soils. *Chemosphere* 30: 1875–1889.

Dugas, W.A., S.A. Prior, and H.H. Rogers. 1997. Transpiration from sorghum and soybean growing under ambient and elevated CO_2 concentrations. *Agric. Forest Meteorol.* 83: 37–48.

Duveen, D.I. 1971. Lavoisier, Antoine Laurent. *Encycl. Brit.* 13: 818–819.

Eamus, D. 1991. The interaction of rising CO_2 and temperatures with water use efficiency. *Plant Cell Environ.* 14: 843–852.

Eamus, D., C.A. Berryman, and G.A. Duff. 1995. The impact of CO_2 enrichment on water relations in *Maranthes corymbosa* and *Eucalyptus tetrodonta*. *Aust. J. Bot.* 43: 273–282.

Ebach, M.C. and C. Holdrege. 2005. DNA barcoding is no substitute for taxonomy. *Nature* 434: 697.

Eckardt, N.A. 2007. Measuring daylength: Pharbitis takes a different approach. *Plant Cell* 19: 2968–2969.

Eckardt, N.A. 2008a. Aquaporins and chloroplast membrane permeability. *Plant Cell* 20: 499.

Eckardt, N.A. 2008b. Grass genome evolution. *Plant Cell* 20: 3–4.

Eckardt, N.A. 2010. Evolution of domesticated bread wheat. *Plant Cell* 22: 993.

Edwards, E.J., C.P. Osborne, C.A.E. Strömberg, S.A. Smith, and the C_4 Grasses Consortium [E.J. Edwards, C.P. Osborne, C.A.E. Strömberg, S.A. Smith, W.J. Bond, P.-A. Christin, A.B. Cousins, M.R. Duvall, D.L. Fox, R.P. Freckleton, O. Ghannoum, J. Hartwell, Y. Huang, C.M. Janis, J.E. Keeley, E.A. Kellogg, A.K. Knapp, A.D.B. Leakey, D.M. Nelson, J.M. Saarela, R.F. Sage, O.E. Sala, N. Salamin, C.J. Still, and B. Tipple]. 2010. The origins of C_4 grasslands: Integrating evolutionary and ecosystem science. *Science* 328: 587–591.

Edwards, G.R. and P.C.D. Newton. 2007. Plant performance and implications for plant population dynamics and species composition in a changing climate. In Newton, P.C.D., Carran, R.A., Edwards, G.R., and Niklaus, P.A., Eds., *Agroecosystems in a Changing Climate*. Boca Raton, FL: CRC, Taylor & Francis Group, pp. 189–210.

Ehleringer, J.R., R.F. Sage, L.B. Flanagan, and R.W. Pearcy. 1991. Climate change and the evolution of C_4 photosynthesis. *Trends Ecol. Evol.* 6: 95–99.

Eissa, A.-G.M. and D.N. Baker. Undated. *Effect of Carbon Dioxide Enrichment and Temperature on Wheat Growth*. Unpublished Report, 91pp. (Obtained from USDA, ARS, Crop Simulation Research Unit, Mississippi State, Mississippi 39762. The report probably was done in the late 1970s.)

El Kohen, A., L. Venet, and M. Mousseau. 1993. Growth and photosynthesis of two deciduous forest species at elevated carbon dioxide. *Funct. Ecol.* 7: 480–486.

Ellsworth, D.S. 1999. CO_2 enrichment in a maturing pine forest: Are CO_2 exchange and water status in the canopy affected? *Plant Cell Environ.* 22: 461–472.

El-Sharkawy, M.A. 2006. International research on cassava photosynthesis, productivity, eco-physiology, and responses to environmental stresses in the tropics. *Photosynthetica* 44: 481–512.

El-Sharkawy, M.A. 2007. Physiological characteristics of cassava tolerance to prolonged drought in the tropics: Implications for breeding cultivars adapted to seasonally dry and semiarid environments. *Braz. J. Plant Physiol.* 19: 257–286.

El-Sharkawy, M.A. 2009a. Pioneering research on C_4 leaf anatomical, physiological, and agronomic characteristics of tropical monocot and dicot plant species: Implications for crop water relations and productivity in comparison to C_3 cropping systems. *Photosynthetica* 47: 163–183.

El-Sharkawy, M.A. 2009b. Pioneering research on C_4 photosynthesis: Implications for crop water relations and productivity in comparison to C_3 cropping systems. *J. Food Agric. Environ.* 7: 132–148.

El-Sharkawy, M. and J. Hesketh. 1965. Photosynthesis among species in relation to characteristics of leaf anatomy and CO_2 diffusion resistances. *Crop Sci.* 5: 517–521.

El-Sharkawy, M.A. and S.M. De Tafur. 2010. Comparative photosynthesis, growth, productivity, and nutrient use efficiency among tall- and short-stemmed rain-fed cassava cultivars. *Photosynthetica* 48: 173–188.

Encylopaedia Britannica (author anonymous). 1971a. Boussingault, Jean Baptiste. *Encycl. Brit.* 4: 27.

Encyclopaedia Britannica (author anonymous). 1971b. Mangrove. *Encycl. Brit.* 14: 781–782.

Engel, V.C., K.L. Griffin, R. Murthy, L. Patterson, C. Klimas, and M. Potosnak. 2004. Growth CO_2 concentration modifies the transpiration response of *Populus deltoides* to drought and vapor pressure deficit. *Tree Physiol.* 24: 1137–1145.

Enoch, H.Z. and B.A. Kimball, Eds. 1986a. *Carbon Dioxide Enrichment of Greenhouse Crops. Volume I. Status and CO_2 Sources.* Boca Raton, FL: CRC Press, 181pp.

Enoch, H.Z. and B.A. Kimball, Eds. 1986b. *Carbon Dioxide Enrichment of Greenhouse Crops. Volume II. Physiology, Yield, and Economics.* Boca Raton, FL: CRC Press, 230pp.

Epstein, P.R. 2005. Climate change and human health. *New Engl. J. Med.* 353: 1433–1436 (October 6, 2005, issue).

Epstein, P.R., H.F. Diaz, S. Elias, G. Grabherr, N.E. Graham, W.J.M. Martens, E. Mosley-Thompson, and J. Susskind. 1998. Biological and physical signs of climate change: Focus on mosquito-borne diseases. *Bull. Am. Meteorol. Soc.* 79: 409–417.

Esau, K. 1965. *Plant Anatomy*, 2nd edn. New York: John Wiley & Sons, 767pp.

Esau, K. 1977. *Anatomy of Seed Plants*, 2nd edn. New York: John Wiley & Sons, 550pp.

Estiarte, M., J. Peñuelas, B.A. Kimball, S.B. Idso, R.L. LaMorte, P.J. Pinter, Jr., G.W. Wall, and R.L. Garcia. 1994. Elevated CO_2 effects on stomatal density of wheat and sour orange trees. *J. Exp. Bot.* 45: 1665–1668.

Fahn, A. and D.F. Cutler. 1992. *Xerophytes: Encyclopedia of Plant Anatomy*, Vol. 13, Part 3. Berlin: Gebrüder Brontraeger, 176pp.

Fang, Y., P. Gundersen, W. Zhang, G. Zhou, J. Riis Christiansen, J. Mo, S. Dong, and T. Zhang. 2009. Soil-atmosphere exchange of N_2O, CO_2 and CH_4 along a slope of an evergreen broad-leaved forest in southern China. *Plant Soil* 319: 37–48.

Farquhar, G.D. 1983. On the nature of carbon isotope discrimination in C_4 species. *Aust. J. Plant Physiol.* 10: 205–226.

Farquhar, G.D. and R.A. Richards. 1984. Isotopic composition of plant carbon correlates with water-use efficiency of wheat genotypes. *Aust. J. Plant Physiol.* 11: 539–552.

Farquhar, G.D., M.H. O'Leary, and J.A. Berry. 1982. On the relationship between carbon isotope discrimination and the intercellular carbon dioxide concentration in leaves. *Aust. J. Plant Physiol.* 9: 121–137.

Farquhar, G.D., J.R. Ehleringer, and K.T. Hubick. 1989. Carbon isotope discrimination and photosynthesis. *Annu. Rev. Plant Physiol. Plant Mol. Biol.* 90: 503–537.

Farrar, J.F. and M.L. Williams. 1991. The effects of increased atmospheric carbon dioxide and temperature on carbon partitioning, source-sink relations and respiration. *Plant Cell Environ.* 14: 819–830.

Farrar, C.D., M.L. Sorey, W.C. Evans, J.F. Howle, B.D. Kerr, B.M. Kennedy, C.-Y. King, and J.R. Southon. 1995. Forest-killing diffuse CO_2 emission at Mammoth Mountain as a sign of magmatic unrest. *Nature* 376: 675–678.

Fehr, W.R. and C.E. Caviness. 1977. *Stages of Soybean Development.* Special Report 80. Ames, IA: Iowa State University Cooperative Extension Service Agriculture and Home Economics Experiment Station, 11pp.

Fernald, M.L. 1950. *Gray's Manual of Botany*, 8th edn. New York: American Book Company, 1,632pp.

Ferretti, D.F., E. Pendall, J.A. Morgan, J.A. Nelson, D. LeCain, and A.R. Mosier. 2003. Partitioning evapotranspiration fluxes from a Colorado grassland using stable isotopes: Seasonable variations and ecosystem implications of elevated atmospheric CO_2. *Plant Soil* 254: 291–303.

Ferris, R. and G. Taylor. 1994a. Elevated CO_2, water relations and biophysics of leaf extension in four chalk grassland herbs. *New Phytol.* 127: 297–307.

Ferris, R. and G. Taylor. 1994b. Increased root growth in elevated CO_2: A biophysical analysis of root cell elongation. *J. Exp. Bot.* 45: 1603–1612.

Feyerherm, A.M. 1977. *Responses of Winter and Spring Wheat Grain Yields to Meteorological Variation.* Final Report Contract NAS9-14282. Houston, TX: National Aeronautics and Space Administration, Johnson Space Center, 64pp + 5 appendices.

Finney, K.F., W.T. Yamazaki, V.L. Youngs, and G.L. Rubenthaler. 1987. Quality of hard, soft, and durum wheats. In Heyne, E.G., Ed., *Wheat and Wheat Improvement,* 2nd edn. Madison, WI: American Society of Agronomy, Crop Science Society of America, and Soil Science Society of America, pp. 677–748.

Fischer, R.A. 1968. Stomatal opening: Role of potassium uptake by guard cells. *Science* 160: 784–785.

Fischer, R.A. and M.I. Aguilar. 1976. Yield potential in a dwarf spring wheat and the effect of carbon dioxide fertilization. *Agron. J.* 68: 749–752.

Fischer, R.A. and N.C. Turner. 1978. Plant productivity in the arid and semiarid zones. *Annu. Rev. Plant Physiol.* 29: 277–317.

Fischer, R.A., T.C. Hsiao, and R.M. Hagan. 1970. After-effect of water stress on stomatal opening potential. I. Techniques and magnitudes. *J. Exp. Bot.* 21: 371–385.

Fisher, M.J., I.M. Rao, M.A. Ayarza, C.E. Lascano, J.I. Sanz, R.J. Thomas, and R.R. Vera. 1994. Carbon storage by introduced deep-rooted grasses in the South American savannas. *Nature* 371: 236–238.

Fitter, A.H., G.K. Self, J. Wolfenden, M.M.I. van Vuuren, T.K. Brown, L. Williamson, J.D. Graves, and D. Robinson. 1996. Root production and mortality under elevated atmospheric carbon dioxide. *Plant Soil* 187: 299–307.

Fitzpatrick, E.A. 1973. A comparison of simple climatological parameters for estimating phasic development of wheat in Western Australia. In Slatyer, R.O., Ed., *Plant Response to Climatic Factors.* Paris: United National Educational, Scientific, and Cultural Organization, pp. 389–397.

Fitzpatrick, E.A. and H.A. Nix. 1969. A model for simulating soil water regime in alternating fallow-crop systems. *Agric. Meteorol.* 6: 303–319.

Fonti, P. and I. García-González. 2004. Suitability of chestnut earlywood vessel chronologies for ecological studies. *New Phytol.* 163: 77–86.

Fortin, M.-C., P. Rochette, and E. Pattey. 1996. Soil carbon dioxide fluxes from conventional and no-tillage small-grain cropping systems. *Soil Sci. Soc. Am. J.* 60: 1541–1547.

Foster, A.S. and E.M. Gifford, Jr. 1959. *Comparative Morphology of Vascular Plants.* San Francisco, CA: W.H. Freeman and Company, 555pp.

Frechilla, S., L.D. Talbott, and E. Zeiger. 2002. The CO_2 response of *Vicia* guard cells acclimates to growth environment. *J. Exp. Bot.* 53: 545–550.

Freidel, F. 1971. Depression, New Deal, and World War II. United States of America. *Encycl. Brit.* 22: 673–685.

Friend, H.J. and D.B. Guralnik, Eds. 1959. *Webster's New World Dictionary of the American Language.* Cleveland, OH: The World Publishing Company, 1,724pp.

Frommer, W.B. 2010. CO_2mmon sense. *Science* 327: 275–276.

Fujino, M. 1967. Adenosinetriphosphate and adenosinetriphosphatase in stomatal movement. *Sci. Bull. Fac. Educ. Nagasaki Univ.* 18: 1–47.

Furr, J.R. and W.W. Aldrich. 1943. Oxygen and carbon-dioxide changes in the soil atmosphere of an irrigated date garden on calcareous very fine sandy loam soil. *Proc. Am. Soc. Hort. Sci.* 42: 46–52.

Garbutt, K., W.E. Williams, and F.A. Bazzaz. 1990. Analysis of the differential response of five annuals to elevated CO_2 during growth. *Ecology* 71: 1185–1194.

Garcia, R.L. and J.J. Hanway. 1976. Foliar fertilization of soybeans during the seed-filling period. *Agron. J.* 68: 653–657.

Garcia, R.L., S.B. Idso, G.W. Wall, and B.A. Kimball. 1994. Changes in net photosynthesis and growth of *Pinus eldarica* seedlings in response to atmospheric CO_2 enrichment. *Plant Cell Environ.* 17: 971–978.

Gardner, W.R. 1973. Internal water status and plant response in relation to the external water régime. In Slatyer, R.O., Ed., *Plant Response to Climatic Factors.* Paris, France: United Nations Educational, Scientific and Cultural Organization, pp. 221–225.

Gardner, W.H. 1977. Historical highlights in American soil physics, 1776–1976. *Soil Sci. Soc. Am. J.* 41: 221–229.

Garratt, J. 1999. Necrologies. Charles H.B. Priestley 1915–1998. *Bull. Am. Meteorol. Soc.* 80: 314, 316.

Garten, C.T., Jr., A.T. Classen, and R.J. Norby. 2009. Soil moisture surpasses elevated CO_2 and temperature as a control on soil carbon dynamics in a multi-factor climate change experiment. *Plant Soil* 319: 85–94.

Gay, A.P. and B. Hauck. 1994. Acclimation of *Lolium temulentum* to enhanced carbon dioxide concentration. *J. Exp. Bot.* 45: 1133–1141.

Gegas, V.C., A. Nazari, S. Griffiths, J. Simmonds, L. Fish, S. Orford, L. Sayers, J.H. Doonan, and J.W. Snape. 2010. A genetic framework for grain size and shape variation in wheat. *Plant Cell* 22: 1046–1056.

Ghannoum, O., S. von Caemmerer, E.W.R. Barlow, and J.P. Conroy. 1997. The effect of CO_2 enrichment and irradiance on the growth, morphology and gas exchange of a C_3 (*Panicum laxum*) and a C_4 (*Panicum antidotale*) grass. *Aust. J. Plant Physiol.* 24: 227–237.

Giese, A.C. 1962. *Cell Physiology*, 2nd edn. Philadelphia, PA: W.B. Saunders Company, 592pp.

Gifford, R.M. 1977. Growth pattern, carbon dioxide exchange and dry weight distribution in wheat growing under differing photosynthetic environments. *Aust. J. Plant Physiol.* 4: 99–110.

Gill, W.R. and R.D. Miller. 1956. A method for study of the influence of mechanical impedance and aeration on the growth of seedling roots. *Soil Sci. Soc. Am. Proc.* 20: 154–157.

Gingrich, J.R. and M.B. Russell. 1956. Effect of soil moisture tension and oxygen concentration on the growth of corn roots. *Agron. J.* 48: 517–520.

Gitz, D.C. and J.T. Baker. 2009. Methods for creating stomatal impressions directly onto archivable slides. *Agron. J.* 101: 232–236.

Gleick, P.H. 1987. Regional hydrologic consequences of increases in atmospheric CO_2 and other trace gases. *Climatic Change* 10: 137–161.

Glickman, T.S. 2000. *Glossary of Meteorology*, 2nd edn. Boston, MA: American Meteorological Society, 855pp.

Gliński, J. and W. Stępniewski. 1985. *Soil Aeration and Its Role for Plants*. Boca Raton, FL: CRC Press, 229pp.

Gomez-Roldan, V., S. Fermas, P.B. Brewer, V. Puech-Pagès, E.A. Dun, J.-P. Pillot, F. Letisse et al. 2008. Strigolactone inhibition of shoot branching. *Nature* 455: 189–194 (plus cover).

Gonzàlez-Meler, M.A., E. Blanc-Betes, C.E. Flower, J.K. Ward, and N. Gomez-Casanovas. 2009. Plastic and adaptive responses of plant respiration to changes in atmospheric CO_2 concentration. *Physiol. Plant.* 137: 473–484.

Goodman, M. 1971. Calvin, Melvin. *Encycl. Brit.* 4: 674.

Gorissen, A., J.H. van Ginkel, J.J.B. Keurentjes, and J.A. van Veen. 1995. Grass root decomposition is retarded when grass has been grown under elevated CO_2. *Soil Biol. Biochem.* 27: 117–120.

Goudriaan, J. and H.H. van Laar. 1978. Relations between leaf resistance, CO_2 concentration and CO_2 assimilation in maize, beans, Lalang grass and sunflower. *Photosynthetica* 12: 241–249.

Gould, S.J. 1994. Pride of place. Science without taxonomy is blind. *Sciences* 34(2): 38–39.

Goutzamanis, J.J. and D.J. Connor. 1977. *A Simulation Model of the Wheat Crop*. Bulletin No. 1. Bundoora, Victoria, Australia: School of Agriculture, La Trobe University, 123pp + bibliography and 8 appendices.

Govindasamy, B., S. Thompson, P.B. Duffy, K. Caldeira, and C. Delire. 2002. Impact of geoengineering schemes on the terrestrial biosphere. *Geophys. Res. Lett.* 29(22), 2061, doi:10.1029/2002GL015911.

Graan, T. and J.S. Boyer. 1990. Very high CO_2 partially restores photosynthesis in sunflower at low water potentials. *Planta* 181: 378–384.

Grant, W.J.R., H.M. Fan, W.J.S. Downton, and B.R. Loveys. 1992. Effects of CO_2 enrichment on the physiology and propagation of two Australian ornamental plants, *Chamelaucium uncinatum* (Schauer) × *Chamelaucium floriferum* (MS) and *Correa schlechtendalii* (Behr). *Sci. Hort.* 52: 337–342.

Grantham, A.E. 1917. *The Tillering of Winter Wheat*. Bulletin 117. Newark, DE: Delaware Agricultural Experiment Station, 119pp.

Grecu, S.J., M.B. Kirkham, E.T. Kanemasu, D.W. Sweeney, L.R. Stone, and G.A. Milliken. 1988a. Root growth in a claypan with a perennial-annual rotation. *Soil Sci. Soc. Am. J.* 52: 488–494.

Grecu, S.J., M.B. Kirkham, E.T. Kanemasu, D.W. Sweeney, L.R. Stone, and G.A. Milliken. 1988b. Penetration resistance, root growth, and water content in a subsoiled claypan. *J. Agron. Crop Sci.* 161: 195–206.

Green, S.R., M.B. Kirkham, and B.E. Clothier. 2006. Root uptake and transpiration: From measurements and models to sustainable irrigation. *Agric. Water Manage.* 86: 165–176.

Gregg, D.C. 1963. *Principles of Chemistry*, 2nd edn. Boston: Allyn and Bacon, 716pp + 62pp. in appendix.

Gregory, P.J., J.A. Palta, and G.R. Batts. 1997. Root system and root:mass ratio—Carbon allocation under current and projected atmospheric conditions in arable crops. *Plant Soil* 187: 221–228.

Gropp, R.E. 2003. Are university natural science collections going extinct? *BioScience* 53: 550.

Gropp, R.E. 2004. Budget cuts affecting natural history. *Science* 306: 811.

Grosse, W., J. Frye, and S. Lattermann. 1992. Root aeration in wetland trees by pressurized gas transport. *Tree Physiol.* 10: 285–295.

Guehl, J.M., C. Picon, G. Aussenac, and P. Gross. 1994. Interactive effects of elevated CO_2 and soil drought on growth and transpiration efficiency and its determinants in two European forest tree species. *Tree Physiol.* 14: 707–724.

Gutiérrez, D., E. Guitérrez, P. Pérez, R. Morcuende, A.L. Verdejo, and R. Martínez-Carrasco. 2009. Acclimation to future atmospheric CO_2 levels increases photochemical efficiency and mitigates photochemistry inhibition by warm temperatures in wheat under field chambers. *Physiol. Plant.* 137: 86–100.

Gyllenhaal, C., D.D. Soejarto, N.R. Farnsworth, and M.J. Huft. 1990. The value of herbaria. *Nature* 347: 704.

Haberlandt, G. 1904. *Physiologische Pflanzenanatomie*. Leipzig: Engelmann (cited by El-Sharkawy, 2009b).

Hall, A.E. and L.H. Ziska. 2000. Crop breeding strategies for the 21st century. In Reddy, K.R. and Hodges, H.F., Eds., *Climate Change and Global Crop Productivity*. Wallingford, Oxon, U.K.: CABI Publishing, CAB International, pp. 407–423.

Halse, N.J. and R.N. Weir. 1974. Effects of temperature on spikelet number of wheat. *Aust. J. Agric. Res.* 25: 687–695.

Ham, J.M., C.E. Owensby, and P.I. Coyne. 1993. Technique for measuring air flow and carbon dioxide flux in large, open-top chambers. *J. Environ. Qual.* 22: 759–766.

Ham, J.M., C.E. Owensby, P.I. Coyne, and D.J. Bremer. 1995. Fluxes of CO_2 and water vapor from a prairie ecosystem exposed to ambient and elevated atmospheric CO_2. *Agric. Forest Meteorol.* 77: 73–93.

Hanba, Y.T., M. Shibasaka, Y. Hayashi, T. Hayakawa, K. Kasamo, I. Terashim, and M. Katsuhara. 2004. Overexpression of the barley aquaporin HvPIP2;1 increases internal CO_2 conductance and CO_2 assimilation in the leaves of transgenic rice plants. *Plant Cell Physiol.* 45: 521–529.

Hanway, J.J. 1963. Growth stages of corn (*Zea mays*, L.). *Agron. J.* 55: 487–492.

Harley, R.M. 1990. In defence of taxonomy. *Nature* 347: 223.

Hartz, T.K. and D.B. Holt. 1991. Root-zone carbon dioxide enrichment in field does not improve tomato or cucumber yield. *HortScience* 26: 1423.

Hartz, T.K., A. Baameur, and D.B. Holt. 1991. Carbon dioxide enrichment of high-value crops under tunnel culture. *J. Am. Soc. Hort. Sci.* 116: 970–973.

Hatch, M.D. 1992. C_4 photosynthesis: An unlikely process full of surprises. *Plant Cell Physiol.* 33: 333–342.

Hatch, M.D. and C.R. Slack. 1966. Photosynthesis by sugar-cane leaves: A new carboxylation reaction and the pathway of sugar formation. *Biochem. J.* 101: 103–111.

Hättenschwiler, S., F.H. Schweingruber, and Ch. Körner. 1996. Tree ring responses to elevated CO_2 and increased N deposition in *Picea abies*. *Plant Cell Environ.* 19: 1369–1378.

Hattersley, P.W. 1982. $\delta^{13}C$ values of C_4 types in grasses. *Aust. J. Plant Physiol.* 9: 139–154.

Hattersley, P.W., L. Watson, and C.B. Osmond. 1977. In situ immunofluorescent labelling of ribulose-1,5-bisphosphate carboxylase in leaves of C_3 and C_4 plants. *Aust. J. Plant Physiol.* 4: 523–539.

Haun, J.R. 1973. Visual quantification of wheat development. *Agron. J.* 65: 116–119.

Havelka, U.D., V.A. Wittenbach, and M.G. Boyle. 1984. CO_2-enrichment effects on wheat yield and physiology. *Crop Sci.* 24: 1163–1168.

He, H., M.B. Kirkham, D.J. Lawlor, and E.T. Kanemasu. 1992. Photosynthesis and water relations of big bluestem (C_4) and Kentucky bluegrass (C_3) under high concentration carbon dioxide. *Trans. Kansas Acad. Sci.* 95: 139–152.

He, J.-S., K.S. Wolfe-Bellin, and F.A. Bazzaz. 2005. Leaf-level physiology, biomass, and reproduction of *Phytolacca americana* under conditions of elevated CO_2 and altered temperature regimes. *Int. J. Plant Sci.* 166: 615–622.

Heath, J., E. Ayres, M. Possell, R.D. Bardgett, H.I.J. Black, H. Grant, P. Ineson, and G. Kerstiens. 2005. Rising atmospheric CO_2 reduces sequestration of root-derived soil carbon. *Science* 309: 1711–1713.

Heath, O.V.S. and T.A. Mansfield. 1969. The movements of stomata. In Wilkins, M.B., Ed., *The Physiology of Plant Growth and Development*. New York: McGraw-Hill, pp. 301–332.

Heimann, M. 2005. Obituary. Charles David Keeling 1928–2005. Pioneer in the Modern Science of Climate Change. *Nature* 437: 331.

Hendry, G.R. and B.A. Kimball. 1994. The FACE program. *Agric. Forest Meteorol.* 70: 3–14.

Hendry, M.J., C.A. Mendoza, R.A. Kirkland, and J.R. Lawrence. 1999. Quantification of transient CO_2 production in a sandy unsaturated zone. *Water Resources Res.* 35: 2189–2198.

Herrick, J.D., H. Maherali, and R.B. Thomas. 2004. Reduced stomatal conductance in sweetgum (*Liquidambar styraciflua*) sustained over long-term CO_2 enrichment. *New Phytol.* 162: 387–396.

Hesketh, J.D. and H. Hellmers. 1973. Floral initiation in four plant species growing in CO_2 enriched air. *Environ. Control Biol.* 11: 51–53.

Heslop-Harrison, J. 1969. Development, differentiation, and yield. In Eastin, J.D., Haskins, F.A., Sullivan, C.Y., and van Bavel, C.H.M., Eds., *Physiological Aspects of Crop Yield*. Madison, WI: American Society of Agronomy, Crop Science Society of America, pp. 291–321.

Hetherington, A.M. and F.I. Woodward. 2003. The role of stomata in sensing and driving environmental change. *Nature* 424: 901–908.

Heyne, E.G. and H.R. Sherman. 1971. Wheat. *Encycl. Brit.* 23: 466–472.

Hicks, R.A., D.D. Briske, C.A. Call, and R.J. Ansley. 1990. Co-existence of a perennial C_3 bunchgrass in a C_4 dominated grassland: An evaluation of gas exchange characteristics. *Photosynthetica* 24: 63–75.

Higgs, R., C. Heidenreich, R. Loberger, R. Cropp, and M. Mitchell. 1981. *Agricultural Mathematics. Problems in Production, Management, Marketing, Mechanization, Environmental Quality.* Danville, IL: The Interstate Printers and Publishers, 297pp.

Hill, P.W., J.F. Farrar, E.L. Boddy, A.M. Gray, and D.L. Jones. 2006. Carbon partitioning and respiration—their control and role in plants at high CO_2. In Nösberger, J., Long, S.P., Norby, R.J., Stitt, M., Hendrey, G.R., and Blum, H., Eds., *Managed Ecosystems and CO_2. Case Studies, Processes, and Perspectives.* Berlin: Springer-Verlag, pp. 271–292.

Hodgman, C.D., Ed. 1959. *Handbook of Chemistry and Physics*, 40th edn. Cleveland, OH: Chemical Rubber Publishing Company, 3,456pp.

Hogan, K.P., A.P. Smith, and L.H. Ziska. 1991. Potential effects of elevated CO_2 and changes in temperature on tropical plants. *Plant Cell Environ.* 14: 763–778.

Högberg, P., A. Nordgren, N. Buchmann, A.F.S. Taylor, A. Ekblad, M.N. Högberg, G. Nygerg, M. Ottosson-Löfvenius, and D.J. Read. 2001. Large-scale forest girdling shows that current photosynthesis drives soil respiration. *Nature* 411: 789–792.

Hook, D.D., C.L. Brown, and P.P. Kormanik. 1971. Inductive flood tolerance in swamp tupelo (*Nyssa sylvatica* var. *biflora* (Walt.) Sarg.). *J. Exp. Bot.* 22: 78–89.

Hopp, R.J. and M.T. Vittum. 1978. Nature's own weather bureau. *Plants Gardens* 33(4): 44–45.

Hovenden, M.J. and L.J. Schimanski. 2000. Genotype differences in growth and stomatal morphology of Southern Beech, *Nothofagus cunninghamii*, exposed to depleted CO_2 concentrations. *Aust. J. Plant Physiol.* 27: 281–287.

Howell, T.A. and S.R. Evett. 2004. The Penman-Monteith Method. Section 3. In *Evapotranspiration: Determination of Consumptive Use in Water Rights Proceedings.* Denver, CO: Continuing Legal Education in Colorado, Inc., 14pp.

Hsiao, T.C. 1975. Relationships among aperture, mass-flow resistance and diffusive resistance. In Kanemasu, E.T., Ed., *Measurement of Stomatal Aperture and Diffusive Resistance.* Bulletin No. 809. Pullman, WA: College of Agriculture Research Center, Washington State University, pp. 24–25.

Hsiao, T.C., E. Acevedo, E. Fereres, and D.W. Henderson. 1976. Stress metabolism. Water stress, growth, and osmotic adjustment. *Phil. Trans. Roy. Soc. London* 273B: 479–500.

Huang, B., J.W. Johnson, and D.S. NeSmith. 1997. Responses to root-zone CO_2 enrichment and hypoxia of wheat genotypes differing in waterlogging tolerance. *Crop Sci.* 37: 464–468.

Huck, M.G. 1970. Variation in taproot elongation rate as influenced by composition of the soil air. *Agron. J.* 62: 815–818.

Hufstetler, E.V., H.R. Boerma, T.E. Carter, Jr., and H.J. Earl. 2007. Genotypic variation for three physiological traits affecting drought tolerance in soybean. *Crop Sci.* 47: 25–35.

Hungate, B.A., E.A. Holland, R.B. Jackson, F.S. Chapin III, H.A. Mooney, and C.B. Field. 1997. The fate of carbon in grasslands under carbon dioxide enrichment. *Nature* 388: 576–579.

Hunt, J.E., F.M. Kelliher, T.M. McSeveny, D.J. Ross, and D. Whitehead. 2004. Long-term carbon exchange in a sparse, seasonally dry tussock grassland. *Global Change Biol.* 10: 1785–1800.

Hunter, C. and E.M. Rich. 1925. The effect of artificial aeration of the soil on *Impatiens balsamia* L. *New Phytol.* 24: 257–271.

Idso, S.B. 1982. *Carbon Dioxide: Friend or Foe?* Tempe, AZ: IBR Press, A division of the Institute for Biospheric Research, Inc., 92pp.

Idso, S.B. 1991. A general relationship between CO_2-induced increases in net photosynthesis and concomitant reductions in stomatal conductance. *Environ. Exp. Bot.* 31: 381–383.

Idso, K.E. and S.B. Idso. 1994. Plant responses to atmospheric CO_2 enrichment in the face of environmental constraints: A review of the past 10 years' research. *Agric. Forest Meteorol.* 69: 153–203.

Idso, S.B. and B.A. Kimball. 1989. Growth response of carrot and radish to atmospheric CO_2 enrichment. *Environ. Exp. Bot.* 29: 135–139.

Idso, S.B. and B.A. Kimball. 1991a. Effects of two and a half years of atmospheric CO_2 enrichment on the root density distribution of three-year-old sour orange trees. *Agric. Forest Meteorol.* 55: 345–349.

Idso, S.B. and B.A. Kimball. 1991b. Doubling CO_2 triples growth rate of sour orange trees. DOE Research Summary No. 13. *Current Research Sponsored by the U.S. Department of Energy's Carbon Dioxide Research Program.* Oakridge, TN: Carbon Dioxide Information Analysis Center, Environmental Sciences Division, Oak Ridge National Laboratory, 4pp.

Idso, S.B. and B.A. Kimball. 1994a. Effects of atmospheric CO_2 enrichment on regrowth of sour orange trees (*Citrus aurantium*; Rutaceae) after coppicing. *Am. J. Bot.* 81: 843–846.

Idso, S.B. and B.A. Kimball. 1994b. Effects of atmospheric CO_2 enrichment on biomass accumulation and distribution in Eldarica pine trees. *J. Exp. Bot.* 45: 1669–1672.

Idso, S.B., B.A. Kimball, and J.R. Mauney. 1987. Atmospheric carbon dioxide enrichment effects on cotton midday foliage temperature: Implications for plant water use and crop yield. *Agron. J.* 79: 667–672.

Idso, S.B., B.A. Kimball, and S.G. Allen. 1991a. CO_2 enrichment of sour orange trees: 2.5 years into a long-term experiment. *Plant Cell Environ.* 14: 351–353.

Idso, S.B., B.A. Kimball, and S.G. Allen. 1991b. Net photosynthesis of sour orange trees maintained in atmospheres of ambient and elevated CO_2 concentration. *Agric. Forest Meteorol.* 54: 95–101.

Idso, S.B., B.A. Kimball, M.G. Anderson, and J.R. Mauney. 1987. Effects of atmospheric CO_2 enrichment on plant growth: The interactive role of air temperature. *Agric. Ecosyst. Environ.* 20: 1–10.

Idso, S.B., S.G. Allen, M.G. Anderson, and B.A. Kimball. 1989. Atmospheric CO_2 enrichment enhances survival of *Azolla* at high temperatures. *Environ. Exp. Bot.* 29: 337–341.

Imai, K. and M. Okamoto-Sato. 1991. Effects of temperature on CO_2 dependence of gas exchanges in C_3 and C_4 crop plants. *Jpn. J. Crop Sci.* 60: 139–145.

Imamura, S. 1943. Unterscuchunger über den Mechanismus der Turgorschwankung der Spaltöffnungssschliesszellen. *Jpn. J. Bot.* 12: 251–346.

Ingram, D.S., G. Ll. Lucas, and R. Huxley. 1994. Botanical collections at risk. *Nature* 371: 9.

Ingvardsen, C. and B. Veierskov. 1994. Response of young barley plants to CO_2 enrichment. *J. Exp. Bot.* 45: 1373–1378.

Ishibashi, M. and I. Terashima. 1995. Effects of continuous leaf wetness on photosynthesis: Adverse aspects of rainfall. *Plant Cell Environ.* 18: 431–438 (plus cover).

Jach, M.E., R. Ceulemans, and M.B. Murray. 2001. Impacts of greenhouse gases on the phenology of forest trees. In Karnosky, D.E., Ceulemans, R., Scarascia-Mugnozza, G.E., and Innes, J.L., Eds., *The Impact of Carbon Dioxide and Other Greenhouse Gases on Forest Ecosystems.* Report No. 3 of the IUFRO Task Force on Environmental Changes. CABI Publishing in association with The International Union of Forestry Research Organizations (IUFRO). Wallingford, Oxon, UK: CAB International, pp. 193–235.

Jackson, R.B., O.E. Sala, C.B. Field, and H.A. Mooney. 1994. CO_2 alters water use, carbon gain, and yield for the dominant species in a natural grassland. *Oecologia* 98: 257–262.

Jaeger, C.H. and H. Hellmers. 1981. An improved medium for nutriculture incorporating arcillite. *HortScience* 16: 176–177.

Jagtap, S.S. and J.W. Jones. 1989. Evapotranspiration model for developing crops. *Trans. Am. Soc. Agric. Eng.* 32: 1342–1350.

Janick, J. 1972. *Horticultural Science*, 2nd edn. San Francisco: W.H. Freeman and Company, 586pp.

Jarvis, P.G. 1985. Specific leaf weight equals 1.0—always! *HortScience* 20: 812. (The same article appears in the *News Bulletin of the British Plant Growth Regulator Group*, Vol. 7, No. 1, October, 1984, pp. 23–24.)

Jarzembowski, E.A. 1990. In defence of taxonomy. *Nature* 347: 222.

Jasoni, R., C. Kane, C. Green, E. Peffley, D. Tissue, L. Thompson, P. Payton, and P.W. Paré. 2004. Altered leaf and root emissions from onion (*Allium cepa* L.) grown under elevated CO_2 conditions. *Environ. Exp. Bot.* 51: 273–280.

Johnston, C.O. and E.C. Miller. 1934. Relation of wheat rust to yield, growth, and water economy of two varieties of wheat. *J. Agric. Res.* 49: 955–981.

Jones, H.G. 1992. *Plants and Microclimate*, 2nd edn. Cambridge: Cambridge University Press, 428pp.

Jones, P., L.R. Allen, Jr., J.W. Jones, and R. Valle. 1985. Photosynthesis and transpiration responses of soybean canopies to short- and longterm CO_2 treatments. *Agron. J.* 77: 119–126.

Jongen, M., M.B. Jones, T. Hebeisen, H. Blum, and G. Hendry. 1995. The effects of elevated CO_2 concentrations on the root growth of *Lolium perenne* and *Trifolium repens* grown in a FACE system. *Global Change Biol.* 1: 361–371.

Joseph, E.Z., Ed. 1985. *Chemical Safety Data Guide.* Washington, DC: The Bureau of National Affairs, 927pp.

Kaddour, A.A. and M.P. Fuller. 2004. The effect of elevated CO_2 and drought on the vegetative growth and development of durum wheat (*Triticum durum* Desf.) cultivars. *Cereal Res. Commun.* 32: 225–232.

Kamakura, M. and A. Furukawa. 2008. Response of individual stomata in *Ipomoea pes-caprae* [sic] to various CO_2 concentrations. *Physiol. Plant.* 132: 255–261.

Kammann, C., L. Grünhage, U. Grüters, S. Janze, and H.-J. Jäger. 2005. Response of aboveground grassland biomass and soil moisture to moderate long-term O_2 enrichment. *Basic Appl. Ecol.* 6: 351–365.

Kanemasu, E.T. and C.K. Hiebsch. 1975. Net carbon dioxide exchange of wheat, sorghum, and soybean. *Can. J. Bot.* 53: 382–389.

Kanemasu, E.T., G.W. Thurtell, and C.B. Tanner. 1969. Design, calibration and field use of a stomatal diffusion porometer. *Plant Physiol.* 44: 881–885.

Kanemasu, E.T., L.R. Stone, and W.L. Powers. 1976. Evapotranspiration model tested for soybean and sorghum. *Agron. J.* 68: 569–572.

Kanemasu, E.T., M.L. Wesely, B.B. Hicks, and J.L. Heilman. 1979. Techniques for calculating energy and mass fluxes. In Barfield, B.J. and Gerber, J.F., Eds., *Modification of the Aerial Environment of Plants* [This is the title given on title page. The title given on cover of the book is "*Modification of the Aerial Environment of Crops*."] St. Joseph, MI: American Society of Agricultural Engineers, pp. 156–182.

Kansas Department of Agriculture. 2009. *Kansas Agricultural Statistics. Crops*, Vol. 9(7), Topeka, KS: Kansas Department of Agriculture, U.S. Department of Agriculture, pp. 1 and 3.

Katul, G.G., S. Palmroth, and R. Oren. 2009. Leaf stomatal responses to vapour pressure deficit under current and CO_2-enriched atmosphere explained by the economics of gas exchange. *Plant Cell Environ.* 32: 968–979.

Kawamitsu, Y., S. Yoda, and W. Agata. 1993. Humidity pretreatment affects the responses of stomata and CO_2 assimilation to vapor pressure difference in C_3 and C_4 plants. *Plant Cell Physiol.* 34: 113–119.

Keeling, C.D. 1970. Is carbon dioxide from fossil fuel changing man's environment? *Proc. Am. Philos. Soc.* 114(1): 10–17.

Keeling, C.D. 1982. The oceans and terrestrial biosphere as future sinks for fossil fuel CO_2. In Reck, R.A. and Hummel, J.R., Eds., AIP Conference Proceedings No. 82. *Interpretation of Climate and Photochemical Models, Ozone and Temperature Measurements.* New York: American Institute of Physics, pp. 47–82.

Keeling, R.F. 2005. Comment on "The Oceanic Sink for Anthropogenic CO_2." *Science* 308: 1743.

Keeling, R.F., S.C. Piper, and M. Heimann. 1996. Global and hemispheric CO_2 sinks deduced from changes in atmospheric O_2 concentration. *Nature* 381: 218–221.

Keller, C.K. 1991. Hydrogeochemistry of a clayey till. 2. Sources of CO_2. *Water Resources Res.* 27: 2555–2564.

Kelliher, F.M., M.B. Kirkham, and J.E. Hunt. 2000. Photosynthesis and stomatal conductance of the New Zealand tree, *Meryta sinclairii*, grown under two watering regimes. *N. Z. J. Bot.* 38: 515–519.

Kellogg, E.A. and C.R. Buell. 2009. Splendor in the grasses. *Plant Physiol.* 149: 1–3 (plus cover).

Kellogg, W.W. and Z.-C. Zhao. 1988. Sensitivity of soil moisture to doubling of carbon dioxide in climate model experiments. Part I. North America. *J. Climate* 1: 348–366.

Kendall, A.C., J.C. Turner, and S.M. Thomas. 1985. Effects of CO_2 enrichment at different irradiances on growth and yield of wheat. I. Effects of cultivar and of duration of CO_2 enrichment. *J. Exp. Bot.* 36: 252–260.

Kenrick, P. 2001. Turning over a new leaf. *Nature* 410: 309–310.

Kerp, H. 2002. Atmospheric CO_2 from fossil plant cuticles. *Nature* 415: 38.

Kerr, R.A. 2003. Inheriting the family science. *Science* 301: 1312–1316.

Kerstiens, G. and C.V. Hawes. 1994. Response to growth and carbon allocation to elevated CO_2 in young cherry (*Prunus avium* L.) saplings in relation to root environment. *New Phytol.* 128: 607–614.

Kerstiens, G., J. Townend, J. Heath, and T.A. Mansfield. 1995. Effects of water and nutrient availability on physiological responses of woody species to elevated CO_2. *Forestry* 68: 303–315.

Kiesselbach, T.A. 1910. *Transpiration Experiments with the Corn Plant*, 23rd Annual Report. Lincoln, NE: Nebraska Agricultural Experiment Station, pp. 125–139 (cited by Tanner and Sinclair, 1983).

Kiesselbach, T.A. 1916. *Transpiration as a Factor in Crop Production.* Research Bulletin No. 6. Lincoln, NE: Nebraska Agricultural Experiment Station (cited by Tanner and Sinclair, 1983).

Kiesselbach, T.A. and E.G. Montgomery. 1911. *The Relation of Climatic Factors to the Water Used by the Corn Plant*, 24th Annual Report. Lincoln, NE: Nebraska Agricultural Experiment Station, pp. 91–107 (cited by Tanner and Sinclair, 1983).

Kim, T.-H., M. Böhmer, H. Hu, N. Nishimura, and J.I. Schroeder. 2010. Guard cell signal transduction network: Advances in understanding abscisic acid, CO_2, and Ca^{2+} signaling. *Annu. Rev. Plant Biol.* 61: 561–591.

Kimball, B.A. 1983. Carbon dioxide and agricultural yield: An assemblage and analysis of 430 prior observations. *Agron. J.* 75: 779–788.

Kimball, B.A. 1985. Adaptation of vegetation and management practices to a higher carbon dioxide world. In Strain, B.R. and Cure, J.D., Eds., *Direct Effects of Increasing Carbon Dioxide on Vegetation.* DOE/ ER-0238. Washington, DC: United States Department of Energy, Office of Energy Research, Office of Basic Energy Sciences, Carbon Dioxide Research Division, pp. 185–204.

Kimball, B.A. 2006. The effects of free-air [CO_2] enrichment of cotton, wheat, and sorghum. *Ecol. Stud.* 187: 47–70.

Kimball, B.A. and C.J. Bernacchi. 2006. Evapotranspiration, canopy temperature, and plant water relations. *Ecol. Stud.* 187: 311–324.

Kimball, B.A., J.R. Mauney, J.W. Radin, F.S. Nakayama, S.B. Idso, D.L. Hendrix, D.H. Akey, S.G. Allen, M.G. Anderson, and W. Hartung. 1986. Effects of increasing atmospheric CO_2 on the growth, water relations, and physiology of plants grown under optimal and limiting levels of water and nitrogen. *Response of Vegetation to Carbon Dioxide.* Research Report No. 039. Washington, DC: U.S. Department of Energy, Carbon Dioxide Research Division and the U.S. Department of Agriculture, Agricultural Research Service, 125pp.

Kimball, B.A., J.R. Mauney, D.H. Akey, D.L. Hendrix, S.G. Allen, S.B. Idso, J.W. Radin, and E.A. Lakatos. 1987. Effects of increasing atmospheric CO_2 on the growth, water relations, and physiology of plants grown under optimal and limiting levels of water and nitrogen. *Response of Vegetation to Carbon Dioxide.* Research Report No. 049. Washington, DC: U.S. Department of Energy, Carbon Dioxide Research Division and the U.S. Department of Agriculture, Agricultural Research Service, 124pp.

Kimball, B.A. (Chair), N.J. Rosenberg, and L.H. Allen, Jr. (Ed. Comm.). 1990. *Impact of Carbon Dioxide, Trace Gases, and Climate Change on Global Agriculture.* Madison, WI: American Society of Agronomy, Crop Science Society of America, Soil Science Society of America, 133pp.

Kimball, B.A., J.R. Mauney, F.S. Nakayama, and S.B. Idso. 1993. Effects of increasing atmospheric CO_2 on vegetation. *Vegetatio* 104/105: 65–75.

Kimball, B.A., R.L. LaMorte, P.J. Pinter, Jr., G.W. Wall, D.J. Hunsaker, F.J. Adamsen, S.W. Leavitt, T.L. Thompson, A.D. Matthias, and T.J. Brooks. 1999. Free-air CO_2 enrichment and soil nitrogen effects on energy balance and evapotranspiration of wheat. *Water Resources Res.* 35: 1179–1190.

Kimball, B.A., K. Kobayashi, and M. Bindi. 2002. Responses of agricultural crops to free-air CO_2 enrichment. *Adv. Agron.* 77: 293–368.

Kimball, B.A., S.B. Idso, S. Johnson, and M.C. Rillig. 2007. Seventeen years of carbon dioxide enrichment of sour orange trees: Final results. *Global Change Biol.* 13: 2171–2183.

Kimball, B.A., M.M. Conley, S. Wang, X. Lin, C. Luo, J. Morgan, and D. Smith. 2008. Infrared heater arrays for warming ecosystem field plots. *Global Change Biol.* 14: 309–320.

King, F.H. 1899. *Irrigation and Drainage.* Principles and Practice of Their Cultural Phases. The Rural Sciences Series. New York: The Macmillan Company, 502pp.

King, F.H. 1901. *Physics of Agriculture*, 2nd edn. Madison, WI: Published by the Author, 604pp.

King, F.H. 1905. *Investigations in Soil Management.* U.S. Department of Agriculture Bureau of Soils Bulletin No. 26. Washington, DC: U.S. Department of Agriculture, 205pp.

King, F.H. 1911. *Farmers of Forty Centuries.* Emmaus, PA: Rodale Press, Inc., 441pp.

King, F.H. 1914. *Soil Management.* New York: Orange Judd Company, 311pp.

Kirkham, D. 1995–1996. *Oral Information from Don Kirkham about His Utah State Work.* Written by M.B. Kirkham from notes taken December 20–22, 1995 with clarifications on September 21–24, 1996. Available from M.B. Kirkham, Department of Agronomy, Kansas State University, Manhattan, Kansas, 14pp.

Kirkham, M.B. 1978. Water potential and turgor pressure as a selection basis for wind-grown winter wheat. *Agric. Water Manage.* 1: 343–349.

Kirkham, M.B. 1983a. Effect of ethephon on the water status of a drought-resistant and a drought-sensitive cultivar of winter wheat. *Z. Pflanzenphysiologie (J. Plant Physiol.)* 112: 103–112.

Kirkham, M.B. 1983b. Effect of ABA on the water relations of winter-wheat cultivars varying in drought resistance. *Physiol. Plant.* 59: 153–157.

Kirkham, M.B. 1986. Water relations of the upper and lower surfaces of maize leaves. *Biol. Plant.* 28: 249–257.

Kirkham, M.B. 1989. Growth and water relations of two wheat cultivars grown separately and together. *Biol. Agric. Hort.* 6: 35–46.

Kirkham, M.B. 1994. Streamlines for diffusive flow in vertical and surface tillage: A model study. *Soil Sci. Soc. Am. J.* 58: 85–93.

Kirkham, M.B. 2005. *Principles of Soil and Plant Water Relations.* Amsterdam: Elsevier, 500pp.

Kirkham, M.B. 2008. Horizontal root growth: Water uptake and stomatal resistance under microgravity. *Vadose Zone J.* 7: 1125–1131.

Kirkham, M.B., W.R. Gardner, and G.C. Gerloff. 1972. Regulation of cell division and cell enlargement by turgor pressure. *Plant Physiol.* 49: 961–962.

Kirkham, M.B., W.R. Gardner, and G.C. Gerloff. 1974. Internal water status of kinetin-treated, salt-stressed plants. *Plant Physiol.* 53: 241–243.

Kirkham, M.B., K. Suksayretrup, C.E. Wassom, and E.T. Kanemasu. 1984. Canopy temperature of drought-resistant and drought-sensitive genotypes of maize. *Maydica* 29: 287–303.

Kirkham, M.B., H. He, T.P. Bolger, D.J. Lawlor, and E.T. Kanemasu. 1991. Leaf photosynthesis and water use of big bluestem under elevated carbon dioxide. *Crop Sci.* 31: 1589–1594.

Kirkham, M.B., D. Nie, H. He, and E.T. Kanemasu. 1993. Responses of plants to elevated levels of carbon dioxide. In *Proceedings of the Symposium on Plant Growth and Environment*, October 16, 1993, Suwon, Korea. Suwon, Korea: Korean Agricultural Chemical Society, pp. 130–161.

Kissel, D.E., Ed. 1988. SSSA Awards—1987. Fellows of SSSA Elected in 1987. Bruce A. Kimball. *Soil Sci. Soc. Am. J.* 52: 567–572.

Klee, H. 2008. Hormones branch out. *Nature* 455: 176–177.

Klepper, B. 2000. Exploring the hidden half: Root systems in the soil. In Rosenzweig, C., Russo, S., and Hillel, D., Eds., *A Spectrum of Achievements in Agronomy: Women Fellows of the Tri-Societies*. Madison, WI: American Society of Agronomy, Crop Science Society of America, Soil Science Society of America, pp. 27–34.

Knapp, A.K. 1993. Gas exchange dynamics in C_3 and C_4 grasses: Consequences of differences in stomatal conductance. *Ecology* 74: 113–123.

Knapp, A.K., E.P. Hamerlynck, and C.E. Owensby. 1993. Photosynthetic and water relations responses to elevated CO_2 in the C_4 grass *Andropogon gerardii*. *Int. J. Plant Sci.* 154: 459–466.

Knapp, A.K., J.T. Fahnestock, and C.E. Owensby. 1994a. Elevated atmospheric CO_2 alters stomatal responses to variable sunlight in a C_4 grass. *Plant Cell Environ.* 17: 189–195.

Knapp, A.K., M. Cocke, E.P. Hamerlynck, and C.E. Owensby. 1994b. Effect of elevated CO_2 on stomatal density and distribution in a C_4 grass and a C_3 forb under field conditions. *Ann. Bot.* 74: 595–599.

Knapp, S., R.M. Bateman, N.R. Chalmers, C.J. Humphries, P.S. Rainbow, A.B. Smith, P.D. Taylor, R.I. Vane-Wright, and M. Wilkinson. 2002. Taxonomy needs evolution, not revolution. *Nature* 419: 559.

Knight, J., J.J. Kennedy, C. Folland, G. Harris, G.S. Jones, M. Palmer, D. Parker, A. Scaife, and P. Stott. 2009. Do global temperature trends over the last decade falsify climate predictions? [in "State of the Climate 2008"]. *Bull. Am. Meteorol. Soc.* 90(8): S22–S23.

Knoll, A.H. 1991. End of the Proterozoic eon. *Sci. Am.* 265(4): 64–73.

Kobayashi, K., M. Okada, H.Y. Kim, M. Lieffering, S. Miura, and T. Hasegawa. 2006. Paddy rice responses to free-air [CO_2] enrichment. In Nösberger, J., Long, S.P., Norby, R.J., Stitt, M., Hendrey, G.R., and Blum, H., Eds., *Managed Ecosystems and CO_2. Case Studies, Processes, and Perspectives*. Berlin: Springer-Verlag, pp. 87–104.

Koch, G.W. and H.A. Mooney. 1996. Response of terrestrial ecosystems to elevated CO_2: A synthesis and summary. In Koch, G.W. and Mooney, H.A., Eds., *Carbon Dioxide and Terrestrial Ecosystems*. San Diego, CA: Academic Press, pp. 415–429.

Koepp, S. (Deputy Managing Ed.). 2005. Notebook. Milestones. Died. Charles David Keeling. *Time* 166(1), July 4, 2005, p. 21.

Kok, B. 1967. Photosynthesis—Physical aspects. In San Pietro, A., Greer, F.A., and Army, T.J., Eds., *Harvesting the Sun. Photosynthesis in Plant Life*. New York: Academic Press, pp. 29–48.

Kok, B. 1976. Photosynthesis: The path of energy. In Bonner, J. and Varner, J.E., Eds., *Plant Biochemistry*, 3rd edn. New York: Academic Press, pp. 845–885.

Kondo, T., R. Kajita, A. Miyazaki, M. Hokoyama, T. Nakamura-Miura, S. Mizuno, Y. Masuda, K. Irie, Y. Tanaka, S. Takada, T. Kakimoto, and Y. Sakagami. 2010. Stomatal density is controlled by a mesophyll-derived signaling molecule. *Plant Cell Physiol.* 51: 1–8.

Kontunen-Soppela, S., J. Riikonen, H. Ruhanen, M. Brosché, P. Somervuo, P. Peltonen et al. 2010. Differential gene expression in senescing leaves of two silver birch genotypes in response to elevated CO_2 and tropospheric ozone. *Plant Cell Environ.* 33: 1016–1028.

Körner, C. 1988. Does global increase of CO_2 alter stomatal density? *Flora* 181: 243–257.

Körner, C., M. Diemer, B. Schäppi, and L. Zimmermann. 1996. Response of alpine vegetation to elevated CO_2. In Koch, G.W and Mooney, H.A., Eds., *Carbon Dioxide and Terrestrial Ecosystems*. San Diego, CA: Academic Press, pp. 177–196.

Körner, C., R. Asshoff, O. Bignucolo, S. Hättenschwiler, S.G. Keel, S. Peláez-Riedl, S. Pepin, R.T.W. Siegwolf, and G. Zotz. 2005. Carbon flux and growth in mature deciduous forest trees exposed to elevated CO_2. *Science* 309: 1360–1362.

Kortschak, H.P., C.E. Hartt, and G.O. Burr. 1965. CO_2 fixation in sugarcane leaves. *Plant Physiol.* 40: 209–213.

Kouwenberg, L.L.R., J.C. McElwain, W.M. Kürschner, F. Wagner, D.J. Beerling, F.E. Mayle, and H. Visscher. 2003. Stomatal frequency adjustment of four conifer species to historical changes in atmospheric CO_2. *Am. J. Bot.* 90: 610–619.

Kramer, P.J. 1969. *Plant and Soil Water Relationships: A Modern Synthesis*. New York: McGraw-Hill Book Company, 482pp.

Krekule, J. 1964. Varietal differences on replacing vernalization by a short day in winter wheat. *Biol. Plant.* 6: 299–305 (in Russian with English summary).

Krenzer, E.G., Jr. and D.N. Moss. 1975. Carbon dioxide enrichment effects upon yield and yield components in wheat. *Crop Sci.* 15: 71–74.

Kürschner, W.M. 2001. Leaf sensor for CO_2 in deep time. *Nature* 411: 247–248.

Kutílek, M. 2008. *Racionálně o globálním oteplování.* (*Rationality on Global Warming*). Prague, Czech Republic: Dokořán, 185pp (in Czech).

Kutílek, M. and D.R. Nielsen. 2010. *Facts about Global Warming*. Reiskirchen, Germany: Catena Verlag 227 pp.

Laforge, F., C. Lussier, Y. Desjardins, and A. Gosselin. 1991. Effect of light intensity and CO_2 enrichment during in vitro rooting on subsequent growth of plantlets of strawberry, raspberry and asparagus in acclimatization. *Sci. Hort.* 47: 259–269.

Lagomarsino, A., P. De Angelis, M.C. Moscatelli, and S. Grego. 2009. The influence of temperature and labile C substrates on heterotrophic respiration in response to elevated CO_2 and nitrogen fertilization. *Plant Soil* 317: 223–234.

Lake, J.A., W.P. Quick, D.J. Beerling, and F.I. Woodward. 2001. Signals from mature to new leaves. *Nature* 411: 154.

Lal, R., J.M. Kimble, R.F. Follett, and C.V. Cole. 1999. *The Potential of U.S. Cropland to Sequester Carbon and Mitigate the Greenhouse Effect*. Boca Raton, FL: Lewis Publishers, 128pp.

Lambers, H., I. Stulen, and A. van der Werf. 1996. Carbon use in root respiration as affected by elevated atmospheric CO_2. *Plant Soil* 187: 251–263.

Lanning, S.P., K. Kephart, G.R. Carlson, J.E. Eckhoff, R.N. Stougaard, D.M. Wichman, J.M. Martin, and L.E. Talbert. 2010. Climatic change and agronomic performance of hard red spring wheat from 1950 to 2007. *Crop Sci.* 50: 835–841.

LaPolla, J.S. 2005. Academy of Natural Sciences: Job cuts. *Science* 307: 1560.

Large, E.C. 1954. Growth stages in cereals. Illustration of the Feekes Scale. *Plant Path.* 3: 128–129.

Latimer, W.M. 1971. Lewis, Gilbert Newton. *Encycl. Brit.* 13: 1008.

Laurie, A. and V.H. Ries. 1950. *Floriculture: Fundamentals and Practices*. New York: McGraw-Hill Book Company, 525pp.

Laurie, A., D.C. Kiplinger, and K.S. Nelson. 1969. *Commercial Flower Forcing*, 7th edn. New York: McGraw-Hill Book Co., 514pp.

Lawlor, D.W. and R.A.C. Mitchell. 1991. The effects of increasing CO_2 on crop photosynthesis and productivity: A review of field studies. *Plant Cell Environ.* 14: 807–818.

Lawlor, D.W. and R.A.C. Mitchell. 2000. Crop ecosystem responses to climatic change: Wheat. In Reddy, K.R. and Hodges, H.F., Eds., *Climate Change and Global Crop Productivity*. Wallingford, Oxon, U.K.: CABI Publishing, CAB International, pp. 57–80.

LeCain, D.R., J.A. Morgan, A.R. Mosier, and J.A. Nelson. 2003. Soil and plant water relations determine photosynthetic responses of C_3 and C_4 grasses in a semi-arid ecosystem under elevated CO_2. *Ann. Bot.* 92: 41–52.

LeCain, D.R., J.A. Morgan, D.G. Milchunas, A.R. Mosier, J.A. Nelson, and D.P. Smith. 2006. Root biomass of individual species, and root size characteristics after five years of CO_2 enrichment on native shortgrass steppe. *Plant Soil* 279: 219–228.

Ledig, F.T., G.E. Rehfeldt, C. Sáenz-Romero, and C. Flores-López. 2010. Projections of suitable habitat for rare species under global warming scenarios. *Am. J. Bot.* 97: 970–987.

Lee, H.S.J. and P.G. Jarvis. 1996. Effects of tree maturity on some responses to elevated CO_2 in Sitka spruce (*Picea sitchensis* Bong. Carr). In Koch, G.W. and Mooney, H.A., Eds., *Carbon Dioxide and Terrestrial Ecosystems*. San Diego, CA: Academic Press, pp. 53–70.

LeGrand, F.E. and H.R. Myers. 1976. *Weather Observation and Use*. OSU Extension Facts No. 9410. Stillwater, OK: Cooperative Extension Service, Oklahoma State University, 4pp.

Letey, J. 1985. Relationship between soil physical properties and crop production. *Adv. Soil Sci.* 1: 277–294.

Levitt, J. 1980. *Responses of Plants to Environmental Stress. Vol. I. Chilling, Freezing, and High Temperature Stresses*, 2nd edn. New York: Academic Press, 497pp.

Levitt, S.D. and S.J. Dubner. 2009. *Superfreakonomics*. New York: William Morrow. An Imprint of HarperCollins Publishers, 270pp.

Lewin, K.F., G.R. Hendrey, J. Nagy, and R.L. LaMorte. 1994. Design and application of a free-air carbon dioxide enrichment facility. *Agric. Forest Meteorol.* 70: 15–29.

Leyton, L. and L.Z. Rousseau. 1958. Root growth of tree seedlings in relation to aeration. In Thimann, K.V., Critchfield, W.B., and Zimmermann, M.H., Eds., *The Physiology of Forest Trees*. New York: The Ronald Press Company, pp. 467–475.

Li, F., S. Kang, and J. Zhang. 2004. Interactive effects of elevated CO_2, nitrogen and drought on leaf area, stomatal conductance, and evapotranspiration of wheat. *Agric. Water Manage.* 67: 221–233.

Liang, N. and K. Maruyama. 1995. Interactive effects of CO_2 enrichment and drought stress on gas exchange and water-use efficiency in *Alnus firma*. *Environ. Exp. Bot.* 35: 353–361.

Liang, N., K. Maruyama, and Y. Huang. 1995. Interactions of elevated CO_2 and drought stress in gas exchange and water-use efficiency in three temperate deciduous tree species. *Photosynthetica* 31: 529–539.

Li-Cor, Inc. 1985. *The LI-1600 Steady State Porometer*. Brochure No. SSP1-685, Lincoln, NE: Li-Cor, Inc., 8pp.

Lide, D.R., Ed. 1994. *CRC Handbook of Chemistry and Physics*, 75th edn. Boca Raton, FL: CRC Press.

Lieth, H., Ed. 1974. *Phenology and Seasonality Modeling*. New York: Springer-Verlag, 444pp.

Lilien, J. 2003. Safe home planned for Iowa herbarium. *Nature* 425: 121.

Livingston, B.E. 1916a. Physiological temperature indices for the study of plant growth in relation to climatic conditions. *Physiol. Res.* 1: 399–420.

Livingston, B.E. 1916b. A single index to represent both moisture and temperature conditions as related to plants. *Physiol. Res.* 1: 421–440.

Livingston, B.E., and G.J. Livingston. 1913. Temperature coefficients in plant geography and climatology. *Bot. Gaz.* 56: 349–375.

Livnè, A. and Y. Vaadia. 1965. Stimulation of transpiration rate in barley leaves by kinetin and gibberellic acid. *Physiol. Plant.* 18: 658–664.

Lockwood, J.G. 1995. The suppression of evapotranspiration by rising levels of atmospheric CO_2. *Weather* 50: 304–308.

Long, S.P. 1991. Modification of the response of photosynthetic productivity to rising temperature by atmospheric CO_2 concentrations: Has its importance been underestimated? Opinion. *Plant Cell Environ.* 14: 729–739.

Long, S.P., E.A. Ainsworth, A.D.B. Leakey, J. Nösberger, and D.R. Ort. 2006. Food for thought: Lower-than-expected crop yield stimulation with rising CO_2 concentrations. *Science* 312: 1918–1921.

Lorenz, A.J., T.J. Gustafson, J.G. Coors, and N. de Leon. 2010. Breeding maize for a bioeconomy: A literature survey examining harvest index and stover yield and their relationship to grain yield. *Crop Sci.* 50: 1–12.

Lowry, W.P. 1969. *Weather and Life: An Introduction to Biometeorology*. New York: Academic Press, 305pp.

Luellen, W.R. (Managing Ed.). 1998. *Publications Handbook Style Manual*. Madison, WI: American Society of Agronomy, Crop Science Society of America, Soil Science Society of America, 154pp.

Luke, H.H. and T.E. Freeman. 1967. Rapid bioassay for phytokinins based on transpiration of excised oat leaves. *Nature* 215: 874–875.

Lundegårdh, H. 1927. Carbon dioxide evolution of soil and crop growth. *Soil Sci.* 23: 417–453.

Luo, Y. and P.S. Nobel. 1993. Growth characteristics of newly initiated cladodes of *Opuntia ficus-indica* as affected by shading, drought and elevated CO_2. *Physiol. Plant.* 87: 467–474.

Luo, Y., S. Wan, D. Hui, and L.L. Wallace. 2001. Acclimatization of soil respiration to warming in a tall grass prairie. *Nature* 413: 622–625.

Luomala, E.-M., K. Laitinen, S. Sutinen, S. Kellomaki, and E. Vapaavuori. 2005. Stomatal density, anatomy and nutrient concentrations of Scots pine needles are affected by elevated CO_2 and temperature. *Plant Cell Environ.* 28: 733–749.

Macallum, A.G. 1905. On the distribution of potassium in animal and vegetable cells. *J. Physiol.* 52: 95–128.

Maherali, H., C.D. Reid, H.W. Polley, H.B. Johnson, and R.B. Jackson. 2002. Stomatal acclimation over a subambient to elevated CO_2 gradient in a C_3/C_4 grassland. *Plant Cell Environ.* 25: 557–566.

Maherali, H., H.B. Johnson, and R.B. Jackson. 2003. Stomatal sensitivity to vapour pressure difference over a subambient to elevated CO_2 gradient in a C_3/C_4 grassland. *Plant Cell Environ.* 26: 1297–1306.

Malkin, R. and K. Niyogi. 2000. Photosynthesis. In Buchanan, B.B., Gruissem, W., and Jones, R.L., Eds., *Biochemistry and Molecular Biology of Plants*. Rockville, MD: American Society of Plant Physiologists, pp. 568–628.

Malone, S.R., H.S. Mayeux, H.B. Johnson, and H.W. Polley. 1993. Stomatal density and aperture length in four plant species grown across a subambient CO_2 gradient. *Am. J. Bot.* 80: 1413–1418.

Manabe, S. and R.T. Wetherald. 1980. On the distribution of climate change resulting from an increase in CO_2 content of the atmosphere. *J. Atmos. Sci.* 37: 99–118.

Manabe, S. and R.T. Wetherald. 1987. Large-scale changes of soil wetness induced by an increase in atmospheric carbon dioxide. *J. Atmos. Sci.* 44: 1211–1235.

Mao, Z.-J., Y.-J. Wang, X.-W. Wang, and P. Yu. Voronin. 2005. Effect of doubled CO_2 on morphology: Inhibition of stomata development in growing birch (*Betula platyphylla* Suk.) leaves. *Russ. J. Plant Physiol.* 52: 171–175.

Marc, J. and R.M. Gifford. 1984. Floral initiation in wheat, sunflower, and sorghum under carbon dioxide enrichment. *Can. J. Bot.* 62: 9–14.

Marks, S. and B.R. Strain. 1989. Effects of drought and CO_2 enrichment on competition between two old-field perennials. *New Phytol.* 111: 181–186.

Marquis Who's Who. 2004. *Who's Who in America 2005*, 59th edn. New Providence, NJ: Marquis Who's Who, p. 2437.

Marquis Who's Who. 2007. *Who's Who in America 2008*, 62nd edn. New Providence, NJ: Marquis Who's Who.

Marron, N., C. Plain, B. Longdoz, and D. Epron. 2009. Seasonable and daily time course of the ^{13}C composition in soil CO_2 efflux recorded with a tunable diode laser spectrophotometer (TDLS). *Plant Soil* 318: 137–151.

Marsh, B.a'B. 1971. Measurement of length in random arrangements of lines. *J. Appl. Ecol.* 8: 265–267.

Martin, D. 2008. "Francis Childs, 68, Dies; Sage of High Corn Yields." *The New York Times*. Sunday, January 20, 2008, p. YT 31.

Martinić, Z. 1973. Vernalization and photoperiodism of common wheat as related to the general and specific adaptability of varieties. In Slatyer, R.O., Ed., *Plant Response to Climatic Factors*. Paris: United Nations Educational, Scientific and Cultural Organization, pp. 153–163.

Matamala, R., M.A. Gonzàlez-Meler, J.D. Jastrow, R.J. Norby, and W.H. Schlesinger. 2003. Impacts of fine root turnover on forest NPP and soil C sequestration potential. *Science* 302: 1385–1387.

Mauney, J.R. and D.L. Hendrix. 1988. Responses of glasshouse grown cotton to irrigation with carbon dioxide-saturated water. *Crop Sci.* 28: 835–838.

Maximov, N.A. 1929. *The Plant in Relation to Water*. London: George Allen and Unwin, Ltd., 451pp.

McElwain, J.C. and W.G. Chaloner. 1995. Stomatal density and index of fossil plants track atmospheric carbon dioxide in the Palaeozoic. *Ann. Bot.* 76: 389–395.

McElwain, J., F.J.G. Mitchell, and M.B. Jones. 1995. Relationship of stomatal density and index to *Salix cinerea* to atmospheric carbon dioxide concentrations in the Holocene. *Holocene* 5: 216–219.

McIntosh, M.S. and S.R. Simmons. 2008. A century of women in agronomy: Lessons from diverse life stories. *Agron. J.* 100(Suppl.): S-53–S-69.

McKinney, H.H. and W.J. Sando. 1935. Earliness of sexual reproduction in wheat as influenced by temperature and light in relation to growth phases. *J. Agric. Res.* 51: 621–641.

McPherson, D.C. 1939. Cortical air spaces in the roots of *Zea mays* L. *New Phytol.* 38: 190–202 plus 2 plates.

Medlyn, B.E., F.W. Badeck, D.G.G. de Pury, C.V.M. Barton, M. Broadmeadow, H. Ceulemans, P. de Angelis et al. 1999. Effects of elevated [CO_2] on photosynthesis in European forest species: A meta-analysis of model parameters. *Plant Cell Environ.* 22: 1475–1485.

Meidner, H. and T.A. Mansfield. 1968. *Physiology of Stomata*. New York: McGraw-Hill, 179pp.

Menzel, A. and T. Sparks. 2006. Temperature and plant development: Phenology and seasonality. In Morison, J.I.L and Morecroft, M.D., Eds., *Plant Growth and Climate Change*. Oxford: Blackwell, pp. 70–95.

Mercader, J. 2009. Mozambican grass seed consumption during the Middle Stone Age. *Science* 326: 1680–1683.

Meserve, R.A. 2008–2009. The President's Report. *Carnegie Institution for Science. Year Book 2008–2009*. Washington, DC: Carnegie Institution of Washington, 92pp.

Meyer, A. 2004. Learning from the Altmeister. *Nature* 428: 897.

Miao, S. 1995. Acorn mass and seedling growth in *Quercus rubra* in response to elevated CO_2. *J. Veg. Sci.* 6: 697–700.

Miao, S.L., P.M. Wayne, and F.A. Bazzaz. 1992. Elevated CO_2 differentially alters the responses of cooccurring birch and maple seedlings to a moisture gradient. *Oecologia* 90: 300–304.

Miglietta, F., A. Raschi, I. Bettarini, R. Resti, and F. Selvi. 1993a. Natural CO_2 springs in Italy: A resource for examining long-term response of vegetation to rising atmospheric CO_2 concentrations. *Plant Cell Environ.* 16: 873–878.

Miglietta, F., A. Raschi, R. Resti, and M. Badiani. 1993b. Growth and onto-morphogenesis of soybean (*Glycine max* Merril [sic]) in an open, naturally CO_2-enriched environment. *Plant Cell Environ.* 16: 909–918.

Milchunas, D.G., J.A. Morgan, A.R. Mosier, and D.R. LeCain. 2005a. Root dynamics and demography in shortgrass steppe under elevated CO_2, and comments on minirhizotron methodology. *Global Change Biol.* 11: 1837–1855.

Milchunas, D.G., A.R. Mosier, J.A. Morgan, D.R. LeCain, J.Y. King, and J.A. Nelson. 2005b. Root production and tissue quality in a shortgrass steppe exposed to elevated CO_2: Using a new ingrowth method. *Plant Soil* 268: 111–122.

Miller, S.E., W.J. Kress, and C. Samper K. 2004. Crisis for biodiversity collections. *Science* 303: 310.

Miller-Rushing, A.J. and D.W. Inouye. 2009. Variation in the impact of climate change on flowering phenology and abundance: An examination of two pairs of closely related wildflower species. *Am. J. Bot.* 96: 1821–1829.

Miller-Rushing, A.J., R.B. Primack, P.H. Templer, S. Rathbone, and S. Mukunda. 2009. Long-term relationships among atmospheric CO_2, stomata, and intrinsic water use efficiency in individual trees. *Am. J. Bot.* 96: 1779–1786.

Mitchell, J.F.B. and D.A. Warrilow. 1987. Summer dryness in northern mid-latitudes due to increased CO_2. *Nature* 330: 238–240.

Monteith, J.L. 1965. Evaporation and environment. In Fogg, G.E., Ed., *Symposium of the Society for Experimental Biology, The State and Movement of Water in Living Organisms*, Vol. 19. New York: Academic Press, Inc., pp. 205–234.

Monteith, J.L. 1973. *Principles of Environmental Physics*. New York: American Elsevier Publishing Company, 241pp.

Monteith, J.L. 1978. Reassessment of maximum growth rates for C_3 and C_4 crops. *Exp. Agric.* 14: 1–5.

Montgomery, E.G. and T.A. Kiesselbach. 1912. *Studies in Water Requirements of Corn*. Bulletin No. 128. Lincoln, NE: Nebraska Agricultural Experiment Station (cited by Tanner and Sinclair, 1983).

Moore, P.D. 1992. A leaf for all seasons. *Nature* 360: 110–111.

Moore, T.R. and M. Dalva. 1993. The influence of temperature and water table position on carbon dioxide and methane emissions from laboratory columns of peatland soils. *J. Soil Sci.* 44: 651–664.

Morgan, J.A. 2002. Perspectives: Ecology. Looking beneath the surface. *Science* 298: 1903–1904.

Morgan, J.A., D.R. LeCain, A.R. Mosier, and D.G. Milchunas. 2001. Elevated CO_2 enhances water relations and productivity and affects gas exchange in C_3 and C_4 grasses of the Colorado shortgrass steppe. *Global Change Biol.* 7: 451–466.

Morgan, J.A., D.E. Pataki, C. Körner, H. Clark, S.J. Del Grosso, J.M. Grünzweig, A.K. Knapp et al. 2004. Water relations in grassland and desert ecosystems exposed to elevated atmospheric CO_2. *Oecologia* 140: 11–25.

Morison, J.I.L. 1987a. Intercellular CO_2 concentration and stomatal response to CO_2. In Zeiger, Z., Farquhar, G.D., and Cowan, I.R., Eds., *Stomatal Function*. Stanford, CA: Stanford University Press, pp. 229–251.

Morison, J.I.L. 1987b. Climatology. Plant growth and CO_2 history. *Nature* 327: 560.

Morison, J.I.L. 1993. Response of plants to CO_2 under water limited conditions. *Vegetatio* 104/105: 193–209.

Morse, S.R., P. Wayne, S.L. Miao, and F.A. Bazzaz. 1993. Elevated CO_2 and drought alter tissue water relations of birch (*Betula populifolia* Marsh.) seedlings. *Oecologia* 95: 599–602.

Mortensen, L.M. 1994a. Effects of carbon dioxide concentration on assimilate partitioning, photosynthesis and transpiration of *Betula pendula* Roth. and *Picea abies* (L.) Karst. seedlings at two temperatures. *Acta Agric. Scand.* 44: 164–169.

Mortensen, L.M. 1994b. Effects of elevated CO_2 concentrations on growth and yield of eight vegetable species in a cool climate. *Sci. Hort.* 58: 177–185.

Mortlock, M.Y. and G.L. Hammer. 1999. Genotype and water limitation effects on transpiration efficiency in sorghum. *J. Crop Prod.* 2: 265–286.

Mortvedt, J.J. (Editor-in-Chief). 1986. Fellows of SSSA Elected in 1985. Elizabeth L. Klepper. *Soil Sci. Soc. Am. J.* 50: 521.

Moses, V. 1997. Obituary. Melvin Calvin (1911–97). *Nature* 385: 586.

Mott, K.A. 2009. Opinion: Stomatal responses to light and CO_2 depend on the mesophyll. *Plant Cell Environ.* 32: 1479–1486.

Mousseau, M., E. Dufrêne, A. El Kohen, D. Epron, D. Godard, R. Liozon, J.Y. Pontailler, and B. Saugier. 1996. Growth strategy and tree response to elevated CO_2: A comparison of beech (*Fagus sylvatica*) and sweet chestnut (*Castanea sativa* Mill.). In Koch, G.W. and Mooney, H.A., Eds., *Carbon Dioxide and Terrestrial Ecosystems*. San Diego, CA: Academic Press, pp. 71–103.

Müller, J. 1993. Stoffbildung, CO_2-Gaswechsel, Kohlenhydrat- und Stickstoffgehalt von Winterweizen bei erhöhter CO_2-Konzentration und Trockenstreß. [Dry matter production, CO_2 exchange, carbohydrate and nitrogen content of winter wheat at elevated CO_2 concentrations and drought stress.] *J. Agron. Crop Sci.* 171: 217–235. [In German, with an English summary.]

Murray, D.R. 1997. *Carbon Dioxide and Plant Responses*. Taunton, Somerset, U.K.: Research Studies Press, Ltd., and New York: John Wiley & Sons, 275pp.

Nachtergaele, F.O. 2005. FAO [Food and Agriculture Organization]. In Hillel, D., Ed., *Encyclopedia of Soils in the Environment* (Four Volumes), Vol. 1. Amsterdam: Elsevier, pp. 216–222.

Nadeau, J.A. and F.D. Sack. 2002. Control of stomatal distribution on the *Arabidopsis* leaf surface. *Science* 296: 1697–1700.

Nakayama, F.S. and D.A. Bucks. 1980. Using subsurface trickle system for carbon dioxide enrichment. In Jensen, M.H. and Oebker, N.F., Eds., *Proceedings of the 15th National Agricultural Plastics Congress*, Tucson, Arizona, April 13–17. Manchester, MO: National Agricultural Plastics Association, pp. 13–18.

Neales, T.F. and A.O. Nicholls. 1978. Growth responses of young wheat plants to a range of ambient CO_2 levels. *Aust. J. Plant Physiol.* 5: 45–59.

Nederhoff, E.M. 1994. *Effects of CO_2 Concentration on Photosynthesis, Transpiration and Production of Greenhouse Fruit Vegetable Crops*. PhD dissertation. Wageningen, the Netherlands: Agricultural University, 213pp.

Nederhoff, E.M. and K. Buitelaar. 1992. Effects of CO_2 on greenhouse grown eggplants (*Solanum melongena* L.). II. Leaf tip chlorosis and fruit production. *J. Hort. Sci.* 67: 805–812.

Nederhoff, E.M. and R. de Graaf. 1993. Effects of CO_2 on leaf conductance and canopy transpiration of greenhouse grown cucumber and tomato. *J. Hort. Sci.* 68: 925–937.

Nederhoff, E.M., A.A. Rijsdijk, and R. de Graaf. 1992. Leaf conductance and rate of crop transpiration of greenhouse grown sweet pepper (*Capsicum annuum* L.) as affected by carbon dioxide. *Sci. Hort.* 52: 283–301.

Negi, J., O. Matsuda, T. Nagasawa, Y. Oba, H. Takahashi, M. Kawai-Yamada, H. Uchimiya, M. Hashimoto, and K. Iba. 2008. CO_2 regulator SLAC1 and its homologues are essential for anion homeostasis in plant cells. *Nature* 452: 483–486.

Nelson, J.A., J.A. Morgan, D.R. LeCain, A.R. Mosier, D.G. Milchunas, and B.A. Parton. 2004. Elevated CO_2 increases soil moisture and enhances plant water relations in a long-term field study in semi-arid shortgrass steppe of Colorado. *Plant Soil* 259: 169–179.

Nepstad, D.C., C.R. de Carvalho, E.A. Davidson, P.H. Jipp, P.A. Lefebvre, G.H. Negreiros, E.D. da Silva, T.A. Stone, S.E. Trumbore, and S. Vieira. 1994. The role of deep roots in the hydrological and carbon cycles of Amazonian forests and pastures. *Nature* 372: 666–669.

Newcomb, E.H., G.C. Gerloff, and W.F. Whittingham. 1964. *Plants in Perspective: A Laboratory Manual of Modern Biology*. San Francisco: W.H. Freeman and Company, 218pp.

Newman, E.I. 1965. A method for measuring the total length of the root. *J. Appl. Ecol.* 2: 139–140.

Newton, P.C.D. 1991. Direct effects of increasing carbon dioxide on pasture plants and communities. *N. Z. J. Agric. Res.* 34: 1–24.

Newton, P.C.D., C.C. Bell, and H. Clark. 1996a. Carbon dioxide emissions from mineral springs in Northland and the potential of these sites for studying the effects of elevated carbon dioxide on pastures. *N. Z. J. Agric. Res.* 39: 33–40.

Newton, P.C.D., H. Clark, C.C. Bell, and E.M. Glasgow. 1996b. Interaction of soil moisture and elevated CO_2 on the above-ground rate, root length density and gas exchange of turves from temperate pasture. *J. Exp. Bot.* 47: 771–779.

Newton, P.C.D., H. Clark, C.C. Bell, E.M. Glasgow, K.R. Tate, D.J. Ross, and G.W. Yeates. 1995. Plant growth and soil processes in temperate grassland communities at elevated CO_2. *J. Biogeogr.* 22: 235–240.

Newton, R.F. 1971. Thermodynamics. *Encycl. Brit.* 21: 1015–1026.

Nicolas, M.E., R. Munns, A.B. Samarakoon, and R.M. Gifford. 1993. Elevated CO_2 improves the growth of wheat under salinity. *Aust. J. Plant Physiol.* 20: 349–360.

Nie, D., H. He, M.B. Kirkham, and E.T. Kanemasu. 1992a. Photosynthesis of a C_3 grass and a C_4 grass under elevated CO_2. *Photosynthetica* 26:189–198.

Nie, D., M.B. Kirkham, L.K. Ballou, D.J. Lawlor, and E.T. Kanemasu. 1992b. Changes in prairie vegetation under elevated carbon dioxide levels and two soil moisture regimes. *J. Veg. Sci.* 3: 673–678.

Nijs, I., I. Impens, and T. Behaeghe. 1989. Effects of long-term elevated CO_2 concentration on *Lolium perenne* and *Trifolium repens* canopies in the course of a terminal drought stress period. *Can. J. Bot.* 67: 2720–2725.

Nilson, S.E. and S.M. Assmann. 2007. The control of transpiration. Insights from Arabidopsis. *Plant Physiol.* 143: 19–27.

Nix, H.A. and E.A. Fitzpatrick. 1969. An index of crop water stress related to wheat and grain sorghum yields. *Agric. Meteorol.* 6: 321–337.

Nobel, P.S. and A.A. Israel. 1994. Cladode development, environmental responses of CO_2 uptake, and productivity for *Opuntia ficus-indica* under elevated CO_2. *J. Exp. Bot.* 45: 295–303.

Nobel, P.S. and J.A. Palta. 1989. Soil O_2 and CO_2 effects on root respiration of cacti. *Plant Soil* 120: 263–271.

Nobel, P.S., M. Cui, P.M. Miller, and Y. Luo. 1994. Influences of soil volume and an elevated CO_2 level on growth and CO_2 exchange for the Crassulacean acid metabolism plant *Opuntia ficus-indica*. *Physiol. Plant.* 90: 173–180.

Norby, R. 1997. Carbon cycle. Inside the black box. *Nature* 388: 522–523.

North, G.B., T.L. Moore, and P.S. Nobel. 1995. Cladode development for *Opuntia ficus-indica* (Cactaceae) under current and doubled CO_2 concentration. *Am. J. Bot.* 82: 159–166.

Northen, H.T. and R.T. Northen. 1973. *Greenhouse Gardening*, 2nd edn. New York: The Ronald Press Company, 388pp.

Novero, R., D.H. Smith, F.D. Moore, J.F. Shanahan, and R. d'Andria. 1991. Field-grown tomato responses to carbonated water application. *Agron. J.* 83: 911–916.

Nowak, R.S., S.F. Zitzer, D. Babcock, V. Smith-Longozo, T.N. Charlet, J.S. Coleman, J.R. Seemann, and S.D. Smith. 2004a. Elevated atmospheric CO_2 does not conserve soil water in the Mojave Desert. *Ecology* 85: 93–99.

Nowak, R.S., D.S. Ellsworth, and S.D. Smith. 2004b. Functional responses of plants to elevated atmospheric CO_2—Do photosynthetic and productivity data from FACE experiments support early predictions? *New Phytol.* 162: 253–280.

Nuttonson, M.Y. 1948. Some preliminary observations of phenological data as a tool in the study of photoperiodic and thermal requirements of various plant material. In Murneek, A.E. and Whyte, R.O., Eds., *Vernalization and Photoperiodism*. Waltham, MA: Chronica Botanica Company, pp. 129–143.

Nuttonson, M.Y. 1955. *Wheat-Climate Relationships and the Use of Phenology in Ascertaining the Thermal and Photo-Thermal Requirements of Wheat*. Washington, DC: American Institute of Crop Ecology, 388pp.

O'Leary, J.W. and G.N. Knecht. 1981. Elevated CO_2 concentration increases stomata [sic] numbers in *Phaseolus vulgaris* leaves. *Bot. Gaz.* 142: 438–441.

O'Leary, J.W. and J. Tarquinio Prisco. 1970. Response of osmotically stressed plants to growth regulators. *Adv. Front. Plant Sci.* 25: 129–139.

Oechel, W.C. and B.R. Strain. 1985. Native species responses to increased atmospheric carbon dioxide concentration. In Strain, B.R. and Cure, J.D., Eds., *Direct Effects of Increasing Carbon Dioxide on Vegetation*. DOE/ER-0238. Washington, DC: United States Department of Energy, Office of Energy Research, Office of Basic Energy Sciences, Carbon Dioxide Research Division, pp. 117–154.

Oesper, R.E. 1971. Henry, William. *Encycl. Brit.* 11: 380.

Oikawa, S., K.-M. Miyagi, K. Hikosaka, M. Okada, T. Matsunami, M. Kokubun, T. Kinugasa, and T. Hirose. 2010. Interactions between elevated CO_2 and N_2-fixation determine soybean yield—a test using a non-nodulated mutant. *Plant Soil* 330: 163–172.

Ort, D.R., E.A. Ainsworth, M. Aldea, D.J. Allen, C.J. Bernacchi, M.R. Berenbaum, G.A. Bollero et al. 2006. SoyFACE: The effects and interactions of elevated $[CO_2]$ and $[O_3]$ on soybean. In Nösberger, J., Long, S.P., Norby, R.J., Stitt, M., Hendrey, G.R., and Blum, H., Eds., *Managed Ecosystems and CO_2. Case Studies, Processes, and Perspectives*. Berlin: Springer-Verlag, pp. 71–86.

Osozawa, S. and S. Hasegawa. 1995. Diel and seasonal changes in carbon dioxide concentration and flux in an Andisol. *Soil Sci.* 160: 117–124.

Ottman, M.J., B.A. Kimball, P.J. Pinter, G.W. Wall, R.L. Vanderlip, S.W. Leavitt, R.L. LaMorte, A.D. Matthias, and T.J. Brooks. 2001. Elevated CO_2 increases sorghum biomass under drought conditions. *New Phytol.* 150: 261–273.

Overdieck, D. and M. Forstreuter. 1994. Evapotranspiration of beech stands and transpiration of beech leaves subject to atmospheric CO_2 enrichment. *Tree Physiol.* 14: 997–1003.

Owensby, C.E., P.I. Coyne, J.M. Ham, L.M. Auen, and A.K. Knapp. 1993. Biomass production in a tallgrass prairie ecosystem exposed to ambient and elevated CO_2. *Ecol. Appl.* 3: 644–653.

Paoletti, E. and R. Gellini. 1993. Stomatal density variation in beech and holm oak leaves collected over the last 200 years. *Acta Oecol.* 14: 173–178.

Parkhurst, D.F. 1994. Diffusion of CO_2 and other gases inside leaves. Tansley Review No. 65. *New Phytol.* 126: 449–479.

Parry, M.A.J., E. Delgado, J. Vadell, A.J. Keys, D.W. Lawlor, and H. Medrano. 1993. Water stress and the diurnal activity of ribulose-1,5-bisphosphate carboxylase in field grown *Nicotiana tabacum* genotypes selected for survival at low CO_2 concentrations. *Plant Physiol. Biochem.* 31: 113–120.

Patiño-Zúñiga, L., J.A. Ceja-Navarro, B. Govaerts, M. Luna-Guido, K.D. Sayre, and L. Dendooven. 2009. The effect of different tillage and residue management practices on soil characteristics, inorganic N dynamics and emissions of N_2O, CO_2 and CH_4 in the central highlands of Mexico: A laboratory study. *Plant Soil* 314: 231–241.

Patterson, D.T. and E.P. Flint. 1990. Implications of increasing carbon dioxide and climate change for plant communities and competition in natural and managed ecosystems. In Kimball, B.A., Rosenberg, N.J., Allen, Jr., L.H., Eds., *Impact of Carbon Dioxide, Trace Gases, and Climate Change on Global Agriculture*. Madison, WI: American Society of Agronomy, Crop Science Society of America, Soil Science Society of America, pp. 83–110.

Pearson, M. and G.L. Brooks. 1995. The influence of elevated CO_2 on growth and age-related changes in leaf gas exchange. *J. Exp. Bot.* 46: 1651–1659.

Pendall, E., A.R. Mosier, and J.A. Morgan. 2004a. Rhizodeposition stimulated by elevated CO_2 in a semiarid grassland. *New Phytol.* 162: 447–458.

Pendall, E., S. Bridgham, P.J. Hanson, B. Hungate, D.W. Kicklighter, D.W. Johnson, B.E. Law et al. 2004b. Research review. Below-ground process responses to elevated CO_2 and temperature: A discussion of observations, measurement methods, and models. *New Phytol.* 162: 311–322.

Penman, H.L. 1948. Natural evaporation from open water, bare soil, and grass. *Proc. Roy. Soc. London* 193A: 120–145.

Pennisi, E. 2004. Museums that made a master. *Science* 305: 37.

Peñuelas, J. and J. Azcón-Bieto. 1992. Changes in leaf $\Delta^{13}C$ of herbarium plant species during the last 3 centuries of CO_2 increase. *Plant Cell Environ.* 15: 485–489.

Peñuelas, J. and R. Matamala. 1990. Changes in N and S leaf content, stomatal density and specific leaf area of 14 plant species during the last three centuries of CO_2 increase. *J. Exp. Bot.* 41: 1119–1124.

Pérez-López, U., A. Robredo, M. Lacuesta, C. Sgherri, A. Muñoz-Rueda, F. Navari-Izzo, and A. Mena-Petite. 2009. The oxidative stress caused by salinity in two barley cultivars is mitigated by elevated CO_2. *Physiol. Plant.* 135: 29–42.

Peterson, T.C. and M.O. Baringer, Eds. 2009a. State of the Climate in 2008. *Bull. Am. Meteorol. Soc.* 90: S1–S196.

Peterson, T.C. and M.O. Baringer. 2009b. Introduction [in "State of the Climate in 2008"]. *Bull. Am. Meteorol. Soc.* 90(8): S13.

Pettersson, R., H.S.J. Lee, and P.G. Jarvis. 1993. The effect of CO_2 concentration on barley. In Rozema, J., Lambers, H., van de Geijn, S.C., and Cambridge, M.L., Eds., *CO$_2$ and Biosphere*. Dordrecht, the Netherlands: Kluwer Academic Publishers, pp. 462–463.

Philip, J.R. 1998. Physics, mathematics, and the environment: The 1997 Priestley lecture. *Aust. Meteorol. Mag.* 47: 273–283.

Pinter, P.J., Jr., B.A. Kimball, R.L. Garcia, G.W. Wall, D.J. Hunsaker, and R.L. LaMorte. 1996. Free-air CO_2 enrichment: Responses of cotton and wheat crops. In Koch, G.W and Mooney, H.A., Eds., *Carbon Dioxide and Terrestrial Ecosystems*. San Diego, CA: Academic Press, pp. 215–249.

Pohl, R.W. 1968. *How to Know the Grasses*, Revised edn. Dubuque, IA: Wm. C. Brown Company Publishers, 244pp.

Polley, H.W., H.B. Johnson, and H.S. Mayeux. 1992. Growth and gas exchange of oats (*Avena sativa*) and wild mustard (*Brassica kaber*) at subambient CO_2 concentrations. *Int. J. Plant Sci.* 153: 453–461.

Polley, H.W., H.B. Johnson, and H.S. Mayeux. 1994. Increasing CO_2: Comparative responses of the C_4 grass *Schizachyrium* and grassland invader *Prosopis*. *Ecol.* 75: 976–988.

Polley, H.W., H.S. Mayeux, H.B. Johnson, and C.R. Tischler. 1997. Viewpoint: Atmospheric CO_2, soil water, and shrub/grass ratios on rangelands. *J. Range Manage.* 50: 278–284.

Poole, I., J.D.B. Weyers, T. Lawson, and J.A. Raven. 1996. Variations in stomatal density and index: Implications for palaeoclimatic reconstructions. *Plant Cell Environ.* 19: 705–712 (plus cover).

Poole, I., T. Lawson, J.D.B. Weyers, and J.A. Raven. 2000. Effect of elevated CO_2 on the stomatal distribution and leaf physiology of *Alnus glutinosa*. *New Phytol.* 145: 511–521.

Porter, J.R. 2005. Rising temperatures are likely to reduce crop yields. *Nature* 436: 174.

Prasad, P.V.V., K.J. Boote, L.H. Allen, Jr., and J.M.G. Thomas. 2002. Effects of elevated temperature and carbon dioxide on seed-set and yield of kidney bean (*Phaseolus vulgaris* L.). *Global Change Biol.* 8: 710–721.

Prasad, P.V.V., K.J. Boote, L.H. Allen, Jr., and J.M.G. Thomas. 2003. Super-optimal temperatures are detrimental to peanut (*Arachis hypogaea* L.) reproductive processes and yield at both ambient and elevated carbon dioxide. *Global Change Biol.* 9: 1775–1787.

Prasad, P.V.V., L.H. Allen, Jr., and K.J. Boote. 2005. Crop responses to elevated carbon dioxide and interaction with temperature: Grain legumes. In Tuba, Z., Ed., *Ecological Responses and Adaptations of Crops to Rising Atmospheric Carbon Dioxide*. New York: Food Products Press. An Imprint of the Haworth Press, pp. 113–155.

Prévost, D., A. Bertaud, C. Juge, and F.P. Chalifour. 2010. Elevated CO_2 induces differences in nodulation of soybean depending on bradyrhizobial strain and method of inoculation. *Plant Soil* 331: 115–127.

Prévost-Bouré, N.C., J. Ngao, D. Berveiller, D. Bonal, C. Damesin, E. Dufrêne, J.-C. Lata et al. 2009. Root exclusion through trenching does not affect the isotopic composition of soil CO_2 efflux. *Plant Soil* 319: 1–13.

Price, H.L. 1911. *The Application of Meteorological Data in the Study of Physiological Constants*. Annual Report 1909–1910. Blacksburg, VA: Virginia Agricultural Experiment Station, pp. 206–212.

Priestley, C.H.B. and R.J. Taylor 1972. On the assessment of surface heat flux and evaporation using large-scale parameters. *Month. Weather Rev.* 100: 81–92.

Prior, S.A. and H.H. Rogers. 1995. Soybean growth response to water supply and atmospheric carbon dioxide enrichment. *J. Plant Nutr.* 18: 617–636.

Prior, S.A., H.H. Rogers, N. Sionit, and R.P. Patterson. 1991. Effects of elevated atmospheric CO_2 on water relations of soya bean. *Agric. Ecosyst. Environ.* 35: 13–25.

Prior, S.A., H.H. Rogers, G.B. Runion, and G.R. Hendrey. 1994a. Free-air CO_2 enrichment of cotton. Vertical and lateral root distribution patterns. *Plant Soil* 165: 33–44.

Prior, S.A., H.H. Rogers, G.B. Runion, and J.R. Mauney. 1994b. Effects of free-air CO_2 enrichment on cotton root growth. *Agric. Forest Meteorol.* 70: 69–86.

Prior, S.A., H.H. Rogers, G.B. Runion, B.A. Kimball, J.R. Mauney, K.F. Lewin, J Nagy, and G.R. Hendry. 1995. Free-air carbon dioxide enrichment of cotton: Root morphological characteristics. *J. Environ. Qual.* 24: 678–683.

Prior, S.A., H.H. Rogers, G.L. Mullins, and G.B. Runion. 2003. The effects of elevated atmospheric CO_2 and soil P placement on cotton root deployments. *Plant Soil* 255: 1709–187.

Prior, S.A., G.B. Runion, H.A. Torbert, and H.H. Rogers. 2004. Elevated atmospheric CO_2 in agroecosystems: Soil physical properties. *Soil Sci.* 169: 434–439.

Pugsley, A.T. 1963. The inheritance of a vernalization response in Australian spring wheats. *Aust. J. Agric. Res.* 14: 622–627.

Pugsley, A.T. 1965. Inheritance of a correlated day-length response in spring wheat. *Nature* 207: 108.

Pugsley, A.T. 1966. The photoperiodic sensitivity of some spring wheats with special reference to the variety Thatcher. *Aust. J. Agric. Res.* 17: 591–599.

Purvis, O.N. 1934. An analysis of the influence of temperature during germination on the subsequent development of certain winter cereals and its relation to the effect of length of day. *Ann. Bot.* 48: 919–956.

Qaderi, M.M., L.V. Kurepin, and D.M. Reid. 2006. Growth and physiological responses of canola (*Brassica napus*) to three components of global climate change: Temperature, carbon dioxide and drought. *Physiol. Plant.* 128: 710–721.

Rabbinge, R., H.C. van Latesteijn, and J. Goudriaan. 1993. Assessing the greenhouse effect in agriculture. In Lake, J.V., Bock, G.R., and Ackrill, K., Eds., *Environmental Health and Human Health*. Ciba Foundation Symposium 175. Chichester, U.K.: John Wiley & Sons, Ltd., pp. 62–79.

Rachidi, F., M.B. Kirkham, L.R. Stone, and E.T. Kanemasu. 1993a. Soil water depletion by sunflower and sorghum under rainfed conditions. *Agric. Water Manage.* 24: 49–62.

Rachidi, F., M.B. Kirkham, E.T. Kanemasu, and L.R. Stone. 1993b. Energy balance comparison of sorghum and sunflower. *Theoret. Appl. Climatol.* 48: 29–39.

Radoglou, K.M. and P.G. Jarvis. 1990a. Effects of CO_2 enrichment on four poplar clones. II. Leaf surface properties. *Ann. Bot.* 65: 627–632.

Radoglou, K.M. and P.G. Jarvis. 1990b. Effects of CO_2 enrichment on four poplar clones. I. Growth and leaf anatomy. *Ann. Bot.* 65: 617–626.

Radoglou, K.M. and P.G. Jarvis. 1993. Effects of atmospheric CO_2 enrichment on early growth of *Vivia* [sic] *faba*, a plant with large cotyledons. *Plant Cell Environ.* 16: 93–98.

Rasulov, B., K. Hüve, M. Välbe, A. Laisk, and U. Niinemets. 2009. Evidence that light, carbon dioxide, and oxygen dependencies of leaf isoprene emissions are driven by energy status in hybrid aspen. *Plant Physiol.* 151: 448–460.

Raven, P.H. 2003. Biodiversity and the future. *Am. Sci.* 91: 382.

Rawson, H.M. 1970. Spikelet number, its control, and relation to yield per ear in wheat. *Aust. J. Biol. Sci.* 23: 1–15.

Rawson, H.M. 1971. An upper limit for spikelet number per ear in wheat as controlled by photoperiod. *Aust. J. Agric. Res.* 22: 537–546.

Rawson, H.M. and J.E. Begg. 1977. The effect of atmospheric humidity on photosynthesis, transpiration and water use efficiency of leaves of several plant species. *Planta* 134: 5–10.

Reddy, K.R., H.F. Hodges, and J.M. McKinion. 1992. Cotton crop response to global climate change. *World Resource Rev.* 4: 348–365.

Reddy, K.R., H.F. Hodges, and B.A. Kimball. 2000. Crop ecosystem responses to climatic change: Cotton. In Reddy, K.R. and Hodges, H.F., Eds., *Climate Change and Global Crop Productivity*. Wallingford, Oxon, U.K.: CABI Publishing, pp. 161–187.

Reddy, K.R., P.V.V. Prasad, and V.G. Kakani. 2005. Crop responses to elevated carbon dioxide and interaction with temperature: Cotton. In Tuba, Z., Ed., *Ecological Responses and Adaptations of Crops to Rising Atmospheric Carbon Dioxide*. New York: Food Products Press. An Imprint of the Haworth Press, pp. 157–191.

Reekie, J.Y.C., P.R. Hicklenton, and E.G. Reekie. 1994. Effects of elevated CO_2 on time of flowering in four short-day and four long-day species. *Can. J. Bot.* 72: 533–538.

Reichman, O.J. 1987. *Konza Prairie: A Tallgrass Natural History*. Lawrence, KS: University Press of Kansas, 226pp.

Reicosky, D.C. and M.J. Lindstrom. 1993. Fall tillage method: Effect on short-term carbon dioxide flux from soil. *Agron. J.* 85: 1237–1243.

Retallack, G.J. 2001. A 300-million-year record of atmospheric carbon dioxide from fossil plant cuticles. *Nature* 411: 287–290.

Retallack, G.J. 2002. Reply. Atmospheric CO_2 from fossil plant cuticles. *Nature* 415: 38.

Richards, L.A., Ed. 1954. *Diagnosis and Improvement of Saline and Alkali Soils*. Agriculture Handbook No. 60. Washington, DC: United States Department of Agriculture, 160pp.

Riddell, J.A. and G.A. Gries. 1958a. Development of spring wheat. II. The effect of temperature on responses to photoperiod. *Agron. J.* 50: 739–742.

Riddell, J.A. and G.A. Gries. 1958b. Development of spring wheat. III. Temperature of maturation and age of seeds as factors influencing their response to vernalization. *Agron. J.* 50: 743–746.

Riddell, J.A., G.A. Gries, and F.W. Stearns. 1958. Development of spring wheat. I. The effect of photoperiod. *Agron. J.* 50: 735–738.

Riley, J.J. and H.E. Korab. 1971. Soft drinks. *Encycl. Brit.* 20: 829–830.

Robbins, W.W., T.E. Weier, and C.F. Stocking. 1964. *Botany: An Introduction to Plant Science*, 3rd edn. New York: John Wiley & Sons, 614pp.

Robertson, G.W. 1968. A biometeorological time scale for a cereal crop involving day and night temperatures and photo-period. *Int. J. Biometeorol.* 12: 191–233.

Robertson, E.J. and R.M. Leech. 1995. Significant changes in cell and chloroplast development in young wheat leaves (*Triticum aestivum* cv Hereward) grown in elevated CO_2. *Plant Physiol.* 107: 63–71.

Robinson, J.M. 1994a. Opinion. Speculations on carbon dioxide starvation, Late Tertiary evolution of stomatal regulation and floristic modernization. *Plant Cell Environ.* 17: 345–354.

Robinson, J.M. 1994b. Atmospheric CO_2 and plants. *Nature* 368: 105–106.

Robinson, A.B., N.E. Robinson, and W. Soon. 2007. Environmental effects of increased atmospheric carbon dioxide. *J. Am. Physicians Surg.* 12: 79–90.

Rogers, H.H., G.E. Bingham, J.D. Cure, J.M. Smith, and K.A. Surano. 1983. Responses of selected plant species to elevated carbon dioxide in the field. *J. Environ. Qual.* 12: 569–574.

Rogers, H.H., C.M. Peterson, J.N. McCrimmon, and J.D. Cure. 1992. Response of plant roots to elevated atmospheric carbon dioxide. *Plant Cell Environ.* 15: 749–752.

Rogers, H.H., G.B. Runion, and S.V. Krupa. 1994. Plant responses to atmospheric CO_2 enrichment with emphasis on roots and the rhizosphere. *Environ. Pollut.* 83: 155–189.

Rogers, H.H., S.A. Prior, G.B. Runion, and R.J. Mitchell. 1996. Root to shoot ratio of crops as influenced by CO_2. *Plant Soil* 187: 229–248.

Rogers, H.H., G.B. Runion, S.V. Krupa, and S.A. Prior. 1997. Plant responses to atmospheric carbon dioxide enrichment: Implications in root-soil-microbe interactions. In Allen, Jr., L.H., Kirkham, M.B., Olszyk, D.M., and Whitman, C., Eds., *Advances in Carbon Dioxide Effects Research*. Madison, WI: American Society of Agronomy, Crop Science Society of America, Soil Science Society of America, pp. 1–34.

Romell, L.G. 1922. Luftväxlingen i marken som ekologisk faktor. *Meddelanden Från Statens Skogsförsökanstalt* 19: 125–359 (in Swedish).

Rose, C.W. 1966. *Agricultural Physics*. Oxford, U.K.: Pergamon Press, 226pp.

Rose, É. 1985. *Water Relations of Winter Wheat. I. Genotypic Differences in Ethylene and Abscisic Acid Production during Drought Stress. II. Effect of a Dwarfing Gene on Root Growth, Shoot Growth, and Water Uptake in the Field.* PhD dissertation. Manhattan, KS: Kansas State University, 206pp + abstract (Dissertation Abstract No. 8510242).

Rosenberg, N.J. 1974. *Microclimate: The Biological Environment.* New York: John Wiley & Sons, 315pp.

Rosenberg, N.J. 1981. The increasing CO_2 concentration in the atmosphere and its implication on agricultural productivity. I. Effects on photosynthesis, transpiration and water use efficiency. *Climatic Change* 3: 265–279.

Rosenberg, N.J. 1982. The increasing CO_2 concentration in the atmosphere and its implication on agricultural productivity. II. Effects through CO_2-induced climatic change. *Climatic Change* 4: 239–254.

Rosenthal, W.D., E.T. Kanemasu, R.J. Raney, and L.R. Stone. 1977. Evaluation of an evapotranspiration model for corn. *Agron. J.* 69: 461–464.

Rosenberg, N.J., M.S. McKenney, and P. Martin. 1989. Evapotranspiration in a greenhouse-warmed world: A review and a simulation. *Agric. Forest Meteorol.* 47: 303–320.

Rosenzweig, C. and D. Hillel. 1998. *Climate Change and the Global Harvest. Potential Impacts of the Greenhouse Effect on Agriculture.* New York: Oxford University Press, 324pp.

Ross, D.J., S. Saggar, K.R. Tate, C.W. Feltham, and P.C.D. Newton. 1996. Elevated CO_2 effects on carbon and nitrogen cycling in grass/clover turves of a Psammaquent soil. *Plant Soil* 182: 185–198.

Ross, D.J., K.R. Tate, P.C.D. Newton, and H. Clark. 2002. Decomposability of C_3 and C_4 grass litter sampled under different concentrations of atmospheric carbon dioxide at a natural CO_2 spring. *Plant Soil* 240: 275–286.

Roth-Nebelsick, A. 2005. Reconstructing atmospheric carbon dioxide with stomata: Possibilities and limitations of a botanical pCO_2-sensor. *Trees* 19: 251–265.

Rowland-Bamford, A.J., C. Nordenbrock, J.T. Baker, G. Bowes, and L.H. Allen, Jr. 1990. Changes in stomatal density in rice grown under various CO_2 regimes with natural solar irradiance. *Environ. Exp. Bot.* 30: 175–180.

Roy, J., J.-L. Guillerm, M.-L. Navas, and S. Dhillion. 1996. Responses to elevated CO_2 in Mediterranean old-field microcosms: Species, community, and ecosystem components. In Körner, C., and Bazzaz, F.A., Eds., *Carbon Dioxide, Populations, and Communities.* San Diego, CA: Academic Press, pp. 123–138.

Rustad, L. 2001. Global change. Matter of time on the prairie. *Nature* 413: 578–579.

Sack, F.D. 2004. Yoda would be proud: Valves for land plants. *Science* 304: 1461–1462.

Sage, T.L. and R.F. Sage. 2009. The functional anatomy of rice leaves: Implications for refixation of photorespiratory CO_2 and efforts to engineer C_4 photosynthesis into rice. *Plant Cell Physiol.* 50: 756–772.

Salisbury, E.J. 1927. On the causes and ecological significance of stomatal frequency, with special reference to woodland flora. *Phil. Trans. Roy. Soc. London* 431B: 1–65.

Salisbury, F.B. and C.W. Ross. 1978. *Plant Physiology*, 2nd edn. Belmont, CA: Wadsworth Publishing Company, 436pp.

Salsman, K.J., D.N. Jordan, S.D. Smith, and D.S. Neuman. 1999. Effect of atmospheric CO_2 enrichment on root growth and carbohydrate allocation of *Phaseolus* spp. *Int. J. Plant Sci.* 160: 1075–1081.

Samarakoon, A.B., W.J. Muller, and R.M. Gifford. 1995. Transpiration and leaf area under elevated CO_2: Effects of soil water status and genotype of wheat. *Aust. J. Plant Physiol.* 22: 33–44.

Sanità de Toppi, L., A. Fabbri, M.A. Favali, T. Ganino, S. Grassi, and A. Raschi. 2003. Elevated CO_2 reduces vessel diameter and lignin deposition in some legume plants grown in mini-FACE rings. *Biol. Plant.* 46: 243–249.

Santner, A. and M. Estelle. 2009. Recent advances and emerging trends in plant hormone signalling. *Nature* 459: 1073–1078.

Šantrůček, J., H. Šantrůčková, J. Květoň, M. Šimková, and K. Roháček. 1994. The effect of elevated CO_2 concentration on photosynthetic CO_2 fixation, respiration and carbon economy of wheat plants. *Rostlinná Výroba* 40: 689–696.

Sator, C., U. Schenk, H.J. Weigel, and U. Menge-Hartmann. 1995. Auswirkung erhöhter CO_2-Konzentrationen auf die Stomatadichte von Kleeblättern (*Trifolium repens*). [Effects of high CO_2-concentrations on stomatal densities of white clover leaves.] *Landbauforshung Volkenrode* 45: 107–112 (in German with English summary).

Schaffer, B., C. Searle, A.W. Whiley, and R.J. Nissen. 1996. Effects of atmospheric CO_2 enrichment and root restriction on leaf gas exchange and growth of banana (*Musa*). *Physiol. Plant.* 97: 685–693.

Schaffer, B., A.W. Whiley, C. Searle, and R.J. Nissen. 1997. Leaf gas exchange, dry matter partitioning, and mineral element concentrations in mango as influenced by elevated atmospheric carbon dioxide and root restriction. *J. Am. Soc. Hort. Sci.* 122: 849–855.

Schellenberger, H. 1896. Beiträge zur Kenntiniss von Bau und Funktion der Spaltöffnungen. *Bot. Zeit.* 1: 169 (cited by Darwin, 1898).

Schnell, R.C. 2005. Carbon dioxide [in "State of the Climate in 2004"]. *Bull. Am. Meteorol. Soc.* 86(6): S20.

Senock, R.S., J.M. Ham, T.M. Loughin, B.A. Kimball, D.J. Hunsaker, P.J. Pinter, G.W. Wall, R.L. Garcia, and R.L. LaMorte. 1996. Sap flow in wheat under free-air CO_2 enrichment. *Plant Cell Environ.* 19: 147–158.

Shantz, H.L. and L.N. Piemeisel. 1927. The water requirement of plants at Akron, Colo. *J. Agric. Res.* 34: 1093–1190.

Shaw, M.R., E.S. Zavaleta, N.R. Chiariello, E.E. Cleland, H.A. Mooney, and C.B. Field. 2002. Grassland responses to global environmental changes suppressed by elevated CO_2. *Science* 298: 1987–1990.

Shimono, H., M. Okada, M. Inoue, H. Nakamura, K. Kobayashi, and T. Hasegawa. 2010. Diurnal and seasonal variations in stomatal conductance of rice at elevated atmospheric CO_2 under fully open-air conditions. *Plant Cell Environ.* 33: 322–331.

Showstack, R. 2005. Effects of drought on rainforest. *Eos* 86: 194.

Sicher, R.C. 2005. Interactive effects of inorganic phosphate nutrition and carbon dioxide enrichment on assimilate partitioning in barley roots. *Physiol. Plant.* 123: 219–226.

Siedow, J.N. and D.A. Day. 2000. Respiration and photorespiration. In: Buchanan, B.B., Gruissem, W., and Jones, R.L., Eds., *Biochemistry and Molecular Biology of Plants.* Rockville, MD: American Society of Plant Physiologists, pp. 676–728.

Singsaas, E.L., D.R. Ort, and E.H. DeLucia. 2003 [see note at end of reference about date]. Elevated CO_2 effects on mesophyll conductance and its consequences for interpreting photosynthetic physiology. *Plant Cell Environ.* 27: 41–50. (A printing error in January issue of this volume of *Plant, Cell and Environment* has all papers in this issue with the year printed as "2003." This is wrong. Volume 27 is year 2004.)

Sionit, N., H. Hellmers, and B.R. Strain. 1980. Growth and yield of wheat under CO_2 enrichment and water stress. *Crop Sci.* 20: 687–690.

Sionit, N., B.R. Strain, and H. Hellmers. 1981a. Effects of different concentrations of atmospheric CO_2 on growth and yield components of wheat. *J. Agric. Sci.* 79: 335–339.

Sionit, N., D.A. Mortensen, B.R. Strain, and H. Hellmers. 1981b. Growth response of wheat to CO_2 enrichment and different levels of mineral nutrition. *Agron. J.* 73: 1023–1027.

Sionit, N., H.H. Rogers, G.E. Bingham, and B.R. Strain. 1984. Photosynthesis and stomatal conductance with CO_2-enrichment of container and field-grown soybeans. *Agron. J.* 76: 447–451.

Sionit, N., B.R. Strain, H. Hellmers, G.H. Riechers, and C.H. Jaeger. 1985. Long-term atmospheric CO_2 enrichment affects the growth and development of *Liquidambar styraciflua* and *Pinus taeda* seedlings. *Can. J. Forest Res.* 15: 468–471.

Sircar, S.M. 1948. Vernalization and photoperiodism in the tropics. In Murneek, A.E. and Whyte, R.O., Eds., *Vernalization and Photoperiodism.* Waltham, MA: Chronica Botanica Company, pp. 121–128.

Slafer, G.A. and H.M. Rawson. 1997. CO_2 effects on phasic development, leaf number and rate of leaf appearance in wheat. *Ann. Bot.* 79: 75–81.

Slavík, B. 1963. The distribution pattern of transpiration rate, water saturation deficit, stoma number and size, photosynthetic and respiration rate in the area of the tobacco leaf blade. *Biol. Plant.* 5: 143–153.

Smith, L.P. 1967. The effect of weather on the growth of barley. *Field Crops Abstr.* 20: 273–278.

Smith, E.L. 1976. The genetics of wheat architecture. In *The Grasses and Grasslands of Oklahoma. Annals of the Oklahoma Academy of Science Publication No. 6.* Ardmore, OK: The Samuel Roberts Noble Foundation, pp. 117–132.

Smith, S.D., D.N. Jordan, and E.P. Hamerlynck. 1999. Effects of elevated CO_2 and temperature stress on ecosystem processes. In Luo, Y. and Mooney, H.A., Eds., *Carbon Dioxide and Environmental Stress.* San Diego, CA: Academic Press, pp. 107–137.

Smith, K.A., T. Ball, F. Conen, K.E. Dobbie, J. Massheder, and A. Rey. 2003. Exchange of greenhouse gases between soil and atmosphere: Interactions of soil physical factors and biological processes. *Eur. J. Soil Sci.* 54: 779–791.

Smith, S.D., T.E. Huxman, S.F. Zitzer, T.N. Charlet, D.C. Housman, J.S. Coleman, L.K. Fenstermaker, J.R. Seemann, and R.S. Nowak. 2000. Elevated CO_2 increases productivity and invasive species success in an arid ecosystem. *Nature* 408: 79–82.

Soil Science Society of America. 1997. *Glossary of Soil Science Terms.* Madison, WI: Soil Science Society of America, 138pp.

Soil Survey Staff. 1975. *Soil Taxonomy. A Basic System of Soil Classification for Making and Interpreting Soil Surveys.* Agriculture Handbook No. 436. Washington, DC: Soil Conservation Service, United States Department of Agriculture, 754pp.

Song, Y., J.M. Ham, M.B. Kirkham, and G.J. Kluitenberg. 1998. Measuring soil water content under turfgrass using the dual-probe heat-pulse technique. *J. Am. Soc. Hort. Sci.* 123: 937–941.

Song, Y., M.B. Kirkham, J.M. Ham, and G.J. Kluitenberg. 1999. Measurement resolution of the dual-probe heat-pulse technique. In van Genuchten, M.Th., Leij, F.J., and Wu, L., Eds., *Characterization and Measurement of the Hydraulic Properties of Unsaturated Porous Media*. Riverside, CA: University of California, pp. 381–386.

Song, Y., M.B. Kirkham, J.M. Ham, and G.J. Kluitenberg. 2000. Root-zone hydraulic lift evaluated with the dual-probe heat-pulse technique. *Aust. J. Soil Res.* 38: 927–935.

Sorauer, P. 1880. Studien über Verdunstung. *Forschungen auf dem Gebiete der Agrikultur-Physik* Bd. 3, pp. 351–490. See p. 467 (cited by Briggs and Shantz, 1913b).

Spencer, E.W. 1962. *Basic Concepts of Historial Geology*. New York: Thomas Y. Crowell Company, 504pp.

Stafford, N. 2007. The other greenhouse effect. *Nature* 448: 526–528.

Ståhle, N.K. 1971. Nobel prizes. *Encycl. Brit.* 16: 548–552.

Stanghellini, C. and J.A. Bunce. 1993. Response of photosynthesis and conductance to light, CO_2, temperature and humidity in tomato plants acclimated to ambient and elevated CO_2. *Photosynthetica* 29: 487–497.

Stanier, R.Y., M. Doudoroff, and E.A. Adelberg. 1963. *The Microbial World*, 2nd edn. Englewood Cliffs, NJ: Prentice-Hall, Inc., 753pp.

Stępniewski, W. and Z. Stępniewska. 1998. Oxygenology as a new discipline in the environmental sciences—a proposal for discussion. *Int. Agrophysics* 12: 53–55.

Stępniewski, W., Z. Stępniewska, R.P. Bennicelli, and J. Gliński. 2005. *Oxygenology in Outline*. EU Fifth Framework Program, QLAM-2001-00428. Lublin, Poland: Centre of Excellence for Applied Physics in Sustainable Agriculture, Institute of Agrophysics, 121pp. (Printed by ALF-GRAF, ul. Kościuszki 4, 20-006, Lublin, Poland.)

Stevens, P.F. 1990. In defence of taxonomy. *Nature* 347: 222–223.

Stirton, C.H., L. Boulos, T.D. MacFarlane, N.P. Singh, and A. Nicholas. 1990. In defence of taxonomy. *Nature* 347: 224.

Stolwijk, J.A.J. and K.V. Thimann. 1957. On the uptake of carbon dioxide and bicarbonate by roots, and its influence on growth. *Plant Physiol.* 32: 513–520.

Stone, J.F. 1983. On Julian day notation for meteorological conditions. *Agric. Meteorol.* 29: 137–140.

Stone, L.R. and M.L. Horton. 1974. Estimating evapotranspiration using canopy temperatures: Field evaluation. *Agron. J.* 66: 450–454.

Stone, L.R., M.B. Kirkham, D.E. Johnson, and E.T. Kanemasu. 1982. *Yield and Water Use of Alfalfa*. Keeping Up With Research No. 58. Manhattan, KS: Kansas State University Agricultural Experiment Station, 6pp.

Stone, L.R., D.E. Goodrum, A.J. Schlegel, M.N. Jaafar, and A.H. Khan. 2002. Water depletion depth of grain sorghum and sunflower in the Central High Plains. *Agron. J.* 94: 936–943.

Stone, L.R., F.R. Lamm, A.J. Schlegel, and N.L. Klocke. 2008. Storage efficiency of off-season irrigation. *Agron. J.* 100: 1185–1192.

Storlie, C.A. and J.R. Heckman. 1996. Bell pepper yield response to carbonated irrigation water. *J. Plant Nutr.* 19: 1477–1484.

Strain, B.R. and J.D. Cure, Eds. 1985a. *Direct Effects of Increasing Carbon Dioxide on Vegetation*. DOE/ER-0238. Washington, DC: United States Department of Energy, Office of Energy Research, Office of Basic Energy Sciences, Carbon Dioxide Research Division, 286pp.

Strain, B.R. and J.D. Cure. 1985b. Status of knowledge and recommendations for future work. In Strain, B.R. and Cure, J.D., Eds., *Direct Effects of Increasing Carbon Dioxide on Vegetation*. DOE/ER-0238. Washington, DC: United States Department of Energy, Office of Energy Research, Office of Basic Energy Sciences, Carbon Dioxide Research Division, pp. 205–213.

Stronach, I.M., S.C. Clifford, A.D. Mohamed, P.R. Singleton-Jones, S.N. Azam-Ali, and N.M.J. Crout. 1994. The effects of elevated carbon dioxide, temperature and soil moisture on the water use of stands of groundnut (*Arachis hypogaea* L.) *J. Exp. Bot.* 45: 1633–1638.

Sun, Y., H. Cao, J.Yin, L. Kang, and F. Ge. 2010. Elevated CO_2 changes the interactions between nematode and tomato genotypes differing in the JA pathway. *Plant Cell Environ.* 33: 729–739.

Syme, J.R. 1968. Ear emergence of Australia, Mexican and European wheats in relation to time of sowing and their response to vernalization and daylength. *Aust. J. Exp. Agric. Anim. Husb.* 8: 578–581.

Syme, J.R. 1973. Quantitative control of flowering time in wheat cultivars by vernalization and photoperiod sensitivities. *Aust. J. Agric. Res.* 24: 657–665.

Syvertsen, J.P., J. Lloyd, C. McConchie, P.E. Kriedemann, and G.D. Farquhar. 1995. On the relationship between leaf anatomy and CO_2 diffusion through the mesophyll of hypostomatous leaves. *Plant Cell Environ.* 18: 149–157.

Tackett, J.L. and R.W. Pearson. 1964a. Oxygen requirements of cotton seedling roots for penetration of compacted soil cores. *Soil Sci. Soc. Am. Proc.* 28: 600–605.

Tackett, J.L. and R.W. Pearson. 1964b. Effect of carbon dioxide on cotton seedling root penetration of compacted soil cores. *Soil Sci. Soc. Am. Proc.* 28: 741–743.

Tal, M., D. Imber, and C. Itai. 1970. Abnormal stomatal behavior and hormonal imbalance in *flacca*, a wilty mutant of tomato. I. Root effect and kinetin-like activity. *Plant Physiol.* 46: 367–372.

Talbott, L.D., E. Rahveh, and E. Zeiger. 2003. Relative humidity is a key factor in the acclimation of the stomatal response to CO_2. *J. Exp. Bot.* 54: 2141–2147.

Tanner, C.B. 1967. Measurement of evapotranspiration. In Hagan, R.M., Haise, H.R., and Edminster, T.W., Eds., *Irrigation of Agricultural Lands*. Madison, WI: American Society of Agronomy, pp. 534–574.

Tanner, C.B. and T.R. Sinclair. 1983. Efficient water use in crop production: Research or re-search? In Taylor, H.M., Jordan, W.R., and Sinclair, T.R., Eds., *Limitations to Efficient Water Use in Crop Production*. Madison, WI: American Society of Agronomy, Crop Science Society of America, Soil Science Society of America, pp. 1–27.

Tanner, C.B. and R.W. Simonson. 1993. Franklin Hiram King—Pioneer scientist. *Soil Sci. Soc. Am.* 57: 286–292.

Taylor, S.A. 1949. Oxygen diffusion in porous media as a measure of soil aeration. *Soil Sci. Soc. Am. Proc.* 14: 55–61.

Taylor, S.A. 1968. Terminology in plant and soil water relations. In Kozlowski, T.T., Ed., *Water Deficits and Plant Growth*, Vol. 1. *Development, Control, and Measurement*. New York: Academic Press, pp. 49–72.

Taylor, S.A. and G.L. Ashcroft. 1972. *Physical Edaphology: The Physics of Irrigated and Nonirrigated Soils*. San Francisco: W.H. Freeman and Company, 533pp.

Taylor, H.M., M.G. Huck, and B. Klepper. 1972. Root development in relation to soil physical conditions. In Hillel, D., Ed., *Optimizing the Soil Physical Environment Toward Greater Crop Yields*. New York: Academic Press, pp. 57–77.

Taylor, G., S. Ranasinghe, C. Bosac, S.D.L. Gardner, and R. Ferris. 1994. Elevated CO_2 and plant growth: Cellular mechanisms and responses of whole plants. *J. Exp. Bot.* 45: 1761–1774.

Teskey, R.O. 1995. A field study of the effects of elevated CO_2 on carbon assimilation, stomatal conductance and leaf and branch growth of *Pinus taeda* trees. *Plant Cell Environ.* 18: 565–573.

Thimann, K.V. 1969. The auxins. In Wilkins, M.B., Ed., *The Physiology of Plant Growth and Development*. New York: McGraw-Hill, pp. 1–45.

Thimann, K.V. 1972a. Ethylene. In Steward, F.C., Ed., *Plant Physiology. A Treatise. Vol. VIB. Physiology of Development: The Hormones*. New York: Academic Press, pp. 213–221.

Thimann, K.V. 1972b. Interactions between hormones. In Steward, F.C., Ed., *Plant Physiology. A Treatise. Vol. VIB. Physiology of Development: The Hormones*. New York: Academic Press, pp. 222–236.

Thimann, K.V. 1977. *Hormone Action in the Whole Life of Plants*. Amherst, MA: The University of Massachusetts Press, 448pp.

Thomas, R.B. and B.R. Strain. 1991. Root restriction as a factor in photosynthetic acclimation of cotton seedlings grown in elevated carbon dioxide. *Plant Physiol.* 96: 627–634.

Thomas, S.C. and F.A. Bazzaz. 1996. Elevated CO_2 and leaf shape: Are dandelions getting toothier? *Am. J. Bot.* 83: 106–111.

Thomas, S.M., D. Whitehead, J.A. Adams, J.B. Reid, R.R. Sherlock, and A.C. Leckie. 1996. Seasonal root distribution and soil surface carbon fluxes for one-year-old *Pinus radiata* trees growing at ambient and elevated carbon dioxide concentration. *Tree Physiol.* 16: 1015–1021.

Thorne, D.W. 1972. Dedication. In Brown, R.W. and Van Haveren, B.P., Eds., *Psychrometry in Water Relations Research*. Logan, UT: Utah Agricultural Experiment Station, Utah State University, pp. vi–viii.

Thornley, J.H.M. and M.G.R. Cannell. 1996. Temperate forest responses to carbon dioxide, temperature and nitrogen: A model analysis. *Plant Cell Environ.* 19: 1331–1348.

Thornthwaite, C.W. 1948. An approach toward a rational classification of climate. *Geogr. Rev.* 38: 55–94.

Tingey, D.T., M.G. Johnson, and D.L. Phillips. 2005. Independent and contrasting effects of elevated CO_2 and N-fertilization on root architecture in *Pinus ponderosa*. *Trees* 19: 43–50.

Tissue, D.T., K.L. Griffin, R.B. Thomas, and B.R. Strain. 1995. Effects of low and elevated CO_2 on C_3 and C_4 annuals. II. Photosynthesis and leaf biochemistry. *Oecologia* 101: 21–28.

Tolley, L.C. and B.R. Strain. 1984a. Effects of CO_2 enrichment on growth of *Liquidambar styraciflua* and *Pinus taeda* seedlings under different irradiance levels. *Can. J. Forest Res.* 14: 343–350.

Tolley, L.C. and B.R. Strain. 1984b. Effects of atmospheric CO_2 enrichment and water stress on growth of *Liquidamabar styraciflua* and *Pinus taeda* seedlings. *Can. J. Bot.* 62: 2135–2139.

Tolley, L.C. and B.R. Strain. 1985. Effects of CO_2 enrichment and water stress on gas exchange of *Liquidambar styraciflua* and *Pinus taeda* seedlings grown under different irradiance levels. *Oecologia* 65: 166–172.

Townend, J. 1993. Effects of elevated carbon dioxide and drought on the growth and physiology of clonal Sitka spruce plants (*Picea sitchensis* (Bong.) (Carr.). *Tree Physiol.* 13: 389–399.

Troughton, J.H. 1969. Plant water status and carbon dioxide exchange of cotton leaves. *Aust. J. Biol. Sci.* 22: 289–301.

Troughton, J.H. and R.O. Slatyer. 1969. Plant water status, leaf temperature, and the calculated mesophyll resistance to carbon dioxide of cotton leaves. *Aust. J. Biol. Sci.* 22: 815–827.

Tschaplinski, T.J., R.J. Norby, and S.D.Wullschleger. 1993. Responses of loblolly pine seedlings to elevated CO_2 and fluctuating water supply. *Tree Physiol.* 13: 282–296.

Tschaplinski, T.J., D.B. Stewart, and R.J. Norby. 1995. Interactions between drought and elevated CO_2 on osmotic adjustment and solute concentration of tree seedlings. *New Phytol.* 131: 169–177.

Tschaplinski, T.J., D.B. Stewart, P.J. Hanson, and R.J. Norby. 1995. Interactions between drought and elevated CO_2 on growth and gas exchange of seedlings of three deciduous tree species. *New Phytol.* 129: 63–71.

Tu, C., F.L. Booker, K.O. Burkey, and S. Hu. 2009. Elevated atmospheric carbon dioxide and O_3 differentially alter nitrogen acquisition in peanut. *Crop Sci.* 49: 1827–1836.

Tuba, Z., K. Szente, Z. Nagy, Z. Csintalan, J. Koch, G. Kemény, E. Laitar, E. Masarovicova, and Z. Takács. 1993. The response of Hungarian loess grassland species to long term elevated CO_2 (ecophysiological responses to the first 9.5 months exposure). *Hung. Agric. Res.* 2(1): 37–40.

Tuba, Z., K. Szente, and J. Koch. 1994. Response of photosynthesis, stomatal conductance, water use efficiency and production to long-term elevated CO_2 in winter wheat. *J. Plant Physiol.* 144: 661–668.

Turner, N.C. 1993. Water use efficiency of crop plants: Potential for improvement. In Buxton, D.R., Shibles, R., Forsberg, R.A., Blad, B.L., Asay, K.H., Paulsen, G.M., and Wilson, R.F., Eds., *International Crop Science I*. Madison, WI: Crop Science Society of America, pp. 75–82.

Uehlein, N., C. Lovisolo, F. Siefritz, and R. Kaldenhoff. 2003. The tobacco aquaporin NtAQP1 is a membrane CO_2 pore with physiological functions. *Nature* 425: 734–737.

Uehlein, N., B. Otto, D.T. Hanson, M. Fischer, N. McDowell, and R. Kaldenhoff. 2008. Function of *Nicotiana tabacum* aquaporins as chloroplast gas pores challenges the concept of membrane CO_2 permeability. *Plant Cell* 20: 648–657.

United States Department of Agriculture. 2009a. *Crops. Kansas Agricultural Statistics*, Vol. 9, No. 7, July 10, 2009. Topeka, KS: United States Department of Agriculture, National Agricultural Statistics Service, Kansas Field Office, 4pp.

United States Department of Agriculture. 2009b. *Agricultural Prices. Kansas Agricultural Statistics*, Vol. 9, No. 5, May 1, 2009. Topeka, KS: United States Department of Agriculture, National Agricultural Statistics Service, Kansas Field Office, 4pp.

United States Department of Agriculture. 2009c. *Crops. Kansas Agricultural Statistics*, Vol. 9, No. 6, June 10, 2009. Topeka, KS: United States Department of Agriculture, National Agricultural Statistics Service, Kansas Field Office, 4pp.

Vaadia, Y. and Y. Waisel. 1967. Physiological processes as affected by water balance. In Hagan, R.M., Haise, H.R., and Edminster, T.W., Eds., *Irrigation of Agricultural Lands*. Madison, WI: American Society of Agronomy, pp. 354–372.

Vahisalu, T., H. Kollist, Y.-F. Wang, N. Nishimura, W.-Y. Chan, G. Valerio et al. 2008. SLAC1 is required for plant guard cell S-type anion channel function in stomatal signalling. *Nature* 452: 487–491.

Valdés, J.B., R.S. Seoane, and G.R. North. 1994. A methodology for the evaluation of global warming impact on soil moisture and runoff. *J. Hydrol.* 161: 389–413.

Valentini, R., G. Matteucci, A.J. Dolman, E.-D. Schulze, C. Rebmann, E.J. Moore, A. Granier et al. 2000. Respiration as the main determinant of carbon balance in European forests. *Nature* 404: 861–865.

van Bavel, C.H.M. 1966. Potential evaporation: The combination concept and its experimental verification. *Water Resources Res.* 2: 455–467.

van Bavel, C.H.M. 1974. Antitranspirant action of carbon dioxide on intact sorghum plants. *Crop Sci.* 14: 208–212.

van Bavel, C.H.M. and L.E. Myers. 1962. An automatic weighing lysimeter. *Agric. Eng.* 43: 580–583, 587–588.

van Bavel, C.H.M., F.S. Nakayama, and W.L. Ehrler. 1965. Measuring transpiration resistance of leaves. *Plant Physiol.* 40: 535–540.

Van de Water, P.K., S.W. Leavitt, and J.L. Betancourt. 1994. Trends in stomatal density and $^{13}C/^{12}C$ ratios of *Pinus flexilis* needles during last glacial-interglacial cycle. *Science* 264: 239–243 (plus cover).

van den Honert, T.H. 1948. Water transport in plants as a catenary process. *Discuss. Faraday Soc.* 3: 146–153.

van der Ploeg, R.R., W. Ehlers, and R. Horn. 2006. Schwerlast auf dem Acker. [Heavy loads on fields.] *Spektrum der Wissenschaft* [*Spectrum of Science*], August issue, pp. 80–88 [in German].

van Ginkel, J.H., A. Gorissen, and J.A. van Veen. 1996. Long-term decomposition of grass roots as affected by elevated atmospheric carbon dioxide. *J. Environ. Qual.* 25: 1122–1128.

Van Klooster, H.S. and J.A. Passmore. 1971. Priestley, Joseph. *Encycl. Brit.* 18: 500–502.

van Vuuren, M.M.I., D. Robinson, A.H. Fitter, S.C. Chasalow, L. Williamson, and J.A. Raven. 1997. Effects of elevated atmospheric CO_2 and soil water availability on root biomass, root length, and N, P and K uptake by wheat. *New Phytol.* 135: 455–465.

Verheyden, A., F. De Ridder, N. Schmitz, H. Beeckman, and N. Koedam. 2005. High-resolution time series of vessel density in Kenyan mangrove trees reveal a link with climate. *New Phytol.* 167: 425–435.

Vivin, P., P. Gross, G. Aussenac, and J.-M. Guehl. 1995. Whole-plant CO_2 exchange, carbon partitioning and growth in *Quercus robur* seedlings exposed to elevated CO_2. *Plant Physiol. Biochem.* 33: 201–211.

Vogel, G. 2004. Berlin's scientific treasure house shakes off the dust. *Science* 305: 35–37.

Von Caemmerer, S., and H. Griffiths. 2009. Stomatal responses to CO_2 during a diel Crassulacean acid metabolism cycle in *Kalanchoe daigremontiana* and *Kalanchoe pinnata*. *Plant Cell Environ.* 32: 567–576.

Waggoner, P.E. 1974a. Modeling seasonality (pp. 301–323), with an appendix by J.-Y. Parlange, Analytic solution to model passages through phenophases (pp. 323–327). In Lieth, H., Ed., *Phenology and Seasonality Modeling*. New York: Springer-Verlag, pp. 301–327.

Waggoner, P.E. 1974b. Using models of seasonality. In Lieth, H., Ed., *Phenology and Seasonality Modeling*. New York: Springer-Verlag, pp. 401–405.

Waggoner, P.E. 1977. Simulation modelling of plant physiological processes to predict crop yields. In Landsberg, J.J. and Cuttings, C.V., Eds., *Environmental Effects on Crop Physiology*. London: Academic Press, pp. 351–359.

Waggoner, P.E. 1984. Agriculture and carbon dioxide. *Am. Sci.* 72: 179–184.

Wagner, F., D.L. Dilcher, and H. Visscher. 2005. Stomatal frequency responses in hardwood-swamp vegetation from Florida during a 60-year continuous CO_2 increase. *Am. J. Bot.* 92: 690–695.

Walker, R.F., D.R. Geisinger, D.W. Johnson, and J.T. Ball. 1997. Elevated atmospheric CO_2 and soil N fertility effects on growth, mycorrhizal colonization, and xylem water potential of juvenile ponderosa pine in a field soil. *Plant Soil* 195: 25–36.

Wall, P.C. and P.M. Cartwright. 1974. Effects of photoperiod, temperature and vernalization on the phenology and spikelet numbers of spring wheats. *Ann. Appl. Biol.* 76: 299–309.

Wall, G.W., R.L. Garcia, B.A. Kimball, D.J. Hunsaker, P.J. Pinter, Jr., S.P. Long, C.P. Osborne et al. 2006. Interactive effects of elevated carbon dioxide and drought on wheat. *Agron. J.* 98: 354–381.

Walthall, C.L. 2009. Managing agriculture in a climate of change. *Agric. Res.* 57(10): 2. [The entire issue of this journal (November/December 2009) describes research that the Agricultural Research Service is doing related to climate change.]

Wang, D., A.R. Portis, Jr., S.P. Moose, and S.P. Long. 2008. Cool C_4 photosynthesis: Pyruvate P_i dikinase expression and activity corresponds to the exceptional cold tolerance of carbon assimilation in *Miscanthus* x *giganteus*. *Plant Physiol.* 148: 557–567 (plus cover).

Ward, J.K. and B.R. Strain. 1997. Effects of low and elevated CO_2 partial pressure on growth and reproduction of *Arabidopsis thaliana* from different elevations. *Plant Cell Environ.* 20: 254–260.

Watson, B. 2008. Obituary. Bert Bolin (1925–2008). Pioneering Climate Scientist and Communicator. *Nature* 451: 642.

Weast, R.C., Ed. 1964. *Handbook of Chemistry and Physics*, 45th edn. Cleveland, OH: Chemical Rubber Company.

Webster's Collegiate Dictionary. 1939, 5th edn. Springfield, MA: G. & C. Merriam Company, 1,276pp.

Wechsung, G., F. Wechsung, G.W. Wall, F.J. Adamsen, B.A. Kimball, R.L. Garcia, P.J. Pinter, Jr., and T. Kartschall. 1995. Biomass and growth rate of a spring wheat root system grown in free-air CO_2 enrichment (FACE) and ample soil moisture. *J. Biogeogr.* 22: 623–634.

Weibel, D.E. 1958. Vernalization of immature winter wheat embryos. *Agron. J.* 50: 267–270.

Weigel, H.J., R. Manderscheid, H.-J. Jäger, and G.J. Mejer. 1994. Effects of season-long CO_2 enrichment on cereals: I. Growth performance and yield. *Agric. Ecosyst. Environ.* 48: 231–240.

Weiss, I., Y. Mizrahi, and E. Raveh. 2009. Synergistic effects of elevated CO_2 and fertilization on net CO_2 uptake and growth of the CAM plant *Hylocereus undatus*. *J. Am. Soc. Hort. Sci.* 134: 364–371.

West, J.G. and B.J. Conn. 1990. In defence of taxonomy. *Nature* 347: 222.

West, N.E. and R.W. Wein. 1971. A plant phenological index technique. *BioScience* 21(3): 116–117.

Wheeler, Q.D., P.H. Raven, and E.O. Wilson. 2004. Taxonomists and conservation. Response. *Science* 305: 1104–1105.

White, J.W.C., P. Ciais, R.A. Figge, R. Kenny, and V. Markgraf. 1994. A high-resolution record of atmospheric CO_2 content from carbon isotopes in peat. *Nature* 367:153–156.

Whitehead, D., J.R. Leathwick, and A.S. Walcroft. 2001. Modeling annual carbon uptake for the indigenous forests of New Zealand. *Forest Sci.* 47: 9–20.

Whitehead, D., A.S. Walcroft, K.L. Griffin, D.T. Tissue, M.H. Turnbull, V. Engel, K.J. Brown, and W.S.F. Schuster. 2004. Scaling carbon uptake from leaves to canopies: Insights from two forests with contrasting properties. In Mencuccini, M., Grace, J., Moncrieff, J., and McNaughton, K.G., Eds., *Forests at the Land–Atmosphere Interface*. Wallingford, U.K.: CAB International, pp. 231–254.

Whyte, R.O. 1948. History of research in vernalization. In Murneek, A.E. and Whyte, R.O., Eds., *Vernalization and Photoperiodism*. Waltham, MA: Chronica Botanica Company, pp. 1–38.

Widodo, W., J.C.V. Vu, K.J. Boote, J.T. Baker, and L.H. Allen, Jr. 2003. Elevated growth CO_2 delays drought stress and accelerates recovery of rice leaf photosynthesis. *Environ. Exp. Bot.* 49: 259–272.

Widtsoe, J.A. 1911. *Dry-Farming. A System of Agriculture for Countries under a Low Rainfall*. New York: The Macmillan Company, 445pp.

Wiegand, C.L. and E.R. Lemon. 1958. A field study of some plant-soil relations in aeration. *Soil Sci. Soc. Am. Proc.* 22: 216–221.

Wielgolaski, F.-E. 1974. Phenology in agriculture. In Lieth, H., Ed., *Phenology and Seasonality Modeling*. New York: Springer-Verlag, pp. 369–381.

Williams, S.N. 1995. Dead trees tell tales. *Nature* 376: 644.

Williams, R.F. and C.N. Williams. 1968. Physiology of growth in the wheat plant. IV. Effects of day length and light energy level. *Aust. J. Biol. Sci.* 21: 835–854.

Willis, W.O., W.E. Larson, and D. Kirkham. 1957. Corn growth as affected by soil temperature and mulch. *Agron. J.* 49: 323–328.

Wittwer, S.H. 1983. Rising atmospheric CO_2 and crop productivity. *HortScience* 18: 667–673.

Wittwer, S.H. 1995. *Food, Climate, and Carbon Dioxide. The Global Environment and World Food Production*. Boca Raton, FL: Lewis Publishers. An imprint of CRC Press, 236pp.

Woodward, F.I. 1987. Stomatal numbers are sensitive to increases in CO_2 from pre-industrial levels. *Nature* 327: 617–618.

Woodward, F.I. 1988. The responses of stomata to changes in atmospheric levels of CO_2. *Plants Today* 1: 132–135.

Woodward, F.I. and F.A. Bazzaz. 1988. The responses of stomatal density to CO_2 partial pressure. *J. Exp. Bot.* 39: 1771–1781.

Woodward, F.I. and C.K. Kelly. 1995. The influence of CO_2 concentration on stomatal density. *New Phytol.* 131: 311–327.

Xiao, C.-W., O.J. Sun, G.-S. Zhou, J.-Z. Zhao, and G. Wu. 2005. Interactive effects of elevated CO_2 and drought stress on leaf water potential and growth in *Caragana intermedia. Trees* 19: 711–720.

Xin, Z., C. Franks, P. Payton, and J.J. Burke. 2008. A simple method to determine transpiration efficiency in sorghum. *Field Crops Res.* 107: 180–183.

Xu, Q. and M.B. Kirkham. 2003. Combined effect of irradiance and water regime on sorghum photosynthesis. *Photosynthetica* 41: 27–32.

Xu, D.-Q., R.M. Gifford, and W.S. Chow. 1994. Photosynthetic acclimation in pea and soybean to high atmospheric CO_2 partial pressure. *Plant Physiol.* 106: 661–671.

Yamashita, T. 1952. Influence of potassium supply upon various properties and movement of the guard cell. *Sieboldia Acta Biol.* 1: 51–70.

Yao, Z., B. Wolf, W. Chen, K. Butterbach-Bahl, N. Brüggemann, M. Wiesmeier, M. Dannenmann, B. Blank, and X. Zheng. 2010. Spatial variability of N_2O, CH_4 and CO_2 fluxes within the Xilin River catchment of Inner Mongolia, China: A soil core study. *Plant Soil* 331: 341–359.

Young, J.J., S. Mehta, M. Israelsson, J. Godoski, E. Grill, and J.I. Schroeder. 2006. CO_2 signaling in guard cells: Calcium sensitivity response modulation, a Ca^{2+}-independent phase, and CO_2 insensitivity of the *gac2* mutant. *Proc. Natl. Acad. Sci.* 103: 7506–7511.

Young, S.L., F.J. Pierce, J.D. Streubel, and H.P. Collins. 2009. Performance of solid-state sensors for continuous, real-time measurement of soil CO_2 concentrations. *Agron. J.* 101: 1417–1420.

Zadoks, J.C., T.T. Chang, and C.F. Konzak. 1974. A decimal code for the growth stages of cereals. *Weed Res.* 14: 415–421.

Zanne, A.E., M. Westoby, D.S. Falster, D.D. Ackerly, S.T. Loarie, S.E.J. Arnold, and D.A. Coomes. 2010. Angiosperm wood structure: Global patterns in vessel anatomy and their relation to wood density and potential conductivity. *Am. J. Bot.* 97: 207–215.

Zelitch, I. 1967. Water and CO_2 transport in the photosynthetic process. In San Pietro, A., Greer, F.A., and Army, T.J., Eds., *Harvesting the Sun. Photosynthesis in Plant Life*. New York: Academic Press, pp. 231–248.

Zeroni, M. and J. Gale. 1989. Response of 'Sonia' roses to salinity at three levels of ambient CO_2. *J. Hort. Sci.* 64: 503–511.

Zhu, X.-G., S.P. Long, and D.R. Ort. 2010. Improving photosynthetic efficiency for greater yield. *Annu. Rev. Plant Biol.* 612: 235–261.

Ziska, L.H. and J.A. Bunce. 1995. Growth and photosynthetic response of three soybean cultivars to simultaneous increases in growth temperature and CO_2. *Physiol. Plant.* 94: 575–584.

Ziska, L.H., B.G. Drake, and S. Chamberlain. 1990. Long-term photosynthetic response in single leaves of a C_3 and C_4 salt marsh species grown in elevated atmospheric CO_2 *in situ*. *Oecologia* 83: 469–472.

Ziska, L.H., M.B. Tomecek, and D.R. Gealy. 2010. Competitive interactions between cultivated and red rice as a function of recent and projected increases in atmospheric carbon dioxide. *Agron. J.* 102: 118–123.

Index